T0212802

# Lecture Notes in Computer Science 8344

Commenced Publication in 1973
Founding and Former Series Editors:
Gerhard Goos, Juris Hartmanis, and Jan van Leeuwen

Sudebkumar Prasant Pal
Kunihiko Sadakane (Eds.)

# Algorithms and Computation

8th International Workshop, WALCOM 2014
Chennai, India, February 13-15, 2014
Proceedings

 Springer

Volume Editors

Sudebkumar Prasant Pal
Indian Institute of Technology Kharagpur
Department of Computer Science and Engineering
Kharagpur 721302, India
E-mail: spp@cse.iitkgp.ernet.in

Kunihiko Sadakane
National Institute of Informatics
2-1-2 Hitotsubashi, Chiyoda-ku, Tokyo 101-8430, Japan
E-mail: sada@nii.ac.jp

ISSN 0302-9743                                    e-ISSN 1611-3349
ISBN 978-3-319-04656-3                            e-ISBN 978-3-319-04657-0
DOI 10.1007/978-3-319-04657-0
Springer Cham Heidelberg New York Dordrecht London

Library of Congress Control Number: 2014930232

CR Subject Classification (1998): F.2, G.2.1-2, G.4, I.1, I.3.5, E.1

LNCS Sublibrary: SL 1 – Theoretical Computer Science and General Issues

*Typesetting:* Camera-ready by author, data conversion by Scientific Publishing Services, Chennai, India

Printed on acid-free paper

Springer is part of Springer Science+Business Media (www.springer.com)

# Preface

The 8th International Workshop on Algorithms and Computation (WALCOM 2014) was held during February 13–15, 2014 at the Indian Institute of Technology Madras, Chennai, India. This event was organized by the Department of Computer Science and Engineering, Indian Institute of Technology Madras. The workshop covered a diverse range of topics on algorithms and computations including computational geometry, approximation algorithms, graph algorithms, parallel and distributed computing, graph drawing, and computational complexity.

This volume contains 29 contributed papers presented during WALCOM 2014. There were 62 submissions from 16 countries. These submissions were rigorously refereed by the Program Committee members with the help of external reviewers. Abstracts of three invited talks delivered at WALCOM 2014 are also included in this volume.

We would like to thank the authors for contributing high-quality research papers to the workshop. We express our heartfelt thanks to the Program Committee members and the external referees for their active participation in reviewing the papers. We are grateful to Kurt Mehlhorn, Ian Munro, and Pavel Valtr for delivering excellent invited talks. We thank the Organizing Committee, chaired by N.S. Narayanaswamy for the smooth functioning of the workshop. We thank Springer for publishing the proceedings in the reputed Lecture Notes in Computer Science series. We thank our sponsors for their support. Finally, we remark that the EasyChair conference management system was very effective in handling the reviewing process.

February 2014

Sudebkumar Prasant Pal
Kunihiko Sadakane

# Organization

## Steering Committee

| | |
|---|---|
| Kyung-Yong Chwa | KAIST, South Korea |
| Costas S. Iliopoulos | King's College London, UK |
| M. Kaykobad | BUET, Bangladesh |
| Petra Mutzel | TU Dortmund, Germany |
| Shin-ichi Nakano | Gunma University, Japan |
| Subhas Chandra Nandy | ISI Kolkata, India |
| Takao Nishizeki | Tohoku University, Japan |
| Md. Saidur Rahman | BUET, Bangladesh |
| C. Pandu Rangan | IIT Madras, India |

## Organizing Committee

| | |
|---|---|
| N.S. Narayanaswamy | Indian Institute of Technology Madras, Chennai, India (Chair) |
| Sasanka Roy | Chennai Mathematical Institute, Chennai, India |
| Sajin Koroth | Indian Institute of Technology Madras, Chennai, India |
| R. Krithika | Indian Institute of Technology Madras, Chennai, India |
| C.S. Rahul | Indian Institute of Technology Madras, Chennai, India |

## Program Committee

| | |
|---|---|
| Hee-Kap Ahn | Pohang University of Science and Technology, Gyeongbuk, South Korea |
| V. Arvind | Institute of Mathematical Sciences, Chennai, India |
| Amitabha Bagchi | Indian Institute of Technology, Delhi, India |
| Giuseppe Battista | Third University of Rome, Italy |
| Arijit Bishnu | Indian Statistical Institute, Kolkata, India |
| Franz Brandenburg | University of Passau, Germany |
| Sumit Ganguly | Indian Institute of Technology, Kanpur, India |
| Subir Ghosh | Tata Institute of Fundamental Research, Mumbai, India |

Sathish Govindarajan            Indian Institute of Science, Bangalore, India
Shuji Kijima                    Kyushu University, Fukuoka, Japan
Ramesh Krishnamurti             Simon Fraser University, Burnaby, BC, Canada
Giuseppe Liotta                 University of Perugia, Italy
Sudebkumar Pal                  IIT Kharagpur, India (Co-chair)
Leonidas Palios                 University of Ioannina, Greece
Rina Panigrahy                  Microsoft Research, Mountain View, CA, USA
Rosella Petreschi               Sapienza University of Rome, Italy
Sheung-Hung Poon                National Tsing Hua University, Hsinchu,
                                    Taiwan
Sohel Rahman                    Bangladesh University of Engineering and
                                    Technology, Dhaka, Bangladesh
Rajeev Raman                    University of Leicester, UK
Abhiram Ranade                  Indian Institute of Technology, Bombay, India
C. Pandu Rangan                 Indian Institute of Technology, Madras, India
Kunihiko Sadakane               National Institute of Informatics, Tokyo, Japan
                                    (Co-chair)
Nicola Santoro                  Carleton University, Ottawa, ON, Canada
Jayalal Sarma                   Indian Institute of Technology, Madras, India
Saket Saurabh                   Institute of Mathematical Sciences, Chennai,
                                    India
Shakhar Smorodinsky             Ben-Gurion University, Be'er Sheva, Israel
Takeshi Tokuyama                Tohoku University, Sendai, Japan
Peter Widmayer                  ETH Zurich, Switzerland
Hsu-Chen Yen                    National Taiwan University, Taipei, Taiwan

## Additional Reviewers

Alam, Muhammad Rashed           Iranmanesh, Ehsan
Angelini, Patrizio              Karmakar, Arindam
Bae, Sang Won                   Khidamoradi, Kamyar
Bari, Md. Faizul                Kindermann, Philipp
Baswana, Surender               Komarath, Balagopal
Bekos, Michael                  Koroth, Sajin
Bohmova, Katerina               Ku, Tsung-Han
Bonifaci, Vincenzo              Lin, Chun-Cheng
Calamoneri, Tiziana             Lin, Jin-Yong
Curticapean, Radu               Lu, Chia Wei
Da Lozzo, Giordano              Mehta, Shashank
Das, Gautam Kumar               Misra, Neeldhara
Di Giacomo, Emilio              Mondal, Debajyoti
Frati, Fabrizio                 Monti, Angelo
Fusco, Emanuele                 Morgenstern, Gila
Ghosh, Arijit                   Nandakumar, Satyadev
Hossain, Md. Iqbal              Narayanaswamy, N.S.

Nekrich, Yakov
Nicosia, Gaia
Nishat, Rahnuma Islam
Nöllenburg, Martin
Ono, Hirotaka
Paul, Subhabrata
Peng, Yuejian
Pröger, Tobias
Rafiey, Arash
Rao B.V., Raghavendra

Roselli, Vincenzo
Savicky, Petr
Sinaimeri, Blerina
Tanigawa, Shin-Ichi
Tewari, Raghunath
Tschager, Thomas
Vatshelle, Martin
Zhang, Guochuan
Zohora, Fatema Tuz

# Table of Contents

# Distributed Computing and Networks

# Graph Algorithms

# Complexity and Bounds

## Graph Embeddings and Drawings

# Algorithms for Equilibrium Prices
# in Linear Market Models

Kurt Mehlhorn

Max-Planck-Institut für Informatik, Saarbrücken, Germany
mehlhorn@mpi-inf.mpg.de

Near the end of the 19th century, Leon Walrus [Wal74] and Irving Fisher [Fis91] introduced general market models and asked for the existence of equilibrium prices. Chapters 5 and 6 of [NRTV07] are an excellent introduction into the algorithmic theory of market models. In Walrus' model, each person comes to the market with a set of goods and a utility function for bundles of goods. At a set of prices, a person will only buy goods that give him maximal satisfaction.[1] The question is to find a set of prices at which the market clears, i.e., all goods are sold and all money is spent. Observe that the money available to an agent is exactly the money earned by selling his goods. Fisher's model is somewhat simpler. In Fisher's model every agent comes with a predetermined amount of money. Market clearing prices are also called *equilibrium prices*. Walrus and Fisher took it for granted that equilibrium prices exist. Fisher designed a hydro-mechanical computing machine that would compute the prices in a market with three buyers, three goods, and linear utilities [BS00].

In the 20th century it became clear that the existence of equilibrium prices requires rigorous proof. Arrow and Debreu [AD54] refined Walras' model and proved the existence of equilibrium prices for general convex utility functions. Their proof is non-constructive and uses a fixed-point theorem in a crucial way. The obvious next question for an algorithmicist is whether market clearing prices can be computed (efficiently)? We discuss the situation for linear markets.

In the linear Fisher market, there are $n$ buyers and $n$ goods. We assume for w.l.o.g that there is one unit of each good. The $i$-th buyer comes with a non-negative budget $b_i$. The utility for buyer $i$ of receiving the full unit of good $j$ is $u_{ij} \geq 0$. Let $x_{ij} \geq 0$ be the fraction of good $j$ that is allocated to buyer $i$. Under this assigment and the assumption of linear additive utilities, the total utility of $i$ is

$$\sum_j u_{ij} x_{ij}.$$

---

[1] Consider the case of linear additive utilites, i.e., two items of the same good give twice the utility of one item and utilities of different goods add. Assume that an agent values an item of good $A$ twice as much as an item of good $B$. If the price of an item of $A$ is less than twice the price of an item of $B$, the agent will only want $A$. If the price is more than twice, the agent will only want $B$. If the price is twice the price of $B$, the agent is indifferent and any combination of $A$ and $B$ is equally good. Linear utilities are a gross simplification.

S.P. Pal and K. Sadakane (Eds.): WALCOM 2014, LNCS 8344, pp. 1–4, 2014.

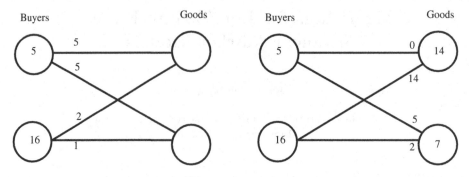

**Fig. 1.** A Fisher market: there are two buyers and two goods: the first buyer has a budget of five and the second buyer has a budget of 16 ($b_1 = 5$ and $b_2 = 16$). The first buyer draws a utility of 5 from both goods and the second buyer draws a utility of 2 from the first good and a utility of 1 from the second good ($u_{11} = u_{12} = 5$, $u_{21} = 2$, and $u_{22} = 1$). A solution is shown on the right. For price vector $p_1 = 14$ and $p_2 = 7$, the first buyer prefers the second good over the first good and therefore is only willing to spend money on the second good, and the second buyer is indifferent and hence is willing to spend money on both goods. For the allocation shown (the first good is allocated completely to the second buyer and the second good is split in the ratio 5:2), the market is in equilibrium.

Let $p_j$ be the (to be determined) price of good $j$. Then the utility of good $j$ for buyer $i$ per unit of money is $u_{ij}/p_j$. Buyers spend their money only on goods that give them maximal utility per unit of money, i.e.,

$$x_{ij} > 0 \quad \Rightarrow \quad \frac{u_{ij}}{p_j} = \alpha_i = \max_j u_{ij}/p_j. \tag{1}$$

$\alpha_i$ is called the bang-per-buck for agent $i$ at price vector $p$.

A price vector $p$ is *market clearing* in the Fisher model if there is an allocation $x = (x_{ij})$ such that (1) and

$$\sum_i x_{ij} = 1 \quad \text{for all } j \qquad\qquad \text{good } j \text{ is completely sold} \tag{2}$$

$$\sum_j p_{ij} x_{ij} = b_i \quad \text{for all } i \qquad \text{buyer } i \text{ spends his complete budget} \tag{3}$$

hold. In the linear Arrow-Debreu market there is the additional constraint

$$b_i = p_i \quad \text{for all } i, \tag{4}$$

i.e., the $i$-th buyer is also the owner of the $i$-th good and his budget is precisely the revenue for this good. Figures 1 and 2 illustrate the market concepts.

Fisher's model is a special case of the Arrow-Debreu model. In the former model, each buyer comes with a budget and money has intrinsic value. In the latter model, money is only used for comparing goods. The former model reduces to the latter by introducing a $n+1$-th good corresponding to money.

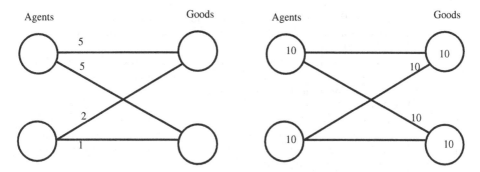

**Fig. 2.** An Arrow-Debreu market: there are two agents and two goods: the first agent owns good one and the second agent owns good two. As in Figure 1, the first buyer draws a utility of 5 from both goods and the second buyer draws a utility of 2 from the first good and a utility of 1 from the second good ($u_{11} = u_{12} = 5$, $u_{21} = 2$, and $u_{22} = 1$). A solution is shown on the right. For price vector $p_1 = 10$ and $p_2 = 10$, the first agent is indifferent between the goods and is willing to spend money on both goods. The second agent prefers the first good and is only willing to spend money on the first good.

In recent years, polynomial time algorithms were found for the computation of equilibrium prices in linear markets. Not surprisingly, Fisher's model was solved first. Already in 1958, Eisenberg and Gale [EG58] characterized equilibrium prices by a convex program. With the advent of the Ellipsoid method, the characterization became a polynomial time algorithm. In 2008, Devanur, Papadimitriou, Saberi, Vazerani [DPSV08] gave the first combinatorial algorithm. It computes market clearing prices by repeated price adjustments and maximum flow computations. A simpler and more efficient algorithm was found by Orlin [Orl10] in 2010. For the Arrow-Debreu market, the first polynomial time algorithms are due to Jain [Jai07] and Ye [Ye07]. Both algorithms give a characterization by a convex program and then use the Ellipsoid and interior point method, respectively, to solve the program. Duan and Mehlhorn [DM13] found a combinatorial algorithm in 2012.

# References

[AD54]   Arrow, K.J., Debreu, G.: Existence of an equilibrium for a competitive economy. Econometrica 22, 265–290 (1954)

[BS00]   Brainard, W.C., Scarf, H.E.: How to compute equilibrium prices in 1891. Cowles Foundation Discussion Papers 1272, Cowles Foundation for Research in Economics, Yale University (August 2000)

[DM13]   Duan, R., Mehlhorn, K.: A Combinatorial Polynomial Algorithm for the Linear Arrow-Debreu Market. In: Fomin, F.V., Freivalds, R., Kwiatkowska, M., Peleg, D. (eds.) ICALP 2013, Part I. LNCS, vol. 7965, pp. 425–436. Springer, Heidelberg (2013)

[DPSV08] Devanur, N.R., Papadimitriou, C.H., Saberi, A., Vazirani, V.V.: Market equilibrium via a primal–dual algorithm for a convex program. J. ACM 55(5), 22:1–22:18 (2008)

[EG58]   Eisenberg, E., Gale, D.: Consensus of Subjective Probabilities: the Parimutuel Method. Defense Technical Information Center (1958)

[Fis91]  Fisher, I.: Mathematical Investigations in the Theory of Value and Prices. PhD thesis, Yale University (1891)

[Jai07]  Jain, K.: A polynomial time algorithm for computing an Arrow-Debreu market equilibrium for linear utilities. SIAM J. Comput. 37(1), 303–318 (2007)

[NRTV07] Nisan, N., Roughgarden, T., Tardos, É., Vazirani, V.V. (eds.): Algorithmic Game Theory. Cambridge University Press (2007)

[Orl10]  Orlin, J.B.: Improved algorithms for computing Fisher's market clearing prices. In: Proceedings of the 42nd ACM Symposium on Theory of Computing, STOC 2010, pp. 291–300. ACM, New York (2010)

[Wal74]  Walrus, L.: Elements of Pure Economics, or the theory of social wealth (1874)

[Ye07]   Ye, Y.: A path to the Arrow-Debreu competitive market equilibrium. Math. Program. 111(1), 315–348 (2007)

# In as Few Comparisons as Possible

J. Ian Munro

Cheriton School of Computer Science, University of Waterloo,
Waterloo, Ontario N2L 3G1, Canada
imunro@uwaterloo.ca

**Abstract.** We review a variety of data ordering problems with the goal of solving them in as few comparisons as possible. En route we highlight a number of open problems, some new, some a couple of decades old, and others open for up to a half century. The first is that of sorting and the Ford-Johnson Merge-Insertion algorithm [8] of 1959, which remains the "best", at least for the "best and worst" values of n. Is it optimal, or are its extra .028..$n$ or so comparisons beyond the information theoretic lower bound necessary?

Moving to selection problems we first examine a special case. The problem of finding the second largest member of a set is fairly straightforward in the worst case. The best expected case method remains the $n + \Theta(\lg \lg n)$ method of Matula from 1973 [10]. It begs the question as to whether the $\lg \lg n$ term is necessary. The status of median finding has remained unchanged for a couple of decades, since the work of Dor and Zwick [4,5]. $(3 - \delta)n$ comparisons are sufficient, while $(2 + \epsilon)n$ are necessary. So the constant isn't an integer, but is it $\log_{4/3} 2$ as conjectured by Paterson [11]? This worst case behavior is in sharp contrast with the expected case of median finding where the answer has been known since the mid-'80's [3,6].

Finally we look at the problem of partial sorting (arranging elements according to a given partial order) and completing a sort given partially ordered data. The latter problem was posed and solved within $n$ or so comparisons of optimal by Fredman in 1975 [7]. The method, though, could use exponential time to determine which comparisons to perform. The more recent approaches of Cardinal et al [2,1] to these problems are based on graph entropy arguments and require only polynomial time to determine the comparisons to be made. Indeed the solution to the partial ordering problem involves a reduction to multiple selection [9]. In both cases the number of comparisons used differs from the information theoretic lower bound by only a lower order term plus a linear term.

## References

1. Cardinal, J., Fiorini, S., Joret, G., Jungers, R.M., Ian Munro, J.: An efficient algorithm for partial order production. SIAM J. Comput. 39(7), 2927–2940 (2010)
2. Cardinal, J., Fiorini, S., Joret, G., Jungers, R.M., Ian Munro, J.: Sorting under partial information (without the ellipsoid algorithm). In: STOC, pp. 359–368 (2010)
3. Cunto, W., Ian Munro, J.: Average case selection. J. ACM 36(2), 270–279 (1989)

S.P. Pal and K. Sadakane (Eds.): WALCOM 2014, LNCS 8344, pp. 5–6, 2014.

4. Dor, D., Zwick, U.: Selecting the median. SIAM J. Comput. 28(5), 1722–1758 (1999)
5. Dor, D., Zwick, U.: Median selection requires $(2 + \epsilon)n$ comparisons. SIAM J. Discrete Math. 14(3), 312–325 (2001)
6. Floyd, R.W., Rivest, R.L.: Expected time bounds for selection. Commun. ACM 18(3), 165–172 (1975)
7. Fredman, M.L.: Two applications of a probabilistic search technique: Sorting x + y and building balanced search trees. In: STOC, pp. 240–244 (1975)
8. Ford Jr., L.R., Johnson, S.B.: A tournament problem. American Mathematical Monthly 66(5), 387–389 (1959)
9. Kaligosi, K., Mehlhorn, K., Ian Munro, J., Sanders, P.: Towards optimal multiple selection. In: Caires, L., Italiano, G.F., Monteiro, L., Palamidessi, C., Yung, M. (eds.) ICALP 2005. LNCS, vol. 3580, pp. 103–114. Springer, Heidelberg (2005)
10. Matula, D.W.: Selecting the the best in average $n + \theta(\log \log n)$ comparisons. Washington University Report, AMCS-73-9 (1973)
11. Paterson, M.: Progress in selection. In: Karlsson, R., Lingas, A. (eds.) SWAT 1996. LNCS, vol. 1097, pp. 368–379. Springer, Heidelberg (1996)

# The Happy End Theorem and Related Results

Pavel Valtr

Department of Applied Mathematics and Institute for Theoretical Computer Science,
Charles University, Faculty of Mathematics and Physics,
Malostranské nám. 25, 118 00 Praha 1, Czech Republic

The Erdős–Szekeres $k$-gon theorem [1] says that for any integer $k \geq 3$ there is an integer $n(k)$ such that any set of $n(k)$ points in the plane, no three on a line, contains $k$ points which are vertices of a convex $k$-gon. It is a classical result both in combinatorial geometry and in Ramsey theory. Sometimes it is called the Happy End(ing) Theorem (a name given by Paul Erdős), since George Szekeres later married Eszter Klein who proposed a question answered by the theorem.

It is still widely open if the minimum possible value of $n(k)$ is equal to $2^{k-2} + 1$, as conjectured by Erdős and Szekeres [2] more than fifty years ago. The conjecture is known to be true for $k \leq 6$. It was verified for $k = 6$ by Szekeres and Peters [3] about ten years ago (the paper appeared in 2006), using an extensive computer search in a clever way.

There are many extensions and modifications of the Happy End Theorem. Some of them will be mentioned in the talk.

## References

[1] P. Erdős, and G. Szekeres, A combinatorial problem in geometry, *Compos. Math.* 2 (1935), 463–470.
[2] P. Erdős, and G. Szekeres, On some extremum problems in elementary geometry, *Ann. Univ. Sci. Bp. Rolan do Eötvös Nomin., Sect. Math.* 3/4 (1960–1961), 53–62.
[3] G. Szekeres and L. Peters, Computer solution to the 17-point Erdős-Szekeres problem, *ANZIAM Journal* 48 (2006), 151164.

S.P. Pal and K. Sadakane (Eds.): WALCOM 2014, LNCS 8344, p. 7, 2014.

# Generalized Class Cover Problem
# with Axis-Parallel Strips

Apurva Mudgal* and Supantha Pandit

Department of Computer Science and Engineering
Indian Institute of Technology Ropar, Nangal Road, Rupnagar, Punjab-140001
{apurva,supanthap}@iitrpr.ac.in

**Abstract.** We initiate the study of a generalization of the class cover problem [1,2], the *generalized class cover problem*, where we are allowed to misclassify some points provided we pay an associated positive penalty for every misclassified point. We study five different variants of generalized class cover problem with axis-parallel strips and half-strips in the plane, thus extending similar work by Bereg et al. [2] on the class cover problem. For each of these variants, we either show that they are in P, or prove that they are NP-complete and give constant factor approximation algorithms.

**Keywords:** class cover problem, axis-parallel strips, approximation algorithms, geometric set cover.

## 1 Introduction

The class cover problem [1,2] is the following : given a set $R$ of red points, a set $B$ of blue points, and a set $\mathcal{O}$ of geometric objects, find a minimum cardinality subset $\mathcal{O}' \subseteq \mathcal{O}$ which covers every blue point, but does not cover any red point. If we identify the blue points as positive examples and the red points as negative examples, the set $\mathcal{O}'$ gives us a classifier of complexity $|\mathcal{O}'|$, since every point $p \in R \cup B$ can be correctly classified as blue or red using the disjunction of $|\mathcal{O}'|$ queries of the form "Does $p$ lie inside object $o \in \mathcal{O}'$ ?".

In this paper, we study the *generalized class cover problem* where the classifier $\mathcal{O}' \subseteq \mathcal{O}$ is allowed to *misclassify* some blue points as red and some red points as blue. We measure the amount of misclassification by a penalty function $\mathscr{P} : R \cup B \to \mathbb{R}^+$ assigning positive penalties to every point, where the penalty of a point is understood to be the cost we pay for misclassifying it using our classifier i.e., for reporting a blue point as red and vice versa. The objective now is to minimize the complexity of the classifier (i.e., the cardinality of the set $\mathcal{O}' \subseteq \mathcal{O}$) plus the sum of penalties of every point incorrectly classified by $\mathcal{O}'$.

Allowing for misclassification in the class cover problem can be useful in several ways. First, a small fraction of the red and blue points in the data may be

* Partially supported by grant No. SB/FTP/ETA-434/2012 under DST-SERB Fast Track Scheme for Young Scientist.

S.P. Pal and K. Sadakane (Eds.): WALCOM 2014, LNCS 8344, pp. 8–21, 2014.

"outliers" and allowing for misclassification can identify these points and lead to a classifier with much smaller complexity. Secondly, there may occur scenarios where no subset $\mathcal{O}' \subseteq \mathcal{O}$ can completely separate blue points from red points (one such case occurs when every object $o \in \mathcal{O}$ contains a red point). In such cases, all classifiers are "approximate" and the generalized class cover problem gives us an approximate classifier which minimizes the sum of classifier complexity and penalty due to misclassification. Finally, not all data points may be equally important and the practitioner can fine-tune the point penalties to reflect the relative importance of a data point as a positive or negative example for the classifier.

In the following, we define $P = R \cup B$, $n = |B|$, $m = |R|$, and for any function $f : U \to \mathbb{R}$ and a subset $S \subseteq U$, we use $f(S)$ to denote the sum $\sum_{x \in S} f(x)$. We assume that no two points have the same $x$- or $y$-coordinates. We consider two versions of generalized class cover problem, *single coverage* and *multiple coverage*, which differ in the way the penalty of misclassified red points is counted:

1. **Generalized Class Cover (Single Coverage).** The penalty of each red point $r \in R$ covered by $\mathcal{O}'$ is counted exactly once. The cost $c_1(\mathcal{O}')$ is defined as $|\mathcal{O}'| + \mathscr{P}(B') + \mathscr{P}(R')$, where $B' \subseteq B$ is the set of blue points that are *not covered* by objects in $\mathcal{O}'$ and $R' \subseteq R$ is the set of red points that are *covered* by objects in $\mathcal{O}'$.
2. **Generalized Class Cover (Multiple Coverage).** The penalty of a red point $r \in R$ is counted once for every object containing $r$ in $\mathcal{O}'$. The cost $c_2(\mathcal{O}')$ of $\mathcal{O}'$ is defined as $|\mathcal{O}'| + \mathscr{P}(B') + \sum_{r \in R} \mathscr{P}(r) \cdot m(r, \mathcal{O}')$, where $B' \subseteq B$ is the set of blue points not covered by any object in $\mathcal{O}'$, and $m(r, \mathcal{O}')$ is the number of objects in $\mathcal{O}'$ which contain point $r$.

Note that if the penalty of each red and blue point is infinity, both the single and multiple coverage versions of generalized class cover problem reduce to the class cover problem.

In general, the set $\mathcal{O}$ of geometric objects may be triangles, circles, axis-parallel squares and rectangles, etc. [1, 2], but in this paper we assume $\mathcal{O}$ to consist of only axis-parallel strips and half-strips in the plane. A horizontal (resp. vertical) strip $(a, b)$ is the set of points $(x, y) \in \mathbb{R}^2$ satisfying the equation $a \leq y \leq b$ (resp. $a \leq x \leq b$). A half-strip $(a, b, c)$ extending to infinity in the southern direction is the set of points satisfying $a \leq x \leq b$ and $y \leq c$. Similarly, we can define half-strips extending to infinity in the northern, eastern, and western directions. We now define the set $\mathcal{O}$ of geometric objects for six different generalized class cover problems with strips and half-strips:

1. *STRIP.* All vertical and horizontal strips.
2. *HS-1D.* All half-strips extending to infinity southwards.
3. *HS-2D-SAME.* All half-strips extending to infinity in two opposite directions.
4. *HS-2D-DIFF.* All half-strips extending to infinity in two mutually orthogonal directions.
5. *HS-3D.* All half-strips extending to infinity in three different directions.
6. *HS-4D.* All half-strips extending to infinity in four different directions.

We denote the single and multiple coverage versions of each of the above six problems by suffixes *SC* and *MC* respectively. For example, *STRIP-SC* and *STRIP-MC* denote the single and multiple coverage versions of generalized class cover problem when the objects are axis-parallel strips.

**Previous Work.** The class cover problem was introduced by Cannon and Cowen [1]. They give a PTAS for the problem of separating the red points from blue points using minimum number of disks of equal radius whose centers are constrained to lie on the blue points. Interestingly, in the conclusion of their paper, they note that the class cover problem is "very sensitive to misclassified points" and ask whether approximation algorithms can be designed for the scenario where some points are allowed to be misclassified.

Bereg et al. [2] study an unconstrained version of the class cover problem where the geometric objects (disks or axis-parallel squares) are allowed to be centered at any location and can have different sizes. They prove that the class cover problem with axis-parallel rectangles or squares is NP-hard. Bereg et al. [2] also show that $O(1)$ approximation algorithms for the class cover problem with disks and axis-parallel squares can be derived from results in the theory of $\epsilon$-nets [3–5]. Further, they study in detail the complexity of class cover problem when the covering objects are axis-parallel strips and half-strips in the plane. Our paper can be looked as a further continuation of this line of research for the generalized class cover problem.

Mustafa and Ray [6] give a PTAS using local search for geometric set cover with disks in the plane. Aschner et al. [7] also describe a PTAS for geometric set cover with disks (or axis-parallel squares) in the plane based on ideas in [6,8]. They further note that this implies a PTAS for the class cover problem with disks or axis-parallel squares. A related problem is the red-blue set cover problem [9], where the objective is to cover all blue points with the given objects while minimizing the number of red elements covered. Recently, Chan et al. [10] show that red-blue set cover with axis-parallel unit squares is NP-complete and give a PTAS for the same.

## Our Contributions

1. We show that *STRIP-MC* is in P, whereas *STRIP-SC* is NP-complete. We give a factor 2 approximation algorithm for *STRIP-SC* (see Section 2).
2. We show that both *HS-1D-SC* and *HS-1D-MC* are in P by giving a dynamic programming algorithm (based on [11]) (see Section 3).
3. We show that *HS-2D-DIFF* (and *HS-3D*) is NP-complete even when all points have penalty infinity. Earlier, NP-completeness was known only for the infinite penalty version of *HS-4D* [2] (see Section 4).
4. We give $O(1)$ approximation algorithms for both single and multiple coverage versions of *HS-2D-DIFF*, *HS-3D*, and *HS-4D* (see Section 5).

We leave open the question of algorithmic complexity of *HS-2D-SAME-SC* and *HS-2D-SAME-MC*.

## 2    Strip Covering Problem

We note that the set $\mathcal{O}$ of geometric objects in this section consists of all possible
horizontal and vertical strips. Since we can contract a strip as long as we remove
only red points, we only consider strips with blue points incident on their two
bounding lines. Further, we can assume that in any solution to *STRIP-MC*
(and *STRIP-SC*), no two horizontal and no two vertical strips overlap. (If two
horizontal or two vertical strips overlap, we can replace them by a single strip
covering their union without increasing the cost of the solution.)

### 2.1    Multiple Coverage

In this section, we show that *STRIP-MC* is in $P$. Let $\mathcal{A}$ be the ordered set
$\{b_1, b_2, \ldots, b_n\}$ of blue points in $B$ ordered by increasing $x$-coordinate. Similarly,
let $\mathcal{B}$ be the ordered set $\{b'_1, b'_2, \ldots, b'_n\}$ of blue points in $B$ ordered by increasing
$y$-coordinate. We define an ordered 1-regular bipartite graph $G = (\mathcal{A}, \mathcal{B}, E)$
whose edge set $E = \{e_1, e_2, \ldots, e_n\}$ forms a perfect matching. The $i$th edge $e_i$
connects $b_i \in \mathcal{A}$ with $b'_{\sigma(i)} \in \mathcal{B}$, where $\sigma(i)$ is the position of point $b_i$ in ordered
set $\mathcal{B}$. The weight $w'(e_i)$ of edge $e_i$ is equal to the penalty of the corresponding
blue point i.e., $w'(e_i) = \mathscr{P}(b_i) = \mathscr{P}(b'_{\sigma(i)})$.

Let $\mathcal{I}$ and $\mathcal{J}$ be the sets of all possible intervals covering points in ordered sets
$\mathcal{A}$ and $\mathcal{B}$ respectively. An interval $I = (i, j) \in \mathcal{I}$ corresponds to the vertical strip
covering blue points from $b_i$ to $b_j$ and has weight $w(I) = 1 + \sum_{k=i}^{j-1} \mathbf{q}_k$, where
$\mathbf{q}_k$ is the total penalty of red points with $x$-coordinates between $b_k$ and $b_{k+1}$.
Similarly, an interval $J = (i, j) \in \mathcal{J}$ corresponds to the horizontal strip from $b'_i$
to $b'_j$ and has weight $w(J) = 1 + \sum_{k=i}^{j-1} \mathbf{q}'_k$, where $\mathbf{q}'_k$ is the total penalty of red
points with $y$-coordinates between $b'_k$ and $b'_{k+1}$.

Finding the optimal solution to *STRIP-MC* is equivalent to finding subsets
$\mathcal{I}' \subseteq \mathcal{I}, \mathcal{J}' \subseteq \mathcal{J}$ such that the quantity $w(\mathcal{I}') + w(\mathcal{J}') + w'(E'(\mathcal{I}' \cup \mathcal{J}'))$ is
minimized, where $E'(\mathcal{I}' \cup \mathcal{J}') \subseteq E$ is the set of edges in $G$ not covered by intervals
in $\mathcal{I}' \cup \mathcal{J}'$ (an edge is covered if at least one of its end points is contained in an
interval). Let us call this equivalent problem on bipartite graph $G = (\mathcal{A}, \mathcal{B}, E)$
as *Prize-collecting Bipartite Interval Vertex Cover Problem (PC-BP-IVCP)*.

We now show that *PC-BP-IVCP* (and hence *STRIP-MC*) is in $P$. The key
property that makes this possible is the following restricted form of submodu-
larity [12]:

**Lemma 1.** *Let $I_1, I_2 \in \mathcal{I}$ be two intervals such that $I_1 \cap I_2 \neq \phi$. Then, $w(I_1) +$
$w(I_2) \geq w(I_1 \cup I_2) + w(I_1 \cap I_2)$. A similar property holds for intervals in $\mathcal{J}$.*

In other words, the set $\mathcal{I}$ of intervals forms an *intersecting family* [12] and $w$
is a submodular[1] function defined on this family.

We will use Lemma 1 to show that *STRIP-MC* is in $P$ by an application of
submodular function minimization [13, 14].

---

[1] Note that our weight function $w$ is actually modular.

**Submodular Function Minimization.** We first extend the weight function defined on intervals in $\mathcal{I}$ to all possible subsets of $\mathcal{A}$. For $S \subseteq \mathcal{A}$, define $\rho_{\mathcal{A}}(S) = min\{w(\mathcal{I}')|\mathcal{I}' \subseteq \mathcal{I}$ and each point in $S$ is covered by some interval in $\mathcal{I}'\}$. Similarly, we can define a function $\rho_{\mathcal{B}}(S)$ for each possible subset $S \subseteq \mathcal{B}$. The following lemma can be proved using results in [12]:

**Lemma 2.** *The functions $\rho_{\mathcal{A}} : 2^{\mathcal{A}} \to \mathbb{R}^+ \cup \{0\}$ and $\rho_{\mathcal{B}} : 2^{\mathcal{B}} \to \mathbb{R}^+ \cup \{0\}$ are submodular and can be computed in polynomial time.*

*PC-BP-IVCP* is now equivalent to finding a set $S \subseteq \mathcal{A} \cup \mathcal{B}$ such that $\rho_{\mathcal{A}}(S \cap \mathcal{A}) + \rho_{\mathcal{B}}(S \cap \mathcal{B}) + w'(E(S))$ is minimized, where $E(S) \subseteq E$ is the set of edges not covered by vertices in $S$. This is an instance of *prize-collecting bipartite submodular vertex cover* (see [13,15,16]), which can be solved in polynomial time using submodular function minimization [13,14]. Hence, we get the following theorem:

**Theorem 3.** *STRIP-MC can be solved in polynomial time by submodular function minimization.*

## 2.2   Single Coverage

In this section, we show that *STRIP-SC* is NP-complete. Let $\mathcal{A}$ be the ordered set $\{p_1, p_2, \ldots, p_{n+m}\}$ of points in $P$ ordered by increasing $x$-coordinate. Similarly, let $\mathcal{B}$ be the ordered set $\{p'_1, p'_2, \ldots, p'_{n+m}\}$ of points in $P$ ordered by increasing $y$-coordinate. We define an ordered 1-regular bipartite graph $G = (\mathcal{A}, \mathcal{B}, E)$ as follows (see Figures 1(a) and 1(b)). The set $E = \{e_1, e_2, \ldots, e_{n+m}\}$ has exactly $n+m$ edges which form a perfect matching of $G$. The $i$th edge $e_i$ connects $p_i \in \mathcal{A}$ with $p'_{\sigma(i)} \in \mathcal{B}$, where $\sigma(i)$ is the position of point $p_i$ in ordered set $\mathcal{B}$. Edge $e_i$ is blue or red according to whether $p_i$ is blue or red. The weight $w'(e_i)$ of edge $e_i$ is equal to the penalty of point $p_i$ i.e., $w'(e_i) = \mathscr{P}(p_i) = \mathscr{P}(p'_{\sigma(i)})$.

Let $\mathcal{I}$ and $\mathcal{J}$ be the sets of all possible intervals covering points in ordered sets $\mathcal{A}$ and $\mathcal{B}$ respectively. A vertical strip covering points from $p_i$ to $p_j$ corresponds to interval $(i, j) \in \mathcal{I}$ of the ordered set $\mathcal{A}$. A similar correspondence exists between horizontal strips and intervals in $\mathcal{J}$.

Finding the optimal solution to *STRIP-SC* is equivalent to finding subsets $\mathcal{I}' \subseteq \mathcal{I}, \mathcal{J}' \subseteq \mathcal{J}$ which minimize the quantity $|\mathcal{I}'| + |\mathcal{J}'| + w'(E_1(\mathcal{I}' \cup \mathcal{J}')) + w'(E_2(\mathcal{I}' \cup \mathcal{J}'))$, where $E_1(\mathcal{I}' \cup \mathcal{J}')$ is the set of red edges covered by $\mathcal{I}' \cup \mathcal{J}'$ and $E_2(\mathcal{I}' \cup \mathcal{J}')$ is the set of blue edges not covered by $\mathcal{I}' \cup \mathcal{J}'$. Let us call this equivalent problem on 1-regular bipartite graph $G(\mathcal{A}, \mathcal{B}, E)$ as *Single Coverage Prize-collecting Bipartite Interval Vertex Cover Problem (PC-BP-IVCP-SC)*.

We now prove that *PC-BP-IVCP-SC* (and hence *STRIP-SC*) is NP-complete by reducing it from 3-SAT [17]. We first define 3-SAT:

**Definition 1. 3-SAT [17].** *We are given a 3-CNF formula $\phi = C_1 \wedge C_2 \wedge \ldots \wedge C_{m'}$ over $n'$ variables $x_1, x_2, \ldots, x_{n'}$, where each clause $C_j$ is a disjunction of exactly three literals from the set $\{x_i, \overline{x}_i | i \in 1, 2, \ldots, n'\}$. Is there a truth assignment to the variables which satisfies the formula?*

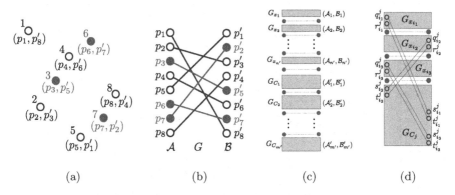

**Fig. 1.** (a) Point configuration (b) Reduced ordered 1-regular bipartite graph (c) Global structure of $G_\phi$ (d) Interaction between variable and clause $(C_j=(\overline{x}_{i_1} \vee x_{i_2} \vee \overline{x}_{i_3}))$ gadgets

The reduction converts $\phi$ to a bipartite graph $G_\phi(\mathcal{A}, \mathcal{B}, E)$ whose vertex set can be partitioned into $m' + n'$ components, $G_{x_i} = (\mathcal{A}_i, \mathcal{B}_i)$ for each variable $x_i, i = 1, 2, \ldots, n'$ and $G_{C_j} = (\mathcal{A}'_j, \mathcal{B}'_j)$ for each clause $C_j, j = 1, 2, \ldots, m'$ (Figure 1(c)). Any two components are separated by a red edge of weight $\infty$ which ensures that each interval picked in the solution lies completely inside some component.

*Variable gadget:* $G_{x_i}$ consists of two vertical columns of vertices. The right (resp. left) column contains two vertices $q_i^j, r_i^j$ for every clause $C_j$ in which $x_i$ occurs as a positive (resp. negative) literal. There are two blue edges at the top and bottom of $G_{x_i}$ with weight $\infty$.

*Clause gadget:* The clause gadgets $G_{C_j}$ for clauses of the form $(x_{i_1} \vee x_{i_2} \vee x_{i_3})$ and $(\overline{x}_{i_1} \vee x_{i_2} \vee x_{i_3})$ are shown in Figures 2(a) and 2(b) respectively. (The gadgets for the remaining 6 types of clauses can be obtained by flipping these two gadgets.) All blue and red edges in the clause gadgets are internal except for three "interacting pairs" $(s_{i_1}^j, t_{i_1}^j), (s_{i_2}^j, t_{i_2}^j), (s_{i_3}^j, t_{i_3}^j)$ of vertices which connect through blue edges to vertex pairs $(q_{i_1}^j, r_{i_1}^j), (q_{i_2}^j, r_{i_2}^j), (q_{i_3}^j, r_{i_3}^j)$ in variable gadgets $G_{x_{i_1}}, G_{x_{i_2}}, G_{x_{i_3}}$ respectively (see Figure 1(d)). Let $E_{C_j}$ denote the blue edges internal to clause gadget $G_{C_j}$ and let $E_{i_1}^j, E_{i_2}^j, E_{i_3}^j$ denote the blue edges going from $G_{C_j}$ to $G_{x_{i_1}}, G_{x_{i_2}}, G_{x_{i_3}}$ respectively.

The following Lemma can be proved by extensive case analysis:

**Lemma 4.** *If the blue edges in at least one of the sets $E_{i_1}^j, E_{i_2}^j, E_{i_3}^j$ are covered by an interval from the corresponding variable gadget, the optimal interval cover for the remaining blue edges in $G_{C_j}$ has cost 4. On the other hand, the optimal interval cover in $G_{C_j}$ for all blue edges in $E_{C_j} \cup E_{i_1}^j \cup E_{i_2}^j \cup E_{i_3}^j$ has cost 4.5.*

**Theorem 5.** *$\phi$ is satisfiable iff there exists a solution to $G_\phi$ with cost at most $n' + 4m'$.*

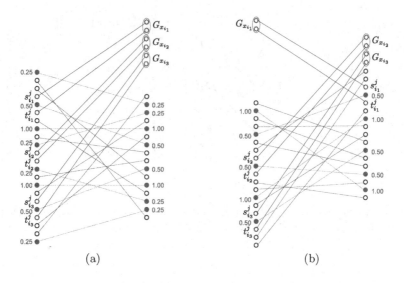

**Fig. 2.** Clause gadgets. (a) $C_j = x_{i_1} \vee x_{i_2} \vee x_{i_3}$ (b) $C_j = \overline{x}_{i_1} \vee x_{i_2} \vee x_{i_3}$.

**Proof:** (if part) Let $\Pi : \{x_1, x_2, \ldots, x_{n'}\} \to \{0, 1\}$ be a satisfying assignment to $\phi$. Pick the right (resp. left) interval in gadget $G_{x_i}$ according to whether $x_i$ is true (resp. false). Since for each clause gadget $G_{C_j}$ at least one of the three interacting pairs is covered by an interval in the corresponding variable gadget, by Lemma 4, the remaining blue edges can be covered at cost 4 per clause, for a total cost of $n' + 4m'$.

(only if part) If we have a solution of cost $n' + 4m'$, by Lemma 4, each variable gadget contributes exactly 1 and each clause gadget contributes exactly 4 to the total cost. Set $\Pi(x_i) = 1$ (resp. 0) if the right (resp. left) interval is picked by the solution in $G_{x_i}$. Since each clause gadget has cost 4, by Lemma 4, at least one interacting pair in each clause gadget must be covered by the corresponding variable gadget, and hence $\Pi$ is a satisfying assignment. ∎

**Factor 2 Approximation Algorithm for STRIP-SC.** The approximation algorithm finds the optimal solution to *STRIP-MC* on the same point set in polynomial time and outputs it as the answer to *STRIP-SC*. Since each red point can be covered by at most two strips, the optimal solution to *STRIP-MC* is within factor 2 of the optimal solution to *STRIP-SC* on the same point set.

## 3    Unidirectional Half-Strip Covering Problem

We now show that an algorithm of Chin et al. [11] for variable-size rectangle covering can be adapted to solve *HS-1D-SC* and *HS-1D-MC* in polynomial time. We now give an outline of the algorithm along with proofs of main theorems.

Without loss of generality we assume that all points in $P$ have positive $y$-coordinates. We number the blue points in $B$ from $1, 2, \ldots, n$ according to

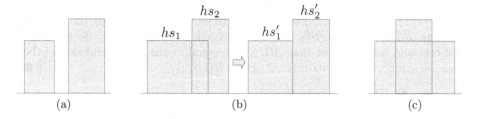

**Fig. 3.** Possible intersections between two half-strips

increasing $x$-coordinate. Let $(x_i, y_i)$ denote the Cartesian coordinates of the $i$th blue point in this order. Let $Y = \{y_1, y_2, \ldots, y_n\}$. We denote any half-strip $hs$ covering points in $B$ by a 3-tuple $(i, j, z)$, where $1 \leq i \leq j \leq n$ and $z \in Y$. $hs$ is the region of the plane given by the equations $x_i \leq x \leq x_j$ and $y \leq z$. Let $\mathcal{H}$ denote the set of all possible half-strips.

Any two half-strips $hs_1, hs_2 \in \mathcal{H}$ are either (i) disjoint (see Figure 3(a)), or (ii) partially intersecting (see Figure 3(b)), or (iii) fully intersecting (see Figure 3(c)).

**Lemma 6.** *There exists an optimal solution $\mathcal{H}^* \subseteq \mathcal{H}$ to HS-1D-SC such that no two half-strips in $\mathcal{H}^*$ are partially intersecting.*

**Proof:** If two half-strips in $\mathcal{H}^*$ are partially intersecting, we can replace them with two disjoint half-strips $hs_1', hs_2'$ (see Figure 3(b)) without increasing the cost of $\mathcal{H}^*$. One can show that in a polynomial number of such replacements, we will get an optimal solution without any partially intersecting half-strips.    ■

Let $S(i, j, z)$ ($1 \leq i \leq j \leq n, z \in Y \cup \{0\}$) denote the subset of points in $P$ with $x$-coordinate in the interval $[x_i, x_j]$ and $y$-coordinate strictly greater than $z$. Let $A(i, j, z)$ denote the cost of optimal solution to *HS-1D-SC* on point set $S(i, j, z)$. Finally, let $\mathcal{H}(i, j, z) = \{(i', j', z') | i \leq i' \leq j' \leq j \text{ and } z' > z\}$ be the set of half-strips covering points in $S(i, j, z)$.

**Lemma 7.** *The costs $A(i, j, z)$, where $1 \leq i \leq j \leq n$ and $z \in Y \cup \{0\}$, satisfy the recurrence:*

$$A(i, j, z) = \min_{(i', j', z') \in \mathcal{H}(i,j,z)} (A(i, i' - 1, z) + A(i', j', z') + 1 + \mathscr{P}(B') + \mathscr{P}(R'))$$

*where $B' \subseteq B$ are the set of blue points in $S(i, j, z)$ with $x$-coordinate greater than $j'$ and $R' \subseteq R$ is the set of red points in the set $S(i, j, z)$ covered by half-strip $(i', j', z')$.*

**Proof:** Let $OPT(i, j, z)$ denote an optimal set of half-strips for *HS-1D-SC* on point set $S(i, j, z)$. Let $j^*$ be rightmost blue point in $S(i, j, z)$ covered by $OPT(i, j, z)$ and let $(i^*, j^*, z^*)$ denote the half-strip with minimum height among all half-strips in $OPT(i, j, z)$ covering $j^*$. Since $OPT(i, j, z)$ is not partially intersecting (Lemma 6), the half-strips in $OPT(i, j, z)$ except $(i^*, j^*, z^*)$ can be partitioned into two sets $S_1 \subseteq \mathcal{H}(i, i^* - 1, z)$ and $S_2 \subseteq \mathcal{H}(i^*, j^*, z^*)$ which cover points in $S(i, i^* - 1, z)$ and $S(i^*, j^*, z^*)$ respectively. Replacing $S_1$ and $S_2$ by the optimal solutions to *HS-1D-SC* on $S(i, i^* - 1, z)$ and $S(i^*, j^*, z^*)$ can only decrease the total cost and hence we get that $A(i, j, z)$ is equal to the right hand side of the above recurrence for half-strip $(i^*, j^*, z^*) \in \mathcal{H}(i, j, z)$. ∎

**Theorem 8.** *HS-1D-SC is in P.*

**Proof:** First, all $A(i, j, z)$'s such that $j < i$ are set to 0. We compute the remaining $A(i, j, z)$'s in increasing order of $k = j - i$. For a given $k$, we compute $A(i, i + k, z)$ using the above recurrence where $i$ goes from 1 to $n - k$ and $z$ goes from highest to the lowest $y$-coordinate in $Y \cup \{0\}$. One can show that for each $A(i, i + k, z)$, the right hand side of the recurrence can be evaluated in $O(n^3)$ time and since there are $O(n^3)$ subproblems, the total time taken by the dynamic program is $O(n^6)$. We note that the running time can be further reduced to $O(n^4)$ [11]. ∎

A minor modification of the above algorithm also shows that *HS-1D-MC* is in P.

## 4  Hardness of *HS-2D-DIFF*

In this section we prove that *HS-2D-DIFF* is NP-complete by a reduction from 3-SAT (see Definition 1). We partition the clauses of the given 3-SAT formula $\phi$ into two types: (i) A Type I clause contains at most one negative literal, whereas (ii) a Type II clause contains at most one positive literal. Given a 3-SAT formula $\phi$, we construct an instance $V_\phi$ of *HS-2D-DIFF* as follows.

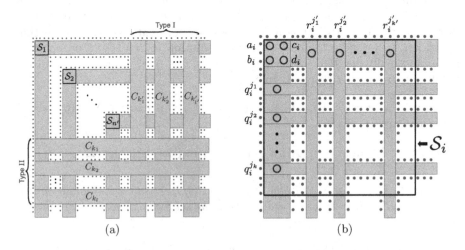

**Fig. 4.** (a) Global structure of $V_\phi$ (b) Variable gadget $V_{x_i}$

Globally, $V_\phi$ consists of $n'$ variable gadgets $V_{x_i}$ (each contained in an axis-parallel square $\mathcal{S}_i$) arranged diagonally with respect to each other (see Figure 4(a)).

We have a vertical strip for each clause of Type I and a horizontal strip for each clause of Type II. These horizontal and vertical strips contain the clause gadgets for their respective clauses. All points in $V_\phi$ have penalty infinity. In the construction below, we use a large number of red points to restrict the set of possible half-strips. However, we note that by carefully choosing the positions of the red points, their number can be reduced to a polynomial in $n', m'$.

*Variable gadget* $V_{x_i}$: The axis-parallel square $\mathcal{S}_i$ containing $V_{x_i}$ has sides of length $m' + 1$ (see Figure 4(b)). Four blue points $a_i, b_i, c_i, d_i$ are placed at the four corners of an axis-parallel unit square such that $a_i$ coincides with the top left corner of $\mathcal{S}_i$. For every clause $C_j$, if $x_i$ is the only negative (resp. positive) literal in $C_j$, we add a blue point $q_i^j$ (resp. $r_i^j$) in the unit width vertical (resp. horizontal) half-strip containing $\{a_i, b_i, c_i, d_i\}$. The blue points $q_i^j$ (resp. $r_i^j$) are placed at unit distance from each other vertically (resp. horizontally). We now add red points to the gadget so that the maximal horizontal and vertical half-strips containing blue points in $V_{x_i}$ are as shown in Figure 4(b).

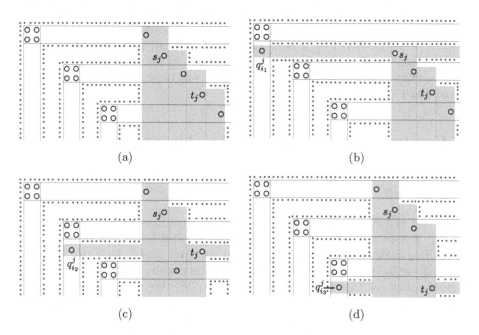

(a)                                      (b)

(c)                                      (d)

**Fig. 5.** Clause gadgets for different Type I clauses. (a) $C_j = (x_{i_1} \vee x_{i_2} \vee x_{i_3})$ (b) $C_j = (\overline{x}_{i_1} \vee x_{i_2} \vee x_{i_3})$ (c) $C_j = (x_{i_1} \vee \overline{x}_{i_2} \vee x_{i_3})$ (d) $C_j = (x_{i_1} \vee x_{i_2} \vee \overline{x}_{i_3})$.

*Clause gadget* $V_{C_j}$: The clause gadgets for the four different clauses of Type I are shown in Figure 5. The gadgets for Type II clauses can be obtained by flipping these gadgets about the diagonal. Two points $s_j, t_j$ in each clause gadget are special and are called "core points".

A little care is needed in the placement of red and blue points in the clause gadgets. We describe here only the construction in Figure 5(b). Point $s_j$ lies in the maximal horizontal half-strip containing the blue point $q_{i_1}^j$. Point $t_j$ is placed so that the horizontal line through it lies below square $\mathcal{S}_{i_2}$, above square $\mathcal{S}_{i_3}$, and does not intersect any other square $\mathcal{S}_i, 1 \leq i \leq n'$. The remaining two points lie in the maximal horizontal half-strips containing $\{a_{i_2}, b_{i_2}, c_{i_2}, d_{i_2}\}$ and $\{a_{i_3}, b_{i_3}, c_{i_3}, d_{i_3}\}$ respectively. We now place the red points so that maximal half-strips containing either $s_j$ or $t_j$ are constrained to lie in the shaded region of Figure 5(b).

**Theorem 9.** $\phi$ *is satisfiable iff there exists a solution to HS-2D-DIFF on $V_\phi$ with cost at most $n' + 2m'$.*

**Proof:** (if part) Let $\Pi : \{x_1, x_2, \ldots, x_{n'}\} \to \{0, 1\}$ be a satisfying assignment. Cover $\{a_i, b_i, c_i, d_i\}$ with a horizontal half-strip if $\Pi(x_i) = 1$ and with a vertical half-strip if $\Pi(x_i) = 0$. The uncovered points in each clause gadget can be covered using 2 half-strips per clause, for a total of $n' + 2m'$ half-strips.

(only if part) Let $S = \{a_i | i \in [1, n']\} \cup \{s_j, t_j | j \in [1, m']\}$ denote a set of $n' + 2m'$ points. Since no two points in $S$ can be covered by the same half-strip, we have a unique half-strip in the solution for every point in $S$. Set $\Pi(x_i) = 1$ (resp. 0) if $a_i$ is covered by a horizontal (resp. vertical) half-strip. If a clause is not satisfied by $\Pi$, then covering the five points in the shaded region of the corresponding clause gadget $V_{C_j}$ (see Figure 5) will require at least three half-strips, which is a contradiction. ∎

## 5   Constant Factor Approximation Algorithms

**Theorem 10.** *There exist $O(1)$ approximation algorithms for HS-2D-DIFF-MC, HS-3D-MC, and HS-4D-MC.*

**Proof:** We first consider the case when all blue points have penalty infinity. Let $\mathcal{H}_H, \mathcal{H}_V$ denote the sets of all horizontal and vertical half-strips respectively. Then, HS-4D-SC is an instance of weighted set cover on $(B, \mathcal{H}_H \cup \mathcal{H}_V)$ where the weight $w(hs)$ of half-strip $hs$ is 1 plus the total penalty of red points covered by it. We now use the technique of Gaur et al. [18]. We solve the natural LP relaxation for weighted set cover on $(B, \mathcal{H}_H \cup \mathcal{H}_V)$ and partition the blue points into two sets $B_1$ and $B_2$ based on whether horizontal or vertical half-strips cover a point to value greater than $\frac{1}{2}$. Since $\mathcal{H}_H$ and $\mathcal{H}_V$ are families of pseudo-disks, by the result of Chan et al. [19], in randomized polynomial time we can compute integral set covers $S_1$ and $S_2$ of $(B_1, \mathcal{H}_H)$ and $(B_2, \mathcal{H}_V)$ with weight at most $O(1)$ times the optimal fractional solutions of the respective set cover LPs. An analysis similar to that in Gaur et al. [18] shows that $S_1 \cup S_2$ is an $O(1)$ solution to HS-4D-SC.

The approximation algorithm for the prize-collecting version (blue points have finite penalties) can be obtained by applying the deterministic rounding technique of Bienstock et al. [20] as follows. Let $(x^*, y^*, z^*)$ be an optimal fractional solution to the following LP for the prize-collecting version:

$$\min \sum_{hs \in \mathcal{H}_V} w(hs)x_{hs} + \sum_{hs \in \mathcal{H}_H} w(hs)y_{hs} + \sum_{p \in B} \mathscr{P}(p)z_p$$

$$\text{s.t.} \sum_{hs|hs \in \mathcal{H}_V \ \& \ p \in hs} x_{hs} + \sum_{hs|hs \in \mathcal{H}_H \ \& \ p \in hs} y_{hs} + z_p \geq 1 \quad \forall \, p \in B$$

(LP)

(1)

$$x_{hs} \geq 0 \ \forall \, hs \in \mathcal{H}_V, y_{hs} \geq 0 \ \forall \, hs \in \mathcal{H}_H \ \& \ z_p \geq 0 \ \forall \, p \in B \qquad (2)$$

Let $\alpha, 0 < \alpha < 1$, be a constant (say $\alpha = 0.5$) and let $B_\alpha$ denote the set of blue points $p \in B$ with $z_p^* \geq \alpha$. Clearly, $\frac{1}{1-\alpha}(x^*, y^*)$ forms a feasible solution to the natural set cover LP on $(B \setminus B_\alpha, \mathcal{H}_H \cup \mathcal{H}_V)$. Hence, by applying the above algorithm for infinite penalties, we can obtain a half-strip cover $S$ for all blue points in $B \setminus B_\alpha$ with cost within $O(1)$ of $\frac{1}{1-\alpha} \cdot (\sum_{hs \in \mathcal{H}_V} w(hs)x_{hs}^* + \sum_{hs \in \mathcal{H}_H} w(hs)y_{hs}^*)$. Since the penalty for not covering the blue points in $B_\alpha$ is at most $\frac{1}{\alpha} \sum_{p \in B} \mathscr{P}(p)z_p^*$, the set $S$ of half-strips gives an $O(1)$ solution to *HS-4D-MC* with arbitrary penalties. ∎

**Lemma 11.** *Let $OPT_M$ and $OPT_S$ denote the optimal solution to HS-1D-MC and HS-1D-SC respectively on point set $P$. Then $OPT_M \leq 2 \cdot OPT_S$.*

**Proof:** For simplicity, assume all blue points are covered and all points in $P$ have positive $y$-coordinates. Then the region of $\mathbb{R}^2$ above the $x$-axis covered by half-strips in $\mathcal{H}^*$, the optimal solution to *HS-1D-SC*, is a disjoint union of $1.5D$ rectilinear terrains.

Let $RT_1, RT_2, \ldots, RT_m$ be these terrains. Let $S_i$ be the set of half-strips in $\mathcal{H}^*$ forming terrain $RT_i$ and let $k_i = |S_i|$. Let $R_i \subseteq R$ be the red points inside terrain $RT_i$. Then $OPT_S = \sum_{i=1}^{m} k_i + \sum_{i=1}^{m} \mathscr{P}(R_i)$. For each terrain $RT_i$, we define a set $T_i$ of disjoint half-strips covering $RT_i$ as follows. For each horizontal edge $e \in RT_i$, we add the half-strip with top edge $e$ to $T_i$. Since each of the half-strips in $T_i$ are disjoint, $\bigcup_{i=1}^{m} T_i$ forms a solution to *HS-1D-MC* of cost $\sum_{i=1}^{m} l_i + \sum_{i=1}^{m} \mathscr{P}(R_i)$, where $l_i = |T_i|$.

One can show using simple arguments, that for each $i$, $l_i \leq 2k_i$. This proves the lemma, as $OPT_M = \sum_{i=1}^{m} l_i + \sum_{i=1}^{m} \mathscr{P}(R_i) \leq 2\sum_{i=1}^{m} k_i + \mathscr{P}(R_i) \leq 2(\sum_{i=1}^{m} k_i + \mathscr{P}(R_i)) = 2 \cdot OPT_S$. ∎

**Theorem 12.** *There exist $O(1)$ approximation algorithms for HS-2D-DIFF-SC, HS-3D-SC, and HS-4D-SC.*

**Proof:** Using Lemma 11, we can show that $OPT\text{-}4D_M \leq 8 \cdot OPT\text{-}4D_S$, where $OPT\text{-}4D_M$ and $OPT\text{-}4D_S$ denote the optimal solutions to *HS-4D-MC* and *HS-4D-SC* on point set $P$. Therefore, the $O(1)$ algorithm to *HS-4D-MC* given by Theorem 10 also gives an $O(1)$ approximation algorithm for *HS-4D-SC*. ∎

**Acknowledgements.** We thank Prakhar Asthana for his help. We also thank the anonymous referees for their valuable comments, in particular on NP-hardness proofs, which helped to improve the presentation of this paper.

# References

1. Cannon, A.H., Cowen, L.J.: Approximation algorithms for the class cover problem. Annals of Mathematics and Artificial Intelligence 40(3-4), 215–223 (2004)
2. Bereg, S., Cabello, S., Díaz-Báñez, J.M., Pérez-Lantero, P., Seara, C., Ventura, I.: The class cover problem with boxes. Computational Geometry 45(7), 294–304 (2012)
3. Brönnimann, H., Goodrich, M.T.: Almost optimal set covers in finite VC-dimension. Discrete & Computational Geometry 14(1), 463–479 (1995)
4. Matoušek, J., Seidel, R., Welzl, E.: How to net a lot with little: small ε-nets for disks and halfspaces. In: Proc. of the 6th Annual Symposium on Computational Geometry, SCG 1990, pp. 16–22. ACM (1990)
5. Clarkson, K.L., Varadarajan, K.: Improved approximation algorithms for geometric set cover. Discrete & Computational Geometry 37(1), 43–58 (2007)
6. Mustafa, N.H., Ray, S.: Improved results on geometric hitting set problems. Discrete & Computational Geometry 44(4), 883–895 (2010)
7. Aschner, R., Katz, M.J., Morgenstern, G., Yuditsky, Y.: Approximation schemes for covering and packing. In: Ghosh, S.K., Tokuyama, T. (eds.) WALCOM 2013. LNCS, vol. 7748, pp. 89–100. Springer, Heidelberg (2013)
8. Gibson, M., Pirwani, I.A.: Algorithms for dominating set in disk graphs: Breaking the $\log n$ barrier. In: de Berg, M., Meyer, U. (eds.) ESA 2010, Part I. LNCS, vol. 6346, pp. 243–254. Springer, Heidelberg (2010)
9. Carr, R.D., Doddi, S., Konjevod, G., Marathe, M.: On the red-blue set cover problem. In: Proc. of 11th Annual ACM-SIAM Symposium on Discrete Algorithms, SODA 2000, pp. 345–353 (2000)
10. Chan, T.M., Hu, N.: Geometric red-blue set cover for unit squares and related problems. In: Proceedings of the 25th Canadian Conference on Computational Geometry, CCCG 2013, pp. 289–293 (2013)
11. Chin, F.Y.L., Ting, H.-F., Zhang, Y.: Variable-size rectangle covering. In: Du, D.-Z., Hu, X., Pardalos, P.M. (eds.) COCOA 2009. LNCS, vol. 5573, pp. 145–154. Springer, Heidelberg (2009)
12. Lovász, L.: Submodular functions and convexity. In: Bachem, A., Korte, B., Grötschel, M. (eds.) Mathematical Programming The State of the Art, pp. 235–257. Springer, Heidelberg (1983)
13. Iwata, S., Nagano, K.: Submodular function minimization under covering constraints. In: 50th Annual IEEE Symposium on Foundations of Computer Science, FOCS 2009, Atlanta, Georgia, USA, October 25-27, pp. 671–680 (2009)
14. Schrijver, A.: A combinatorial algorithm minimizing submodular functions in strongly polynomial time. Journal of Combinatorial Theory, Series B 80(2), 346–355 (2000)
15. Delong, A., Veksler, O., Osokin, A., Boykov, Y.: Minimizing sparse high-order energies by submodular vertex-cover. Advances in Neural Information Processing Systems 25, 971–979 (2012)
16. Goel, G., Karande, C., Tripathi, P., Wang, L.: Approximability of combinatorial problems with multi-agent submodular cost functions. In: FOCS 2009: Proceedings of 50th Annual IEEE Symposium on Foundations of Computer Science (2009)

17. Garey, M.R., Johnson, D.S.: Computers and Intractability; A Guide to the Theory of NP-Completeness. W. H. Freeman & Co., New York (1990)
18. Gaur, D.R., Ibaraki, T., Krishnamurti, R.: Constant ratio approximation algorithms for the rectangle stabbing problem and the rectilinear partitioning problem. Journal of Algorithms 43(1), 138–152 (2002)
19. Chan, T.M., Grant, E., Könemann, J., Sharpe, M.: Weighted capacitated, priority, and geometric set cover via improved quasi-uniform sampling. In: Proc. of the 13th Annual ACM-SIAM Symposium on Discrete Algorithms, SODA 2012, pp. 1576–1585 (2012)
20. Bienstock, D., Goemans, M., Simchi-Levi, D., Williamson, D.: A note on the prize collecting traveling salesman problem. Mathematical Programming 59(1-3), 413–420 (1993)

# Top-$k$ Manhattan Spatial Skyline Queries*

Wanbin Son[1], Fabian Stehn[2], Christian Knauer[2], and Hee-Kap Ahn[1]

[1] Department of Computer Science and Engineering, POSTECH, Pohang, Republic of Korea
{mnbiny,heekap}@postech.ac.kr
[2] Institute of Computer Science, Universität Bayreuth, 95440 Bayreuth, Germany
{fabian.stehn,christian.knauer}@uni-bayreuth.de

**Abstract.** Efficiently retrieving relevant data from a huge spatial database is and has been the subject of research in fields like database systems, geographic information systems and also computational geometry for many years. In this context, we study the retrieval of relevant points with respect to a query and a scoring function: let $P$ and $Q$ be point sets in the plane, the *skyline* of $P$ with respect to $Q$ consists of points $P$ for which no other point of $P$ is closer to all points of $Q$. A skyline of a point set $P$ with respect to a query set $Q$ can be seen as the most "relevant" or "desirable" subset of $P$ with respect to $Q$. As the skyline of a set $P$ can be as large as the set $P$ itself, it is reasonable to *filter* the skyline using a scoring function $f$, only reporting the $k$ best skyline points with respect to $f$.

In this paper, we consider the top-$k$ Manhattan spatial skyline query problem for monotone scoring functions, where distances are measured in the $L_1$ metric. We present an algorithm that computes the top-$k$ skyline points in near linear time in the size of $P$. The presented strategy improves over the direct approach of first using the state-of-the-art algorithm to compute the Manhattan spatial skyline [1] and then filtering it by the scoring function by a $\log(|P|)$ factor. The improvement has been validated in experiments that show a speedup of up to an order of magnitude.

## 1 Introduction

The amount of data collected by various devices and applications has significantly increased over the last years. There is no reason to believe that this trend will change in the near future. In the face of ever increasing amount of data, the task of efficiently receiving information of high relevance with respect to a given query and separating that data from irrelevant information becomes more and more important.

Throughout this paper, we assume the data to be given as points in the $L_1$ plane, that is, distances are measured using the $L_1$ metric. One concept of defining the relevance of a point in a point set $P$ with respect to a query set $Q$ is to consider the so-called *skyline* of $P$ with respect to $Q$. The relevance of a point in $P$ is correlated to its distances to the individual points of the query: the skyline of a point set $P$ with respect to a query set $Q$ consists of all points $s \in P$ so that no other point of $P$ is closer to all points of $Q$ than $s$; skylines will be formally introduced in the next section.

---

* This research was supported by the National Research Foundation of Korea(NRF) grant funded by the Korea government(MSIP) (No. 2011-0030044).

S.P. Pal and K. Sadakane (Eds.): WALCOM 2014, LNCS 8344, pp. 22–33, 2014.

Computing the skyline of a set $P$ of $n$ points with respect to a query set $Q$ of $m$ points in the $L_1$ plane is known as the Manhattan Spatial Skyline Query problem [1]. Unfortunately, the entire skyline for a specific query can consist of (too) many points; actually, the skyline may be $P$ itself. A common strategy to further reduce the number of relevant points is to apply a (monotone) scoring function and to report $k$ points of the skyline with the highest scoring function value, where usually $k$ is much smaller than $n$.

In this paper, we hence study the top-$k$ Manhattan spatial skyline query problem. In a top-$k$ Manhattan spatial skyline query we ask for the $k$ *best* skyline points of $P$ with respect to $f$, where the scoring function $f$ and the parameter $k$ are part of the input.

A straightforward algorithm to compute the top-$k$ skyline points would be as follows: first, compute the entire skyline, then apply the function to the resulting set and finally report $k$ points with the highest scores.

The best known algorithm to answer a Manhattan spatial skyline query problem takes $O(n(\log n + \log m) + m \log m)$ time [1], implying that the straightforward algorithm for the problem considered here takes $O(n(\log n + \log m + t_f) + m \log m)$ time, where $t_f$ is the time to evaluate the scoring function $f$. This approach, however, might be very time consuming, as one needs to test for each data point $p \in P$ whether or not $p$ actually is part of the skyline by comparing it with all other points of $P$, which itself is a time consuming operation. In this paper, we propose efficient algorithms that avoid this inefficiency.

## Contribution

In this paper, we present an algorithm for answering top-$k$ Manhattan spatial skyline queries for an arbitrary monotone scoring function $f$ in time $O(n(\log m + \log k + t_f) + m \log m)$ using $O(n + m)$ space after spending $O(n \log n)$ time for a preprocessing. This improves over the direct (and the previously only known) approach mentioned above by a factor of $\log n$.

In many practical applications, $m$ and $k$ are much smaller than $n$. For example, a Geographic Information System (GIS) stores a huge number of spatial data, but a typical query consist only of a few locations of interest chosen by a user. And as the scoring function already reflects the users preference, only a small number of reported points is usually desired.

The time $t_f$ to evaluate the monotone scoring function $f$ depends on the function itself. For some point $p$, $f(p)$ is defined by relations of distances from query points to $p$. Let $\|p - p'\|_1$ denote the $L_1$ distance between two points $p$ and $p'$. Typical monotone scoring functions are

$$\text{SDIST}(p) := \sum_{q \in Q} \|p - q\|_1, \ \text{MAX}(p) := \max_{q \in Q} \|p - q\|_1, \text{ and MIN}(p) := \min_{q \in Q} \|p - q\|_1.$$

For these three functions, $t_f = O(\log m)$ after a preprocessing time of $O(m \log m)$. The computational complexity of SDIST will be investigated and presented later. In case of MAX and MIN, the (farthest) Voronoi diagram of $Q$ can be used to compute the score of a point.

If $m$ and $k$ are much smaller than $n$, and $t_f$ is also small enough, our algorithm computes top-$k$ Manhattan spatial skyline points in a time that is almost linear in $n$.

## 2 Problem Definition

In the Manhattan Spatial Skyline Query problem [1], we have a set $P$ of $n$ data points in the $L_1$ plane. For a given set $Q$ of $m$ query points, a point $p \in P$ is a spatial skyline point if and only if it is not *spatially dominated* by any other point of $P$ with respect to $Q$. The notion of *spatial dominance* of a point over another point with respect to $Q$ is defined as follows:

**Definition 1 (spatial dominance).** *Let $p, p'$ be points in the $L_1$ plane and let $Q$ be a point set in the $L_1$ plane. We say that $p$ spatially dominates $p'$ with respect to $Q$ if and only if*

$$\forall q \in Q : \|p - q\|_1 \leq \|p' - q\|_1 \quad and \quad \exists q' \in Q : \|p - q'\|_1 < \|p' - q'\|_1.$$

**Definition 2 (skyline).** *Let $P$ and $Q$ be point sets in the $L_1$ plane. A point $p \in P$ is the spatial skyline point of $P$ with respect to $Q$ if and only if $p$ is not dominated by any other point of $P$ with respect to $Q$.*

We want to compute top-$k$ spatial skyline points in the $L_1$ plane with respect to $Q$, where $k$ and a monotone scoring function $f$ are given as part of the input. In this context, a scoring function is called monotone, if for any two points $p$ and $p'$

$$\forall q \in Q : \|p - q\|_1 \leq \|p' - q\|_1 \implies f(p) \leq f(p').$$

The top-$k$ Manhattan spatial skyline problem can be formalized as

*Problem 1 (Top-k Manhattan Spatial Skyline Query).*
Given:
at preprocessing time:
$$P = \{p_1, ...., p_n\} \quad \text{a set of data points in the } L_1 \text{ plane}$$
at query time:
$$Q = \{q_1, ..., q_m\} \quad \text{a set of query points}$$
$$f : P \to \mathbb{R}^+ \quad \text{a monotone scoring function}$$
$$k > 0 \quad \text{a parameter}$$
Task:
Compute a set $S_k \subseteq P$ of $k$ points in the plane, so that the points of $S_k$ are the $k$-smallest (with respect to $f$) skyline points of $P$ with respect to $Q$.

## 3 Related Work

**Skyline Computation.** The problem of computing the skyline was known as the maximal vector problem [2,3] where the goal is to find the subset of the vectors that are not dominated by any of the vectors from the set. Recently, the maximal vector problem was rediscovered by Börzsönyi et al. [4] who introduced *skyline queries* in database context. Since then, various algorithms for skyline queries have been introduced including the linear elimination-sort for skyline (LESS) algorithm [5], the sort-filter-skyline (SFS) algorithm [6], and the branch and bound skyline (BBS) algorithm [7].

*Spatial skyline queries* were introduced by Sharifzadeh and Shahabi [8] to support skyline queries over spatial data. Given a set $P$ of data points and a set $Q$ of query points in a $d$-dimensional space, spatial skyline queries retrieve skyline points among data points with respect to $Q$. For each data point, its spatial attribute is defined by its distance to query points in $Q$. Spatial dominance between any two data points is determined by their spatial attributes. Sharifzadeh and Shahabi presented two algorithms for the problem in the Euclidean space. Later, Lee et al. [9] improved the results of Sharifzadeh and Shahabi. Son et al. [1] recently considered the problem in the $L_1$ plane, which is also called *the Manhattan spatial skyline queries*, and proposed an efficient algorithm.

**Top-$k$ Skyline Computation.** Skyline queries provide a useful method to get desirable objects from massive data, but the resulting skyline objects can still be too big in size to process efficiently in practice. So it is more desirable to require users to specify some ranking function and to retrieve the most ideal $k$ skyline objects among skylines for small $k$. This problem is called *top-$k$ skyline queries*. Goncalves and Vidal [10] presented an algorithm to solve top-$k$ skyline queries using an arbitrary scoring function. Their algorithm maintains a sorted list of $n$ objects for each attribute and additionally for the scoring function, and takes $O(n^2d)$ time, where $d$ is the number of attributes.

Recently, Xu and Gao [11] proposed an algorithm for top-$k$ skyline queries that uses a reference object specified by the user to evaluate user preference. Meanwhile, there have been works to use subspace domination relationships in scoring skyline objects [12,13].

There have been another interesting line of research to retrieve top-$k$ objects for $k$ larger than the number of skyline objects [14,15]. They report $k$ objects such that each of them is either a skyline objects or is dominated only by the reported objects.

## 4   Algorithm

In this section, we propose an algorithm to compute top-$k$ Manhattan spatial skyline points involving a given monotone scoring function $f$.

We first observe some properties of the dominance relation and monotone scoring functions on a point.

**Lemma 1.** *If a point $p$ is not a skyline point then there is a skyline point $p'$ that dominates $p$.*

*Proof.* Among the points that dominate $p$, at least one of them is a skyline point by the transitivity of the dominance relation. By Definition 1, the remaining part is obvious.                                                                    ☐

For a monotone scoring function $f$ the following lemma holds.

**Lemma 2.** *[6] For $p$, $p' \in P$, if $p'$ dominates $p$, then $f(p') \leq f(p)$.*

Let $\mathcal{L}_q = [p_1, p_2, ..., p_n]$ be a sorted list of points in $P$ in the ascending order of distance from a query point $q$. For the moment we assume that no two points have the same distance from $q$. We will discuss later how to handle the case that more than one point of $P$ have the same distance from $q$. By Lemma 1 and Lemma 2 we get the following corollary.

**Corollary 1.** *If $p_j \in \mathcal{L}_q$ is not a skyline point, then there is a point $p_i \in \mathcal{L}_q$ that satisfies the following conditions (by Lemma 1 and Lemma 2):*

1. *$p_i$ is a skyline point, and $i < j$*
2. *$f(p_i) \leq f(p_j)$*

The basic idea of our algorithm is as follows: we test the points of $\mathcal{L}_q$ in order until we find the first $k$ skyline points and store them in $S_k$. To be specific, for a point $p$, we perform the dominance test only with points currently stored in $S_k$. So if $p$ is not dominated by any point of $S_k$, we insert $p$ to $S_k$ and update $V_k = \max_{s \in S_k} f(s)$.

After finding first $k$ skyline points, we only consider points $p \in \mathcal{L}_q$ that satisfy $f(p) < V_k$, as we already have $k$ skyline points that have a better score than $p$. For a point $p$ that satisfies $f(p) < V_k$, we perform the dominance test with the points in $S_k$. All skyline points that dominate $p$ should be in $S_k$ by Corollary 1, so testing with points in $S_k$ is sufficient. Whenever we find a new skyline point with a score function value smaller than the worst point of $S_k$, we update $S_k$ and $V_k$ by inserting the new point and removing the worst point. Details of our algorithm are as follows.

1 **Algorithm:** Top-$k$-MSSQ

**Input**: Two lists of points in $P$, one sorted along the line $y = x$ and the other sorted along the line $y = -x$, a set $Q$ of query points, a scoring function $f$ and a parameter $k$.

**Output**: A set $S_k$ of the $k$ best (with respect to $f$) Manhattan spatial skyline points of $P$ on queries of $Q$.

2 **begin**

3     Compute a sequence $p_1, p_2, \ldots, p_n$ of points in $P$ ordered increasingly by distance from a query point $q \in Q$;

4     $V_k := -\infty$, $S_k := \emptyset$;

5     **for** $i \to 1$ *to* $n$ **do**

6        **if** $p_i$ *is not dominated by any point in $S_k$* **then**

7           insert $p_i$ to $S_k$;

8           $V_k := \max \{V_k, f(p_i)\}$;

9           **if** $|S_k| \geq k$ **then**

10              break;

11     **for** $j \to i + 1$ *to* $n$ **do**

12        **if** $f(p_j) \geq V_k$ **then**

13           continue;

14        **if** $p_j$ *is not dominated by any point in $S_k$* **then**

15           insert $p_j$ to $S_k$;

16           delete a point $p'$ in $S_k$ such that $f(p') = V_k$;

17           update $V_k$;

18     **return** $S_k$;

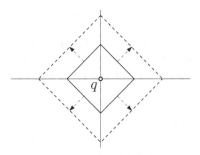

**Fig. 1.** $L_1$ circle centered at $q$, and expanding with increasing radius

In the following, we explain how to sort $P$ efficiently and how to break the tie for two data points that have the same distance from the chosen query $q$.

### 4.1   Sorting Data Points from a Query Point

To compute $\mathcal{L}_q$ efficiently, we preprocess data points of $P$ and maintain two lists of $P$, one sorted along the line $y = x$ and the other sorted along the line $y = -x$. It is possible that more than one data point have the same order along the line $y = x$ or $y = -x$. If this is the case, we sort them by their $x$-coordinates.

Once $Q$ is given, we choose an arbitrary query point $q$ from $Q$. Consider the subdivision of the plane into quadrants defined by the horizontal line and vertical line passing through $q$ (Figure 1). We sort points lying in each quadrant separately, and then merge the four sorted lists into one. In the following, we explain how to sort points of $P$ contained in the top left quadrant only. Data points contained in other quadrants can be sorted analogously.

Let $I(p)$ be the $y$-intercept value of the line passing through $p$ with slope 1. We construct the list $\mathcal{L}_I$ of data points of $P$ sorted in the ascending order of $I$ in the preprocessing time. By using $\mathcal{L}_I$, we sort data points of $P$ lying in the top left quadrant in the ascending order of $I$ as follows. We compute $I(q)$ and locate the data point $p$ in $\mathcal{L}_I$ whose $I$ value is the smallest among all data points $p'$ with $I(p') \geq I(q)$. This can be done using a binary search algorithm. Then we scan the list $\mathcal{L}_I$ starting from $p$ in the ascending order of $I$ and report each data point in $\mathcal{L}_I$ if it lies in the top left quadrant. This gives us the list of data points contained the top left quadrant sorted in the order of the distance from $q$. We also compute the sorted lists for the other three quadrants analogously, and merge the four sorted lists into one to get $\mathcal{L}_q$.

**Lemma 3.** *We can sort $n$ data points in $P$ in the ascending order of distance from a query point $q \in Q$ in $O(n)$ time after $O(n \log n)$ time preprocessing.*

### 4.2   Sorting Data Points of the Same Distance from $q$

Consider now the case that two or more data points have the same distance from $q$. Then Corollary 1 may not hold for $\mathcal{L}_q$: For two points $p_i, p_j$ in $\mathcal{L}_q$ with $i < j$ and $\|p_i - q\|_1 = \|p_j - q\|_1$, it is possible that none of $p_1, p_2, \ldots, p_{i-1}$ of $\mathcal{L}_q$ spatially dominates $p_i$ but

$p_j$ spatially dominates $p_i$. For example, consider two query points $q_1 = (0,0)$ and $q_2 = (5,0)$ and three data points $p_1 = (0,2)$, $p_2 = (2,2)$, and $p_3 = (3,1)$. Then $\mathcal{L}_{q_1} = [p_1, p_2, p_3]$, and $p_1$ is a skyline point. But $p_2$ is spatially dominated only by $p_3$.

Therefore we have to refine the ordering in a suitable way: we first sort by distance from $q$, and if more than one point have the same distance from $q$ we sort them by sum-of-distances from $Q$. Then Corollary 1 holds for the resulting list $\mathcal{L}_q$. Note that two points of the same distance from $q$ may also have the same sum-of-distances from $Q$. If this is the case, they do not dominate each other (Definition 1) and we use any order of them. In the following, we first propose a method for computing the sum-of-distances of a data point from $Q$. Then we will explain how to use this method for sorting data points of the same distance from $q$ efficiently.

Let $\text{SDIST}(p)$ denote the sum-of-distances of $p$ from $Q$. In general it takes $O(m)$ time to compute $\text{SDIST}(p)$ for a data point $p$ and therefore $O(nm)$ time for all data points of $P$. We propose another method that computes $\text{SDIST}(p)$ for each data point $p$ in $O(\log m)$ time after $O(m \log m)$ time preprocessing.

Consider any two data points $p$ and $p'$. Then

$$\text{SDIST}(p') =$$
$$\text{SDIST}(p) + \sum_{q \in Q}(|p'.x - q.x| - |p.x - q.x|) + \sum_{q \in Q}(|p'.y - q.y| - |p.y - q.y|)$$
$$\tag{1}$$

Equation (1) suggests that $\text{SDIST}(p')$ can be computed immediately once we know the values of $\text{SDIST}(p)$ and the last two terms. In the following we will show how to evaluate the second term of Equation (1) efficiently. The third term can also be evaluated in the same way.

We can rewrite the second term as follows. Without loss of generality, we assume that $p.x \leq p'.x$. Let $Q_L = \{q \mid q.x < p.x, \forall q \in Q\}$, $Q_M = \{q \mid p.x \leq q.x \leq p'.x, \forall q \in Q\}$, and $Q_R = \{q \mid q.x > p'.x, \forall q \in Q\}$.

$$\sum_{q \in Q}(|p'.x - q.x| - |p.x - q.x|)$$
$$= \sum_{q \in Q_L}(p'.x - q.x) + \sum_{q \in Q_M}(p'.x - q.x) + \sum_{q \in Q_R}(q.x - p'.x)$$
$$- \left(\sum_{q \in Q_L}(p.x - q.x) + \sum_{q \in Q_M}(q.x - p.x) + \sum_{q \in Q_R}(q.x - p.x)\right)$$
$$= |Q_L|(p'.x - p.x) + |Q_M|(p'.x + p.x) - 2\sum_{q \in Q_M}q.x + |Q_R|(p.x - p'.x)$$

The last equation can be evaluated if we maintain a data structure on $Q$ that supports the following two queries: (1) given a real value $r$, return the number of queries $q$ satisfying $q.x \leq r$ (or $q.x \geq r$), (2) given two real values $r_1$ and $r_2$ with $r_1 \leq r_2$, return the sum of $x$-coordinates of queries $q$ satisfying $r_1 \leq q.x \leq r_2$.

We can support a query of type (1) by using a binary search algorithm and a query of type (2) by using a partial sum query structure [16] after sorting the query points in the ascending order of their $x$-coordinates. For a set of elements stored in an array, a partial sum query structure returns the sum of the elements that lie in an orthogonal query range [16]. Chan and Pătrașcu showed that the partial sum query for one-dimensional array of size $m$ can be answered in $O(\log m / \log \log m)$ time after $O(m \log^{0.5+\epsilon} m)$ time preprocessing for an arbitrarily small constant $\epsilon > 0$ [16].

**Lemma 4.** *For any two data points $p$ and $p'$, we can compute* $\mathrm{SDIST}(p')$ *in* $O(\log m)$ *time if we know the value of* $\mathrm{SDIST}(p)$.

Now we show how to sort points that have the same distance from $q$ in the ascending order of SDIST. Let $P_t$ denote the set of points of $P$ at distance $t$ from $q$. Assume that $P_t$ contains more than one data point for some fixed $t > 0$. Then the points in $P_t$ all lie on a $L_1$ circle $C$ centered at $q$ with radius $t$, each lying on one of four segments of $C$.

Let $p_L$ be the point of $P$ with the smallest $x$-coordinate, and assume that we have already computed $\mathrm{SDIST}(p_L)$ by taking the sum-of-distances of $p_L$ from $Q$. We can compute SDIST values of points in $P_t$ by using the value of $\mathrm{SDIST}(p_L)$ as stated in Lemma 4. For these SDIST values, the following lemma holds.

**Lemma 5.** *The function* SDIST *is a convex along a line.*

*Proof.* Consider any line $\ell$. Clearly the distance function from a query point $q$ is convex along $\ell$. Since SDIST is the sum of these convex distance functions, SDIST is also convex [17]. □

Consider the set of points in $P_t$ that lie on a line segment of $C$. These points are already sorted by their $x$-coordinates in the preprocessing phase. Because of convexity of SDIST in Lemma 5, we sort them in the ascending order of SDIST in time linear to the number of points. We do this for the remaining points of $P_t$ in a similar way, and get the final list of points in $P_t$ sorted in the ascending order of SDIST in time $O(|P_t|)$.

**Lemma 6.** *We can sort data points having the same distance from $q$ in the ascending order of* SDIST *in time linear to the number of the points.*

### 4.3  Analysis

We outline our algorithm as follows. For a given set $Q$ of $m$ query points, we sort $n$ data points of $P$ in the order of distance from a query point as described in Sections 4.1 and 4.2. For the sorted list we can find top-$k$ skyline points by using Algorithm Top-$k$-MSSQ in Section 4.

Let us analyze the running time of the algorithm. In the preprocessing phase, we spend $O(n \log n)$ time to compute two lists of points in $P$, one sorted along the line $y = x$ and the other sorted along the line $y = -x$.

For Algorithm Top-$k$-MSSQ, the time complexity is as follows. For a query set $Q$, we spend $O(n)$ time to sort $P$ in the ascending order of distance from a query point $q \in Q$ and $O(n \log m)$ time to break ties.

We evaluate $f$ for each point in lines 8 and 12, so we spend $O(nt_f)$ time to evaluate $f$ for all points. In lines 6 and 14, we check whether a data point is dominated by any point in $S_k$. In a straightforward way, it would take $O(km)$ time, but in the $L_1$ plane we can do it more efficiently by using an idea similar to the one used in Son et al. [1] as follows.

Let $C(p, q)$ be the $L_1$ disk centered at $q$ with radius $\|p - q\|_1$. Let $R(p)$ denote the common intersection $\cap_{q \in Q} C(p, q)$. Definition 1 implies that a data point $p'$ dominates $p$ if and only if $p'$ lies in $R(p)$. Therefore, it suffices to check whether $R(p)$ contains any point of $S_k$ or not. To do this, we use a dynamic orthogonal range query structure of $S_k$ [18]. This structure allows us to determine whether an orthogonal range contains any point of $S_k$. We construct this data structure for $S_k$ and update it when $S_k$ is updated.

To use the dominance test described above, we compute $R(p)$ in $O(\log m)$ time after constructing a data structure in $O(m \log m)$ time [1]. The dynamic orthogonal range query structure for $S_k$ uses $O(k)$ space and takes $O(\log k)$ time for a query and $O(\log k)$ amortized time for an update [18]. Therefore the total time complexity of our algorithm is as follows.

**Theorem 1.** *Given n data points and m query points in the $L_1$ plane, we can compute the top-k Manhattan spatial skyline points for a monotone scoring function f in time $O(n(\log m + \log k + t_f) + m \log m)$ using $O(n + m)$ space after $O(n \log n)$- time preprocessing, where $t_f$ is the time to evaluate f.*

In many practical applications, $k$ and $m$ are much smaller than $n$, and $t_f$ is sublinear to $m$. In this case our algorithm spends near linear time to compute top-$k$ Manhattan spatial skyline points.

## 5    Experimental Evaluation

In this section, we outline our experimental settings and show evaluation results to validate the efficiency and effectiveness of our algorithms. We carry out our experiments on Linux with Intel Q6600 CPU and 3GB memory. The algorithm is coded in C++.

We compare our algorithm Top-$k$-MSSQ with a straightforward implementation of reporting the $k$ best ones (with respect to $f$) from the skylines returned by MSSQ [1]. We use both synthetic datasets and a real-world dataset of points of interest (POI) in California [1].

A synthetic dataset contains up to a million uniformly distributed locations in the plane. We use five datasets with 200k, 400k, 600k, 800k and 1M data points. Data points in the synthetic datasets are independent, and uniformly distributed in the unit square. Query points are distributed in a randomly selected bounding square contained in the unit square. We also investigate the effect of $m$, $k$ and $f$. The parameters are summarized in Table 1.

We also use a POI dataset to validate our algorithm. This dataset has 104,770 locations in 63 different categories. Except dataset cardinality, we use the same parameters in Table 1.

We first show the efficiency of our algorithm for synthetic datasets. Figure 2 shows the efficiency of the two algorithms for synthetic datasets when $f(p) = \text{SDIST}(p)$.

---

[1] Available at http://www.cs.fsu.edu/~lifeifei/SpatialDataset.htm

**Table 1.** Parameters used for synthetic datasets

| Parameter | Setting (default value is underlined) |
|---|---|
| Dataset cardinality | 200k, 400k, **600k**, 800k, 1M |
| The number of query points | 4, 8, **12**, 16, 20 |
| $k$ | 4, 8, **12**, 16, 20 |
| $f(p)$ | SDIST, MAX |

We observe that our proposed algorithm is faster than MSSQ. The performance gap increases as $n$ increases. We also show the response times of the algorithms for $f(p) =$ MAX$(p)$ in Figure 3.

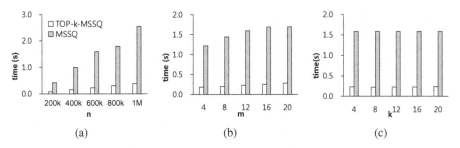

**Fig. 2.** Response time over (a) $n$ (b) $m$ (c) $k$, for synthetic dataset, when $f(p) = $ SDIST$(p)$

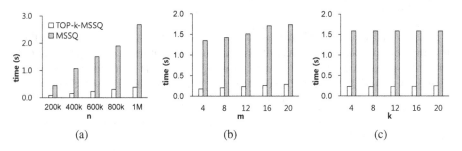

**Fig. 3.** Response time over (a) $n$ (b) $m$ (c) $k$, for synthetic dataset, when $f(p) = $ MAX$(p)$

Now we show the experimental results for the POI dataset.

Figure 4 shows experimental results for $f(p) = $ SDIST$(p)$. We observe that our proposed algorithm is faster than MSSQ. We observe similar trends for $f(p) = $ MAX$(p)$ as shown in Figure 5.

All the experimental results show that Top-$k$-MSSQ outperforms MSSQ for both synthetic datasets and the POI dataset.

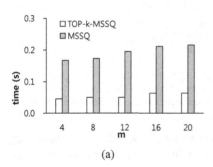
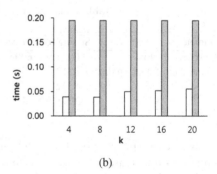

(a)                                    (b)

**Fig. 4.** Response time over (a) $m$ (b) $k$ for the POI dataset when $f(p) = \mathrm{SDIST}(p)$

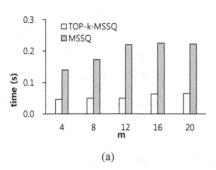
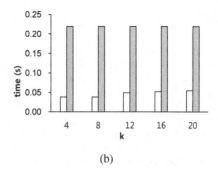

(a)                                    (b)

**Fig. 5.** Response time over (a) $m$ (b) $k$ for the POI dataset when $f(p) = \mathrm{MAX}(p)$

## 6   Conclusion

We have studied the top-$k$ Manhattan spatial skyline query problem. Our algorithm can compute top-$k$ skyline points with respect to a monotone scoring function $f$ in time $O(n(\log m + \log k + t_f) + m \log m)$ using $O(n + m)$ space after $O(n \log n)$-time preprocessing. In addition our extensive experiments validate the efficiency and effectiveness of our proposed algorithms.

## References

1. Son, W., Hwang, S.-w., Ahn, H.-K.: MSSQ: Manhattan Spatial Skyline Queries. In: Pfoser, D., Tao, Y., Mouratidis, K., Nascimento, M.A., Mokbel, M., Shekhar, S., Huang, Y. (eds.) SSTD 2011. LNCS, vol. 6849, pp. 313–329. Springer, Heidelberg (2011)
2. Kung, H.T., Luccio, F., Preparata, F.: On finding the maxima of a set of vectors. Journal of the Association for Computing Machinery 22(4), 469–476 (1975)
3. Preparata, F., Shamos, M.: Computational Geometry: An Introduction. Springer-Verlag (1985)
4. Börzsönyi, S., Kossmann, D., Stocker, K.: The skyline operator. In: ICDE 2001: Proc. of the 17th International Conference on Data Engineering, pp. 421–430. IEEE Computer Society (2001)

5. Godfrey, P., Shipley, R., Gryz, J.: Maximal vector computation in large data sets. In: VLDB 2005: Proc. of the 31st International Conference on Very Large Data Bases, pp. 229–240. IEEE Computer Society (2005)
6. Chomicki, J., Godfery, P., Gryz, J., Liang, D.: Skyline with presorting. In: ICDE 2003: Proc. of the 19th International Conference on Data Engineering, pp. 717–816. IEEE Computer Society (2003)
7. Papadias, D., Tao, Y., Fu, G., Seeger, B.: An optimal and progressive algorithm for skyline queries. In: SIGMOD 2003: Proc. of the 2003 ACM SIGMOD International Conference on Management of Data, pp. 467–478. ACM (2003)
8. Sharifzadeh, M., Shahabi, C.: The spatial skyline queries. In: VLDB 2006: Proc. of the 32nd International Conference on Very Large Data Bases, pp. 751–762, VLDB Endowment (2006)
9. Lee, M.W., Son, W., Ahn, H.K., Hwang, S.W.: Spatial skyline queries: exact and approximation algorithms. GeoInformatica 15(4), 665–697 (2011)
10. Goncalves, M., Vidal, M.-E.: Reaching the top of the skyline: An efficient indexed algorithm for top-k skyline queries. In: Bhowmick, S.S., Küng, J., Wagner, R. (eds.) DEXA 2009. LNCS, vol. 5690, pp. 471–485. Springer, Heidelberg (2009)
11. Xu, C., Gao, Y.: A novel approach for selecting the top skyline under users' references. In: ICIME 2010: Proc. of the 2nd IEEE International Conference on Information Management and Engineering, pp. 708–712. IEEE (2010)
12. Vlachou, A., Vazirgiannis, M.: Ranking the sky: Discovering the importance of skyline points through subspace dominance relationships. Data & Knowledge Engineering 69(9), 943–964 (2010)
13. Lee, J., You, G.W., Hwang, S.W.: Personalized top-k skyline queries in high-dimensional space. Information Systems 34(1), 45–61 (2009)
14. Brando, C., Goncalves, M., González, V.: Evaluating top-k skyline queries over relational databases. In: Wagner, R., Revell, N., Pernul, G. (eds.) DEXA 2007. LNCS, vol. 4653, pp. 254–263. Springer, Heidelberg (2007)
15. Goncalves, M., Vidal, M.-E.: Top-k skyline: A unified approach. In: Meersman, R., Tari, Z. (eds.) OTM-WS 2005. LNCS, vol. 3762, pp. 790–799. Springer, Heidelberg (2005)
16. Chan, T.M., Pătraşcu, M.: Counting inversions, offline orthogonal range counting, and related problems. In: SODA 2010: Proc. of the 21st Annual ACM-SIAM Symposium on Discrete Algorithms, pp. 161–173. Society for Industrial and Applied Mathematics (2010)
17. Rockafellar, R.T.: Convex Analysis. Princeton University Press (1996)
18. Blelloch, G.E.: Space-efficient dynamic orthogonal point location, segment intersection, and range reporting. In: SODA 2008: Proc. of the 19th Annual ACM-SIAM Symposium on Discrete Algorithms, pp. 894–903. Society for Industrial and Applied Mathematics (2008)

# On Generalized Planar Skyline
# and Convex Hull Range Queries

Nadeem Moidu, Jatin Agarwal, Sankalp Khare,
Kishore Kothapalli, and Kannan Srinathan

Center for Security, Theory and Algorithmic Research (CSTAR),
IIIT Hyderabad (India)

**Abstract.** We present output sensitive techniques for the generalized reporting versions of the planar range maxima problem and the planar range convex hull problem. Our solutions are in the pointer machine model, for orthogonal range queries on a static point set. We solve the planar range maxima problem for two-sided, three-sided and four-sided queries. We achieve a query time of $O(\log n + c)$ using $O(n)$ space for the two-sided case, where $n$ denotes the number of stored points and $c$ the number of colors reported. For the three-sided case, we achieve query time $O(\log^2 n + c \log n)$ using $O(n)$ space while for four-sided queries we answer queries in $O(\log^3 n + c \log^2 n)$ using $O(n \log n)$ space. For the planar range convex hull problem, we provide a solution that answers queries in $O(\log^2 n + c \log n)$ time, using $O(n \log^2 n)$ space.

## 1 Introduction

Generalized intersection searching was introduced by Janardan and Lopez [14]. Since then there has been a considerable amount of work on generalized searching and reporting problems [9][10][11][3][12][13][1][7][22][20]. A comprehensive survey of developments in the area can be found in [8]. In the generalized version of a problem, points are associated with colors. Colors capture the idea of membership, dividing objects into groups based on some common property. Such categorization has practical applications in databases, spatial information systems and other areas where objects are separable into classes and queries involve membership checking in these classes.

We present here solutions for the generalized (colored) range-query versions of two classic problems in computational geometry – *Convex Hull* and *Skyline*[1] – in the two dimensional setting, for a static set of points. Both these problems, in our knowledge, have not been tackled before in literature. Generalized intersection searching problems are broadly divided into two kinds, *reporting* and *counting*. In the former, the goal is to report the distinct colors whose points fall in the query range, while in the latter the goal is to count the number of such colors. Our solutions are for the reporting versions of both problems.

---

[1] We will use the terms *maxima* and *skyline* interchangeably in this paper.

S.P. Pal and K. Sadakane (Eds.): WALCOM 2014, LNCS 8344, pp. 34–43, 2014.

## 1.1  Generalized Planar Range Maxima Queries

A point $p = (x, y)$ is said to *dominate* another point $p' = (x', y')$ if both $x \geq x'$ and $y \geq y'$ are true. Given a set of points $P$ in the plane, the set $m \subseteq P$ of all points which are not dominated by any other point in $P$ is known as the maximal set. In the planar range maxima problem, given a range query $q$, we are asked to report the maximal set of the points in $P \cap q$. Note that range maxima queries are sometimes referred to as range skyline queries. We present a solution for the *reporting* version of the problem, where the points must be preprocessed in a way such that given an orthogonal range query $q$, we can efficiently report the distinct colors on the maximal set of $P \cap q$.

## 1.2  Generalized Planar Range Convex Hull Queries

In the planar convex hull range query problem, given a set of points $P$ and a query region $q$, the goal is to find the convex hull of the points in $P \cap q$. We present a solution for generalized *reporting* of the convex hull in a orthogonal range $q$, for which we must preprocess the points in a way such that we can report the distinct colors that constitute the convex hull of the points in $P \cap q$.

# 2  Previous Work

Kalavagattu et al. [15] studied the problem of counting and reporting points belonging to the maxima in an orthogonal range query on a static set of points. Brodal et al. [5] presented a solution for the dynamic version of the same problem. Rahul et al. [21] solved a similar problem where the maxima and the range query are based on different sets of dimensions. The colored version of the range maxima problem has not been studied before.

Brass et al. [4] studied the range convex hull problem and presented a solution using range trees and a method similar to the gift wrapping algorithm [19]. Moidu et al. [18] presented a more efficient solution, using a novel approach using a modified version of the range tree. The colored version of the range convex hull problem has also not been studied before.

# 3  Preliminaries

We assume that all points are distinct and have integer co-ordinates. The more general setting can be transformed to one with integer co-ordinates by reduction to rank space using standard methods [6][2]. Let $\mathcal{C}$ be the set of all colors. Let $n = |P|$. Clearly, $|\mathcal{C}| \leq n$ (if $|\mathcal{C}| = n$, then the problem becomes a case of standard intersection searching, without colors). We encode the colors as integers from 1 to $n$ for notational and operational convenience.

For both the problems under consideration, let $c$ be the number of distinct colors intersected by the query range. Our answer, therefore, will have size $O(c)$. To be output sensitive, we would therefore like our solutions to take time which is a function of $c$ (and not a function of the total number of points on the maxima or convex hull). This rules out any method involving the computation of all maximal or convex hull points followed by some processing on them.

All queries studied in the following sections are orthogonal and axis-parallel, unless explicitly specified otherwise. All results are in the pointer machine model.

### 3.1   Generalized Reporting in One Dimension

Given a set of points in one dimension where each point has a color (not necessarily unique) associated with it, we want to preprocess the point-set such that given a query range we can efficiently report the distinct colors in that range.

Gupta et al. [11] showed a transformation which reduces the generalized one dimensional range reporting problem into the standard grounded range reporting problem in two dimensions and solved the problem using a *priority search tree* (PST) [17]. Thus, for a static set of points in one dimension, the $c$ distinct colors intersected by a query range can be reported using a $O(n)$ space and $O(n \log n)$ preprocessing time data structure $\mathcal{D}$ which answers queries in $O(\log n + c)$ time per query. We use this result in our solutions for both maxima and convex hull.

### 3.2   Heavy-light Decomposition

Heavy-light decomposition is a technique which allows us to break down a rooted tree into a set of mutually disjoint paths. It was first used in literature by Tarjan [25] while the exact phrase was coined by Sleator and Tarjan when they used the technique in their analysis of link-cut trees [23][24].

Let $T$ be a rooted $n$-ary tree. Let size($v$) be the number of nodes in the subtree rooted at a node $v$. An edge $(p, q)$, where $q$ is a child of $p$, is labeled *heavy* if size($q$) $> \frac{1}{2} \cdot$ size($p$) and *light* otherwise. A tree with edges labeled in this manner is said to be *decomposed*. The following properties can be shown easily for a heavy-light decomposed tree:

- At most 1 edge from a node to its children can be heavy.
- Each connected set of heavy edges forms a vertex-disjoint path. We call such a path a *heavy path*.
- A path $(v, u)$ in the tree, where $u$ is an ancestor of $v$, will consist of $O(\log n)$ light edges and $O(\log n)$ heavy paths.

## 4   Generalized Range Maxima Queries

Due to the nature of information conveyed by the maximal chain of a set of points, it is common to perform orthogonal range queries that are unbounded in one or two directions. We present separate solutions to three representative types of range queries for the maxima problem.

## 4.1 Two Sided Queries

We begin with the simplest kind of maxima queries, where the query range is unbounded in two directions. We solve for query ranges of the type $[x_l, \infty) \times [y_l, \infty)$. Notice that in such queries, the range maxima is nothing but a continuous subset of the maximal chain of the entire point-set. Therefore for this type of two-sided query, reporting the generalized range maxima is equivalent to the generalized reporting problem in one dimension. This can be solved using the method described in Section 3.1. The remaining three types of two-sided queries, however, cannot be solved using this approach.

**Theorem 1.** *Let $P$ be a set of colored points in two dimensions. $P$ can be preprocessed into a $O(n)$ space and $O(n \log n)$ preprocessing time data structure such that given a two-sided query $q$ unbounded in the positive $x$ and positive $y$ directions, the $c$ distinct colors in the maxima of $P \cap q$ can be reported in $O(\log n + c)$ time.*

## 4.2 Three Sided Queries

In a three sided query, the maximal chain need not be a continuous subset of the maximal chain for the entire point set. We solve the problem for three sided queries unbounded in the positive $x$ direction, i.e. queries of the type $[x_l, \infty) \times [y_l, y_h]$, where $y_l < y_h$.

We construct a one dimensional range tree on the $x$ co-ordinates of the points in $P$. Let us call this tree $T_x$. Let $S(v)$ be the subtree of an internal node $v$ in $T_x$. At each internal node $v$, we store a pointer to the point in $S(v)$ having the maximum $y$ co-ordinate.

For a point $p = (p_x, p_y)$, let next $(p)$ be the point with maximum $y$ co-ordinate in its south-east quadrant, i.e. in the range $[p_x, \infty) \times (-\infty, p_y]$. We construct a graph where each point in $P$ is a vertex and there is an edge from each $p_i$ to $next(p_i)$. We create a dummy vertex $p_{null}$. For all such points $p_i$ for which no $next(p_i)$ value exists in $P$, we add an edge $(p_i, p_{null})$ to the graph. The next(.) values can be computed in $O(n \log n)$ time using the tree $T_x$ described above. Each vertex in this graph, except $p_{null}$ has exactly one outgoing edge (edge pointing towards its next(.) point) and there are no cycles. Therefore, this graph is a tree. We call it $\mathcal{T}$. A maximal chain, as reported by a three sided query, will be a path in $\mathcal{T}$.

We now decompose $\mathcal{T}$ using heavy-light decomposition. We preprocess all heavy paths with the data structure $\mathcal{D}$ of Section 3.1. For light edges we do not perform any preprocessing.

**Theorem 2.** *Let $P$ be a set of colored points in two dimensions. $P$ can be processed into a $O(n)$ space and $O(n \log n)$ preprocessing time data structure such that given a three sided query $q$, unbounded on the right, the $c$ distinct colors in the maxima of $P \cap q$ can be reported in $O(\log^2 n + c \log n)$ time per query.*

*Proof.* The preprocessing stage involves:

1. The building of the tree $T_x$, which is a one dimensional range tree. It takes $O(n)$ space and can be built in $O(n \log n)$ time.
2. The building of the tree $\mathcal{T}$, which takes $O(n \log n)$ time. The number of nodes in $\mathcal{T}$ is the same as the number of points in $P$, therefore $\mathcal{T}$ has size $O(n)$. In addition, for each heavy path in $\mathcal{T}$ we build the data structure $\mathcal{D}$ of section 3.1. If the length of a heavy path is $l_h$, it takes $O(l_h)$ space and $O(l_h \log l_h)$ time to build the required data structure. Clearly the overall space requirement is $O(n)$ while the time required is $O(n \log n)$.

Thus, for preprocessing, the total space requirement is $O(n)$ and the total time required is $O(n \log n)$.

Given a query $[x_l, \infty) \times [y_l, y_h]$, let $p_a \in P$ be the point with the maximum $y$ co-ordinate lying in the range $[x_l, \infty) \times [-\infty, y_h]$. This point can be found in $O(\log n)$ time by an orthogonal range successor query on the tree $T_x$. The point $p_a$ will be one end point of the maximal chain in $P \cap q$. Let $p_b$ be the other end point and $\mathcal{P}_{mc}$ be the path in $\mathcal{T}$ from $p_a$ to $p_b$. It can be shown that $p_b$ will be an ancestor of $p_a$ in the tree $\mathcal{T}$. Therefore, it follows from Section 3.2 that the path $\mathcal{P}_{mc}$ will consist of $O(\log n)$ light edges and $O(\log n)$ heavy paths.

For heavy paths, we can report the distinct colors using the preprocessed data structure $\mathcal{D}$ of Section 3.1. This takes $O(\log n + c)$ time per heavy path. Since there are $O(\log n)$ heavy paths, the total time required is $O(\log^2 n + c \log n)$ For the remaining points, i.e. those that are not part of a heavy path, we simply report the colors when such a point is encountered. This takes a total of $O(\log n)$ time since there are $O(\log n)$ light edges. Thus reporting the colors on the maximal chain for a three sided query can be done in $O(\log^2 n + c \log n)$. However, there still remains the issue of duplicates.

To ensure that each color in the maxima is reported only once, we leverage the fact that colors are encoded as integers from 1 to $\mathcal{C}$. For every query we initialize a bit-array $B$, of size $\mathcal{C}$, with all $B[i]$ set to 0. As colors are reported from $\mathcal{T}$, for every reported color $k$ we check the value of $B[k]$. If $B[k] = 0$, we output color $k$ and set $B[k]$ to 1, else we do not output color $k$ and move on. Since there are a total of $O(c)$ colors output by $\mathcal{T}$, the aggregation process will not be a dominating factor in the query time.                     □

### 4.3   Four Sided Queries

Four sided (rectangular) orthogonal range queries can be answered using a range tree together with the structure for three sided queries (Section 4.2).

**Theorem 3.** *Let $P$ be a set of colored points in two dimensions. $P$ can be processed into a $O(n \log n)$ space and $O(n \log^2 n)$ preprocessing time data structure such that given a four sided query $q$, the $c$ distinct colors in the maxima of $P \cap q$ can be reported in $O(\log^3 n + c \log^2 n)$ time per query.*

*Proof.* Let $T_x$ be a one dimensional range tree on the $x$ co-ordinates of all points in $P$. Let $S(v)$ be the subtree of an internal node $v$ in $T_x$. At each internal node $v$, we populate the structures described in Section 4.2 for three sided queries, using the points in $S(v)$ as the input. This takes $O(|S(v)|)$ space and $O(|S(v)| \log |S(v)|)$ time per internal node $v$.

Now consider the set $S_d$ of all nodes lying at depth $d$ in the primary tree $T_x$. The total time required to preprocess all nodes in $S_d$ will be:

$$\sum_{v \in S_d} O(|S(v)| \log |S(v)|) \leq \sum_{v \in S_d} O(|S(v)| \log n) = O(\log n \sum_{v \in S_d} |S(v)|) = O(n \log n)$$
(1)

Likewise, the total space required to store the preprocessed data structures on each node in $S_d$ will be:

$$\sum_{v \in S_d} O(|S(v)|) = O(\sum_{v \in S_d} |S(v)|) = O(n)$$
(2)

There will be $O(\log n)$ levels in the tree $T_x$. From equations (1) and (2) it follows that preprocessing the entire tree will take $O(n \log^2 n)$ time and $O(n \log n)$ space.

Given a query $q = [x_l, x_h] \times [y_l, y_h]$, we first query $T_x$ with the range $[x_l, x_h]$. This gives us $O(\log n)$ canonical nodes. We process these canonical nodes from right to left, starting with the canonical node whose $x$ range has upper boundary $x_h$. On the first canonical node, we do a three sided query $[x_l, \infty] \times [y_l, y_h]$. Let $y_m$ be the $y$ co-ordinate of the point with maximum $y$ on the maximal chain of the points covered by this canonical node. Points whose $y$ co-ordinate is less than $y_m$ in the remaining canonical nodes cannot lie on the maximal chain of the query region. So on the next canonical node, we do a three sided query $[x_m, \infty] \times [y_m, y_h]$, and so on. Each of these queries will return at most $c$ colors. The time taken per canonical node will be $O(\log^2 n + c \log n)$. Since there are $O(\log n)$ canonical nodes, the total time per query will be $O(\log^3 n + c \log^2 n)$. Note that even though the colors output by each canonical node will be distinct within themselves, the overall set may have duplicates. To remove them, we will once again use the bit array method described in the proof of Theorem 2.    □

## 5    Generalized Range Convex Hull Queries

We present an output sensitive algorithm to report the distinct colors on the convex hull of the points lying in a query range. We use the modified two dimensional range tree proposed by Moidu et al. [18] together with the generalized one dimensional range reporting structure of Section 3.1.

### 5.1    Preprocessing

We build a modified range tree $\mathcal{R}$, as described in [18], on the set of points $P$ and supplement it with additional preprocessed data structures to support generalized range reporting of the convex hull.

Constructing $\mathcal{R}$ takes $O(n \log n)$ time. Within $\mathcal{R}$, we call the primary range tree, built using the $x$ co-ordinates of the points, $T_x$. Each vertex $v_i$ of $T_x$ has a secondary tree associated with it, which is built using the $y$ co-ordinates of the points rooted at $v_i$. We call this tree $T_y(v_i)$. At each canonical node in every secondary tree $T_y(v_i)$, we precompute the convex hull of the points rooted at $T_y(v_i)$.

**Lemma 1.** *The convex hulls of the points rooted at the canonical nodes in $\mathcal{R}$ can be computed in a total of $O(n \log^2 n)$ time.*

*Proof.* Consider a node $v_x$ in the primary tree $T_x$. Let the number of points rooted at $v_x$ be $n_{v_x}$. Therefore the total number of points in the tree $T_y(v_x)$ will also be $n_{v_x}$.

Instead of computing the convex hull of the points rooted at each node of $T_y(v_x)$ ab initio, we will proceed in a bottom-up fashion starting at the deepest (lowermost) level, merging the convex hulls of the children of each canonical node to get the convex hull of the points at the canonical node itself. Merging of every pair of child hulls to form the parent hull takes $O(\log(n_1 + n_2))$ time using the method described by Kirkpatrick and Snoeyink [16] to compute outer tangents for disjoint convex polyhedra. Here $n_1$ and $n_2$ are the number of points in the respective child convex hulls. Clearly, even in the worst case, $n_1 + n_2 \leq n_{v_x}$ (for tree $T_y(v_x)$). Therefore each merge step takes at most $O(\log n_{v_x})$ time. Notice that the total number of merge operations required to compute the convex hulls for all canonical nodes in $T_y(v_x)$ is the same as the total number of canonical nodes (non-leaf nodes) present in $T_y(v_x)$, i.e. $O(n_{v_x})$. Therefore, the total pre-processing time required to compute the convex hull of each canonical node in a secondary tree $T_y(v_x)$ will be $O(n_{v_x} \log n_{v_x})$, where $n_{v_x} = |T_y(v_x)|$.

Consider the set $S_d$ of all nodes lying at depth $d$ in the primary tree $T_x$. The time required to preprocess the secondary trees corresponding to each node $s \in S_d$ will be $O(n_s \log n_s)$, where $n_s$ is the number of nodes rooted at $s$. The total time required to preprocess all nodes in $S_d$ will be

$$\sum_{s \in S_d} O(n_s \log n_s) \leq \sum_{s \in S_d} O(n_s \log n) = O\left(\log n \sum_{s \in S_d} n_s\right) = O(n \log n) \quad (3)$$

There are a total of $\log n$ levels in the tree $T_x$. Therefore, preprocessing the entire tree at the cost of $O(n \log n)$ per level will take $O(n \log^2 n)$ time. $\qquad \square$

Once the convex hulls of all canonical nodes have been computed, we preprocess each canonical convex hull (convex hull of the points rooted at the canonical node) for generalized range reporting. To do this, we first linearize the list of convex hull points and then preprocess it using the data structure $\mathcal{D}$ described in Section 3.1. At each canonical node, we store the convex hull points in counter-clockwise order in an array. We store a pointer from each convex hull point to its index in the array, allowing lookups in both directions (array index to convex hull point and vice-versa). Using the array indices as one dimensional co-ordinates, we preprocess the array for generalized one dimensional range reporting using the data structure $\mathcal{D}$.

**Lemma 2.** *The canonical convex hulls in $\mathcal{R}$ can be preprocessed for generalized range reporting in $O(n \log^3 n)$ time.*

*Proof.* The data structure $\mathcal{D}$ takes $O(n \log n)$ time to build. We build one instance of $\mathcal{D}$ for every canonical node in $\mathcal{R}$. By a similar analysis as was performed in proving Lemma 1, it can be shown that the preprocessing of all canonical convex hulls in $\mathcal{R}$ will take $O(n \log^3 n)$ time. □

**Lemma 3.** *The data structure $\mathcal{R}$ occupies $O(n \log^2 n)$ space.*

*Proof.* The modified range tree of [18] utilizes $O(n \log n)$ space, as it only stores a constant amount of extra information per node. Our data structure $\mathcal{R}$, however, stores two additional items at each canonical node – the canonical convex hull and the data structure for generalized reporting. Both these structures require $O(n)$ space. Analysis similar to what is shown in the proof of Lemma 1 will show that the total space requirement of the data structure $\mathcal{R}$ is $O(n \log^2 n)$. □

## 5.2 Query

Moidu et al., in [18], show that using their modified range tree, given a query region $q$, in $O(\log^2 n)$ time we can

1. Identify $O(\log n)$ canonical nodes whose convex hulls, when merged, form the convex hull of the points in $P \cap q$. They call these nodes *candidate* nodes (or blocks).
2. Find, for each candidate node, the start and end points of a continuous segment of its convex hull which, after merging, becomes part of the convex hull of $P \cap q$. Let us call these points $p_s$ and $p_e$.

For every candidate node $n_c$, once we obtain the points $p_s$ and $p_e$, we can find the corresponding indices $l$ and $r$, using the arrays populated in the preprocessing stage for generalized reporting (refer Section 5.1). We then do the query $[l, r]$ on the preprocessed generalized reporting data structure $\mathcal{D}$. For each of the $O(\log n)$ candidate nodes, the corresponding instance of $\mathcal{D}$ outputs the distinct colors present in the points which that node contributes to the convex hull of $P \cap q$. The set of colors returned by querying each candidate node can at most be of size $c$, where $c$ is the total number of distinct colors in the convex hull of $P \cap q$. However, the same color can be output by more than one candidate node. To ensure that each color is reported only once, we once again use the bit-array method of Section 4.2 which ensures that there are no duplicates in the final output.

**Lemma 4.** *The data structure $\mathcal{R}$ answers range queries in $O(\log^2 n + c \log n)$ time.*

*Proof.* For every range query of the form $[x_l, x_h] \times [y_l, y_h]$, our query algorithm performs a one dimensional generalized range reporting query on the instance of the data structure $\mathcal{D}$ stored in each of the $O(\log n)$ candidate canonical nodes. $\mathcal{D}$ has a query time of $O(\log n + c)$, where $c$ is the number of colors reported. Hence our method requires $O(\log^2 n + c \log n)$ time per query. □

Thus we have the following result:

**Theorem 4.** *Let $P$ be a set of colored points in two dimensions. $P$ can be preprocessed into a $O(n \log^2 n)$ space and $O(n \log^3 n)$ preprocessing time data structure such that given an orthogonal range query $q$, the $c$ distinct colors in the convex hull of $P \cap q$ can be reported in $O(\log^2 n + c \log n)$ time.*

## 6   Future Work

Designing output sensitive algorithms for generalized intersection searching problems is an challenging problem since the query time must depend on the number of colors and not on the number of points in the result. Both problems discussed in the preceding sections present interesting possibilities in the design of solutions for generalized geometric range aggregate query problems. An immediate open question is that of proposing more efficient solutions for the problems introduced here. Improvements in the runtime by a logarithmic factor should be possible.

Furthermore, we have not tackled the counting versions of the generalized range maxima and convex hull where the result is simply the number of colors. Developing efficient algorithms for the counting versions remains an open line of pursuit. All results here are for a static set of points. Developing solutions for the dynamic case is still an open problem. Solutions to these problems in higher dimensions are also yet to be proposed.

## References

1. Agarwal, P.K., Govindarajan, S., Muthukrishnan, S.M.: Range searching in categorical data: Colored range searching on grid. In: Möhring, R.H., Raman, R. (eds.) ESA 2002. LNCS, vol. 2461, pp. 17–28. Springer, Heidelberg (2002)
2. Alstrup, S., Brodal, G.S., Rauhe, T.: New data structures for orthogonal range searching. In: FOCS, pp. 198–207. IEEE Computer Society (2000)
3. Bozanis, P., Kitsios, N., Makris, C., Tsakalidis, A.K.: New upper bounds for generalized intersection searching problems. In: Fülöp, Z. (ed.) ICALP 1995. LNCS, vol. 944, pp. 464–474. Springer, Heidelberg (1995)
4. Brass, P., Knauer, C., Shin, C.S., Smid, M., Vigan, I.: Range-aggregate queries for geometric extent problems. In: CATS: 19th Computing: Australasian Theory Symposium (2013)
5. Brodal, G.S., Tsakalidis, K.: Dynamic planar range maxima queries. In: Aceto, L., Henzinger, M., Sgall, J. (eds.) ICALP 2011, Part I. LNCS, vol. 6755, pp. 256–267. Springer, Heidelberg (2011)
6. Chazelle, B.: A functional approach to data structures and its use in multidimensional searching. SIAM J. Comput. 17(3), 427–462 (1988)
7. Gagie, T., Kärkkäinen, J., Navarro, G., Puglisi, S.J.: Colored range queries and document retrieval. Theor. Comput. Sci. 483, 36–50 (2013)
8. Gupta, P., Janardan, R., Smid, M.: Computational geometry: Generalized intersection searching. In: Mehta, D., Sahni, S. (eds.) Handbook of Data Structures and Applications, Chapman & Hall/CRC, Boca Raton, FL, pp. 1–17. CRC Press (2005)
9. Gupta, P., Janardan, R., Smid, M.H.M.: Efficient algorithms for generalized intersection searching on non-iso-oriented objects. In: Symposium on Computational Geometry, pp. 369–378 (1994)

10. Gupta, P., Janardan, R., Smid, M.H.M.: Algorithms for generalized halfspace range searching and other intersection searching problems. Comput. Geom. 5, 321–340 (1995)

11. Gupta, P., Janardan, R., Smid, M.H.M.: Further results on generalized intersection searching problems: Counting, reporting, and dynamization. J. Algorithms 19(2), 282–317 (1995)

12. Gupta, P., Janardan, R., Smid, M.H.M.: Algorithms for generalized halfspace range searching and other intersection searching problems. Comput. Geom. 6, 1–19 (1996)

13. Gupta, P., Janardan, R., Smid, M.H.M.: A technique for adding range restrictions to generalized searching problems. Inf. Process. Lett. 64(5), 263–269 (1997)

14. Janardan, R., Lopez, M.A.: Generalized intersection searching problems. Int. J. Comput. Geometry Appl. 3(1), 39–69 (1993)

15. Kalavagattu, A.K., Agarwal, J., Das, A.S., Kothapalli, K.: On counting range maxima points in plane. In: Smyth, B. (ed.) IWOCA 2012. LNCS, vol. 7643, pp. 263–273. Springer, Heidelberg (2012)

16. Kirkpatrick, D., Snoeyink, J.: Computing common tangents without a separating line. In: Sack, J.-R., Akl, S.G., Dehne, F., Santoro, N. (eds.) WADS 1995. LNCS, vol. 955, pp. 183–193. Springer, Heidelberg (1995)

17. McCreight, E.M.: Priority search trees. SIAM J. Comput. 14(2), 257–276 (1985)

18. Moidu, N., Agarwal, J., Kothapalli, K.: Planar convex hull range query and related problems. In: CCCG, Carleton University, Ottawa, Canada (2013)

19. O'Rourke, J.: Computational Geometry in C. Cambridge University Press (1998), http://cs.smith.edu/~orourke/books/compgeom.html

20. Rahul, S., Bellam, H., Gupta, P., Rajan, K.: Range aggregate structures for colored geometric objects. In: CCCG. pp. 249–252 (2010)

21. Rahul, S., Janardan, R.: Algorithms for range-skyline queries. In: Cruz, I.F., Knoblock, C.A., Kröger, P., Tanin, E., Widmayer, P. (eds.) SIGSPATIAL/GIS, pp. 526–529. ACM (2012)

22. Shi, Q., JáJá, J.: Optimal and near-optimal algorithms for generalized intersection reporting on pointer machines. Inf. Process. Lett. 95(3), 382–388 (2005)

23. Sleator, D.D., Tarjan, R.E.: A data structure for dynamic trees. In: STOC, pp. 114–122. ACM (1981)

24. Sleator, D.D., Tarjan, R.E.: A data structure for dynamic trees. J. Comput. Syst. Sci. 26(3), 362–391 (1983)

25. Tarjan, R.E.: Applications of path compression on balanced trees. J. ACM 26(4), 690–715 (1979)

# Boundary Labeling
# with Flexible Label Positions*

Zhi-Dong Huang[1], Sheung-Hung Poon[1,**], and Chun-Cheng Lin[2]

[1] Dept. of Computer Science, National Tsing Hua University, Hsinchu, Taiwan
autohuang0930@gmail.com, spoon@cs.nthu.edu.tw
[2] Dept. of Industrial Engineering and Management,
National Chiao Tung University, Hsinchu, Taiwan
cclin321@nctu.edu.tw

**Abstract.** *Boundary labeling* connects each point site in a rectangular map to a label on the sides of the map by a *leader*, which may be a straight-line segment or a polyline. In the conventional setting, the labels along a side of the map form a single stack of labels in which labels are placed consecutively one by one in a sequence, and the two end sides of a label stack must respect the sides of the map. However, such a setting may be in conflict with generation of a better boundary labeling, measured by the total leader length or the number of bends of leaders. As a result, this paper relaxes this setting to propose the boundary labeling with *flexible* label positions, in which labels are allowed to be placed at any non-overlapping location along the sides of the map so that they do not necessarily form only one single stack, and the two end sides of label stacks do not need to respect the sides of the map. In this scenario, we investigate the total leader length minimization problem and the total bend minimization problem under several variants, which are parameterized by the number of sides to which labels are attached, their label size, port types, and leader types. It turns out that almost all of the total leader length minimization problems using nonuniform-size labels are NP-complete except for one case, while the others can be solved in polynomial time.

# 1 Introduction

One of the basic requirements for *map labeling* [12,13] is to make all labels in the map pairwise disjoint. However, such a requirement is difficult to be achieved in the case that large labels are placed on dense points. Especially in practice, large labels are usually used in technical drawings or medical atlases. To address this problem, Bekos et al. [4] proposed the so-called *boundary labeling* (see Figure 1(a)), where all labels are attached to the boundary (four sides) of a rectangular map enclosing all point sites, and each point site is connected to a

---

* Supported in part by grants 97-2221-E-007-054-MY3 and 100-2628-E-007-020-MY3 of the National Science Council (NSC), Taiwan, R.O.C.
** Corresponding author.

S.P. Pal and K. Sadakane (Eds.): WALCOM 2014, LNCS 8344, pp. 44–55, 2014.

unique label by a *leader*, which may be a rectilinear or a straight-line segment. In such a setting, they assumed that there are no two sites with the same $x$- or $y$-coordinates, and investigated how to place labels and leaders in a drawing such that there are no crossings among leaders and either the total leader length or the total number of bends of leaders is minimized under a variety of constraints. Bekos et al. [3] later investigated a similar problem for labeling polygonal sites under the framework of boundary labeling. Lin et al. [9] investigated the *multi-site-to-one-label boundary labeling*, in which more than one site is allowed to be connected to a common label, and Lin [8] and Bekos et al. [1] further proposed crossing-free multi-site-to-one-label boundary labeling.

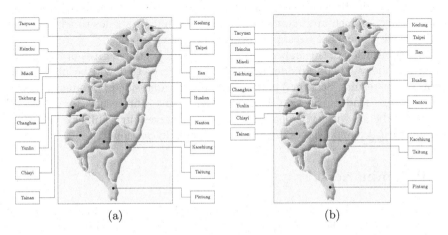

(a)                              (b)

**Fig. 1.** Boundary labeling for Taiwan by (a) the conventional boundary labeling, and (b) boundary labeling with flexible label positions

In the conventional boundary labeling, the labels along a side of the map form a single *stack* of labels, as shown in Figure 1(a), in which labels are placed consecutively one by one in a sequence along a side of the map. Note that conventionally there is a fixed equal gap between adjacent labels in a label stack. This paper proposes the so-called *boundary labeling with flexible label positions*, as shown in Figure 1(b), in which labels are allowed to be placed at any non-overlapping location along the sides of the map. They do not necessarily form only one single stack, and the upper and lower ends of each label stack do not need to respect the upper and lower boundaries of the map area. In the Taiwan map of Figure 1, the counties of Taiwan are labeled using the conventional boundary labeling and our proposed boundary labeling in Figures 1(a) and 1(b), respectively. Compared to Figure 1(a), the boundary labeling in Figure 1(b) contains more straight leaders and thus has much clearer connection correspondences: Figure 1(b) contains 14 straight leaders and two non-straight leaders whereas Figure 1(a) contains no straight leader.

It should be noticed that in the conventional boundary labeling, a fixed equal gap is assumed to lie between adjacent labels, as shown in Figure 1(a). In fact, the gap can also be zero-sized conventionally. In the situation that there are no gaps or equal gaps between labels, the conventional boundary labeling cannot avoid the occurrence of many leader bends, and the bends result in longer leader lengths. In contrast, using flexible label positions can increase the number of straight-line leaders, thus shortening the total leader length. This results in increasing the readability of the relationships between the map regions and their corresponding labels.

In light of the above, it is of interest and of importance to investigate how to generate a good boundary labeling with flexible label positions, measured by either the total leader length or the total bend number. As a result, this paper considers, from a computational complexity viewpoint, the *total leader length minimization* (*TLLM*) problem and the *total bend minimization* (*TBM*) problem for variants of one-sided and two-sided boundary labeling with flexible label positions, which are parameterized by the number of sides to which labels are attached, their label sizes, port types, and leader types.

The organization of this paper is given as follows. Section 2 gives preliminaries to our boundary labeling model and a basic property that will be used in the proofs of our results. Section 3 gives our results for one-sided boundary labeling with type-*opo* leaders. Section 4 and Section 5 give our results for two-sided boundary labeling with type-*opo* leaders. Section 6 gives our results for boundary labeling with type-*po* leaders. Lastly, we conclude in Section 7.

## 2   Preliminaries

In this section, we first introduce the model of the boundary labeling with flexible label positions, and then a basic property that will be used in the proofs for our results.

### 2.1   The Model of the Boundary Labeling with Flexible Label Positions

The conventional boundary labeling [4,5,2] is characterized as *k-sided labeling with type-t leaders* (where $k \in \{1, 2\}$ and $t \in \{opo, po\}$) if the labels are allowed to attach to the $k$ sides of the enclosing rectangle $R$ by only type-$t$ leaders. When $k = 2$, we mean that the labels are placed on two opposite sides of $R$. The parameter $t$ specifies the way a leader is drawn to connect a site to a label. The types *opo* and *po* stand for *orthogonal-parallel-orthogonal* and *parallel-orthogonal* leader types, respectively, which can easily be understood from the examples given in Figures 2(a)(b) and Figures 2(c)(d). It is assumed that the parallel (i.e., '*p*') segment associated with a type-*opo* leader lies in a *track routing area* located between $R$ and the label stack (e.g., see Figure 2(a)). Such a track routing area is given a small fixed width in the input model. We remark that in some labeling models in [2], more labeling sides and more types of leaders are used.

Observe the conventional boundary labeling in Figure 2(a)(b), in which the positions of labels are constrained, i.e., the top and bottom sides of the stack of labels have to respect the top and bottom sides of the map, respectively. Such a constraint may be in conflict with the objectives of general boundary labeling problems, i.e., to minimize the total leader length or the total number of bends of leaders. In this paper, we allow the two end sides of the label stack (sandwiching equal gaps in between adjacent labels) to exceed the upper or lower boundaries the map area, and each label to be placed at any location along a side of the map only if there is no overlapping between labels. Such a setting is said to use *flexible label positions* (see Figures 2(b)(d)), while the original setting is said to use *fixed label positions* (see Figures 2(a)(c)). It is observable from Figures 2(b) and 2(d) that if flexible label positions is allowed, more horizontal straight leaders can be produced so that the total leader length is shorter and the total number of bends is potentially fewer.

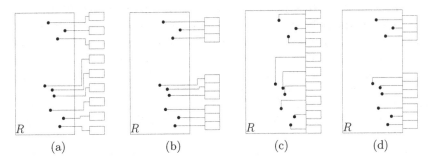

(a)  (b)  (c)  (d)

**Fig. 2.** (a)(b) using type-*opo* leaders; (c)(d) using type-*po* leaders. (a)(c) using fixed label positions; (b)(d) using flexible label positions.

We consider the following labeling problem. Given a rectangular area $R$ of height $h$ and width $w$ whose left lower corner resides at the origin of the $xy$-plane (i.e., $R = [0, w] \times [0, h]$), and a set of $n$ points (called *sites*) $p_i = (x_i, y_i), 1 \leq i \leq n$, located inside $R$ (i.e., $0 \leq x_i \leq w, 0 \leq y_i \leq h$), each of which is associated with a rectangular label $\lambda_i$ of width $w_i$ and height $l_i$, the *boundary labeling* problem is to place the labels along one, two, or four sides of the boundary of $R$, and connect $p_i$ to $\lambda_i, 1 \leq i \leq n$ using *rectilinear leaders* (or leaders, for short) so that a certain criterion is met. As illustrated in Figure 2, a *rectilinear leader* consists of horizontal and/or vertical line segments connecting a site to its corresponding label. Throughout the rest of this paper, we assume that there are no two sites with the same $x$- or $y$-coordinate, and sites are labeled as $p_1, p_2, \ldots, p_n$ in the increasing order of their $y$-coordinates.

In what follows, we give a formal model for the problem of boundary labeling with flexible label positions. Our setting assumes sites to be points of zero size located on the plane. A label placement for all labels is called *legal* if the labels do not overlap, and the leaders are crossing-free. Our objective is to find a legal label placement, such that either the *total leader length is minimum* (*TLLM*) or *the total bend number is minimum* (*TBM*).

By an analogy of Bekos et al.'s work [3], we extend their boundary labeling model to express our model as a 5-tuple (*Leader*, *Side*, *LabelSize*, *LabelPort*, *Objective*), where:

**Leader:** All the leaders are of type *opo* or *po* in this paper.

**Side:** Labels can be placed on the East and West sides of the enclosing rectangle $R$, which are denoted by $E$ and $W$, respectively.

**LabelSize:** Each label $\lambda_i$ is associated with a height $l_i$ and a width $w_i$. As each leader is connected to the left or right side of a label box, without loss of generality, we may assume that $w_i = w_j, \forall 1 \le i, j \le n$. Labels are of *uniform* size if $l_i = l_j, \forall 1 \le i, j \le n$; otherwise, they are called *nonuniform* labels.

**LabelPort:** Depending on the location where a leader touches a label, consider the following two types:

- *Fixed port*: there exists a constant $0 \le \alpha \le 1$, such that the $i$-th leader touches the point of height $\alpha l_i$, from the bottom of the $i$-th label. For boundary labeling with fixed ports, $\alpha$ is usually assumed to be $\frac{1}{2}$, i.e., each leader touches the middle of the corresponding label.
- *Sliding port*: As the name suggests, the contact point of a leader can *slide* along the corresponding label edge.

**Objective:** We aim to find a legal label placement, such that either the total leader length is minimum (*TLLM*) or the number of bends is minimum (*TBM*).

According to the above model, the main contributions of this paper are summarized in Table 1. We note that in Theorems 1, 2, 3, 4, 5 and 7, the results hold for both optimization objectives, *TLLM* and *TBM*. However, due to the space limit, we mostly only provide the proofs for the objective *TLLM* in these theorems, and some of the proofs for the other objective *TBM* are omitted.

## 2.2 Basic Property

An *upward (resp., downward) bending leader* is a type-*opo* leader or type-*po* leader which bends upward (resp., downward) from its associated site. See Figure 3 for examples on upward and downward type-*opo* leaders.

**Table 1.** Time complexity for a variety of boundary labeling models for flexible labels

| (Leader, | Side, | LabelSize, | LabelPort, | Objective) | Time* | Reference |
|---|---|---|---|---|---|---|
| (opo, | E, | uniform/nonuniform, | fixed/sliding, | TLLM/TBM) | $O(n^3)$ | Thm 1 |
| (opo, | EW, | uniform, | fixed, | TLLM/TBM) | $O(n^5)$ | Thm 2 |
| (opo, | EW, | uniform, | sliding, | TLLM/TBM) | $O(n^5)$ | Thm 3 |
| (opo, | EW, | nonuniform, | fixed, | TLLM/TBM) | NPC | Thm 4 |
| (opo, | EW, | nonuniform, | sliding, | TLLM/TBM) | NPC | Thm 5 |
| (po, | E, | uniform, | fixed, | TLLM) | $O(n \log n)$ | [11] |
| (po, | E, | uniform, | sliding, | TLLM) | $O(n^3)$ | Thm 6 |
| (po, | EW, | uniform, | fixed/sliding, | TLLM) | $O(n^5)$ | Thm 6 |
| (po, | E/EW, | uniform, | fixed/sliding, | TBM) | open | – |
| (po, | E/EW, | nonuniform, | fixed/sliding, | TLLM/TBM) | NPC | Thm 7 |

* Note that NPC denotes NP-completeness and time denotes time complexity.

A *stack* of labels contains a maximal set of labels in a solution such that they are placed consecutively one by one in a sequence. Note that throughout the rest of this paper, if labels form a stack, then there are no gaps in the label stack. We now make some observations on

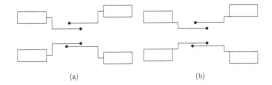

**Fig. 3.** (a) Upward and downward bending type-*opo* leaders with fixed ports; (b) Upward and downward bending type-*opo* leaders with sliding ports.

the shapes and the locations of type-*opo* leaders in the optimal solution. First, it is easy to see that in the optimal solution for the sliding-port model, upward (resp., downward) bending type-*opo* leaders touch the bottom (resp., top) sides of their labels. See Figure 3(b) for illustration. The reason is that otherwise, we can push the connection port between the leader and the label downward (resp., upward) so that we obtain another solution with shorter total leader length for the *TLLM* problem. On the other hand, for the *TBM* problem under the sliding-port model, although it is not compulsory to have this property, by enforcing such a property there is no hurt on the optimality of the solution. Next, we obtain a general property on the stacks of labels for any labeling model. Its proof is omitted.

**Lemma 1.** *There exists an optimal solution for the one-sided or two-sided boundary labeling problem (no matter whether leaders are type-opo or type-po, and no matter whether it is TLLM or TBM) such that each stack of labels in the solution always contains a label with a horizontal straight leader.*

## 3   One-Sided Boundary Labeling with Type-*opo* Leaders

In this section, we propose a polynomial time algorithm for one-sided boundary labeling with type-*opo* leaders, regardless of uniform or nonuniform label sizes, in either fixed- or sliding-port case. In the following, we only consider the fixed-port case; the sliding-port case is similar.

Our dynamic programming based algorithm for the *TLLM* problem is described as follows:

**Step 1.** Let $H = \sum_{i=1}^{n} l_i$. Construct two dummy sites $p_0 = (w, y_1 - H)$ and $p_{n+1} = (w, y_n + H)$ connected to two one-unit-height dummy labels $\lambda_0$ and $\lambda_{n+1}$, respectively.

**Step 2.** For $0 \leq i \leq n + 1$, let $d_i$ denote the length of the leader connected to $p_i$ when the leader is a horizontal straight line. Let $T(i, j)$ denote the minimal total leader length of the boundary labeling for the map with sites $p_i, p_{i+1}, \ldots, p_j$ and labels $\lambda_i, \lambda_{i+1}, \ldots, \lambda_j$ when both labels $\lambda_i$ and $\lambda_j$ are connected with horizontal straight leaders. Use dynamic programming to compute table $T$ according to the following recursive formula:

$$T(i, j) = \min \left\{ \begin{array}{l} \min_{i < k < j}(T(i, k) + T(k, j) - d_k), \\ \min_{i \leq k < j}(I_{down}(i, k) + I_{up}(k + 1, j)) \end{array} \right\}$$

where $I_{down}(i, k)$ (resp., $I_{up}(k + 1, j)$) denotes the total leader length when labels $\lambda_i, \lambda_{i+1}, \ldots, \lambda_k$ (resp., $\lambda_{k+1}, \lambda_{k+2}, \ldots, \lambda_j$) form a stack and label $\lambda_i$ (resp., $\lambda_j$) is connected by a horizontal straight leader.

**Step 3.** Output $T(0, n + 1) - d_0 - d_{n+1}$.

The explanation for the above algorithm is given as follows. By Lemma 1, there is an optimal solution consisting of a number of stacks of labels in which each stack always contains at least a label with a horizontal straight leader. Hence, our dynamic programming formula uses horizontal straight leaders (from outermost to innermost) to divide problems into subproblems. Step 1 adds dummy sites $p_0$ and $p_{n+1}$ (and their corresponding dummy labels) far away from other sites so that both of them must be horizontal straight leaders at the outermost of the map in the optimal labeling, in which one is the bottommost, and the other is the topmost. By doing so, the final solution can be computed by calling $T(0, n+1)$, and at Step 3, the two outermost leader lengths are deducted from $T(0, n+1)$.

To see why the above dynamic programming formula correctly characterizes $T(i, j)$, consider the following cases:

(i) There is a label $\lambda_k$ in $\{\lambda_{i+1}, \ldots, \lambda_{j-1}\}$ with a horizontal straight leader: Recall that $T(i, j)$ assumes both labels $\lambda_i$ and $\lambda_j$ to be connected with horizontal straight leaders. Hence, the minimum $T(i, j)$ may be $T(i, k) + T(k, j) - d_k$, in which the deducted term is due to that $T(i, k)$ and $T(k, j)$ count $d_k$ twice.

(ii) In opposite to the above case, except for labels $\lambda_i$ and $\lambda_j$, all the other labels in $T(i, j)$ are connected with bending leaders: In this case, since there are only two horizontal straight leaders in $T(i, j)$, there are two stacks of labels by Lemma 1: one is led by $\lambda_i$; the other is led by $\lambda_j$. Hence, each label must belong to either stack. As a result, we calculate the sum of the leader lengths for all possible pairs of stacks: $\min_{i \leq k < j-1}(I_{down}(i, k) + I_{up}(k + 1, j))$.

It is easy to see that the above algorithm runs in $O(n^3)$ time, and thus we have the following theorem.

**Theorem 1.** *Regardless of uniform or nonuniform label sizes, the TLLM and TBM problems for one-sided boundary labeling using type-opo leaders can be solved in $O(n^3)$ time, in either fixed or sliding label port case.*

## 4 Two-Sided Boundary Labeling with Type-*opo* Leaders and Uniform Labels

In this section, we investigate the *TLLM/TBM* problems for boundary labeling with type-*opo* leaders and uniform labels in the fixed-port and sliding-port cases, respectively.

### 4.1 Fixed-Port Case

For the fixed-port case, we obtain the following result.

**Theorem 2.** *The two-sided boundary labeling for the model (opo, EW, uniform, fixed, TLLM / TBM ) can be solved in $O(n^5)$ time.*

**Proof:** By Lemma 1, there exists some optimal solution such that each stack $\tau$ of labels on either side of the map must contain a label $\lambda$ with a straight leader. Since all the labels are of the same height, the locations of other labels in stack $\tau$ are predictable with respect to label $\lambda$. That is, with respect to label $\lambda$, there are at most $(2n-2)$ possible label position candidates: $(n-1)$ consecutive label positions above and below label $\lambda$, respectively. As we do not know beforehand which labels have straight leaders in an optimal solution, we treat each of the $n$ labels of our concerned problem as being a possible candidate with a straight leader.

Hence, there are $n(2n-2)$ possible label position candidates in total on either side of the map. Among these $n(2n-2)$ label position candidates, we then proceed to select the appropriate label positions for an optimal solution by the following dynamic programming strategy.

First, we give an order for the $n(2n-2)$ label position candidates for each side of the map: for $1 \le k \le n(2n-2)$, let $\lambda_k$ (resp.,$\gamma_k$) be the $k$-th lowest label position candidate on the left (resp., right) side of the map. Our objective is to find $n$ label position candidates (among the $2n(2n-2)$ positions) for the connection of $n$ sites, so that either the total leader length or the total number of bends is minimum.

Let $T(i,l,r)$ denote the optimal total leader length or total bend number for the boundary labeling problem with sites $p_1,\dots,p_i$, label positions $\lambda_1,\dots,\lambda_l$ on the left side, and label positions $\gamma_1,\dots,\gamma_r$ on the right side. For $i = n, n-1,\dots, 1$, we determine the leader placement of site $p_i$ by the following recursive formula:

$$T(i,l,r) = \min \left\{ \begin{array}{l} T(i-1,l-n_b(\lambda_l)-1,r) + Left(p_i,l), \\ T(i-1,l,r-n_b(\gamma_r)-1) + Right(p_i,r), \\ T(i,l-1,r-1) \end{array} \right\}$$

where $n_b(\ell)$ denotes the number of label position candidates lower than and intersecting label $\ell$; $Left(p,k)$ (resp., $Right(p,k)$ ) denotes the leader length or bend number of the type-*opo* leader from site $p$ to label position $\lambda_k$ (resp., $\gamma_k$). The first term of the above formula corresponds to the case that site $p_i$ links to label $\lambda_l$; the second term corresponds to the case that site $p_i$ links to label $\gamma_r$; the last term corresponds the case that site $p_i$ does not link to label $\lambda_l$ or label $\gamma_r$. Hence, the optimal solution can be found by calculating $T(n, n(2n-2), n(2n-2))$. Since each entry in $T$ can be calculated in $O(1)$ time, we thus obtain an $O(n^5)$-time algorithm.                                   □

## 4.2 Sliding-Port Case

Before showing our result for the sliding-port case, we need the following lemma to bound the number of candidate label positions. Its proof is omitted.

**Lemma 2.** *There exists an optimal solution for the model (opo, E/EW, uniform, sliding, TLLM/TBM) such that each stack of labels in the solution contains a label with its straight leader touching either the top or bottom side of the label.*

Due to this lemma, in any stack $\tau$ of the optimal solution, we know that there is a label $\lambda$ with a straight leader touching either the top or bottom side of the label. In other words, there are exactly two candidate positions for label $\lambda$. Consequently, the candidate label positions for the other labels in $\tau$ are the $n-1$ consecutive label positions above $\lambda$, and the other $n-1$ positions below $\lambda$. Hence, there are totally at most $O(n^2)$ possible candidate label positions on either side of the map area. Then, using a similar dynamic programming formulation as for Theorem 2, we can solve our problem for the sliding case in $O(n^5)$ time. We summarize our result in the following theorem.

**Theorem 3.** *The two-sided boundary labeling for the model (opo, EW, uniform, sliding, TLLM/TBM) can be solved in $O(n^5)$ time.*

## 5   Two-Sided Boundary Labeling with Type-*opo* Leaders and Nonuniform Labels

In this section, we show that two-sided boundary labeling with type-*opo* leaders and nonuniform labels for both fixed-port and sliding-port cases are NP-complete.

### 5.1   Fixed-Port Case

In the following theorem, the NP-completeness of the *TLLM* problem for the fixed-port case is proved by performing a reduction from a single-machine scheduling problem, called *total discrepancy problem* [6]. Such a problem has also been used to show the NP-completeness of other variants of boundary labeling [2,10].

**Theorem 4.** *The boundary labeling problem for the model (opo, EW, nonuniform, fixed, TLLM/TBM) is NP-complete.*

**Proof (Sketch):** We guess an optimal solution with non-zero probability, for which, due to Lemma 1, the label positions must be limited and cannot be arbitrary, and then the decision version of this problem is in NP. In order to show the NP-hardness of our problem, we establish a linear-time reduction to our problem from the total discrepancy problem, which is described as follows.

On one machine, we plan to arrange the schedule for the non-preemptive execution of a set $J$ of $2n + 2$ jobs $J_0, J_1, \ldots, J_{2n}, J_{2n+1}$. Each job $J_i$ has an execution time length $l_i \in \mathbb{Z}^+$ such that $l_0 < l_1 < \ldots < l_{2n}$ and $l_{2n+1} = 2$. For a planned schedule $\sigma$, the actual execution midtime for job $J_i$ is denoted by $m_i(\sigma)$. Each job has a *preferred midtime*, which corresponds to the time at which we would like the first half of the job to be completed. We assume that all the jobs except $J_{2n+1}$ share a single preferred midtime $M = \sum_{i=0}^{2n} l_i/2$, whereas job $J_{2n+1}$ has its own preferred midtime $M' = 2M + l_{2n+1}/2 = 2M + 1$. The *penalty* of job $J_i$ for a schedule $\sigma$ is defined as the absolute difference of its midtime to its preferred midtime, i.e., $|m_i(\sigma) - M|$ for $0 \le i \le 2n$ and $|m_{2n+1}(\sigma) - M'|$. The *cost* incurred in a schedule $\sigma$ is then defined to be the total penalties incurred by all jobs. Hence, the objective of the total discrepancy problem is to determine a schedule $\sigma$ such that the total cost of the schedule, i.e., $\sum_{i=0}^{2n} |m_i(\sigma) - M| + |m_{2n+1}(\sigma) - M'|$, is minimized. Garey, Tarjan and Wilfong [6] showed the following properties for an optimal schedule $\sigma_{opt}$ of the first $2n + 1$ jobs $J_0, J_1, \ldots, J_{2n}$, but the decision problem whether such a schedule exists is NP-complete.

(i)   $\sigma_{opt}$ does not have any gaps between the jobs.
(ii)  The midtime of the shortest job $J_0$ is $M$, i.e., $m_0(\sigma_{opt}) = M$.
(iii) Jobs $J_1, J_2, \ldots, J_{2n}$ are divided into two groups, $A(\sigma_{opt}) = \{J_i : m_i(\sigma_{opt}) < M\}$ and $B(\sigma_{opt}) = \{J_i : m_i(\sigma_{opt}) > M\}$, such that $|A(\sigma_{opt})| = |B(\sigma_{opt})| = n$.
(iv)  Suppose the sequence of the jobs in schedule $\sigma_{opt}$ is $A_n, A_{n-1}, \ldots, A_1, J_0,$ $B_1, B_2, \ldots, B_n$. Then $\{A_i, B_i\} = \{J_{2i-1}, J_{2i}\}$.
(v)   The optimal cost is equal to $\sum_{i=1}^{n} (l_{2i} + l_{2i-1})(n - i + 1/2) + nl_0$.

We will see how this scheduling problem can be reduced to our problem in the following.

Recall the map $R = [0, w] \times [0, h]$. Assume the width of the routing area to be $\delta$, which is a very small constant. We will put all point sites in the map area along the vertical line $x = \frac{w}{2}$; see also Figure 4. For job $J_i, i = 0, \ldots, 2n$, we create its corresponding point site $p_i$ and

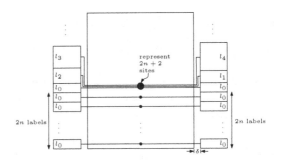

**Fig. 4.** Reduction in Theorem 4

place it at location $(\frac{w}{2}, (2n + \frac{1}{2})l_0 + (i+1)\epsilon)$, in which $\epsilon$ is set to be a very small constant value less than, say for example $\min\{l_0/100n^3, \min_{i=0}^{2n-1}(l_{i+1} - l_i)/(100n^3)\}$. Suppose the leader for $p_i$ connects to the middle position of label $\lambda_i$ with height $l_i$. In addition, a site locates at $(\frac{w}{2}, (2n + \frac{1}{2})l_0)$, and is connected to the middle position of a label with height $l_0$. The $(2n+2)$ point sites are represented by the big dot in Figure 4. Then, we add $4n$ labels of height $l_0$ respectively connected with the $4n$ sites at different positions: $(\frac{w}{2}, (\lfloor j/2 \rfloor + \frac{1}{2})l_0 + (j \bmod 2)\epsilon)$, $\forall j = 0, \ldots, 4n - 1$ (see the $4n$ point sites below the big dot in Figure 4). In all, given $2n + 1$ jobs, we construct $6n + 2$ labels. First, we observe that the total length of horizontal segments of all leaders is $(6n + 2)(\frac{w}{2} + \delta)$.

Moreover, we set $k = \sum_{i=1}^{n} (l_{2i} + l_{2i-1})(n - i + \frac{1}{2}) + nl_0$. Since $\epsilon$ is very small compared to the height of any label, we can nearly treat the locations of point sites $p_i$ to be lying exactly at $(\frac{w}{2}, (2n + \frac{1}{2})l_0)$. The difference between the total leader length under such a scenario and that under the actual scenario is at most

$\frac{l_0}{n}$. Thus, we can show that there is a scheduling with cost at most $k$ if and only if there is a legal labeling with total leader length at most $(6n+2)(\frac{w}{2}+\delta)+k+\frac{l_0}{n}$ (see the label positions in Figure 4 for a hint). This complete the NP-completeness proof for the *TLLM* problem

On the other hand, the NP-completeness for the *TBM* problem under the current labeling model can be proved by performing a polynomial-time reduction from the Partition Problem [7]. The proof details are omitted.   □

## 5.2  Sliding-Port Case

In this subsection, we show that it is NP-complete to find an optimal solution for the boundary labeling problem for the model (*opo*, *EW*, *nonuniform*, *sliding*, *TLLM/TBM*). Our NP-completeness proof is also based on the reduction from the Partition Problem [7]. The proof is omitted due to lack of space.

**Theorem 5.** *The boundary labeling problem for the model (opo, EW, nonuniform, sliding, TLLM/TBM) is NP-complete.*

# 6  Boundary Labeling with Type-*po* Leaders

For boundary labeling with type-*po* leaders, we also obtain some polynomial-time algorithms and NP-completeness results. See the following two theorems. Their proofs are omitted.

**Theorem 6.** *One-sided and two-sided boundary labeling for the model (po, E/EW, uniform, fixed/sliding, TLLM) can be solved in $O(n^3)$ and $O(n^5)$ time, respectively.*

We remark that the one-sided boundary labeling for the model (*po*, *E*, *uniform*, *fixed*, *TLLM*) can be computed in $O(n \log n)$ time [11]. However, our $O(n^3)$-time algorithm here works for both models with fixed-port labels or with sliding-port labels.

**Theorem 7.** *Both one-sided and two-sided boundary labeling under the model (po, E/EW, nonuniform, fixed/sliding, TLLM/TBM) are NP-complete.*

# 7  Conclusions

We have investigated the tractability and algorithms for the one-sided and two-sided boundary labeling problems using flexible label positions for the objective of total leader length minimization (*TLLM*) and the total bend number minimization (*TBM*) under several variants, which are parameterized by the number of sides to which labels are attached, their label size, port types, and leader types. It turns out that except for the model (*po*, *E/EW*, *uniform*, *fixed/sliding*, *TBM*), almost all of the concerned *TLLM/TBM* problems are solved, some of which are solvable in polynomial time, while the others are intractable in general.

# References

1. Bekos, M.A., Cornelsen, S., Fink, M., Hong, S.-H., Kaufmann, M., Nöllenburg, M., Rutter, I., Symvonis, A.: Many-to-one boundary labeling with backbones. In: Wismath, S., Wolff, A. (eds.) GD 2013. LNCS, vol. 8242, pp. 244–255. Springer, Heidelberg (2013)
2. Bekos, M., Kaufmann, M., Nöllenburg, M., Symvonis, A.: Boundary labeling with octilinear leaders. Algorithmica 57(3), 436–461 (2010)
3. Bekos, M., Kaufmann, M., Potina, K., Symvonis, A.: Area-feature boundary labeling. The Computer Journal 53(6), 827–841 (2009)
4. Bekos, M., Kaufmann, M., Symvonis, A., Wolff, A.: Boundary labeling: models and efficient algorithms for rectangular maps. Computational Geometry: Theory and Applications 36(3), 215–236 (2006)
5. Benkert, M., Haverkort, H.J., Kroll, M., Nöllenburg, M.: Algorithms for multi-criteria boundary labeling. Journal of Graph Algorithms and Applications 13(3), 289–317 (2009)
6. Garey, M., Tarjan, R., Wilfong, G.: One-processor scheduling with symmetric earliness and tardiness penalties. Mathematics of Operations Research 13, 330–348 (1988)
7. Garey, M.R., Johnson, D.S.: Computers and Interactability. A Guide to the Theory of NP-Completeness. A Series of Books in the Mathematical Sciences, Freemann and Company (1979)
8. Lin, C.C.: Crossing-free many-to-one boundary labeling with hyperleaders. In: Proc. of 3rd IEEE Pacific Visualization Symposium (PacificVis 2010), pp. 185–192. IEEE Press (2010)
9. Lin, C.C., Kao, H.J., Yen, H.C.: Many-to-one boundary labeling. Journal of Graph Algorithms and Applications 12(3), 319–356 (2008)
10. Lin, C.-C., Poon, S.-H., Takahashi, S., Wu, H.-Y., Yen, H.-C.: One-and-a-half-side boundary labeling. In: Wang, W., Zhu, X., Du, D.-Z. (eds.) COCOA 2011. LNCS, vol. 6831, pp. 387–398. Springer, Heidelberg (2011)
11. Nöllenburg, M., Polishchuk, V., Sysikaski, M.: Dynamic one-sided boundary labeling. In: GIS, pp. 310–319 (2010)
12. Wagner, F.: Approximate map labeling is in $\omega(n \log n)$. Information Processing Letters 52(3), 161–165 (1994)
13. Wagner, F., Wolff, A.: Map labeling heuristics: Provably good and practically useful. In: Proc. of the 11th Annual ACM Symposium on Computational Geometry (SoCG 1995), pp. 109–118. ACM Press (1995)

# Approximating the Bipartite TSP
# and Its Biased Generalization

Aleksandar Shurbevski[1], Hiroshi Nagamochi[1], and Yoshiyuki Karuno[2]

[1] Department of Applied Mathematics and Physics, Kyoto University,
Yoshida-Honmachi, Sakyo-ku, Kyoto 606-8501, Japan
{shurbevski,nag}@amp.i.kyoto-u.ac.jp
[2] Department of Mechanical and System Engineering, Kyoto Institute of Technology,
Matsugasaki, Sakyo-ku, Kyoto 606-8585, Japan
karuno@kit.ac.jp

**Abstract.** We examine a generalization of the symmetric bipartite traveling salesman problem (TSP) with quadrangle inequality, by extending the cost function of a Hamiltonian tour to include a bias factor $\beta \geq 1$. The bias factor is known and given as a part of the input. We propose a novel heuristic procedure for building Hamiltonian cycles in bipartite graphs, and show that it is an approximation algorithm for the generalized problem with an approximation ratio of $1 + \frac{1+\lambda}{\beta+\lambda}$, where $\lambda$ is a real parameter dependent on the problem instance. This expression is bounded above by a constant 2, for any positive real $\lambda$ and $\beta \geq 1$, which improves a previously reported approximation ratio of 16/7. As a part of a composite heuristic, the proposed procedure can contribute to an approximation ratio of $1 + \frac{2}{\zeta+\beta(2-\zeta)}$, where $\zeta$ is an approximation ratio for the metric TSP.

**Keywords:** combinatorial optimization, approximation algorithm, matroid intersection, material handling robot, bipartite TSP, biased cost.

## 1 Introduction

The traveling salesman problem (TSP) is a landmark problem in combinatorial optimization (e.g., Cook [7]). Its bipartite analogue is as follows. Given a bipartite graph $G = (B, W; E)$ with an edge weight function $w : E \rightarrow \mathbb{R}_+$, find a shortest (w.r.t. $w$) alternating tour which visits every point of $B \cup W$ exactly once. We assume that the weight function $w$ is symmetric and satisfies the quadrangle inequality (the bipartite analogue of the triangle inequality, see Eqs. (6) and (7)). We do so not only because do the above conditions suffice in many cases based on real world scenarios, but also because just like the TSP, it is hopeless to approximate the bipartite TSP within a constant factor in the general case, assuming that $P \neq NP$ [10, 15].

The bipartite TSP has justly attracted attention due to its applicability in typical industrial settings where pick and place or grasp and delivery robots are employed with some material handling tasks [3–5, 10, 11, 18]. For the symmetric

S.P. Pal and K. Sadakane (Eds.): WALCOM 2014, LNCS 8344, pp. 56–67, 2014.

case, the best known approximation factor 2 has been independently reported by Chalasani et al. [5] and Frank et al. [10]. With a specific industrial scenario in mind, the bipartite TSP has been extended to account for additional transportation effort [17]. The motivation behind this generalization is to assign certain "difficulty" when transporting an item versus simply moving through space. This has been achieved by the means of a *bias* factor $\beta \geq 1$. The bias factor extends the weight function $w$ as follows

$$\widetilde{w}(u, v) = \begin{cases} \beta w(u, v), & u \in B, v \in W, \\ w(u, v), & u \in W, v \in B. \end{cases} \tag{1}$$

To the best of our knowledge, Shurbevski et al. [17] gave the first account examining the presence of a bias factor, and at the same time, demonstrated a constant 16/7-factor approximation algorithm. The previously reported approximation ratio of 16/7 has been achieved by a composite heuristic (see, e.g., [14] for terminology relating to composite heuristics). In this paper, we present a novel heuristic procedure for building Hamiltonian cycles in bipartite graphs and show that for the biased case it is an approximation algorithm with an approximation ratio of

$$1 + \frac{1 + \lambda}{\beta + \lambda} \tag{2}$$

where $\lambda$ is a real parameter which depends on the problem instance and cannot be known upfront. On one hand, the above expression is bounded by a constant 2 for any positive real $\lambda$ and $\beta \geq 1$, thus the proposed algorithm has a constant factor approximation ratio, improving the one from [17]. On the other hand, for a finite $\lambda$, the above expression approaches 1 as $\beta$ grows larger.

The presented approach by itself does not rely on approximating the metric TSP, however it can be used as a part of a composite heuristic to achieve an approximation ratio of

$$1 + \frac{2}{\zeta + \beta(2 - \zeta)}, \tag{3}$$

where $1 < \zeta \leq 2$ is an approximation ratio for the metric TSP. The expression from Eq. (3) is also bounded above by a constant 2, but it is not dependent on an instance-specific parameter, and has a clear relationship with the bias $\beta$ for a fixed $\zeta < 2$.

## 2   Preliminaries

The set of reals (resp., nonnegative reals) is denoted by $\mathbb{R}$ (resp., $\mathbb{R}_+$).

In general, for a minimization problem $\mathcal{P}$, let $P^*$ be the value of an optimal solution. An approximation algorithm $ALG$ is such that for any instance of $\mathcal{P}$, it can produce a feasible solution of value $P'$. We call the value

$$\alpha_{ALG} = \sup \left\{ \frac{P'}{P^*} \right\} \tag{4}$$

the approximation factor of algorithm $ALG$, and usually say that $ALG$ is an $\alpha_{ALG}$-approximation algorithm.

We use standard notation from graph theory; the ordered pair $G = (V, E)$ is a connected undirected graph. The vertex set and the edge set of $G$ are denoted by $V(G)$ and $E(G)$, respectively. We allow for parallel edges, or think of $G = (V, E)$ as a multigraph. Thus, $E(G)$ is a multiset of elements in $V \times V$. (We will make use of the multiset sum function, denoted by the symbol $\uplus$, as well as the shorthand $k \cdot E$ for $\uplus_{i=1}^{k} E$.) We use $\{u, v\}$, $u, v \in V(G)$ to reference any and all $e \in E(G)$ such that $e$ is incident with $u$ and $v$. For $u \in V(G)$, $d_G(u)$ denotes the degree of the node $u$ in the graph $G$. A graph is weighted if we are given some weight function $w : E(G) \rightarrow \mathbb{R}_+$ over the graph's edges. For any subset of edges $E' \subseteq E$, $w(E')$ denotes $\sum_{e \in E'} w(e)$. Similarly, for a subgraph $G'$ of $G$, $w(G')$ denotes $\sum_{e \in E(G')} w(e)$. A subgraph $G'$ of $G$ is spanning if $V(G') = V(G)$. We assume that all parallel edges are of the same weight, and $\forall e \in E(G)$, $e = \{u, v\}$, we equate the expressions $w(e)$ and $w(u, v)$. The weight function $w$ is said to be symmetric if

$$w(u, v) = w(v, u), \quad \forall e = \{u, v\} \in E(G), \tag{5}$$

and that it satisfies the triangle inequality if

$$w(u, v) \leq w(u, q) + w(q, v), \quad \forall q, u, v \in V(G). \tag{6}$$

A complete bipartite graph $G = (B, W; E)$ is such that $V(G) = B \cup W$, $B \cap W = \emptyset$, and $E(G) = B \times W$. A property similar to the triangle inequality can be extended over complete bipartite graphs, into the quadrangle inequality

$$w(u, v) \leq w(u, q) + w(q, y) + w(y, v), \quad \forall u, y \in B, q, v \in W. \tag{7}$$

For a complete graph induced by a set of vertices $B$, we write $G[B]$. By definition, $V(G[B]) = B$ and $E(G[B]) = B \times B$. Let $G = (B, W; E)$ be a given bipartite graph with an edge weight function $w : E(G) \rightarrow \mathbb{R}_+$, and $G[B]$ is exactly the complete graph induced by the partition $B$. Often in practice the vertex sets are in fact points from some metric space and the distance in this space serves as an edge weight function. In such a case, the edge weight function of $G[B]$ is defined by the distance function in the metric space. However, if we are only given a bipartite graph $G = (B, W; E)$ with an edge weight function $w : E(G) \rightarrow \mathbb{R}_+$, we can extend the edge weight function over the induced graph $G[B]$ as such

$$w(u, y) = \min_{q \in W}\{w(u, q) + w(q, y)\} \quad \forall u, y \in B. \tag{8}$$

**Lemma 1.** *For a given complete bipartite graph $G(B, W; E)$ with a symmetric edge weight function $w : E(G) \rightarrow \mathbb{R}_+$ satisfying the quadrangle inequality, let $G[B]$ be the complete graph induced by the vertex partition $B$. The extension of $w$ as an edge weight function of $G[B]$ of Eq. (8) is symmetric and satisfies the triangle inequality.*

Given a graph $G = (V, E)$, a Hamiltonian cycle $H$ is a connected spanning subgraph of $G$ such that

$$d_H(u) = 2, \quad \forall u \in V(G). \tag{9}$$

The problem of finding a Hamiltonian cycle $H$ of minimum $w(H)$ is commonly referred to as the traveling salesman problem (TSP).

For a complete bipartite graph $G = (B, W; E)$, with $|B| = |W|$, let $n := |B|(= |W|)$ and let $\sigma$ and $\tau$ be permutations on the points of $B$ and $W$, respectively. A traversal of a Hamiltonian cycle $H$ in $G$ is of the form

$$\sigma(1) \to \tau(1) \to \sigma(2) \to \cdots \to \tau(n-1) \to \sigma(n) \to \tau(n) \to \sigma(1). \tag{10}$$

We term Hamiltonian cycles in bipartite graphs *alternating*, for points in $B$ and $W$ appear alternately. When using an indexing device $i = 1, \ldots, n$, we allow it to wrap around, i.e.

$$i := \begin{cases} i + n, & i \leq 0, \\ i - n, & i > n. \end{cases}$$

As subgraphs of $G$, Hamiltonian cycles are undirected. However, once we settle for a way to traverse them, they assume an *orientation*.

In addition to the edge weight $w$, we are concerned with a bias factor $\beta \geq 1$. The bias factor impacts bipartite graphs as in Eq. (1). Assuming a traversal orientation as in Eq. (10), we introduce the biased *cost* $L$ for alternating cycles

$$L(H) = \beta \sum_{i=1}^{n} w(\sigma(i), \tau(i)) + \sum_{i=1}^{n} w(\tau(i), \sigma(i+1)). \tag{11}$$

We are now prepared to state the bipartite analogue of the metric TSP in face of the bias factor $\beta \geq 1$.

**The biased bipartite traveling salesman problem – BBTSP**
*Instance:* A complete bipartite graph $G = (B, W; E)$, with $|B| = |W|$, a symmetric weight function $w : E(G) \to \mathbb{R}_+$ which satisfies the quadrangle inequality, and a bias factor $\beta \geq 1$.
*Task:* Find an alternating Hamiltonian cycle $H^*$ in $G$ such that $L(H^*)$ is minimized.

In this paper we focus exclusively on the version of the BBTSP where the edge weight function $w$ is symmetric and satisfies the quadrangle inequality. We settle for this limitation because it has been shown [1, 10, 13, 15] that the bipartite TSP is not only NP-hard to solve, but also that in the general case, there is no constant factor approximation under the assumption that $P \neq NP$.

## 3 Building Blocks

In this section we will exhibit some of the known lower bounds on the value of an optimal solution for the BBTSP, as well as add a few new insights into their correlations. The presented lower bounds are structures well known in combinatorial optimization, and will serve as building blocks for a new procedure for constructing alternating Hamiltonian cycles in bipartite graphs.

## 3.1  Known Lower Bounds of the BBTSP

We present some of the observations made in [17] concerning the lower bounds of an optimal solution for the BBTSP. Our analysis mainly concerns two combinatorial structures in bipartite graphs; perfect matchings, and alternating spanning trees. We will just briefly state their definitions. Let $G = (B, W; E)$ be a (weighted) complete bipartite graph with an edge weight function $w : E(G) \to \mathbb{R}_+$ and $|B| = |W| =: n$. The edge weight function $w$ is assumed symmetric and satisfying the quadrangle inequality (Eq. (7)). A perfect matching $M \subset E(G)$ is such that there is exactly one edge in $M$ incident with any $u \in V(G)$. An alternating spanning tree $T$ (illustrated in Fig. 1(a)) is a connected acyclic spanning subgraph of $G$ such that

$$d_T(u) \leq 2, \quad \forall u \in B. \tag{12}$$

Both perfect matchings and alternating spanning trees are well studied combinatorial structures, e.g., [12, 16], and there exist polynomial time algorithms for computing perfect matchings and alternating spanning trees (of minimum weight) in bipartite graphs. Henceforth, let $M^*$ denote a perfect matching in $G$ of minimum weight $w(M^*)$, and $T^*$ an alternating spanning tree with minimum $w(T^*)$.

Given an instance of the BBTSP, let $H^*$ be an optimal solution, which minimizes the biased cost $L(H^*)$. The edges of $E(H^*)$ can be decomposed into two disjoint perfect matchings, $\overrightarrow{H^*}$ and $\overleftarrow{H^*}$, as in Fig. 1(b). Without loss of generality, we assume $H^*$ is to be traversed as indicated by arrows in Fig. 1(b), and $\overrightarrow{H^*}$ solely accounts for the bias term. The biased path cost $L(H^*)$ is given by

$$L(H^*) = \beta w(\overrightarrow{H^*}) + w(\overleftarrow{H^*}). \tag{13}$$

It surely holds

$$w(M^*) \leq w(\overrightarrow{H^*}) \leq w(\overleftarrow{H^*}). \tag{14}$$

Concerning alternating spanning trees in $G$, note that $w(T^*)$ is a lower bound of the weight of an alternating Hamiltonian cycle disregarding the bias factor, i.e.,

$$w(T^*) \leq w(\overrightarrow{H^*}) + w(\overleftarrow{H^*}). \tag{15}$$

Observing the graph $G[B]$ induced by the vertex partition $B$, we can see that an alternating Hamiltonian cycle in $G$ does in fact visit each vertex in $B$ exactly once, and can be shortcut to a Hamiltonian cycle of $G[B]$. We will use the extended $w$ from Eq. (8) for $G[B]$. For an optimal alternating Hamiltonian cycle $H^*$, let $\hat{C}$ be the resulting shortcut, as given in Fig. 1(b). Due to Eq. (8) we have

$$w(\hat{C}) \leq w(\overrightarrow{H^*}) + w(\overleftarrow{H^*}). \tag{16}$$

Consequently, for an optimal (w.r.t. the extended $w$) Hamiltonian cycle $C^*$ in $G[B]$ it holds

$$w(C^*) \leq w(\hat{C}) \leq w(\overrightarrow{H^*}) + w(\overleftarrow{H^*}). \tag{17}$$

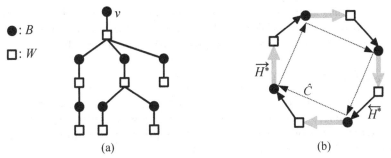

**Fig. 1.** (a) An alternating spanning tree $T$. (b) A minimum cost alternating Hamiltonian path $H^*$ of $G$. The subsets of edges $\overrightarrow{H^*}$ (bold gray arrows) and $\overleftarrow{H^*}$ (slender black arrows) form two disjoint perfect matchings. The shortcut $\hat{C}$ on $G[B]$ is given in dashed lines.

### 3.2  Further Observations

We would like to bring a special attention to an observation with respect to the structures presented above, alternating spanning trees and perfect matchings. Let $M^*$ and $T^*$ be a minimum weight perfect matching and a minimum weight alternating spanning tree in a given bipartite graph $G$, respectively. Owing to its special structure any alternating spanning tree in $G$ contains a perfect matching. Therefore, let $T^M \subset E(T^*)$ denote the edge set forming a perfect matching, and $T^\top$ the remaining edges of the alternating tree, i.e. $T^\top = E(T^*)\backslash T^M$. It simply holds

$$w(T^*) = w(T^M) + w(T^\top). \tag{18}$$

We present our view of the structure of an optimal solution, $H^*$, with $L(H^*) = \beta w(\overrightarrow{H^*}) + w(\overleftarrow{H^*})$, (see Eq. (13)). We introduce a parameter $\lambda \in \mathbb{R}_+$ as

$$\lambda = \frac{w(\overleftarrow{H^*})}{w(\overrightarrow{H^*})}. \tag{19}$$

Then, for the cost of an optimal tour $H^*$ we can write

$$L(H^*) = (\beta + \lambda)w(\overrightarrow{H^*}). \tag{20}$$

For a given instance of the BBTSP, the value of the parameter $\lambda$ cannot be known without solving it exactly. However, for the purpose of our exposition, it suffices that $\lambda \in \mathbb{R}_+$.

## 4   A New Approximation Algorithm

In this section we present a procedure for building an alternating Hamiltonian cycle in a given bipartite graph $G = (B, W; E)$ with $|B| = |W|$. We show that if the graph $G$ is endowed with a positive symmetric edge weight function $w$ which satisfies the quadrangle inequality, this procedure can be used as an approximation algorithm for the BBTSP. The procedure for building an alternating Hamiltonian cycle does not rely on approximating the metric TSP.

## 4.1  Construction

Let $G = (B, W; E)$, be a bipartite graph with $|B| = |W| =: n$. Let $w : E(G) \to \mathbb{R}_+$ be a symmetric edge weight function satisfying the quadrangle inequality. Let $M^*$ and $T^*$ be a perfect matching and an alternating spanning tree in $G$ of minimum $w(M^*)$ and $w(T^*)$, respectively.

We bring to attention the union of $M^*$ and $T^*$. As observed in Section 3.1, the alternating tree $T^*$ contains a perfect matching, $T^M$. The union of $T^M$ and $M^*$ forms a cycle cover of $G$. Let there be $k \le n$ individual cycles, which we will denote by $\mathcal{R} := \{R_i : i = 1, 2, \ldots, k\}$. We can think of elements of $\mathcal{R}$ as nodes, and define a graph $G_{\mathcal{R}} = (V(G_{\mathcal{R}}), E(G_{\mathcal{R}}))$, where $V(G_{\mathcal{R}}) = \mathcal{R}$. For brevity, for a subset $E'$ of $E(G)$, we will use $E'$ for $E(G_{\mathcal{R}})$ to denote that

$$E(G_{\mathcal{R}}) = \{\{i, j\} \mid \exists \{u, v\} \in E', u \in R_i \wedge v \in R_j\}, \quad 1 \le i, j \le k. \quad (21)$$

Since $T^*$ is an alternating spanning tree, thus all vertices in $V[G]$ are connected, the individual cycles $R_i$ must be connected with each other as well, i.e., the graph $G_{\mathcal{R}} = (\mathcal{R}, T^\top)$ is connected. We can choose an inclusion wise minimal $T^\perp \subseteq T^\top$, such that the graph $T_{\mathcal{R}} = (\mathcal{R}, T^\perp)$ remains connected, i.e., $T_{\mathcal{R}}$ is a spanning tree of $G_{\mathcal{R}}$, as in Fig. 2(a).

We term the procedure for constructing alternating Hamiltonian cycles *2APX*. Next, we give a brief summary of the construction procedure *2APX*

**Step 1:** Compute a minimum weight perfect matching $M^*$ and a minimum weight alternating spanning tree $T^*$ in $G$;

**Step 2:** Let $\mathcal{R} := \{R_i : i = 1, 2, \ldots, k\}$ be the cycle cover of $G$ given by $M^* + T^M$;

**Step 3:** Choose an inclusion wise minimal $T^\perp \subseteq T^\top$ such that $T_{\mathcal{R}} = (\mathcal{R}, T^\perp)$ is a spanning tree;

**Step 4:** Construct a multigraph $\mathcal{E}_{2APX} = (V(\mathcal{E}_{2APX}), E(\mathcal{E}_{2APX}))$, where $V(\mathcal{E}_{2APX}) = V(G)$, and $E(\mathcal{E}_{2APX}) = M^* \uplus T^M \uplus 2 \cdot T^\perp$ (Fig. 2(b));

**Step 5:** Shortcut an Eulerian walk of $\mathcal{E}_{2APX}$ to an alternating Hamiltonian cycle $H_{2APX}$, preserving the edges from $M^*$.

The multigraph $\mathcal{E}_{2APX}$ over the vertex set $V(G) = B \cup W$ in Fig. 2(b), has as its edge set a multiset sum of $M^*$, $T^M$ and two copies of $T^\perp$. We need to show that this structure can be used to obtain a valid alternating cycle. As a first step, we will elaborate that there is an Eulerian walk.

**Lemma 2.** *The multigraph $\mathcal{E}_{2APX}$ is Eulerian.*

*Proof.* We need to show that $\mathcal{E}_{2APX}$ is connected, and every vertex has even degree w.r.t. $\mathcal{E}_{2APX}$. Connectedness follows from the fact that we sought the structure $T_{\mathcal{R}} = (\mathcal{R}, T^\perp)$ to be a spanning tree, where $\mathcal{R}$ is a cycle cover of the vertex set $V(G) = B \cup W$. Every vertex in $V(G)$ is of degree 2 w.r.t. the cycle cover $\mathcal{R}$. Finally, we have added two copies of $T^\perp$, hence the claim follows.  □

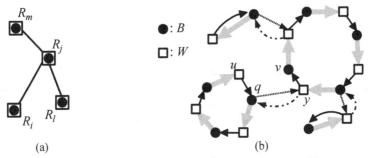

(a)                              (b)

**Fig. 2.** (a) A representation of $T_{\mathcal{R}} = (\mathcal{R}, T^{\perp})$. Nodes of $\mathcal{R}$ (●), are individual cycles over $V(G) = B \cup W$. (b) The resulting multigraph $\mathcal{E}_{2APX}$, arrows added to aid the image of traversing. The perfect matching $M^*$ is given in bold gray lines, $T^M$ in slender black, and the two copies of $T^{\perp}$ in dashed lines.

Next we show how $\mathcal{E}_{2APX}$ can be shortcut to give an alternating Hamiltonian cycle.

**Lemma 3.** *The Eulerian graph $\mathcal{E}_{2APX}$ can always be shortcut to an alternating Hamiltonian cycle $H_{2APX}$, preserving the edges from $M^*$.*

*Proof.* We will prove this claim by induction over the number of cycles $k$ in the cycle cover $\mathcal{R}$
  - *Case $k = 1$:* Trivial, this is $H_{2APX}$;
  - *Case $k > 1$:* Start from the observation that $T^{\perp}$ is bipartite. Therefore there must exist a certain $q \in B$ connected to some $y \in W$ by an arc $\{q, y\} \in T^{\perp}$. Let $q \in R_i$ and $y \in R_j$. Now, let $u \in W$ (also $u \in R_i$) such that $\{u, q\} \in T^M$, and let $v \in B$ (also $v \in R_j$), such that $\{y, v\} \in T^M$ (Fig. 2(b)). We shortcut $\{\{u, q\}, \{q, y\}, \{y, v\}\}$ by $\{u, v\}$, thus merging the two cycles $R_i$ and $R_j$ and decreasing the number of cycles by one.

Note, all of the shortcut edges, $\{u, q\}$, $\{q, y\}$ and $\{y, v\}$ belong to $T$ (either in $T^{\perp} \subseteq T^{\top}$ or $T^M$), thus edges in $M^*$ are preserved intact. Lastly, due to the quadrangle inequality from Eq. (7), this shortcutting will not increase the total weight $w(\mathcal{E}_{2APX})$.                                                      □

In the end, we will have obtained an alternating Hamiltonian cycle $H_{2APX}$.

## 4.2   Approximation Ratio

Next, we investigate the applicability of the $2APX$ procedure as an approximation algorithm.

**Lemma 4.** *For a given instance of the metric BBTSP, let $H^*$ be an alternating Hamiltonian cycle of minimal cost $L(H^*)$. Let the edge set $\overrightarrow{H^*} \subset E(H^*)$ be traversed in the direction from $B$ to $W$, so that the value $L(H^*)$ is parameterized by some $\lambda \in \mathbb{R}_+$ as $L(H^*) = (\beta + \lambda)w(\overrightarrow{H^*})$. For $H_{2APX}$ as the result from the $2APX$ procedure it holds*

$$L(H_{2APX}) \leq \frac{\beta + 2\lambda + 1}{\beta + \lambda} L(H^*). \tag{22}$$

*Proof.* In order to derive an upper bound of the cost $L(H_{2APX})$, we will retrace the steps from the construction process, and recall some of the bounds presented in Section 3, especially Subsection 3.2.

First, recall that we chose a $T^\perp \subseteq T^\top$, therefore $w(T^\perp) \leq w(T^\top)$. It readily follows (see Eq. (18))

$$w(T^\perp) \leq w(T^*) - w(T^M). \tag{23}$$

Let us partition $E(H_{2APX})$ into two disjoint matchings, $\overrightarrow{H}_{2APX}$ and $\overleftarrow{H}_{2APX}$, in such a way that $\overrightarrow{H}_{2APX} = M^*$ and $\overleftarrow{H}_{2APX}$ is a shortcut through $T^M \uplus T^\perp \uplus T^\perp$, as in Lemma 3. We choose a traversal orientation such that exactly the edges of $\overrightarrow{H}_{2APX}$ are traversed in the direction from $B$ to $W$. From the bias factor $\beta$ of Eqs. (1), (11) and (13)

$$L(H_{2APX}) = \beta w(\overrightarrow{H}_{2APX}) + w(\overleftarrow{H}_{2APX})$$
$$\leq \beta w(M^*) + w(T^M) + 2w(T^\perp). \tag{24}$$

Recall the partition of a minimum cost alternating spanning tree from Eq. (18) and the related bounds from Eq. (23) and substitute them in Eq. (24). From this, and the fact that $w(M^*) \leq w(T^M)$, we get

$$L(H_{2APX}) \leq \beta w(M^*) + 2w(T^*) - w(T^M)$$
$$\leq 2w(T^*) + (\beta - 1)w(M^*). \tag{25}$$

Next we substitute for $M^*$ and $T^*$ the bounds given with Eqs. (14) and (15)

$$L(H_{2APX}) \leq 2(w(\overrightarrow{H^*}) + w(\overleftarrow{H^*})) + (\beta - 1)w(\overrightarrow{H^*})$$
$$= 2(1 + \lambda)w(\overrightarrow{H^*}) + (\beta - 1)w(\overrightarrow{H^*}). \tag{26}$$

Finally, following Eq. (20), the expression above leads to the claim. □

Lemma 4 gives the result announced in the Introduction, Eq. (2)

$$\frac{L(H_{2APX})}{L(H^*)} \leq 1 + \frac{1 + \lambda}{\beta + \lambda}.$$

The result from Lemma 4 and the definition of an approximation ratio of Eq. (4) give the following result

$$\alpha_{2APX} = 2,$$

which holds true for any $\beta \geq 1$ and $\lambda \in \mathbb{R}_+$. However, Eq. (2) does provide us with insight of the behavior of $L(H_{2APX})$ for increasing values of $\beta$, and some reasonable finite upper bound on $\lambda$.

## 4.3   As a Part of a Composite Heuristic

As the previously known approximation ratio of $16/7$ described in [17] relies on a composite heuristic, i.e., on a trade-off between two different procedures for building an alternating Hamiltonian path, we investigate a similar approach. For that purpose, we only briefly review a well known procedure for constructing an alternating Hamiltonian cycle in a given complete bipartite graph $G = (B, W; E)$, with $|B| = |W| =: n$. We term this procedure as procedure $SWAP$ (the same procedure has been termed a matching based heuristic in [17].) The $SWAP$ procedure has been described as a heuristic method for the swapping problem [2], and adopted to the bipartite TSP [3]. Briefly described, it is as follows

**Step 1:** Find a minimum cost perfect matching $M^*$ in $G = (B, W; E)$;
**Step 2:** Build a $\zeta$-approximate Hamiltonian cycle $C'$ in $G[B]$;
**Step 3:** Make an Eulerian multigraph $\mathcal{E}_{SWAP} = (V(\mathcal{E}_{SWAP}), E(\mathcal{E}_{SWAP}))$, where $V(\mathcal{E}_{SWAP}) = V(G)$ and $E(\mathcal{E}_{SWAP}) = E(C') \uplus 2 \cdot M^*$;
**Step 4:** Appropriately shortcut an Eulerian walk in $\mathcal{E}_{SWAP}$ to get an alternating Hamiltonian cycle $H_{SWAP}$ in $G$, preserving one copy of $M^*$.

The correctness and validity of the $SWAP$ procedure is argued in more detail in, e.g., [2, 3, 17].

For the purpose of arriving to a suitable expression for a composite heuristic relying on the $2APX$ and $SWAP$ procedures, we will present our bounds on $L(H_{SWAP})$. Analogous to Eq. (17), for a $\zeta$-approximate $C'$ of an optimal $C^*$ we get

$$w(C') \le \zeta w(C^*) \le \zeta \left( w(\overrightarrow{H^*}) + w(\overleftarrow{H^*}) \right). \tag{27}$$

Since we can shortcut an Eulerian walk in $\mathcal{E}_{SWAP}$ to obtain $H_{SWAP}$ in such a way that one copy of $M^*$ is preserved, we can orient the traversal of $H_{SWAP}$ so that exactly the edges in $M^*$ are traversed in the direction from $B$ to $W$. Following Eqs. (14), (20) and (27)

$$\begin{aligned} L(H_{SWAP}) &\le \zeta(1+\lambda)w(\overrightarrow{H^*}) + (\beta+1)w(M^*) \\ &\le \frac{\zeta(1+\lambda) + \beta + 1}{\beta + \lambda} L(H^*). \end{aligned} \tag{28}$$

Since from Lemma 1 we have that the extension of $w$ over the edges of $G[B]$ is symmetric and satisfies the triangle inequality, we can use, e.g., Christofides' heuristic [6] to build a $C'$ with $\zeta = 3/2$.

We propose a simple procedure which will compute both $H_{2APX}$ and $H_{SWAP}$ according to their respective construction procedures, and choose the one of lower cost. Let us term this procedure $COMP$ and the resulting alternating Hamiltonian cycle $H_{COMP}$. From Eqs. (22) and (28) we get

$$L(H_{COMP}) \leq \min\left\{\frac{\beta + 2\lambda + 1}{\beta + \lambda}L(H^*), \frac{\zeta(1 + \lambda) + \beta + 1}{\beta + \lambda}L(H^*)\right\}$$

$$\leq \left(1 + \frac{2}{\zeta + \beta(2 - \zeta)}\right)L(H^*). \qquad (29)$$

The trade-off in Eq. (3) is achieved for $\lambda = \frac{\zeta}{2-\zeta}$, therefore, it only makes sense to be called when $\zeta < 2$. It readily follows

$$\alpha_{COMP} = 1 + \frac{2}{\zeta + \beta(2 - \zeta)},$$

which is not dependent on a hidden instance-specific parameter, such as $\lambda$.

### 4.4 Computational Complexity

Without much deliberation we will state that all procedures undertaken to obtain an alternating Hamiltonian cycle have well known polynomial time implementations. An excellent source of information concerning the presented combinatorial structures as well as their algorithmic implementations can be found in [12, 16], as well as [8]. We will just state that the bottleneck procedure in the computation is finding a minimum cost alternating spanning tree $T^*$ in the bipartite graph $G = (B, W; E)$ ($|B| = |W| =: n$), since it requires a call to a general matroid intersection algorithm, which in turn requires $O(n^7)$ time ([3, 8, 9, 11, 12, 16, 18]). As a consequence, we can state the following

**Theorem 1.** *The biased bipartite traveling salesman problem with a symmetric edge weight function satisfying the quadrangle inequality and a bias $\beta \geq 1$ can be approximated within a constant factor $\alpha = 2$, in polynomial time complexity.*

## 5   Conclusion

We formalized the biased bipartite TSP (BBTSP) as a generalization of the symmetric bipartite TSP with quadrangle inequality by introducing a bias term $\beta \geq 1$, which introduces asymmetry in the cost of an alternating Hamiltonian path. This generalization had been introduced as a means to better capture some features of industrial material handling scenarios.

We presented a novel heuristic for building alternating Hamiltonian cycles in complete bipartite graphs. With that, obtained a first nontrivial approximation algorithm which improves the approximation factor of previously known approaches to a constant 2, and showed that this approximation ratio holds for any value of the bias $\beta \geq 1$. We also analyzed the performance of the proposed procedure for building alternating Hamiltonian cycles as a part of a composite heuristic, and derived an approximation ratio which benefits of both a better approximation for the metric TSP, and an increased value for the bias $\beta$.

It is a standing question whether the constant bound 2 of the approximation ratio presented in this paper can be further improved by some algorithms similar to existing approaches for the standard metric TSP [7].

# References

1. Akiyama, T., Nishizeki, T., Saito, N.: NP-completeness of the Hamiltonian cycle problem for bipartite graphs. Journal of Information Processing 3(2), 73–76 (1980)
2. Anily, S., Hassin, R.: The swapping problem. Networks 22(4), 419–433 (1992)
3. Baltz, A., Srivastav, A.: Approximation algorithms for the Euclidean bipartite TSP. Operations Research Letters 33(4), 403–410 (2005)
4. Chalasani, P., Motwani, R.: Approximating capacitated routing and delivery problems. SIAM Journal on Computing 28(6), 2133–2149 (1999)
5. Chalasani, P., Motwani, R., Rao, A.: Algorithms for robot grasp and delivery. In: Proceedings of 2nd International Workshop on Algorithmic Foundations of Robotics, pp. 347–362 (1996)
6. Christofides, N.: Worst-case analysis of a new heuristic for the travelling salesman problem. Technical Report 388, Graduate School of Industrial Administration, Carnegie-Mellon University, Pittsburgh (1976)
7. Cook, W.J.: In Pursuit of the Traveling Salesman: Mathematics at the Limits of Computation. Princeton University Press (2012)
8. Frank, A.: Connections in Combinatorial Optimization. Oxford Lecture Series in Mathematics and Its Applications. OUP Oxford (2011)
9. Frank, A.: A weighed matroid intersection algorithm. Journal of Algorithms 2, 328–336 (1981)
10. Frank, A., Triesch, E., Korte, B., Vygen, J.: On the bipartite traveling salesman problem. Technical Report 98866-OR, University of Bonn (1998)
11. Karuno, Y., Nagamochi, H., Shurbevski, A.: An approximation algorithm with factor two for a repetitive routing problem of grasp-and-delivery robots. Journal of Advanced Computational Intelligence and Intelligent Informatics 15(8), 1103–1108 (2011)
12. Korte, B., Vygen, J.: Combinatorial Optimization: Theory and Algorithms, 3rd edn. Springer (2005)
13. Krishnamoorthy, M.: An NP-hard problem in bipartite graphs. SIGACT News 7, 26–26 (1975)
14. Langston, M.A.: A study of composite heuristic algorithms. The Journal of the Operational Research Society 38(6), 539–544 (1987)
15. Sahni, S., Gonzalez, T.: P-complete approximation problems. Journal of the ACM 23(3), 555–565 (1976)
16. Schrijver, A.: Combinatorial Optimization - Polyhedra and Efficiency. Springer (2003)
17. Shurbevski, A., Nagamochi, H., Karuno, Y.: Heuristics for a repetitive routing problem of a single grasp-and-delivery robot with an asymmetric edge cost function. In: 10th International Conference of the Society for Electronics, Telecommunications, Automatics and Informatics (ETAI 2011), CD-ROM Proceedings, pp. A1–1 (2011)
18. Srivastav, A., Schroeter, H., Michel, C.: Approximation algorithms for pick-and-place robots. Annals of Operations Research 107(3), 321–338 (2001)

# A $(k + 1)$-Approximation Robust Network Flow Algorithm and a Tighter Heuristic Method Using Iterative Multiroute Flow

Jean-François Baffier[1] and Vorapong Suppakitpaisarn[2]

[1] The University of Tokyo, Université Paris-Sud, JFLI, CNRS
[2] NII, JST, ERATO, Kawarabayashi Large Graph Project

**Abstract.** We consider two variants of a max-flow problem against $k$ edge failures, each of which can be both approximated by a multiroute flow algorithm. The maximum $k$-robust flow problem is to find the minimum max-flow value among $\binom{m}{k}$ networks that can be obtained by deleting each set of $k$ edges. The maximum $k$-balanced flow problem is to find a max-flow of the network such that the flow value is maximum against any set of $k$ edge failures, when deleting the corresponding flow to those $k$ edges in the original flow. We prove $C_M \leqslant C_{M'} \leqslant C_B \leqslant C_R \leqslant (k + 1) \cdot C_M$, where $C_M$ is the max-$(k + 1)$-route flow value, $C_{M'}$ is the effectiveness of the max-$(k + 1)$-route flow after $k$ attacks, $C_B$ is the max-$k$-balanced flow value, and $C_R$ is the max-$k$-robust flow value. Also, we develop a polynomial-time heuristic algorithm for both cases, called the iterative multiroute flow. Our experimental results show that the average improvement made by our heuristic method can be up to 10% better than the multiroute flow algorithm. Compared to the optimal max-$k$-robust flow solutions – obtained by a brute-force algorithm – there is an average gap of 2% at most.

## 1 Introduction

Since its introduction by Ford and Fulkerson [1], the maximum flow problem (max-flow) has been widely studied due to its many theoretical and practical applications. Given a capacity on each link in a network, the problem is to optimize the amount of flow we can send from a given source node to a given sink node. There are many polynomial-time algorithms with which to solve this problem. Recently, Orlin proposed an algorithm to solve it in $O(nm)$ time [2], where $n$ is the number of vertices and $m$ is the number of edges.

The amount of flow we get from a solution to those algorithms can drop significantly if some of the links in the network are attacked or fail. Since the best attacks are equivalent to the worst failures, we will use these two terms interchangeably throughout this paper. A simple way to guarantee the amount of flow sent between a source and a sink is to oversize the capacity of each link in our network [3]. However, higher-capacity links can have a significantly higher maintenance cost [4]. Because of that, several formulations have been introduced

S.P. Pal and K. Sadakane (Eds.): WALCOM 2014, LNCS 8344, pp. 68–79, 2014.

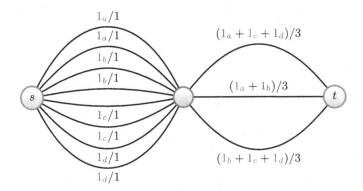

**Fig. 1.** Graph with distinct $C_M$, $C_{M'}$, $C_B$, and $C_R$. Capacity of each edge is shown in black, and the max-2-route flow is shown as four unitary 2-route flows, $a$, $b$, $c$, and $d$ (each pair are denoted by a red and a green member).

to find a robust network. Those include $k$-EDP, $k$-DFP [5] and $k$-EDPCOL [6,7], which show the usage of the $k$ edge-disjoint paths against a specific type of attack.

In this work, we study the robust network problem through the use of $k$-route flow, that is a nonnegative linear combination of $k$-edge disjoint paths. This notion was introduced by Kishimoto and Takeuchi in [8,9], where they extend the max-flow/min-cut duality property to the multiroute flow context and provide an algorithm to compute a max-$k$-route flow based on at most $k$ iterations of a classical max-flow algorithm. A simpler proof of duality can be found in [10]. Later, the algorithm was improved by Aggarwal and Orlin [11] to an algorithm using at most min $\{\log(nU), k\}$ iterations, when $U$ is the maximum capacity on each edge and $n$ is the number of vertices in the network. Also, a simpler proof for the correctness and complexity of this algorithm was given by Du and Chandrasekaran in [12].

We tackled two natural variants of the max-flow problem against $k$ edge failures, referred to as the *maximum $k$-robust flow* and *maximum $k$-balanced flow* problems. The maximum $k$-robust flow problem (max-$k$-robust flow) is to find the minimum max-flow value among $\binom{m}{k}$ networks obtained by deleting each set of $k$ edges. The maximum $k$-balanced flow problem (max-$k$-balanced flow) is to find a max-flow of the network such that the flow value is maximum against any set of $k$ edge failures, when deleting the corresponding flow to those $k$ edges in the original flow. The max-$k$-robust flow is firstly introduced in this paper, but the max-$k$-balanced flow for $k = 1$ is introduced and solved in [13]. For the case when $k \geqslant 2$, we can infer from the discussion in [14] that the problem can be significantly harder. Recent results [15,16] also indicate that several problems involving $k \geqslant 2$ link failures are NP-hard.

Define the $k$-effectiveness of a flow as a minimum value of this flow that remains after $k$ edge failures. We denote $C_M$ the max-$(k + 1)$-route flow value, $C_{M'}$ its $k$-effectiveness, $C_B$ the max-$k$-balanced flow value, and $C_R$ the $k$-robust flow value. Figure 1 is an instance of a network subject to one edge attack, where each one of $C_M$, $C_{M'}$, $C_B$, and $C_R$ are different. More precisely, $C_M = 4$, $C_{M'} = 5$, $C_B = \frac{16}{3}$, and $C_R = 6$. The max-2-route flow is made of the summation of four unitary 2-route flows, $a$, $b$, $c$, and $d$ (each pair are denoted by a red and a green member).

This paper is organized as follow. In Section 3, we introduce the formal definitions and the properties of max-$k$-robust flow, max-$k$-balanced flow, and $k$-effectiveness.

In Section 4, we prove that $C_M \leqslant C_{M'} \leqslant C_B \leqslant C_R \leqslant (k+1) \cdot C_M$. We deduce from this result that the method for finding the max-$(k + 1)$-route flow in [9] is a $(k + 1)$-approximation algorithm for both maximum $k$-robust and maximum $k$-balanced flow problems. This guarantees that the multiroute flow algorithm is useful for our problem when $k$ is relatively small. When $k$ becomes larger, the algorithm can give a solution that is much less efficient than the optimal one.

To improve this, in Section 5, we develop a heuristic algorithm called iterative multiroute flow. The algorithm uses $O(\lambda^2)$ iterations of max-flow, when $\lambda$ is the source-sink edge connectivity of our network.

Our experimental results, in Section 6, show that the average improvement by the iterative multiroute flow can be up to 10% better than the result from the multiroute flow algorithm. Compared to the maximum $k$-robust flow, there is an average gap of 2% at most. As the value of the maximum $k$-balanced flow cannot be larger than the maximum $k$-robust flow, the average gap is even less than 2% for the maximum $k$-balanced flow problem.

## 2  Preliminaries

In this section, we provide the notation that we will use throughout this article. The definition and properties of the multiroute flow are provided in Subsection 2.1.

Let $G = (V, E, c)$ be a network, where $V$ is a set of nodes, $E$ is a set of links, and $c : E \to \mathbb{R}^+$ is a capacity function.

Let $s, t \in V$ be a source node and a sink node, respectively. Throughout this paper, we will consider single-commodity flows from $s$ to $t$. All terminologies are based on that setting unless otherwise specified. The set $\mathscr{C}$ (resp., $\mathscr{F}$) refers to the set of all $s$-$t$ cuts (resp., the set all possible $s$-$t$ flows) of $G$. $\lambda$ refers to the $s$-$t$ edge connectivity of $G$, and $k$ refers to the number of edges that the attacker can remove.

**Definition 2.1 ($k$-robust capacity [17]).** *Given a cut $X \in \mathscr{C}$, let $\{e_0, e_1, \ldots, e_p\}$ be the cut-set of $X$, where $c(e_i) \geq c(e_{i+1})$ for any $0 \leqslant i < p$. For $0 \leqslant k \leqslant p$, we define the $k$-robust capacity of $X$ as follows:*

$$\alpha_k(X) = \sum_{i=k}^{p} c(e_i).$$

## 2.1 Multiroute Flow

In this subsection, we will briefly describe the $k$-route flow introduced in [9], which, for $k \geq 2$, is also called a *multiroute flow*.

**Definition 2.2 ($k$-route flow).** *A $k$-route flow is a nonnegative linear combination of $k$ edge-disjoint s-t paths with unitary flow, in which the value on each edge does not exceed the edge capacity. The value of a $k$-route flow is the summation of coefficients in that linear combination.*

**Definition 2.3 (max-$k$-route flow).** *A maximum $k$-route flow is a $k$-route flow such that its value is at least as large as the value of any other $k$-route flow.*

The max-$k$-route flow can be efficiently found by an algorithm proposed by Kishimoto and Takeuchi [9]. The running time of that algorithm is $O(kT)$, where $T$ is the computation time of the max-flow problem. Let $G^p = (V, E, c_p)$, where $c_p(e) = \min(c(e), p)$. It is shown in that paper that for some $p^* \in \mathbb{R}$ the max-flow of $G^{p^*}$ is the max-$k$-route flow of $G$. Kishimoto and Takeuchi also propose an effective method to search for that $p^*$ based on the max-flow value of $G^p$ for at most $k$ distinct values of $p$.

**Definition 2.4 ($k$-capacity [17]).** *The $k$-capacity of a cut $X$ is given by*

$$\beta_k(X) = \min_{0 \leqslant i \leqslant k-1} \left( \frac{1}{k-i} \cdot \alpha_i(X) \right).$$

A min-$k$-route cut is a cut minimizing the $k$-capacities over all the cuts in the network. Now, we can state the $k$-route duality theorem.

**Theorem 2.1 ($k$-route duality [9,10]).** *The value of a max-$k$-route flow is equal to the $k$-capacity of a min-$k$-route cut. Formally, if we denote $M$ as a max-$k$-route flow solution, its value is $\min_{X \in \mathscr{C}} \beta_k(X)$. We use $C_M$ to represent the value of max-$(k+1)$-route flow.*

## 3 $k$-robust and $k$-balanced Flow

In this section, we will give a precise definition for the maximum $k$-robust flow and the maximum $k$-balanced flow. Also, we will show some of their properties that will be useful for proving our main result in the following section.

For the maximum $k$-robust flow, the attacker will choose $k$ edges to break before we determine the flow. Our optimal choice is to choose the max-flow from the network in which the broken edges are removed. An optimal choice for the attacker is to minimize the value of that max-flow.

The attacker can try all the subsets of links with size $k$, check the max-flow value after those $k$ edges are removed, and select the subset that minimizes that value. By doing this, we can obtain the max-$k$-robust flow in $\binom{m}{k}$ iterations of max-flow, when $m$ is the number of links in our network.

Recall the definition of $k$-robust capacity in Section 2, and define $\phi_S$ as the value of the max-flow of the network $(V, E \backslash S)$. The formal definition of the max $k$-robust flow is as follows.

**Definition 3.1 (max $k$-robust flow).** *Let* $S^* = \underset{S \subseteq E:|S|=k}{\text{argmin}} \phi_S$. *The maximum $k$-robust flow is the max-flow of the network* $(V, E\backslash S^*)$. *We use* $C_R$ *to represent its value.*

**Definition 3.2 (min $k$-robust cut).** *The minimum $k$-robust cut problem is to find a s-t cut with the minimum $k$-robust capacity. The solution to this problem, $X_k^*$, is formally expressed by* $X_k^* = \underset{X \in \mathscr{C}}{\text{argmin}} \, \alpha_k(X)$.

**Lemma 3.1.** *The $k$-robust capacity of a minimum $k$-robust cut,* $\underset{X \in \mathscr{C}}{\min} \alpha_k(X)$, *is equal to the value of a maximum $k$-robust flow,* $C_R$.

*Proof.* Let $\mu_S$ be the capacity of a minimum cut of $(V, E\backslash S)$. We note $\mu_S(X)$ the capacity of a cut $X$. By the max-flow/min-cut duality, we know that

$$C_R = \underset{S \subseteq E:|S|=k}{\min} \phi_S = \underset{S \subseteq E:|S|=k}{\min} \mu_S = \underset{S \subseteq E:|S|=k}{\min} \left( \underset{X \in \mathscr{C}}{\min} \mu_S(X) \right).$$

Let consider an edge-set $S^*$ and a cut $X^*$ such that

$$S^* := \underset{S \subseteq E:|S|=k}{\text{argmin}} \left( \underset{X \in \mathscr{C}}{\min} \mu_S(X) \right), \text{ and } X^* := \underset{X \in \mathscr{C}}{\text{argmin}} \left( \underset{S \subseteq E:|S|=k}{\min} \mu_S(X) \right).$$

Let the cut-set of $X^*$ be $E^* = \{e_0, \ldots, e_p\}$, where $c(e_0) \geq c(e_1) \geq \cdots \geq c(e_p)$. Then we get $\mu_{S^*}(X^*) = \sum_{e \in E^* \backslash S^*} c(e)$. We can consider two cases, $E^* \subseteq S^*$ and $E^* \not\subseteq S^*$. When $E^* \subseteq S^*$, $\mu_{S^*}(X^*) = \alpha_k(X^*) = 0$ and $C_R = \underset{X^* \in \mathscr{C}}{\min} \alpha_k(X^*) = 0$. Next, we consider the case when $E^* \not\subseteq S^*$. Assume $S^* \not\subseteq E^*$. Let $e, e'$ be edges such that $e \in S^* \backslash E^*$ and $e' \in E^* \backslash S^*$. If $S' := S^* \cup \{e'\} - \{e\}$, then $\mu_{S'}(X^*) = \mu_{S^*}(X^*) - c(e') < \mu_{S^*}(X^*)$. This contradicts the assumption that $\mu_{S^*}(X^*)$ is the minimum value among $X \in \mathscr{C}$ and $S \subseteq E$ for $|S| = k$.

Hence, $S^* \subseteq E^*$. To minimize $\mu_{S^*}(X^*) = \sum_{e \in E^* \backslash S^*} c(e)$, it is obvious that $S^*$ must be a set of edges in $E^*$ with the largest capacity, $\{e_0, \ldots, e_{k-1}\}$. Then $\mu_{S^*}(X^*) = \alpha_k(X^*)$, and $C_R = \underset{X \in \mathscr{C}}{\min} \alpha_k(X)$.    □

Next, we will consider the maximum $k$-balanced flow. In this setting, we will choose the flow $F \in \mathscr{F}$ before the attacker selects the edges to attack. Assume that our choice is $F$. We will call the value of the flow that remains after the attack the $k$-effectiveness of $F$, and we define it as follows.

**Definition 3.3 (Effectiveness of a Flow).** *Let $F$ be a valid flow, and let $\phi_S^F$ be a max-flow of a network $G' = (V, E\backslash S, f)$, where $f(e)$ is a value of the flow $F$ on edge $e$. We define the $k$-effectiveness of $F$, $C_{F'}$ as*

$$C_{F'} = \underset{S \subseteq E:|S|=k}{\min} \phi_S^F.$$

$\phi_S^F$, defined in Definition 3.3, is actually the amount of flow $F$ that remains after all the edges in $S$ have been removed. In a way similar to what we did for $C_R$, we can compute $C_{F'}$ by calculating $\phi_S^F$ for all possible $S$, and then take the smallest one. Since the value of the flow on any edge cannot be larger than its capacity, it is obvious that $\phi_S^F \leqslant \phi_S$ for any $S$ and $F$. From this, we know that $C_{F'} \leqslant C_R$ for any flow $F$.

Let $M$ be a max-$(k+1)$-route flow, and let $C_{M'}$ be its $k$-effectiveness. Recall the multiroute flow value $C_M$ defined in Theorem 2.1. While $C_R$, $C_{F'}$, and $C_{M'}$ denote the values of flows in the traditional sense, $C_M$ denotes the value of a $(k+1)$-route flow. As discussed in [11], the $k$-effectiveness of the $(k+1)$-route flow $C_{M'}$ is guaranteed to be greater than or equal to the value of $C_M$.

The formal definition of the maximum $k$-balanced flow is as follow.

**Definition 3.4 (max $k$-balanced flow).** *The maximum $k$-balanced flow can be defined as*

$$F^* = \underset{F \in \mathscr{F}}{\operatorname{argmax}} C_{F'}.$$

*We will use $C_B$ to denote the value of $F^*$.*

## 4   Relations between the Different Kinds of Flow

In this section, we will give various bounds that justify the use of multiroute flow to approximate both max-$k$-balanced flow and max-$k$-robust flow. We will compare a $k$-robust flow with a $(k+1)$-route flow.

Recall the notation that we defined in Sections 2 and 3. We use $C_M$ to represent the value of a max-$(k+1)$-route flow, $C_{M'}$ to represent the effectiveness of that max-$(k+1)$-route flow, $C_B$ to represent the value of a max-$k$-balanced flow, and $C_R$ to represent the value of the max-$k$-robust flow. From those definitions, we can formulate Proposition 4.1.

**Proposition 4.1.** $C_M \leqslant C_{M'} \leqslant C_B \leqslant C_R$.

We know that $C_M \leqslant C_{M'}$ and $C_B \leqslant C_R$ from the discussion in Section 3. It is obvious from the definition of the maximum $k$-balanced flow in Section 2.1 that $C_B$ is the maximum effectiveness that can be obtained from our network. Hence, the effectiveness of the maximum multiroute flow $C_{M'}$ cannot be larger than $C_B$.

**Lemma 4.1.** *For a given cut $X$, $\alpha_k(X) \leqslant (k+1) \cdot \beta_{k+1}(X)$.*

*Proof.* By Definition 2.4, $\beta_{k+1}(X) = \min\limits_{0 \leqslant i \leqslant k} \left( \frac{1}{k+1-i} \cdot \alpha_i(X) \right)$. Hence,

$$
\begin{aligned}
\frac{\alpha_k(X)}{\beta_{k+1}(X)} &= \frac{\alpha_k(X)}{\min\limits_{0 \leqslant i \leqslant k} \left( \frac{1}{k+1-i} \cdot \alpha_i(X) \right)} \\
&= \alpha_k(X) \cdot \max\limits_{0 \leqslant i \leqslant k} \left( \frac{k+1-i}{\alpha_i(X)} \right) \\
&\leqslant \alpha_k(X) \cdot \max\limits_{0 \leqslant i \leqslant k} \left( \frac{k+1}{\alpha_i(X)} \right) = k+1.
\end{aligned}
$$

$\square$

**Corollary 4.1.** $C_R \leqslant (k+1) \cdot C_M$.

*Proof.* Let us assume that $X$ is the cut for a max-$k$-robust flow, and $Y$ is the cut for a max-$(k+1)$-route flow. Hence, $C_R = \alpha_k(X)$, and $C_M = \beta_{k+1}(Y)$. We know from the previous lemma that $\frac{\alpha_k(X)}{\beta_{k+1}(X)} \leqslant k+1$ and $\frac{\alpha_k(Y)}{\beta_{k+1}(Y)} \leqslant k+1$.

By Theorem 2.1 and Definition 3.1, $\beta_{k+1}(Y) \leqslant \beta_{k+1}(X)$ and $\alpha_k(X) \leqslant \alpha_k(Y)$. We will prove this by contradiction and assume that $\frac{C_R}{C_M} = \frac{\alpha_k(X)}{\beta_{k+1}(Y)} > k+1$. Since $\frac{\alpha_k(Y)}{\beta_{k+1}(Y)} \leqslant k+1$, $\frac{\alpha_k(X)}{\beta_{k+1}(Y)} > \frac{\alpha_k(Y)}{\beta_{k+1}(Y)}$. By multiplying $\beta_{k+1}(Y)$ to both sides of the inequality, we get $\alpha_k(X) > \alpha_k(Y)$. That contradicts the starting assumption that $X$ is a min-$k$-robust cut. □

Now we will give and prove a set of propositions to show that $k+1$ is an exact approximation ratio, and this leads to the main theoretical result of this work, Theorem 4.1.

**Proposition 4.2.** *There exists an infinite graph such that* $C_{M'} = (k+1) \cdot C_M$.

**Proposition 4.3.** *There exists an infinite graph such that* $C_B = (k+1) \cdot C_{M'}$.

**Proposition 4.4.** *There exists an infinite graph such that* $C_R = (k+1) \cdot C_B$.

*Proof (4.2).* Let us consider the graph from Figure 2a and $M$, a max-$(k+1)$-route flow solution between $s$ and $r$. On that graph, $C_M = \frac{m}{k+1}$ and $C_{M'} = C_R = m-k$. We have the result we wanted when $m$ approaches infinity. □

*Proof (4.3).* Let us consider the graph from Figure 2a, this time with $s$ and $t$ as the source and sink, respectively. Based on Definition 2.1, for any flow solution $M$ of the max-$(k+1)$-route flow, $C_M = \frac{m}{k+1}$. We choose $M$ such that on the right cut, the flow of $M$ goes through only $k+1$ edges. Hence $C_{M'} = \frac{m}{k+1}$. Obviously $C_B = m - k$. We have the result we wanted when $m$ approaches infinity. □

*Proof (4.4).* Let us consider the graph from Figure 2b. Here, the max-flow value is $m + k$. Naturally, $C_B = \frac{m+k}{k+1}$ and $C_R = m$. We have the result we wanted when $m$ approaches infinity. □

**Theorem 4.1.** *The algorithm for the max-$(k+1)$-route flow problem in [8,9] is also a $(k+1)$-approximation algorithm for the maximum $k$-robust flow and the maximum $k$-balanced flow problems.*

## 4.1   Approximation Ratio and Graph Connectivity

In this subsection, we further improve the upper bound discussed in this section, for a graph with small source-sink edge connectivity $\lambda$. We will show in Theorem 4.2 that the approximation ratio of the maximum multiroute flow can be significantly smaller than $k + 1$.

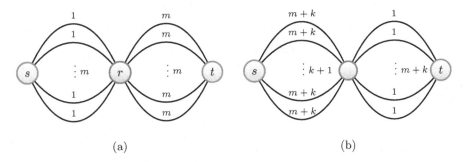

**Fig. 2.** The network used in the proof of Propositions 4.2-4.4

**Lemma 4.2.** *Let $X$ be a minimum $k$-robust cut, and let $Y$ be a minimum $k$-route cut. The ratio between the $k$-robust capacity of $X$ and $k$-capacity of $Y$ cannot be larger than $\frac{p-k}{p}(k+1)$ when $p = \min(p_X, p_Y)$, $p_X$ is the cardinality of the cut-set of $X$, and $p_Y$ is the cardinality of the cut-set of $Y$.*

*Proof.* By Definitions 2.4 and 3.1,

$$\frac{C_R}{C_M} = \frac{\alpha_k(X)}{\min_{0 \leqslant i \leqslant k}\left[\frac{\alpha_i(Y)}{k+1-i}\right]} = \alpha_k(X) \cdot \max_{0 \leqslant i \leqslant k} \frac{k+1-i}{\alpha_i(Y)} \leqslant \alpha_k(Y) \cdot \max_{0 \leqslant i \leqslant k} \frac{k+1-i}{\alpha_i(Y)}.$$

We know that the average value of the capacity $\frac{\alpha_i(Y)}{p_X - i}$ should be at least $\frac{\alpha_j(Y)}{p_X - j}$ if $i \leqslant j$. Hence, $\frac{\alpha_k(Y)}{p_X - k} \leqslant \frac{\alpha_i(Y)}{p_X - i}$. Then,

$$\alpha_k(Y) \cdot \max_{0 \leqslant i \leqslant k} \frac{k+1-i}{\alpha_i(Y)} \leqslant \alpha_k(Y) \cdot \max_{0 \leqslant i \leqslant k} \frac{k+1-i}{\frac{p_X-i}{p_X-k}\alpha_k(Y)}$$
$$\leqslant \max_{0 \leqslant i \leqslant k}\left[(p_X - k)\frac{k+1-i}{p_X-i}\right]$$
$$\leqslant \frac{p_X-k}{p_X}(k+1)$$

Using a similar proof, we know that $\dfrac{\alpha_k(X)}{\min_{0 \leqslant i \leqslant k}\left[\frac{\alpha_i(Y)}{k+1-i}\right]} \leqslant \frac{p_Y-k}{p_Y}(k+1)$. Hence,

$$\frac{\alpha_k(X)}{\min_{0 \leqslant i \leqslant k}\left[\frac{\alpha_i(Y)}{k+1-i}\right]} \leqslant \frac{p-k}{p}(k+1). \qquad \square$$

**Theorem 4.2.** *Let $U$ be the ratio between the largest and smallest capacities of the edges in our network. The maximum multiroute flow algorithm is a $\frac{U(\lambda-k)}{U(\lambda-k)+k}(k+1)$-approximation algorithm for the maximum $k$-robust flow and the maximum $k$-balanced flow problems.*

*Proof.* Let $X$ be a minimum $k$-robust cut, and let $Z$ be a cut in which its cut-set has minimum cardinality. Here, $\alpha_k(Z)$ is the sum of the $\lambda - k$ edge capacities. $\alpha_k(X)$ is the sum of the $p - k$ edge capacities, when $p$ is the cardinality of the cut-set of $X$. If we let $U$ be the ratio between the largest and smallest capacities, then we know that $\frac{\alpha_k(Z)}{\alpha_k(X)} \leqslant U \cdot \frac{\lambda-k}{p-k}$. Since $\alpha_k(Z) \geqslant \alpha_k(X)$ by the definition of the minimum $k$-robust cut, $1 \leqslant \frac{\alpha_k(Z)}{\alpha_k(X)} \leqslant U \cdot \frac{\lambda-k}{p-k}$. Then, $p - k \leqslant U \cdot (\lambda - k)$, $p \leqslant U \cdot (\lambda-k)+k$. We update our upper bound to $\frac{p-k}{p} \cdot (k+1) \leqslant \frac{U(\lambda-k)}{U(\lambda-k)+k}(k+1)$. $\square$

**Corollary 4.2.** *If the cut-set of min-$(k + 1)$-route cut has cardinality $k + 1$, $C_R = C_B = C_M$.*

*Proof.* By Lemma 4.2, we know that $\frac{C_R}{C_M} \leqslant \frac{p-k}{p}(k + 1)$. For this corollary, $p = k + 1$. Hence, $\frac{C_R}{C_M} \leqslant \frac{k+1-k}{k+1}(k + 1) = 1$. $\square$

# 5    Balancing with Iterative Multiroute Flow

In the previous section, we showed that there can be significantly large differences between the values of the solutions to the max-$(k+1)$-route flow and to the max-$k$-balanced flow. We will close that gap in this section by proposing a tighter heuristic algorithm called the *iterative multiroute flow* algorithm. This method is presented as Algorithm 1.

Next, we will discuss some of the intuitive ideas behind the algorithm. We know from Subsection 4.1 that with larger values of $k$, the $k$-route flow tends to be more well balanced, because its effectiveness tends to be close to the effectiveness of a $k$-balanced flow. However, the value of $k$-route flows can be much smaller, as it is seen in [11]. The iterative multiroute flow algorithm is proposed as a way to find a well-balanced flow with a large value. To find that flow, we begin our algorithm by finding a $k$-route flow when $k$ is assigned to $\lambda$, which is the largest value of $k$ for which the value of the $k$-route flow is more than 0. As shown in Corollary 4.2, the flow we obtain in this step tends to be very well balanced, but the value of the flow is small. We then proceed to the next step and consider the capacity that remains after that $\lambda$-route flow. We reduce $k$ by one to $\lambda - 1$, and find a $k$-route flow for that remaining capacity. By doing that, the well-balanced property that we can get from the $\lambda$-route flow is conserved, while the value of our flow increases. From there, to maximize the value of our flow, we iteratively reduce the value of $k$ by one and take the $k$-route flow for the remaining capacity until $k = 1$.

**Theorem 5.1.** *The running time of the iterative multiroute flow algorithm is $O(\lambda^2 T)$, where $\lambda$ is the source-sink edge connectivity of an input network, and $T$ is the computation complexity of the max-flow algorithm.*

*Proof.* As we iterate over $k$ from $\lambda$ down to 1, the number of executions of the max-flow algorithm is $\sum_{k=1}^{\lambda}(k + 1) = \frac{(\lambda+1)(\lambda+2)}{2} = O(\lambda^2)$. $\square$

# 6    Experimental Results

In this section, we evaluate the iterative multiroute flow algorithm proposed in Section 5 for several different network conditions. In each of these settings, we determined the average result from 100 experiments. Let $C_{I'}$ be the $k$-effectiveness

---

**input** : A network graph $G = (V, E, c)$, where $c : E \to \mathbb{R}^+$ is the capacity of
    each edge

**output**: The flow on each edge $F : E \to \mathbb{R}^+$

1   Let $\lambda$ be the source-sink edge connectivity of graph $G$.
2   Assign $F(e) \leftarrow 0$ for all $e \in E$.
3   **for** $k \leftarrow \lambda$ **to** 1 **do**
4      Find the $k$-route flow of graph $G$.
5      Let $F_k : E \to \mathbb{R}^+$ be the flow on each edge in that $k$-route flow.
6      $F(e) = F(e) + F_k(e)$.
7      $c(e) = c(e) - F_k(e)$.
8   **end**

---

**Algorithm 1:** Iterative multiroute flow algorithm

of an iterative multiroute flow. We can show the improvement due to our proposed method by comparing its results $C_{I'}$ with those of the previous method for the effectiveness of a max-$(k + 1)$-route flow, $C_{M'}$. We also perform an experiment to compare our result $C_{I'}$ with the value of a maximum $k$-robust flow $C_R$ to show the difference between our solution and the optimal solution.

In Figures 3a and 3b, we present the results of experiments on a network with $|V| = 10$ and $|E| = 40$. To have a network with higher source-sink connectivity, each node is chosen with probability $\frac{1}{|V|} = 0.1$ to be the tail endpoint of an edge. Exceptions to this are the source node and sink node, where the probabilities are $\frac{2}{|V|} = 0.2$ and 0, respectively. Similarly, the probabilities that the source node, the sink node, and the other nodes are chosen to be a head endpoint are 0, $\frac{2}{|V|} = 0.2$, and $\frac{1}{|V|} = 0.1$, respectively. For this setting, the capacity of each edge was picked at random between 0 and 20. Figure 3a shows the improvement due to the multiroute flow. When $k = 3$, the effectiveness of our iterative multiroute flow is 9.8% larger than the effectiveness of the maximum multiroute flow. We can also see from the figure that the improvement increases with increasing $k$.

As shown in Figure 3b, the average gap between our result $C_{I'}$ and the optimal solution $C_R$ is less than 1.2% for any $k \leqslant 3$. For all of the network conditions considered in this work, this average gap was seldom larger than 2%. Currently, there is no efficient algorithm that finds the maximum $k$-balanced flow $C_B$. However, we can guarantee from these experimental results that $C_{I'}$ is an accurate approximation of $C_B$, since we know from Section 3 that the value cannot be larger than $C_R$. Therefore, the gap between $C_{I'}$ and $C_B$ must be even smaller than 2%. We also performed experiments on the graph where a small number of edges had capacities that were larger than those of others. Let $G = (V, E, c)$ be a network used for the experimental results shown in Figure 3a. We modified the network to $G' = (V, E, c')$, where $c'(e) = \frac{c(e)^h}{20^{h-1}}$ for $1 \leqslant h \leqslant 3$. By doing this, the capacity will follow a power-law distribution with $h$ as the exponent [18]. In Figure 3c, we can see that our improvement significantly increases in this graph setting. That improvement is even larger when the exponent of the distribution is increased.

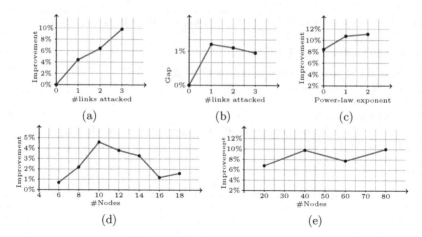

**Fig. 3.** Comparison between iterative multiroute flow, $(k+1)$-route flow, and $k$-robust flow: (a) improvement in the different number of links attacked; (b) gap between the result and the upper bound; (c) improvement when the capacity is not chosen uniformly and at random (a larger exponent represents more bias); (d) improvement in the different number of nodes when the number of edges is fixed; (e) improvement in the different number of nodes when the number of edges is changed proportionally to the number of nodes

Although the multiroute flow and iterative multiroute flow algorithms are fast and efficient, no polynomial-time algorithm exists to find $C_{M'}$ or $C_{I'}$. In other words, we can efficiently find the flows, but it is hard to evaluate how well balanced the flows are, especially in large networks. Thus, we have to perform the experiments on a comparatively small graph in which the number of nodes ranges from 5 to 20, and the number of edges ranges from 20 to 80. In Figures 3d and 3e, we show random networks that were generated in the same way as under the previous settings, and we plot the relationship between the improvements we made and the number of nodes. We are unable to find any associations between the improvement we made and the number of nodes and/or edges for either the case where $|E| = 40$ (Figure 3d) or the case where $|E| = 4|V|$ (Figure 3e). Because of this, experiments on relatively small graphs are sufficient to show the differences between our algorithm, the previous one, and the upper bound.

## 7   Conclusion

In this work, we make two main contributions to the problem of finding an optimal flow for a network in which some links have failed. The first one is a theoretical result showing that the classic maximum multiroute flow algorithm can solve the problem effectively. The other is a new algorithm that can, in practice, solve the problem even more efficiently. Currently, we are developing algorithms that can solve the problem with a smaller approximation ratio. We are also interested in developing a faster algorithm for the maximum $k$-robust cut, a nontrivial exact algorithm for the maximum $k$-balanced flow, and a theoretical model for the iterative multiroute flow.

**Acknowledgement.** The authors would like to thank Mr. Kei Kimura, Dr. Norie Fu, Prof. Hiroshi Imai, and anonymous reviewers for giving several useful comments and feedbacks for this research.

# References

1. Ford, L.R., Fulkerson, D.R.: Flows in Networks. Princeton University Press (1955)
2. Orlin, J.B.: Max flows in $O(nm)$ time, or better. In: Proc. STOC 2013, pp. 765–774 (2013)
3. Bley, A., Grötschel, M., Wessäly, R.: Design of broadband virtual private networks: Model and heuristics for the B-WiN. DIMACS Series in Discrete Mathematics and Theoretical Computer Science 53, 1–16 (1998)
4. Schwartz, M.: Telecommunication Networks: Protocols, Modeling and Analysis, vol. 7. Addison-Wesley Reading (1987)
5. Bagchi, A., Chaudhary, A., Scheideler, C., Kolman, P.: Algorithms for fault-tolerant routing in circuit switched networks. In: Proc. SPAA 2002, pp. 265–274 (2002)
6. Bagchi, A., Chaudhary, A., Goodrich, M.T., Xu, S.: Constructing disjoint paths for secure communication. In: Fich, F.E. (ed.) DISC 2003. LNCS, vol. 2848, pp. 181–195. Springer, Heidelberg (2003)
7. Bagchi, A., Chaudhary, A., Kolman, P.: Short length Menger's theorem and reliable optical routing. Theoretical Computer Science 339(2), 315–332 (2005)
8. Kishimoto, W., Takeuchi, M.: On $m$-route flows in a network. IEICE Trans. J-76-A, 1185–1200 (1993)
9. Kishimoto, W.: A method for obtaining maximum multi-route flows in a network. Networks 27(4), 279–291 (1996)
10. Bagchi, A., Chaudhary, A., Kolman, P., Sgall, J.: A simple combinatorial proof of duality of multiroute flows and cuts. Technical report, Charles Univ. (2004)
11. Aggarwal, C., Orlin, J.B.: On multi-route maximums flows in networks. Networks 39(1), 43–52 (2002)
12. Du, D., Chandrasekaran, R.: The multiroute maximum flow problem revisited. Networks 47(2), 81–92 (2006)
13. Kishimoto, W.: Reliable flow with failures in a network. IEEE Transactions on Reliability 46(3), 308–315 (1997)
14. Lee, P.P., Misra, V., Rubenstein, D.: Distributed algorithms for secure multipath routing. In: Proc. INFOCOM 2005, vol. 3, pp. 1952–1963 (2005)
15. Chuzhoy, J., Makarychev, Y., Vijayaraghavan, A., Zhou, Y.: Approximation algorithms and hardness of the $k$-route cut problem. In: Proc. SODA 2012, pp. 780–799 (2012)
16. Chekuri, C., Khanna, S.: Algorithms for 2-route cut problems. In: Aceto, L., Damgård, I., Goldberg, L.A., Halldórsson, M.M., Ingólfsdóttir, A., Walukiewicz, I. (eds.) ICALP 2008, Part I. LNCS, vol. 5125, pp. 472–484. Springer, Heidelberg (2008)
17. Chandrasekaran, R., Nair, K., Anejac, Y., Kabadib, S.: Multi-terminal multipath flows: synthesis. Discrete Applied Mathematics 143, 182–193 (2004)
18. Faloutsos, M., Faloutsos, P., Faloutsos, C.: On power-law relationships of the internet topology. ACM SIGCOMM Computer Communication Review 29, 251–262 (1999)

# Simple Linear Comparison
# of Strings in $V$-Order[*]
## (Extended Abstract)

Ali Alatabbi[1], Jackie Daykin[1,2],
M. Sohel Rahman[1,3,**], and William F. Smyth[1,***]

[1] Department of Informatics
King's College London, UK
ali.alatabbi@kcl.ac.uk
[2] Department of Computer Science
Royal Holloway, University of London, UK
J.Daykin@cs.rhul.ac.uk
[3] A$\ell$EDA Group
Department of Computer Science and Engineering
Bangladesh University of Engineering and Technology
Dhaka-1000, Bangladesh
msrahman@cse.buet.ac.bd
[4] Algorithms Research Group
Department of Computing & Software
McMaster University, Canada
smyth@mcmaster.ca

**Abstract.** In this paper we focus on a total (but non-lexicographic) ordering of strings called V-order. We devise a new linear-time algorithm for computing the $V$-comparison of two finite strings. In comparison with the previous algorithm in the literature, our algorithm is both conceptually simpler, based on recording letter positions in increasing order, and more straightforward to implement, requiring only linked lists.

**Keywords:** algorithm, array, comparison, complexity, data structure, lexicographic order, linear, linked-list, $V$-order, Lyndon word, string, total order, word.

## 1 Introduction

An important task required in many combinatorial computations is deciding the relative order of two members of a totally ordered set [KS-98, R-03], for instance

---

[*] Part of this research was carried out when Rahman was visiting King's College as a Commonwealth Fellow. This research work is also partially supported by a CodeCrafters-Investortools Research Grant for CSE BUET.
   (http://www.codecraftersintl.com/researchgrant.html)
[**] Partially Supported by a Commonwealth Fellowship and an ACU Titular Award.
[***] Supported in part by the Natural Sciences & Engineering Research Council of Canada.

S.P. Pal and K. Sadakane (Eds.): WALCOM 2014, LNCS 8344, pp. 80–89, 2014.
© Springer International Publishing Switzerland 2014

organizing words in a natural language dictionary. Binary comparison of finite strings (words) thus arises as a primitive operation, a building block, in more complex procedures, which therefore requires efficient implementation.

In this paper we first discuss some known techniques for totally ordering sets, and then introduce our contribution: a new linear string comparison algorithm using $V$-order.

Given an integer $n \geq 1$ and a nonempty set of symbols $\Sigma$ (bounded or unbounded), a string of length $n$ over $\Sigma$ takes the form $x = x_1 \ldots x_n$ with each $x_i \in \Sigma$. The classic and commonly used method for organizing sets of strings is lexicographic (dictionary) order. Formally, if $\Sigma$ is a totally ordered alphabet then *lexicographic ordering* (lexorder) $u < v$ with $u, v \in \Sigma^+$ is defined if and only if either $u$ is a proper prefix of $v$, or $u = ras$, $v = rbt$ for some $a, b \in \Sigma$ such that $a < b$ and for some $r, s, t \in \Sigma^*$.

Lexorder is a very natural method for deciding precedence and organizing information which also finds many uses in computer science, typically in constructing data structures and related applications:

- Building indexes for information retrieval, particularly self-indexes which replace the text and support almost optimal space and search time [NM-07].
- Constructing suffix arrays, which record string suffix starting positions in the lexorder of the suffixes, and thus support binary search [KA-03, KSB-06, NZC-09].
- The Burrows-Wheeler Transform (BWT), which applies suffix sorting, and exhibits data clustering properties, hence is suitable for preprocessing data prior to compression activities [ABM-08, CDP-05].
- The application of automata for bioinformatics sequence alignment. The BWT is extended for finite automata representing the multiple alignment problem - the paths in the automaton are sorted into lexorder thus extending the suffix sorting framework related to the classic BWT [SVM-11].
- An important class in the study of combinatorics on words is *Lyndon words* [L-83] - strings (words) which are lexicographically least amongst the cyclic rotations of their letters (characters) – see also [S-03]; furthermore, any string can be uniquely factored into Lyndon words [CFL-58] - Duval's algorithm cleverly detects the lexicographic order between factors in linear time [Du-83, D-11]. The Lyndon decomposition allows for efficient 'divide-and-conquer' of a string into patterned factors; numerous applications include: periodic musical structures [C-04], string matching [BGM-11, CP-91], and algorithms for digital geometry [BLPR-09].
- Hybrid Lyndon structures, introduced in [DDS-13], based on two methods of ordering strings one of which is lexorder.

Naively, lexorder $u < v$ can be decided in time linear in the length of the shorter string, and space linear in the length of the longer string; various data structures may be used for enhancing this string comparison. In [DIS-94] the Four Russians technique [IS-92] is proposed to compare strings of length $n$ on a bounded alphabet in $O(1)$ time, while for an unbounded alphabet the parallel construction of a merged suffix tree using the CRCW PRAM model [IS-92] is proposed that can be constructed in $O(\log n \log \log n)$ time using $O(n/\log n)$ processors; using this tree, sequential comparison requires $O(\log \log n)$ time.

A class of lexorder-type total orders is easily obtained from permuting the usual order $1, 2, \ldots n$ of pairwise comparison of letters, along with interchanging $<$ with $>$ and so on; for example *relex order* (reverse lexicographic) [R-03], and *co-lexorder* (lexorder of reversed strings) studied and applied to string factorization in [DDS-09].

Non-lexicographic methods include deciding precedence by minimal change such as Gray's *reflected binary code*, where two successive values differ in only one bit, hence well-suited for error correction in digital communications [G-53, S-97]. A more recent example is *V-order* [D-85, DaD-96, DaD-97] which is the focus of this paper: we first introduce this technical method for comparing strings and then consider it algorithmically.

Let $\Sigma$ be a totally ordered alphabet, and let $\boldsymbol{u} = u_1 u_2 \ldots u_n$ be a string over $\Sigma$. Define $h \in \{1, \ldots, n\}$ by $h = 1$ if $u_1 \leq u_2 \ldots \leq u_n$; otherwise, by the unique value such that $u_{h-1} > u_h \leq u_{h+1} \leq u_{h+2} \leq \ldots \leq u_n$. Let $\boldsymbol{u}^* = u_1 u_2 \ldots u_{h-1} u_{h+1} \ldots u_n$, where the star * indicates deletion of the letter $u_h$. Write $\boldsymbol{u}^{s*}$ for $(\ldots(\boldsymbol{u}^*)^* \ldots)^*$ with $s \geq 0$ stars [1]. Let $g = \max\{u_1, u_2, \ldots, u_n\}$, and let $k$ be the number of occurrences of $g$ in $\boldsymbol{u}$. Then the sequence $\boldsymbol{u}, \boldsymbol{u}^*, \boldsymbol{u}^{2*}, \ldots$ ends $g^k, \ldots, g^2, g^1, g^0 = \varepsilon$. In the *star tree* each string $\boldsymbol{u}$ over $\Sigma$ labels a vertex, and there is a directed edge from $\boldsymbol{u}$ to $\boldsymbol{u}^*$, with $\varepsilon$ as the root.

**Definition 1.** *We define V-order $\prec$ between distinct strings $\boldsymbol{u}, \boldsymbol{v}$. First $\boldsymbol{v} \prec \boldsymbol{u}$ if $\boldsymbol{v}$ is in the path $\boldsymbol{u}, \boldsymbol{u}^*, \boldsymbol{u}^{2*}, \ldots, \varepsilon$. If $\boldsymbol{u}, \boldsymbol{v}$ are not in a path, there exist smallest $s, t$ such that $\boldsymbol{u}^{(s+1)*} = \boldsymbol{v}^{(t+1)*}$. Put $\boldsymbol{c} = \boldsymbol{u}^{s*}$ and $\boldsymbol{d} = \boldsymbol{v}^{t*}$; then $\boldsymbol{c} \neq \boldsymbol{d}$ but $|\boldsymbol{c}| = |\boldsymbol{d}| = m$ say. Let $j$ be the greatest $i$ in $1 \leq i \leq m$ such that $\boldsymbol{c}[i] \neq \boldsymbol{d}[i]$. If $\boldsymbol{c}[j] < \boldsymbol{d}[j]$ in $\Sigma$ then $\boldsymbol{u} \prec \boldsymbol{v}$. Clearly $\prec$ is a total order.*

*Example 1.* Over the binary alphabet with $0 < 1$: in lexorder, $0101 < 01110$; in V-order, $0101 \prec 01110$.

Over the naturally ordered integers: in lexorder, $123456 < 2345$; in V-order, $2345 \prec 123456$.

Over the naturally ordered Roman alphabet: in lexorder, *eabecd* < *ebaedc*; in V-order, *ebaedc* $\prec$ *eabecd*.

String comparison in V-order $\prec$ was first considered algorithmically in [DDS-11, DDS-13] - the dynamic longest matching suffix of the pair of input strings, together with a doubly-linked list which simulated letter deletions and hence paths in the star tree, enabled deciding order; these techniques achieved V-comparison in worst-case time and space proportional to string length - thus asymptotically the same as naive comparison in lexorder.

Currently known applications of V-order, utilizing linear-time V-comparison, and generally derived from lexorder or Lyndon cases are as follows:

---

[1] Note that this star operator, as defined in [DaD-96], [DD-03] etc, is distinct from the Kleene star operator: Kleene star is applied to sets, while this V-star is applied to strings.

- A $V$-order structure, an instance of a hybrid Lyndon word and known as a $V$-word [DD-03], similarly to the classic Lyndon case, gives an instance of an African musical rhythmic pattern [CT-03].
- Linear factorization of a string into factors ($V$-words) sequentially [DDS-11] and in parallel [DDIS-13] - yielding factors which are distinct from the Lyndon factorization of the given string [DDS-13].
- Modification of a linear suffix array construction [KA-03] from lexorder to $V$-order [DS-13] thus allowing efficient $V$-ordering of the cyclic rotations of a string.
- Applying the above suffix array modification to compute a novel Burrows-Wheeler transform ($V$-BWT) using, not the usual lexorder, but rather $V$-order [DS-13] - achieving instances of enhanced data clustering.

These initial avenues suggest that further uses of $V$-order, analogous to the practical functions listed for lexorder and Lyndon words, will continue to arise, including for instance those for suffix trees - thus necessitating efficient implementations of the primitive $V$-comparison.

We introduce here a new algorithm for computing the $V$-comparison of two finite strings - the advantage is that it is both conceptually simpler, based on recording letter positions in increasing order, and more straightforward to implement, requiring only linked lists. The time complexity is $O(n + |\Sigma|)$ and similarly the space complexity is $O(n + |\Sigma|)$. However, in computational practice the alphabet, like the input, can be assumed to be finite - at most $O(n)$ - and so the algorithm runs in essentially linear time.

## 2    $V$-Order String Comparison Algorithm

In this section, we present a novel linear-time algorithm for $V$-order string comparison. Before going into the algorithmic details, we present relevant definitions and results from the literature useful in describing and analyzing our algorithm, starting with a unique representation of a string.

**Definition 2.** *([DD-03, DDS-11, DDS-13]) The **V-form** of a string $x$ is defined as*

$$V_k(x) = x = x_0 g x_1 g \cdots x_{k-1} g x_k$$

*for possibly empty $x_i$, $i = 0, 1, \ldots, k$, where $g$ is the largest letter in $x$ – thus we suppose that $g$ occurs exactly $k$ times.*

The following lemma is the key to our algorithm.

**Lemma 1.** *([DaD-96, DD-03, DDS-11, DDS-13]) Suppose we are given distinct strings $v$ and $x$ with the corresponding V-forms as follows:*

$$v = v_0 \mathcal{L}_v v_1 \mathcal{L}_v v_2 \cdots v_{j-1} \mathcal{L}_v v_j$$

$$x = x_0 \mathcal{L}_x x_1 \mathcal{L}_x x_2 \cdots x_{k-1} \mathcal{L}_x x_k$$

Let $h \in \{0 \ldots \max(j, k)\}$ be the least integer such that $v_h \neq x_h$. Then $v \prec x$ if, and only if, one of the following is true:

- $\mathcal{L}_v < \mathcal{L}_x$
- $\mathcal{L}_v = \mathcal{L}_x$ and $j < k$
- $\mathcal{L}_v = \mathcal{L}_x$, $j = k$ and $v_h \prec x_h$.

**Lemma 2.** *([DDS-11, DDS-13]) Suppose we are given distinct strings $v$ and $x$. If $v$ ($x$) is a subsequence of $x$ ($v$) then $v \prec x$ ($x \prec v$).*

We will use some simple data structures, which are initialized by preprocessing steps. We use $Map_u(a)$ to store, in increasing order, the positions of the character $a$ in a string $u$. $Map_u(\Sigma)$ records the 'maps' of all $a \in \Sigma$. To construct $Map_u(\Sigma)$ we take an array of size $\Sigma$. For each $a \in \Sigma$, we construct a linked list that stores the positions $i \in [1..|u|]$ in increasing order such that $u[i] = a$.

*Example 2.* Suppose we have a string $u$ as follows:

$$1\ 2\ 3\ 4\ 5\ 6\ 7\ 8\ 9\ 10\ 11$$
$$u = 8\ 5\ 8\ 2\ 1\ 8\ 7\ 6\ 5\ \ 4\ \ 3$$

$Map_u(\Sigma)$ is shown below for the string $u$ defined above.
$$1\ 2\ 3\ \ 4\ \ 5\ 6\ 7\ 8$$
$$\downarrow \downarrow \downarrow\ \ \downarrow\ \downarrow \downarrow \downarrow \downarrow$$
$$5\ 4\ 11\ 10\ 2\ 8\ 7\ 1$$
$$\qquad\quad \downarrow \qquad \downarrow$$
$$\qquad\quad 9 \qquad\ 3$$
$$\qquad\qquad\qquad\ \downarrow$$
$$\qquad\qquad\qquad\ 6$$

This leads to the following lemma.

**Lemma 3.** *Given a string $u$ of length $n$ we can build $Map_u(\Sigma)$ in $O(n + |\Sigma|)$ time and space.*

*Proof.* Proof will be provided in the journal version.                    □

We will now prove a number of new lemmas that will be used in the string comparison algorithm – first we will introduce some notation. Let $firstMiss(u, v)$ denote the first mismatch entry between $u, v$. More formally, we say $\ell = firstMiss(u, v)$ if and only if $u[\ell] \neq v[\ell]$ and $u[i] = v[i]$, for all $1 \leq i < \ell$. In what follows, the notion of a global mismatch and a local mismatch is useful in the context of two strings $u, v$ and their respective substrings $u', v'$. In particular, $firstMiss(u, v)$ would be termed as the global mismatch in this context and $firstMiss(u', v')$ would be termed as a local mismatch, i.e., local to the corresponding substrings. For this global/local notion, the context $\mathcal{C}$ is important and is defined with respect to the two strings and their corresponding substrings, i.e., the context here would be denoted by $\mathcal{C}\langle(u, u'), (v, v')\rangle$. Also, for the $V$-form of a string $u$ we will use the following convention: $\mathcal{L}_{u,\ell}$ denotes the $\ell$-th $\mathcal{L}_u$

in the $V$-form of $u$ and $pos(\mathcal{L}_{u,\ell})$ will be used to denote its index/position in $u$. With this extended notation, the $V$-form of $u$ can be rewritten as follows:

$$u = u_0 \; \mathcal{L}_{u,1} \; u_1 \; \mathcal{L}_{u,2} \; u_2 \; \cdots \; u_{j-1} \; \mathcal{L}_{u,j} \; u_j.$$

Moreover, within the context $\mathcal{C}$, the strings $u, v$ are referred to as the *superstrings* and $u', v'$ as the *substrings*.

**Lemma 4.** *Suppose we are given distinct strings $v$ and $x$ with the corresponding $V$-forms as follows:*

$$v = v_0 \mathcal{L}_v v_1 \mathcal{L}_v v_2 \cdots v_{j-1} \mathcal{L}_v v_j$$

$$x = x_0 \mathcal{L}_x x_1 \mathcal{L}_x x_2 \cdots x_{k-1} \mathcal{L}_x x_k$$

*Assume that $\mathcal{L}_v = \mathcal{L}_x$ and $j = k$. Let $h \in \{0 \ldots \max(j,k)\}$ be the least integer such that $v_h \neq x_h$. Now assume that $\ell_h = firstMiss(v_h, x_h)$ and $\ell_f = firstMiss(v, x)$. In other words, $\ell_h$ is the index of the first mismatch entry between the substrings $v_h, x_h$, whereas $\ell_f$ is the index of the first mismatch entry between the two strings $v$ and $x$. Then we must have $\ell_f = \sum_{i=0}^{h-1}(|v_i| + 1) + \ell_h$. (Or equivalently, $\ell_f = \sum_{i=0}^{h-1}(|x_i| + 1) + \ell_h$.)*

*Proof.* Proof will be provided in the journal version. □

**Corollary 1.** *If in Case 2 of Lemma 4 we have $\ell_f = pos(\mathcal{L}_{v,\ell})$, then $v_h$ is a proper prefix of $x_h$.*

Interestingly, we can extend Lemma 4 further if we consider the (inner) contexts within (outer) contexts as the following lemma shows. In other words $V$-form can be applied recursively and independently as shown in [DDS-11]. In what follows, for given distinct strings $v$ and $x$ with corresponding $V$-forms, the condition that $\mathcal{L}_v = \mathcal{L}_x$, $j = k$ will be referred to as $Cond\text{-}I(v, x)$.

**Lemma 5.** *Suppose we are given distinct strings $v$ and $x$ with corresponding $V$-forms, and assume that $Cond\text{-}I(v, x)$ holds. Now consider the (outer) context $\mathcal{C}^0 \langle (v, v_{h^0}), (x, x_{h^0}) \rangle$, where $h^0$ is the least integer such that $v_{h^0} \neq x_{h^0}$.*

*Now similarly consider the $V$-forms of $v_{h^0}$ and $x_{h^0}$ and assume that $Cond\text{-}I(v_{h^0}, x_{h^0})$ holds. Further, consider the (inner) context $\mathcal{C}^1 \langle (v_{h^0}, v_{h^1}), (x_{h^0}, x_{h^1}) \rangle$, where $h^1$ is the least integer such that $v_{h^1} \neq x_{h^1}$.*

*Then the global mismatch of the context $\mathcal{C}^0$ coincides with the local mismatch of the context $\mathcal{C}^1$.*

*Proof.* Proof will be provided in the journal version. □

**Corollary 2.** *Given nested contexts $\mathcal{C}^i, 0 \leq i \leq k$ satisfying the hypotheses of Lemma 5, the global mismatch of context $\mathcal{C}^0$ coincides with the local mismatch of context $\mathcal{C}^k$.*

Corollary 2 establishes that the first global mismatch will always be the first mismatch as we go further within inner contexts through the chain of outer and inner contexts.

Now we can focus on the string comparison algorithm: Algorithm $CompareV$. Suppose we are given two distinct strings $p$ and $q$, then the algorithm performs the following steps.

Step 1:  **Preprocessing Step.** Compute $Map_p(\Sigma)$ and $Map_q(\Sigma)$. We also compute the first mismatch position $\ell_f$ between $p$ and $q$. This will be referred to as the global mismatch position and will be independent of any context within the iterations of the algorithm. Then we repeat the following sub-steps in Step 2. During different iterations of the execution of these stages we will be considering different contexts by proceeding from outer to inner contexts. Initially, we will start with the outermost context, i.e., $C^0\langle(p, p_{h^0}), (q, q_{h^0})\rangle$, where $h^0$ is the least integer such that $p_{h^0} \neq q_{h^0}$. At each iteration, we will be considering the largest $\alpha \in \Sigma$ that is present within one of the superstrings in the context. In other words, if the current context is $C^0$, as is the case during the initial iteration, we will consider the largest $\alpha$ such that $\alpha \in p$ or $\alpha \in q$.

Step 2:  Throughout this step we will assume that the current context is $C\langle(v, v_h), (x, x_h)\rangle$, where $h$ is the least integer such that $v_h \neq x_h$. So, initially we have $C = C^0$. Suppose we are now considering $\alpha \in \Sigma$, then it must be the largest $\alpha \in \Sigma$ such that either $\alpha \in v$ or $\alpha \in x$. We proceed to the following sub-steps:

Step 2.a:  We compute $Map_v(\alpha)$ from $Map_p(\alpha)$ where $Map_v(\alpha)$ contains the positions that are only within the range of $v$ in the current context $C$. Similarly, we compute $Map_x(\alpha)$ from $Map_q(\alpha)$ where $Map_x(\alpha)$ contains the positions that are only within the range of $x$ in the current context $C$. Now we compare $Map_v(\alpha)$ and $Map_x(\alpha)$, which yields two cases.

Step 2.a.(i):  In this case, $Map_v(\alpha) = Map_x(\alpha)$. This means that within the current context $C$, considering the $V$-form of the superstrings $v$ and $x$, we must have $\mathcal{L}_v = \mathcal{L}_x$ and $j = k$. So, we need to check Condition 3 of Lemma 1. We identify $h$ such that $h$ is the least integer with $v_h \neq x_h$. By Lemmas 4, 5 and Corollary 2 we know that this $h$ can be easily identified because it is identical to the global mismatch position $\ell_f$.

Then we iterate to Step 2 again with the inner context $C^1\langle(v_h, v_{h^1}), (x_h, x_{h^1})\rangle$, where $h^1$ is the least integer such that $v_{h^1} \neq x_{h^1}$. In other words, we assign $C = C^1$ and then repeat Step 2 for $\beta \in \Sigma$ where $\beta < \alpha$.

Step 2.a.(ii):  In this case, $Map_v(\alpha) \neq Map_x(\alpha)$. [C1] If $Map_v(\alpha) = \emptyset$ $(Map_x(\alpha) = \emptyset)$, we have Condition 1 of Lemma 1 satisfied ($\varepsilon$ is the least string in $V$-order) and hence we return $v \prec x$ ($x \prec v$). Note that this effectively decides $p \prec q$ ($q \prec p$) and the algorithm terminates.

[C2] If $|Map_v(\alpha)| < |Map_x(\alpha)|$ ($|Map_x(\alpha)| < |Map_v(\alpha)|$), we have Condition 2 of Lemma 1 satisfied and hence we return $v \prec x$ ($x \prec v$). Similarly, this effectively decides $p \prec q$ ($q \prec p$) and the algorithm terminates.

[C3] Otherwise, we have $\mathcal{L}_v = \mathcal{L}_x$ and $j = k$. So, we need to check Condition 3 of Lemma 1, and identify $h$ such that $h$ is the least integer such that $v_h \neq x_h$. By Lemmas 4, 5 and Corollary 2 we know that $h$ can be easily identified because it is identical to the global mismatch position $\ell_f$. Now we do a final check as to whether $v_h$ is a subsequence (in fact, a prefix ) of $x_h$ according to Corollary 1. If so, then by Lemma 2 we return $v \prec x$ ($x \prec v$), which decides that $p \prec q$ ($q \prec p$) and the algorithm terminates. Otherwise, we return to Step 2 with the inner context $\mathcal{C}^1 \langle (v_h, v_{h^1}), (x_h, x_{h^1}) \rangle$, where $h^1$ is the least integer such that $v_{h^1} \neq x_{h^1}$. In other words, we assign $\mathcal{C} = \mathcal{C}^1$ and then repeat Step 2 again.

To prove the correctness of the algorithm we need the following lemmas.

**Lemma 6.** *Step 2 of Algorithm CompareV can be realized through a loop that considers each character $\alpha \in \Sigma$ in decreasing order, skipping the ones that are absent in both $v$ and $x$ or in the current context.*

*Proof.* Proof will be provided in the journal version. ◻

**Lemma 7.** *Algorithm CompareV terminates at some point.*

*Proof.* Note that Algorithm *CompareV* can terminate only by conditions [C1] and [C2] of Step 2.a.(ii). Also recall that the input of the algorithm is two distinct strings. Furthermore, we have computed a global mismatch position $\ell_f$. Hence, clearly at some point we will reach either [C1] or [C2] of Step 2.a.(ii). Therefore, the algorithm will definitely terminate. ◻

The correctness of the algorithm follows immediately from Lemmas 1, 2, 6 and 7. Finally we analyze the running time of Algorithm *CompareV* as follows.

**Lemma 8.** *Algorithm CompareV runs in $O(n + |\Sigma|)$ time and space.*

*Proof.* Proof will be provided in the journal version. ◻

Note that a naive $O(n^2)$ rendition of this algorithm was proposed by a reviewer in [DDS-13].

# 3  Conclusion

Lexicographic orderings have also been considered in the case of parallel computations: for instance, an optimal algorithm for lexordering $n$ integers is given in [I-86], and parallel Lyndon factorization in [DIS-94, DDIS-13]. Analogously, we propose future research into parallel forms of $V$-ordering strings.

# References

[ABM-08]   Adjeroh, D., Bell, T., Mukherjee, A.: The Burrows-Wheeler trans- form: data compression, suffix arrays, and pattern matching, p. 352. Springer (2008)

[BGM-11]   Breslauer, D., Grossi, R., Mignosi, F.: Simple real-time constant-space string matching. In: Giancarlo, R., Manzini, G. (eds.) CPM 2011. LNCS, vol. 6661, pp. 173–183. Springer, Heidelberg (2011)

[BLPR-09]  Brlek, S., Lachaud, J.-O., Provençal, X., Reutenauer, C.: Lyndon + Christoffel = digitally convex. Pattern Recognition 42(10), 2239–2246 (2009)

[C-04]     Chemillier, M.: Periodic musical sequences and Lyndon words. Soft Computing - A Fusion of Foundations, Methodologies and Applications 8(9), 611–616 (2004) ISSN: 1432-7643 (Print), 1433-7479 (Online)

[CT-03]    Chemillier, M., Truchet, C.: Computation of words satisfying the "rhythmic oddity property" (after Simha Arom's works). Inf. Proc. Lett. 86, 255–261 (2003)

[CFL-58]   Chen, K.T., Fox, R.H., Lyndon, R.C.: Free differential calculus, IV - The quotient groups of the lower central series. Ann. Math. 68, 81–95 (1958)

[CDP-05]   Crochemore, M., Désarménien, J., Perrin, D.: A note on the Burrows-Wheeler transformation. Theor. Comput. Sci. 332(1-3), 567–572 (2005)

[CP-91]    Crochemore, M., Perrin, D.: Two-way string-matching. J. Assoc. Comput. Mach. 38(3), 651–675 (1991)

[D-85]     Daykin, D.E.: Ordered ranked posets, representations of integers and inequalities from extremal poset problems. In: Rival, I. (ed.) Graphs and Order, Proceedings of a Conference in Banff, Canada. NATO Advanced Sciences Institutes Series C: Mathematical and Physical Sciences, vol. 147, pp. 395–412. Reidel, Dordrecht-Boston (1984, 1985)

[D-11]     Daykin, D.E.: Algorithms for the Lyndon unique maximal factorization. J. Combin. Math. Combin. Comput. 77, 65–74 (2011)

[DaD-96]   Danh, T.-N., Daykin, D.E.: The structure of $V$-order for integer vectors. In: Hilton, A.J.W. (ed.) Congr. Numer., vol. 113, pp. 43–53. Utilas Mat. Pub. Inc., Winnipeg (1996)

[DaD-97]   Danh, T.-N., Daykin, D.E.: Ordering integer vectors for coordinate deletions. J. London Math. Soc. 55(2), 417–426 (1997)

[DD-03]    Daykin, D.E., Daykin, J.W.: Lyndon-like and $V$-order factorizations of strings. J. Discrete Algorithms 1, 357–365 (2003)

[DD-08]    Daykin, D.E., Daykin, J.W.: Properties and construction of unique maximal factorization families for strings. Internat. J. Found. Comput. Sci. 19(4), 1073–1084 (2008)

[DDIS-13]   Daykin, D.E., Daykin, J.W., Iliopoulos, C.S., Smyth, W.F.: Generic algorithms for factoring strings. In: Aydinian, H., Cicalese, F., Deppe, C. (eds.) Ahlswede Festschrift. LNCS, vol. 7777, pp. 402–418. Springer, Heidelberg (2013)

[DDS-09]    Daykin, D.E., Daykin, J.W., (Bill) Smyth, W.F.: Combinatorics of unique maximal factorization families (UMFFs). In: Janicki, R., Puglisi, S.J., Rahman, M.S. (eds.) Fund. Inform, vol. 97-3, pp. 295–309 (2009); Special Issue on Stringology

[DDS-11]    Daykin, D.E., Daykin, J.W., Smyth, W.F.: String comparison and lyndon-like factorization using V-order in linear time. In: Giancarlo, R., Manzini, G. (eds.) CPM 2011. LNCS, vol. 6661, pp. 65–76. Springer, Heidelberg (2011)

[DDS-13]    Daykin, D.E., Daykin, J.W., Smyth, W.F.: A linear partitioning algorithm for Hybrid Lyndons using $V$-order. Theoret. Comput. Sci. 483, 149–161 (2013)

[DIS-94]    Daykin, J.W., Iliopoulos, C.S., Smyth, W.F.: Parallel RAM algorithms for factorizing words. Theoret. Comput. Sci. 127, 53–67 (1994)

[DS-13]     Daykin, J.W., Smyth, W.F.: A bijective variant of the Burrows-Wheeler transform using V-Order (submitted)

[Du-83]     Duval, J.P.: Factorizing words over an ordered alphabet. J. Algorithms 4, 363–381 (1983)

[G-53]      F. Gray, Pulse code communication, U.S. patent no. 2,632,058 (March 17, 1953)

[I-86]      Iliopoulos, C.S.: Optimal cost parallel algorithms for lexicographical ordering, Purdue University, Tech. Rep. 86-602 (1986)

[IS-92]     Iliopoulos, C.S., Smyth, W.F.: Optimal algorithms for computing the canonical form of a circular string. Theoret. Comput. Sci. 92(1), 87–105 (1992)

[KSB-06]    Kärkkäinen, J., Sanders, P., Burkhardt, S.: Linear work suffix array construction. J. ACM 53(6), 918–936 (2006)

[KA-03]     Ko, P., Aluru, S.: Space efficient linear time construction of suffix arrays. In: Baeza-Yates, R., Chávez, E., Crochemore, M. (eds.) CPM 2003. LNCS, vol. 2676, pp. 200–210. Springer, Heidelberg (2003)

[KS-98]     Kreher, D.L., Stinson, D.R.: Combinatorial Algorithms: Generation, Enumeration, and Search. CRC Press (1998)

[L-83]      Lothaire, M.: Combinatorics on Words. Addison-Wesley, Reading (1983); 2nd edn. Cambridge University Press, Cambridge (1997)

[NM-07]     Navarro, G., Mäkinen, V.: Compressed full-text indexes. ACM Computing Surveys - CSUR 39(1), 2-es (2007)

[NZC-09]    Nong, G., Zhang, S., Chan, W.H.: Linear suffix array construction by almost pure induced-sorting. In: Proc. 2009 Data Compression Conf., pp. 193–202 (2009)

[R-03]      Ruskey, F.: Combinatorial Generation (Unpublished book). CiteSeerX: 10.1.1.93.5967, on combinatorics (2003)

[S-97]      Savage, C.: A survey of combinatorial Gray codes. SIAM Rev. 39(4), 605–629 (1997)

[SVM-11]    Sirén, J., Välimäki, N., Mäkinen, V.: Indexing finite language representation of population genotypes. In: Przytycka, T.M., Sagot, M.-F. (eds.) WABI 2011. LNCS, vol. 6833, pp. 270–281. Springer, Heidelberg (2011)

[S-03]      Smyth, B.: Computing patterns in strings, p. 423. Pearson (2003)

# SAHN Clustering in Arbitrary Metric Spaces Using Heuristic Nearest Neighbor Search

Nils Kriege*, Petra Mutzel, and Till Schäfer

Dept. of Computer Science, Technische Universität Dortmund, Germany
{nils.kriege,petra.mutzel,till.schaefer}@cs.tu-dortmund.de

**Abstract.** Sequential agglomerative hierarchical non-overlapping (SAHN) clustering techniques belong to the classical clustering methods that are applied heavily in many application domains, e.g., in cheminformatics. Asymptotically optimal SAHN clustering algorithms are known for arbitrary dissimilarity measures, but their quadratic time and space complexity even in the best case still limits the applicability to small data sets. We present a new pivot based heuristic SAHN clustering algorithm exploiting the properties of metric distance measures in order to obtain a best case running time of $\mathcal{O}(n \log n)$ for the input size $n$. Our approach requires only linear space and supports median and centroid linkage. It is especially suitable for expensive distance measures, as it needs only a linear number of exact distance computations. In extensive experimental evaluations on real-world and synthetic data sets, we compare our approach to exact state-of-the-art SAHN algorithms in terms of quality and running time. The evaluations show a subquadratic running time in practice and a very low memory footprint.

**Keywords:** SAHN clustering, nearest neighbor heuristic, data mining.

## 1 Introduction

Clustering is a generic term for methods to identify homogeneous subsets, so-called *clusters*, in a set of objects. It is a key technique in exploratory data analysis and widely applied in many fields like drug discovery, storage and retrieval, network analysis and pattern recognition [4,12]. A wealth of different clustering algorithms have emerged with varying definition of homogeneity. Typically this definition is based on a symmetric dissimilarity measure for pairs of objects.

A special class of clustering algorithms are hierarchical methods, which provide additional information on the relationship between clusters and can reveal nested cluster structures. A prominent example are sequential agglomerative hierarchical non-overlapping clustering techniques (SAHN) [12]. These approaches start with singleton clusters and iteratively merge two clusters with minimal dissimilarity until only one cluster remains. The inter-cluster dissimilarity is determined by a *linkage* strategy and based on the dissimilarity of the

---

* Nils Kriege was supported by the German Research Foundation (DFG), priority programme "Algorithm Engineering" (SPP 1307).

S.P. Pal and K. Sadakane (Eds.): WALCOM 2014, LNCS 8344, pp. 90–101, 2014.

objects contained in the clusters. The single, complete, average, median, centroid, and Ward linkage methods are well-studied and widely used [11]. A unique advantage of hierarchical methods is that the result can naturally be visualized as a *dendrogram*, a rooted binary tree where each node is linked to a merge operation with a certain dissimilarity. Cutting the dendrogram horizontally at a specific height leads to a set of subtrees where each root is associated with a subcluster. Thus, the result of SAHN clustering allows for iterative refinement of clusters making these methods especially suitable for an interactive exploration process, even for very large data sets [3].

We motivate further requirements for our clustering algorithm by a concrete example arising in cheminformatics, although similar constraints apply in other application areas: (1) Data sets in cheminformatics are often large containing tens of thousands of molecules. (2) A hierarchical method is needed since the whole similarity structure of the data is important. Furthermore, SAHN clustering methods are well-known and studied in cheminformatics [4] and users may be accustomed to dendrogram representations. (3) Support for arbitrary metric distance measures is required, since chemical compounds are complex structures, which are typically represented as graphs or bit vectors, so-called *fingerprints*. (4) Distance measures between these objects may be expensive, e.g., based on the maximum common subgraph of two molecular graphs. Thus, we desire a low dependence on the computational complexity of the distance measure.

A major drawback of hierarchical clustering algorithms is their high time and space complexity. The best exact algorithms known for arbitrary dissimilarity measures have a worst-case running time of $\mathcal{O}(n^2)$ [6] and are optimal since the general problem requires time $\Omega(n^2)$ [13]. Exact approaches are typically based on a symmetric distance matrix, which leads to quadratic memory requirements and a quadratic number of distance computations. However, quadratic time and space complexity is prohibitive when applied to large data sets in practice.

*Related Work.* Several exact algorithms with quadratic worst-case running time are known, some of which are limited to specific linkage methods, e.g., the NN-Chain algorithm [11], the single linkage minimum spanning tree algorithm [16] and methods based on dynamic closest pairs [6]. Some SAHN algorithms (e.g., NNChain) can avoid the quadratic distance matrix when using representatives, e.g., centroids, for cluster representation. However, this approach is limited to vector space and leads to an increased amount of exact distance computations.

Several methods to speed up clustering have been proposed. Data summarization is a common acceleration technique. An easy approach is to draw a random sample and cluster it instead of the whole data set. However, using random sampling leads to distortions in the clustering results. The kind of distortion is influenced by the used linkage method and because of this, many sophisticated summarization techniques are only suitable for special linkages. For example Patra et al. [15] proposed to use an accelerated leaders algorithm to draw a better sampling for average linkage. Another example is the Data Bubble summarization technique [2,21], which was originally developed for OPTICS clustering [1], but is also suitable for single linkage SAHN clustering.

Further acceleration is possible when using heuristic methods. Koga et al. [8] proposed Locality Sensitive Hashing (LSH) for a single linkage like algorithm. Its time complexity is reduced to $\mathcal{O}(nB)$, where $B$ is practically a constant factor. Although the runtime is very promising, it relies on vector data and is limited to single linkage, which is rarely used in cheminformatics.

Using the properties of metric distance functions is a common approach to accelerate different clustering techniques. Pivot based approaches have been proposed to reduce the number of exact distance computations for hierarchical clustering [14] and to speedup $k$-means [5]. To accelerate OPTICS a pivot based approach for heuristic $k$-close neighbor rankings was proposed by Zhou and Sander [19,20]. They also introduced a pivot tree data structure that enhances the effectiveness of the pivots for close neighbor rankings. SAHN clustering algorithms often rely on nearest neighbor (NN) queries (e.g., NNChain, Generic Clustering [13], Conga Line data structure [6]), which can be accelerated for metric distance functions [18]. However, the reduction of the NN search complexity does not necessarily reduce the asymptotic runtime of the clustering algorithms (see Sect. 3 for more details).

*Our Contribution.* We propose a new SAHN clustering algorithm for centroid and median linkage that benefits from sublinear NN queries and combine it with a pivot based indexing structure to obtain subquadratic running time in practice. The theoretical time complexity of our algorithm for clustering $n$ objects is $\mathcal{O}(n^2 \log n)$ in the worst case and $\mathcal{O}(n \log n)$ in the best case. Our approach is broadly applicable since it is not limited to the Euclidean vector space and many dissimilarity measures actually are metrics. Moreover, the new method requires only linear space and a linear number of distance computations and therefore allows to cluster large data sets even when distance computations are expensive. Our extensive experimental evaluation on a real-world data set from cheminformatics and on two synthetic data sets shows that the new method yields high-quality results comparable to exact algorithms, in particular when the data sets indeed contain a nested cluster structure. To our knowledge there are no other competing heuristic SAHN algorithms for general metric space supporting centroid or median linkage.

## 2   Preliminaries

A *clustering* of a set of objects $\mathcal{X} = \{x_1, \ldots, x_n\}$ is a partition $\mathcal{C} = \{C_1, \ldots, C_k\}$ of $\mathcal{X}$. A hierarchical clustering of $n$ objects yields $n$ distinct clusterings obtained from cutting the associated dendrogram at different heights. We refer to a clustering that results from such a cut and contains $i$ clusters as the clustering at level $i \in \{1, \ldots, n\}$. SAHN clustering is performed based on a distance function $d : \mathcal{X} \times \mathcal{X} \to \mathbb{R}^{\geq 0}$ between the objects and an inter-cluster distance $D : \mathcal{P}(\mathcal{X}) \times \mathcal{P}(\mathcal{X}) \to \mathbb{R}^{\geq 0}$ which is also called *linkage*.

Let $P \subset \mathcal{X}$ be a set of *pivots*. The triangle inequality in combination with the symmetric property fulfilled by metric distance functions yields lower and upper bounds for the distances between any two objects based on the exact distances between the objects and the pivots:

$$\forall x_i, x_j \in \mathcal{X}, p \in P : |d(x_i, p) - d(x_j, p)| \leq d(x_i, x_j) \leq d(x_i, p) + d(x_j, p) \quad (1)$$

In the following we are using the tilde and the hat to denote heuristic methods, e.g., $\widetilde{NN}$ and $\widehat{NN}$ refer to a heuristic NN search.

# 3   A Heuristic SAHN Clustering Algorithm

We present a heuristic SAHN (HSAHN) algorithm that is based on a variation of the generic clustering algorithm from [13] and utilizes an index structure for NN search. To efficiently determine heuristic NNs we adopt the approach of [20,19] based on best frontier search combined with a pivot tree and generalize it to support the specific requirements for use with SAHN clustering.

## 3.1   Generic Clustering

The original generic clustering algorithm has a complexity of $\Omega(n \ (\log n + k + m))$ and $\mathcal{O}(n^2 \ (\log n + k))$ for geometric linkages in vector space and $\Omega(n^2)$ for arbitrary linkages and distance measures. The value $k$ is the complexity of the NN search and $m$ the complexity of the merge process. Although other SAHN clustering algorithms have a quadratic upper bound, the practical runtime of the generic clustering competes with the other algorithms [13] and is orientated towards the lower bound. Our modified version of the generic clustering algorithm (Alg. 1) achieves $\Omega(n \ (\log n + k + m))$ for arbitrary linkage and distance measures and therefore the lower bound directly depends on the complexity of the NN search. This is also the case for the NNChain algorithm, but it requires the reducibility property, which is not guaranteed for heuristic NN searches. Note that HSAHN requires a metric distance function and is therefore limited to median and centroid linkage, but this is due to the NN search and not a limitation of the clustering algorithm.

Besides the runtime improvement, we modified the generic clustering algorithm in order to minimize distance distortions caused by our heuristic NN search. Our heuristic NN search $\widehat{NN}(C)$ also calculates a non-symmetric, heuristic distance $\widehat{D}(C, \widehat{NN}(C))$ (see Sect. 3.3 for more details). Since we use the lower bound of (1), we know that:

$$D(C, \widehat{NN}(C)) \geq \max\{\widehat{D}(C, \widehat{NN}(C)), \widehat{D}_{\mathrm{rev}} := \widehat{D}(\widehat{NN}(C), C)\}$$

It is possible to detect some cases where $\widehat{D}$ is smaller than $\widehat{D}_{\mathrm{rev}}$ without recalculating the distance over all pivots. For symmetric distance measures the minimal pairwise distance implies a reciprocal NN pair. Although this assumption does not hold for the used heuristic NN search, it does hold with a high probability. In such a case, we can use the already computed reverse distance and improve the quality of our heuristic by reinserting the tuple $(C, \widehat{NN}(C))$ in the priority

```
 1: function GENERICCLUSTERING(𝒳)
 2:     currentLevel ← singletonClusters(𝒳)                    ▷ clusters of the actual level
 3:     for all C ∈ currentLevel do                            ▷ initialization of Q
 4:         Q.insert(C, N̂N(C), D̂(C, N̂N(C)))                   ▷ Q is sorted by D̂(C, N̂N(C))
                                                                 (value at the time of insertion)
 5:     while currentLevel.size() > 1 do                       ▷ main loop
 6:         (Cᵢ, Cⱼ) ← Q.extractMin()
 7:         while not currentLevel.contains(Cᵢ) or not currentLevel.contains(Cⱼ) do
                                                                 ▷ invalid entry → recalculation of NN
 8:             if currentLevel.contains(Cᵢ) then
 9:                 Q.insert(Cᵢ, N̂N(Cᵢ), D̂(Cᵢ, N̂N(Cᵢ)))
10:             (Cᵢ, Cⱼ) ← Q.extractMin()
11:         if N̂N(Cⱼ) = Cᵢ and D̂(Cᵢ, Cⱼ) < D̂(Cⱼ, N̂N(Cⱼ)) then
                                                                 ▷ using already calculated N̂N(Cⱼ)
12:             Q.insert(Cᵢ, Cⱼ, D̂(Cⱼ, N̂N(Cⱼ)))
13:             continue
14:         Cₖ ← mergeCluster(Cᵢ, Cⱼ)                          ▷ (Cᵢ, Cⱼ) minimal N̂N pair
15:         currentLevel ← currentLevel \ {Cᵢ, Cⱼ} ∪ Cₖ
16:         Q.insert(Cₖ, N̂N(Cₖ), D̂(Cₖ, N̂N(Cₖ)))
17:     return currentLevel.get(0)                             ▷ return root node of the dendrogram
```

**Algorithm 1.** Modified Generic Clustering Algorithm

queue with the distance $\max\{\hat{D}, \hat{D}_{\mathrm{rev}}\}$ (lines $11-13$ of Alg. 1). Our benchmarks have proven that this approach is faster than recalculating the distance over all pivots and it does not harm the observed clustering quality.

## 3.2   Pivot Tree

As mentioned before, we are using the lower bound of (1) for heuristic distance approximations:

$$\tilde{D}(C_i, C_j) = \max_{p \in P} |D(\{p\}, C_i) - D(\{p\}, C_j)| \tag{2}$$

To increase the effectiveness of the pivots for close or NN queries Zhou and Sander proposed a pivot tree data structure [20]. The main idea behind this structure is that the heuristic distance between close objects must be more precise than between further objects to calculate the correct close or nearest neighbors. In our case we determine NNs according to:

$$\widetilde{NN}(C_i) = \mathrm{argmin}_{C_j}\{\tilde{D}(C_i, C_j)\} \tag{3}$$

The original pivot tree is a static data structure. In contrast SAHN clustering merges clusters and therefore we extended the data structure to allow deletion and insertion of objects and clusters, respectively. Additionally we used a different strategy to calculate the heuristic distances within the pivot tree and a simplified notion.

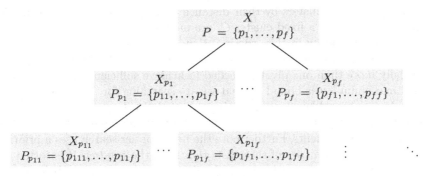

**Fig. 1.** Pivot Tree

As shown in Fig. 1 each node of the pivot tree is linked to a set of singleton clusters $X$ and a set of pivots $P \subseteq X$. The set of pivots is randomly chosen from $X$. One child node is created for each pivot. The set $X$ belongs to the root node and contains all clusters, while the set $X_{pi}$ of any child node contains all nodes from $X$ which are closest to $p_i$. Therefore all clusters in $X_{pi}$ are relatively close to each other. The calculation of the heuristic distance $\widetilde{D}(C_i, C_j)$ is performed according to (2) based on the common pivots $P_{i \cup j}$:

$$P_{i \cup j} = \{p \in P_k \mid C_i, C_j \in X_k\}$$

It is computationally cheap to compute $P_{i \cup j}$ since we know that each cluster is present only on the direct path from a leaf to the root node. To find the leaf node in constant time we store this relationship in a hashing data structure during the construction process. The relevant nodes for $P_{i \cup j}$ are all ancestors of the lowest common ancestor (LCA) of the leaf nodes for $C_i$ and $C_j$ and the LCA itself.

The construction process starts with the root node followed by a series of split operations until a maximum number of leaf nodes is reached. Each split operation creates the child nodes for a leaf node with maximum number of clusters. At construction time the data structure contains all singleton clusters.

The deletion of a cluster $C$ from the pivot tree is simply the removal of $C$ from all $X_i$. Inserting a merged cluster $C_{i \cup j}$ into the data structure is a bit more complicated since we cannot compute the exact distances to all pivots efficiently. It can be done efficiently with the Lance Williams Update Formula [9] for all ancestors of the LCA of $C_i$ and $C_j$, because we know the distance of both clusters to the nodes pivots and we can use $\widetilde{D}$ as distance between the merged clusters. This approach has the drawback that the depth of the pivot tree will decrease over a series of merge operations. However, this will happen relatively late in the merge process because close clusters will be merged first during SAHN clustering. Also the locality property of clusters in $X$ will not be violated.

### 3.3 Best Frontier Search

The best frontier search was already suggested to accelerate the OPTICS clustering algorithm in [19,20]. We will briefly describe the main idea below.

When ordering all clusters by their distance to a single pivot $p$, the heuristic distances $\widetilde{D}(C_i, C_j)$ to a fixed cluster $C_i$ are monotonically increasing when $C_i$ and $C_j$ are further away in the ordering. Hence, it is sufficient to consider the neighbors of $C_i$ (in the ordering) to find $C_i$'s NN with respect to a single pivot.

Typically more than one pivot is needed to achieve sufficient quality and the minimal maximum lower bound is what we are searching for. The best frontier search solves this problem by calculating a sorted list $L_p$ of clusters for each pivot $p$. To find the position of a cluster in $L_p$ in constant time, the clusters are linked to their list elements. Furthermore the best frontier search uses a priority queue which contains entries of these lists that form the *frontier*. It is sorted by the lower bound with respect to the single pivot to which the entry belongs.

When searching an NN of $C$, the queue initially contains all clusters next to $C$ in a list $L_p$ for some $p \in P$. Then the clusters with the lowest bounds are successively extracted from the queue and it is counted how often a certain cluster is retrieved. After the retrieval of each cluster $C_x$ the frontier is *pushed* by adding the cluster to the queue that is next to $C_x$ in the list $L_i$ from which $C_x$ was added to the queue. The cluster $C_j$ that is first counted $|P|$ times is the heuristic NN with respect to (3). The rationale is that the lower bounds induced by the clusters retrieved from the queue are monotonically increasing. That means all lower bounds which will be obtained in the future are greater than the heuristic distance $\widetilde{D}(C_i, C_j)$ and therefore $C_j$ must be the heuristic NN of $C_i$.

To combine the pivot tree with the best frontier search, Zhou and Sander run a $k$-close neighbor ranking for each node of the pivot tree and join the different close neighbors for each object afterwards. This approach is not feasible for SAHN clustering since we cannot efficiently determine which of the heuristic NNs found for each node is the best. Furthermore it is possible that the heuristic NNs found for each node are not correct with respect to (3), while the best frontier search in general can guarantee to find the correct heuristic NN. For that reason we need to use a different technique which will be described below.

Our integration of the best frontier search into the pivot tree runs an NN query for cluster $C_i$ over all pivots $p \in P_x$ where $C_i \in X_x$. While searching for the NN of $C_i$ the cardinality of $P_{i \cup j}$ is not fixed for an arbitrary cluster $C_j$. Therefore it must be calculated for each cluster $C_j$ that is retrieved from the frontier separately. The value can be calculated by finding the LCA of $C_i$ and $C_j$ in the pivot tree and counting the number of all pivots on the path between the LCA and the root node. To avoid unnecessary calculations, it is computed on demand and cached for the time of the NN query.

Because the asymptotic worst case complexity of an NN query with the best frontier search is not better than linear (see Sect. 3.4), a search depth bound $s$ is used. After $s$ clusters are retrieved from the priority queue the best frontier search is stopped and the cluster that is counted most often is returned as an NN. We use the terms $\widehat{NN}$ and $\widehat{D}$ for the search bounded NN search and distance. Note that the search bound is also the reason for the asymmetric behavior of $\widehat{D}$.

### 3.4  Theoretical Complexity

*Time Complexity.* To initialize the pivot tree data structure $n$ clusters are assigned to $f$ pivots on each level of the tree. The construction of the pivot tree therefore needs $\Theta(d\ f\ n)$ time where $d$ represents the tree depth. Since the number of required pivots for achieving a certain quality does not depend on the input size (Sect. 4), $f$ and $d$ can be considered constant and the overall construction time is linear.

As mentioned before, the modified generic clustering algorithm has a runtime of $\Omega(n\ (\log n + k + m))$ and $\mathcal{O}(n^2\ (\log n + k))$, if $k$ is the complexity of the NN search and $m$ the complexity of the merge process. The merging consists of two deletions from and one insertion into the pivot tree. Therefore, the complexity for merging is $\mathcal{O}(d\ \log n)$ as it takes $\mathcal{O}(\log n)$ time to insert a cluster into a sorted list $L_i$. The search of an NN is bounded by $\mathcal{O}(p\ s)$ if $p$ is the number of the used pivot elements and $s$ the search depth bound. The runtime includes $\mathcal{O}(s)$ extractions from the frontier queue with length $\mathcal{O}(p)$. Pushing the frontier and finding the initial elements in $L_i$ for each pivot takes constant time. With the same rationale as before, $p$ can be considered constant. It is shown in the experimental evaluation in Sect. 4 that $s$ can be chosen as a constant value, too.

The overall time complexity for HSAHN clustering is therefore bounded by $\mathcal{O}(n^2 \log n)$ in the worst case and $\mathcal{O}(n \log n)$ in the best case.

*Space Requirements.* Since the tree depth $d$ is a constant factor and therefore the number of nodes is also a constant, the pivot tree needs only linear space. Each node stores a constant number of sorted lists $L_i$ which store at most $n$ clusters. The hash table (to find the leaf nodes), the priority queue in the generic clustering algorithm and the frontier queue require $\mathcal{O}(n)$ space. Therefore the overall space requirements are linear with respect to the input size.

## 4  Experimental Results

This section will cover performance and quality measurements of our Java implementation. All tests were performed on an Intel Core i7 CPU 940 (2.93GHz) with a Linux operating system and a Java HotSpot virtual machine (version 1.6), which was limited to 5 GB of heap space. The implementation as well as the evaluation framework is publicly available at the Scaffold Hunter Website[1] and licensed under the GPLv3.

We used real-world as well as synthetic data sets[2] for the evaluation. The real-world data set (SARFari kinase) stems from the ChEMBL[3] database and contains a set of $\approx 50\,000$ molecules. The Euclidean and Tanimoto distances are utilized for this data set. The latter is applied to Daylight bit fingerprints (1024 bits), which represent structural information of the molecules and were generated with the toolkit CDK[4]. Two synthetic euclidean data sets are used

---

[1] http://scaffoldhunter.sourceforge.net
[2] https://ls11-www.cs.tu-dortmund.de/staff/schaefer/publication_data
[3] https://www.ebi.ac.uk/chembldb
[4] http://cdk.sourceforge.net

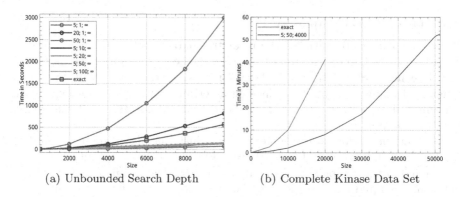

(a) Unbounded Search Depth          (b) Complete Kinase Data Set

**Fig. 2.** Runtime / Kinase Data Set / Tanimoto Distance

to analyze the impact of the clusterability on the quality. The first contains uniformly distributed data and the second 32 normally distributed and well-separated clusters.

All quality measurements are done by comparing each level of the HSAHN results with the exact results. The Fowlkes Mallows Index (FMI) [7] and the Normalized Variation of Information (NVI) [10] measurement are employed to measure each single level. The first produces very differentiated results for different settings while the second is less sensitive to the number of clusters. All test results are averaged over three runs, because the random selection of pivots results in a non-deterministic behavior of the algorithm. Without an exception the differences were very small and the results were stable over different runs. In the plots the parameters of the best frontier search are noted as $(f; l; s)$, where $f$ is the number of pivots per node, $l$ is the number of leaf nodes and $s$ is the search depth bound. Note that $l = 1$ means that the pivot tree is deactivated.

*Speed.* As shown in Fig. 2(a) the performance of the algorithm scales linear with the number of pivots. For the unbounded search depth the empirical runtime behavior is clearly quadratic and the absolute runtime for a high pivot count even exceeds the runtime of the exact algorithm. It is noteworthy that the number of leaves in the pivot tree does not have a major influence on the overall performance. With a reasonable set of parameters (the rationale follows in the quality evaluation) the time to cluster the whole kinase data set (Fig. 2(b)) is much lower than for the exact case. The heuristic curve flattens in the higher levels. Therefore the observed behavior is subquadratic.

It is important to know that we were unable to cluster a data set with 30 000 structures with the exact algorithm due to memory constraints. On the contrary, the heuristic clustering algorithm used less than 1 GB of memory to cluster the whole kinase data set.

*Quality - Pivot Count and Pivot Tree.* From the theoretical point of view more pivots should result in a better quality of the best frontier search. However, our test results do not show significant differences in quality if the number of pivots

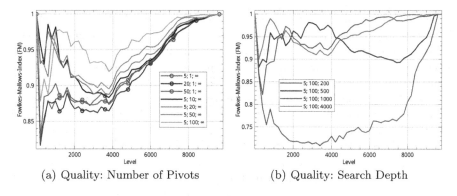

(a) Quality: Number of Pivots            (b) Quality: Search Depth

**Fig. 3.** Kinase Data Set / Euclidean Distance (5 Dimensions)

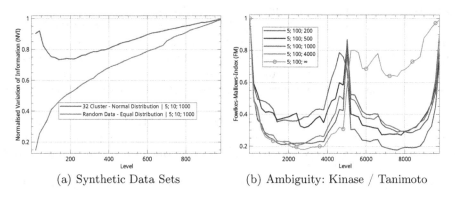

(a) Synthetic Data Sets            (b) Ambiguity: Kinase / Tanimoto

**Fig. 4.** Influence of the Data Distribution

are increased over a certain threshold. From our observations the main aspect that influences this threshold is the intrinsic dimensionality of the data and not the input size. Figure 3(a) clearly shows that there is no significant difference in quality if no pivot tree and 5, 20 or 50 pivots are used. Anyway, when using the pivot tree data structure, the quality can be enhanced further. This is a remarkable result, as the runtime of the setting $(50; 1; \infty)$ is more than 10 times higher than the runtime of the setting $(5; 100; \infty)$.

*Quality - Search Depth and Ambiguity.* The search depth is limiting the runtime of the best frontier search. Therefore it is very important to know, if the search depth can be chosen in a sublinear relation to the input size, while retaining a constant quality. Our tests revealed that this search bound can be chosen constant. For this we calculated an average quality score over all levels but the lowest 10% (e.g., level 9 001 to 10 000 for the input size 10 000) and compared these values for different fixed search depths over a series of different input sizes. Also for very low search depths the quality was constant over all input sizes. The reason to not use the lowest 10% is that these levels are strongly influenced by the starting situation where both clusterings contain only singletons.

The experimental evaluation showed that the search depth can be chosen about 500 in the low dimensional euclidean space (Fig. 3(b)). Lower values significantly harm the quality of the results. The search depth seems to be sensitive to the number of pivots used and the dimensionality of the data. The first observations are not surprising, since an increased amount of pivots increases the exit condition in the best frontier search loop. The second observation can be explained by the distribution of the distances. For high dimensional space the distances become more equal to each other. Equal distance means that even a small deviation of the lower distance bound (1) results in a higher probability, that this item is retrieved falsely from the frontier.

This observation also explains why the quality of the Tanimoto measurement (Fig. 4(b)) is lower than the quality of the Euclidean measurement and why the limitation of the search depth has such a huge impact. The number of different distances for the Tanimoto distance is limited by the length of the Farey sequence which is $\frac{3n^2}{\pi^2}$, where $n$ is the bit count. For a data set size of 10 000 this means that the number of object pairs is about 300 times higher than the number of distinct distance values. This leads to a high ambiguity in the clustering process and makes the results even unstable when comparing two exact algorithms. It is a general problem, that alternative equally good clusterings are not adequately scored by such a measure.

The peak in Fig. 4(b) at level 5 000 is also explainable when having a look at Fig. 4(a). If we have separated clusters inside the data set, the quality is good at exactly the level which corresponds to the number of clusters. This implies that we are able to identify real clusters in a data set with the HSAHN algorithm albeit the clustering quality might not be very well over all levels. This conclusion is very important because it proves the practicability of our approach.

## 5    Conclusions

Our tests show that the HSAHN algorithm can greatly expand the size of data sets which can be clustered in a reasonable amount of time. Furthermore the memory usage is lowered dramatically, which often sets a hard limit to the data set size. The linear dependence on exact distance calculations makes it possible to use computationally expensive distance measures even on huge data sets.

Our approach was integrated in the software Scaffold Hunter [17], a tool for the analysis and exploration of chemical space, and has been proven a valuable alternative to exact approaches in practice.

## References

1. Ankerst, M., Breunig, M.M., Kriegel, H.P., Sander, J.: OPTICS: Ordering Points To Identify the Clustering Structure. SIGMOD Rec. 28(2), 49–60 (1999)
2. Breunig, M.M., Kriegel, H.P., Kröger, P., Sander, J.: Data bubbles: quality preserving performance boosting for hierarchical clustering. SIGMOD Rec. 30(2), 79–90 (2001)

3. Chen, J., MacEachren, A.M., Peuquet, D.J.: Constructing overview + detail dendrogram-matrix views. TVCG 15(6), 889–896 (2009)
4. Downs, G.M., Barnard, J.M.: Clustering Methods and Their Uses in Computational Chemistry, pp. 1–40. John Wiley & Sons, Inc., New Jersey (2003)
5. Elkan, C.: Using the triangle inequality to accelerate k-means. In: ICML 2003, pp. 147–153. AAAI Press, Menlo Park (2003)
6. Eppstein, D.: Fast hierarchical clustering and other applications of dynamic closest pairs. Exp. Algorithmics 5(1) (2000)
7. Fowlkes, E.B., Mallows, C.L.: A method for comparing two hierarchical clusterings. Journal of the American Statistical Association 78(383), 553–569 (1983)
8. Koga, H., Ishibashi, T., Watanabe, T.: Fast agglomerative hierarchical clustering algorithm using locality-sensitive hashing. Knowledge and Information Systems 12(1), 25–53 (2007)
9. Lance, G.N., Williams, W.T.: A general theory of classificatory sorting strategies 1. hierarchical systems. The Computer Journal 9(4), 373–380 (1967)
10. Meilă, M.: Comparing clusterings—an information based distance. JMVA 98(5), 873–895 (2007)
11. Murtagh, F.: Multidimensional clustering algorithms. In: COMPSTAT Lectures 4. Physica-Verlag, Wuerzburg (1985)
12. Murtagh, F., Contreras, P.: Algorithms for hierarchical clustering: an overview. WIREs Data Mining Knowl. Discov. 2(1), 86–97 (2012)
13. Müllner, D.: Modern hierarchical, agglomerative clustering algorithms, arXiv:1109.2378v1 (2011)
14. Nanni, M.: Speeding-up hierarchical agglomerative clustering in presence of expensive metrics. In: Ho, T.-B., Cheung, D., Liu, H. (eds.) PAKDD 2005. LNCS (LNAI), vol. 3518, pp. 378–387. Springer, Heidelberg (2005)
15. Patra, B.K., Hubballi, N., Biswas, S., Nandi, S.: Distance based fast hierarchical clustering method for large datasets. In: Szczuka, M., Kryszkiewicz, M., Ramanna, S., Jensen, R., Hu, Q. (eds.) RSCTC 2010. LNCS, vol. 6086, pp. 50–59. Springer, Heidelberg (2010)
16. Rohlf, F.J.: Hierarchical clustering using the minimum spanning tree. Computer Journal 16, 93–95 (1973)
17. Wetzel, S., Klein, K., Renner, S., Rauh, D., Oprea, T.I., Mutzel, P., Waldmann, H.: Interactive exploration of chemical space with Scaffold Hunter. Nature Chemical Biology 5(8), 581–583 (2009)
18. Zezula, P., Amato, G., Dohnal, V., Batko, M.: Similarity Search: The Metric Space Approach. In: Advances in Database Systems, vol. 32. Springer (2006)
19. Zhou, J.: Efficiently Searching and Mining Biological Sequence and Structure Data. Ph.D. thesis, University of Alberta (2009)
20. Zhou, J., Sander, J.: Speedup clustering with hierarchical ranking. In: Sixth International Conference on Data Mining, ICDM 2006, pp. 1205–1210 (2006)
21. Zhou, J., Sander, J.: Data Bubbles for Non-Vector Data: Speeding-up Hierarchical Clustering in Arbitrary Metric Spaces. In: Proceedings of the 29th International Conference on Very Large Data Bases, VLDB 2003, vol. 29, pp. 452–463, VLDB Endowment (2003)

# Optimal Serial Broadcast of Successive Chunks

Satoshi Fujita

Department of Information Engineering, Hiroshima University
Kagamiyama 1-4-1, Higashi-Hiroshima, 739-8527, Japan

**Abstract.** This paper studies the problem of broadcasting successive chunks to all nodes in a network under an assumption such that each node can upload a chunk to exactly one node at a time. This models the delivery of a video stream in a Peer-to-Peer (P2P) network in which chunks of a stream are successively given by the media server. We propose two schemes for solving the problem. The first scheme attains an optimum broadcast time of $\lceil \lg n \rceil$ steps, where $n$ is the number of nodes, using an overlay network in which each node has $\mathcal{O}(\lg^2 n)$ children. The second scheme reduces the number of children by slightly increasing the broadcast time. More concretely, it completes each broadcast in $\lg n + o(\lg n)$ steps when the number of children is bounded by $O((f(n))^2)$ for any function satisfying $f(n) = \omega(1)$ and $f(n) = o(n)$.

**Keywords:** Peer-to-Peer video streaming, scheduling, optimum broadcast time.

## 1 Introduction

Consider a distributed system consisting of $n$ homogeneous computers called nodes. Let $V$ denote the set of nodes. All nodes in $V$ are synchronized to a global clock and can directly communicate with each other by using an appropriate routing protocol such as OSPF (Open Shortest Path First) and BGP (Border Gateway Protocol). Each time slot of the global clock is called a **step**. Let $t$ denote the step number of the global clock starting from $t = 0$. In each step, each node can send a chunk of data stream to exactly one receiver in $V$, whereas it is allowed to receive any number of chunks from other nodes in the same step, i.e., we assume that *the download capacity of each node is sufficiently large compared with the upload capacity and that the simultaneous upload to several nodes is not allowed.* Such a restricted model of communication is referred to as the serial communication model (or the single-port communication model [4, 5]) in the literature.

Let $s$ ($\notin V$) be a special node called **source**. The source issues several chunks (of a data stream) which should be received by all nodes in $V$, i.e., we are considering a situation in which the source is a publisher and the other nodes are subscribers. The **broadcast time** of chunk $c_i$ is defined as $\beta_i - \alpha_i$ where $\alpha_i$ is the time step at which $c_i$ is given to the first node in $V$ and $\beta_i$ is the earliest time step at which $c_i$ is received by all nodes in $V$. The problem we will consider

S.P. Pal and K. Sadakane (Eds.): WALCOM 2014, LNCS 8344, pp. 102–113, 2014.

in the current paper is to minimize the maximum broadcast time (MBT, for short) over all chunks provided that $c_i$ is given to the source in the $i^{th}$ step for all $i \geq 0$. We do not mind the order of chunks received by each node, e.g., we allow each node to receive $c_{i+1}$ earlier than $c_i$. The reader should note that if we attain an MBT of $M$ steps, then any node will receive $c_i$ no later than $M - 1$ steps after receiving $c_{i+1}$, i.e., the size of input buffer can be bounded by $M$.

The problem of minimizing MBT was proposed by Liu as a part of the study of efficient dissemination of video streams in Peer-to-Peer (P2P) environment [9]. Liu proved that there exists a broadcasting scheme which attains MBT of $\lceil \lg n \rceil$ steps, where $\lg n$ is the logarithm of $n$ to base two. Note that this bound is optimum for the MBT of successive broadcast since under the serial communication model, the number of nodes which received a chunk becomes at most double in each step and the source can spend exactly one step for each chunk. More recently, a variant of the problem in which the number of possible children of each node is bounded by a constant was extensively investigated by Bianchi et al. [1]. Note that in Liu's model, there is no such limitation on the number of possible children, while it is trivially bounded by at most $n$. If the number of possible children of each node is bounded by one, i.e., if the successor of each node is fixed to a node, we can easily have a tight bound of $n - 1$ steps since the propagation of chunks along one-dimensional chain of length $n$ is the unique solution. On the other hand, if each node can have at most two children, where of course, it can send a chunk to at most one child at a time, a chunk given to the source in the $0^{th}$ step is received by at most $F_i$ nodes in the $i^{th}$ step, where $F_i$ is the $i^{th}$ Fibonacci number[1]. Bianchi at al. generalized the above result to the cases in which the possible number of children is bounded by a constant greater than two, and derived an optimum broadcast scheme for such cases. Some other researches pointed out that the structure of the overlay plays an important role in realizing a short broadcast time in P2P streaming systems [2, 3, 6–8, 10–15].

In this paper, we propose two schemes to broadcast successive chunks with small MBT. The first scheme attains optimum MBT of $\lceil \lg n \rceil$ steps for any $n \geq 1$ using an overlay network in which each node has $\mathcal{O}(\lg^2 n)$ children. This scheme is an explicit implementation of the snow ball scheme suggested by Liu in [9], where the reader should note that in his paper, Liu merely pointed out the *existence* of such scheme and did not give a concrete way to realize MBT of $\lceil \lg n \rceil$ steps. In [1], Bianchi et al. also pointed out that such a bound could be attained if the number of possible children of each node is not bounded, but they did not give a concrete scheme as well. In this sense, our scheme is the first scheme which explicitly gives a way to realize an optimum MBT. Our second scheme reduces the number of children of each node by slightly increasing the broadcat time. More concretely, it completes each broadcast in $\lg n + o(\lg n)$

---

[1] We can prove the claim by mathematical induction, i.e., in the $i^{th}$ step, nodes which received the chunk in the $(i - 1)$st and $(i - 2)$nd steps can send the chunk to the next nodes.

steps when the number of children is bounded by $O((f(n))^2)$ for any function satisfying $f(n) = \omega(1)$ and $f(n) = o(n)$.

The remainder of this paper is organized as follows. Section 2 introduces several definitions used in the proposed schemes. Sections 3 and 4 describe our first and second schemes, respectively. Finally, Section 5 concludes the paper with future work.

## 2     Preliminaries

In this paper, we assume $n = 2^m + \alpha$ for some $m \geq 2$ and $0 \leq \alpha < 2^m$. Note that the case of "$\alpha = 0$" corresponds to the most critical situation for "optimal" broadcast schemes, since in order to attain MBT of $m$ ($= \lceil \lg n \rceil$) steps, every node must be busy all the time. Let $x_0, x_1, \ldots, x_{m-1}$ be a sequence of integers defined as follows: 1) $x_0 := 1$, and 2) $x_i := 2x_{i-1}$ for each $1 \leq i \leq m - 2$. Note that by definition, $x_{m-2} = |V|/4$ and $\sum_{i=0}^{j} x_i = 2x_j - 1$ hold for any $j$.

Let $B_i$ denote a **block** consisting of $x_i$ nodes. As will be described later, our first scheme statically partitions $2^m$ nodes in $V$ into several copies of blocks, in the following manner:

1. $2m$ copies of block $B_0$.
2. $m - i - 1$ copies of block $B_i$ for each $1 \leq i \leq m - 2$.

*Remark 1.* The total number of nodes in those copies of blocks is $2^m$.

*Proof.* See Appendix A.

An *i*-thread $T_i$ is a collection of blocks defined as follows (in contrast to blocks, threads will be dynamically constructed in the proposed scheme):

- $T_1$ is a collection of two copies of $B_0$.
- For each $i \geq 2$, $T_i$ consists of two copies of $B_0$ and one copy of $B_j$ for each $1 \leq j \leq i - 1$.

In the following we distinguish two copies of $B_0$ in each thread and call them as left and right copies, respectively. It is worth noting that for any $1 \leq i \leq m - 1$, $T_i$ consists of $2^i$ nodes.

Note that $T_i$ is associated with a set of nodes which received a chunk within $i$ steps under the normal broadcast scheme. More concretely, in an *i*-thread $T_i$ consisting of left and right copies of $B_0$, say $b_0^1$ and $b_0^2$, and one copy of $B_i$ for each $i$, say $b_i$, the broadcast of a chunk to all nodes in $T_i$ proceeds as follows:

1. suppose $b_0^1$ receives the chunk at time $t = 0$;
2. in the first step, $b_0^1$ sends the chunk to $b_0^2$;
3. in the second step, $b_0^1$ and $b_0^2$ send the received chunk to two nodes in $b_1$ in parallel;
4. in the third step, $b_0^1, b_0^2$ and $b_1$ send the received chunk to four nodes in $b_2$ in parallel; and so on.

In other words, in the $j^{th}$ step, all blocks which have received the chunk in a previous step forward it to $b_{j-1}$ in parallel. By conducting such a parallel forwarding of chunks in a step, an $i$-thread grows to an $(i + 1)$-thread. Note that all nodes belonging to such a "growing" $i$-thread are busy to disseminate the chunk to the nodes which will become a new member of the same thread in succeeding steps.

## 3    First Scheme

In this section, we describe the details of our first broadcast scheme which attains an MBT of $m$ steps for $\alpha = 0$ and that of $m + 1$ steps for $0 < \alpha < 2^m$. In the following, after describing the basic scheme for $\alpha = 0$ in Section 3.1, we will extend it to the cases of $2^{m-1} \le \alpha < 2^m$ and $0 < \alpha < 2^{m-1}$ in Sections 3.2 and 3.3, respectively.

### 3.1    Basic Scheme for $\alpha = 0$

The broadcast of chunk $c_i$ is initiated by the source by *generating* a 1-thread corresponding to $c_i$ and by sending $c_i$ to the left copy of block $B_0$ in the thread. Let $u_i$ be the node corresponding to the left copy, and without loss of generality, let us assume that $u_i$ receives chunk $c_i$ at time $t = 0$. The received chunk is forwarded to the right copy of $B_0$ in the same thread at time $t = 1$, and in the succeeding steps, $u_i$ disseminates chunk $c_i$ to all nodes in $V$ by successively *growing* the thread, in such a way that the size of the thread exactly doubles in each step. More concretely, the 1-thread grows to a 2-thread at time $t = 2$, the resulting 2-thread grows to a 3-thread at time $t = 3$, and so on. Such a grow of thread is controlled by node $u_i$. After becoming an $(m - 1)$-thread consisting of $2^{m-1} = \frac{|V|}{2}$ nodes at time $t = m - 1$, the thread *disappears* in the next step. Of course, to complete the broadcast of chunk $c_i$, nodes in the $(m - 1)$-thread must send $c_i$ to the remaining $\frac{|V|}{2}$ nodes in parallel at time $t = m$, but they can receive chunks from other nodes at the same time, i.e., the word "disappear" means that nodes contained in the $(m-1)$-thread will be used as a part of other threads as a receiver while serving as a sender of chunk $c_i$.

Such a dynamic behavior of the scheme, i.e., generation of a 1-thread, grow of an $i$-thread, and disappear of an $(m - 1)$-thread, is realized by using the notion of "pool of blocks." Let $S$ be a variable representing a pool of (copies of) blocks. $S$ is initialized to the collection of all copies of blocks obtained by partitioning $V$. The generation of a 1-thread is easily done by picking up two copies of $B_0$ from the pool and by regarding them as a 1-thread. The disappear of an $(m - 1)$-thread is also easily done by returning all blocks in the $(m - 1)$-thread to $S$. Finally, in order to grow an $i$-thread to an $(i + 1)$-thread for $i < m - 1$, we first pick up a copy of $B_i$ from the pool and then add it to the tail of the $i$-thread as the receivers of the nodes in the thread. More concretely, in each step of the scheme, the following three operations are executed sequentially:

1. Return of all blocks contained in an $(m-1)$-thread $T_{m-1}$. Recall that $T_{m-1}$ consists of two copies of $B_0$ and one copy of $B_i$ for each $1 \le i \le m-2$. Note that this operation is not executed in the initial $m$ steps.
2. Pickup of two copies of $B_0$ to generate a new 1-thread.
3. Pickup of a copy of $B_i$ to grow an $i$-thread to an $(i+1)$-thread for each $0 \le i \le m-2$.

The correctness of the scheme is clear. In addition, it takes $m$ steps to broadcast a given chunk to all nodes in $V$. Thus the following theorem holds.

**Theorem 1.** *The basic scheme attains an MBT of $m$ steps provided that $n = 2^m$.*

As for the possible number of children of each node, we have the following theorem.

**Theorem 2.** *During the execution of the basic scheme, each node sends chunks to at most $\frac{(m-2)(m-1)}{2} + 2$ different nodes for any $m \ge 3$.*

*Proof.* See Appendix B.

## 3.2    Extended Scheme for $2^{m-1} \le \alpha < 2^m$

The basic scheme proposed in the last subsection can be extended to the case of $|V| = 2^m + \alpha$ for some $2^{m-1} \le \alpha < 2^m$, in the following manner. The reader should note that since $\lceil \lg |V| \rceil = m+1$, we can use one more step after completing the broadcast to $2^m$ nodes.

The extended scheme proceeds as follows. In the $m^{th}$ step of the broadcast of a chunk, nodes in the $(m-1)$-thread $T_{m-1}$ corresponding to the chunk tries to grow (a part of) $m$-thread instead of completing the broadcast to $2^m$ nodes as was done in the basic scheme. The key idea is to conduct the forward of the chunk to (a part of) the remaining $2^{m-1}$ nodes at the same time with the partial grow of the $m$-thread. Since there are $\alpha - 2^{m-1}$ nodes which should receive the chunk from $T_{m-1}$ in the $m^{th}$ step, $2^m - \alpha$ nodes in $m-1$ threads organized for the succeeding chunks can receive the chunk from $T_{m-1}$. Thus, in the $(m+1)$st step of the broadcast, it is enough to forward the chunk to the remaining $\alpha$ nodes, which is always possible since the partial $m$-thread organized in the $m^{th}$ step contains $\alpha$ nodes.

As for the analysis of value $\nu(u, t)$, a similar argument to the case of $|V| = 2^m$ holds for $1 \le t \le m-1$ and $t = m+1$. When $t = m$, $\alpha - 2^{m-1}$ nodes in $T_{m-1}$ send the chunk to $\alpha - 2^{m-1}$ nodes to grow a partial $m$-thread, which can be done in the same way to the basic scheme, and the remaining $2^m - \alpha$ nodes in $T_{m-1}$ send the chunk to $2^m - \alpha$ nodes in $m-1$ threads organized for succeeding chunks, which can be done by fixing the type of block of the receiver for each sender, i.e., $\nu(u, m) \le m$ for such $u$.

Hence we have the following claim.

**Corollary 1.** *When $2^{m-1} \le \alpha < 2^m$, a variant of the basic scheme attains an MBT of $m+1$ steps and during the execution of the scheme, each node sends chunks to at most $\mathcal{O}(m^2)$ different nodes for any $m \ge 3$.*

## 3.3  Extended Scheme for $0 < \alpha < 2^{m-1}$

This subsection outlines a way of extending the basic scheme to the case of $0 < \alpha < 2^{m-1}$. Recall that in the basic scheme, after completing the $(m-1)$st step, the chunk to be broadcast is held by $2^{m-1}$ nodes in an $(m-1)$-thread $T_{m-1}$ and those nodes are used to forward the chunk to the remaining $2^{m-1}$ nodes in a single step. The key point of our extension is to separate the parallel forwarding of the chunk in the $m^{th}$ step into two types, i.e., one-step forwarding and two-step forwarding. More concretely, $2^{m-1} - \alpha$ nodes in $T_{m-1}$ forward the chunk to the nodes in the succeeding $m$ threads in a single step, whereas the remaining $\alpha$ nodes forward the chunk through $\alpha$ additional nodes which are not contained in any of $m+1$ threads including $T_{m-1}$. In other words, such $\alpha$ nodes are used merely to relay chunks to $\alpha$ nodes in the succeding $m$ threads. Such a modification of the scheme does not increase the maximum number of children of nodes in the resulting broadcast trees. Hence we have the following claim.

**Corollary 2.** *When $0 < \alpha < 2^{m-1}$, a variant of the basic scheme attains an MBT of $m + 1$ steps and during the execution of the scheme, each node sends chunks to at most $\mathcal{O}(m^2)$ different nodes for any $m \geq 3$.*

# 4   Second Scheme

By allowing more steps, we could significantly reduce the number of children of each node in the overlay. In this section, we describe the details of our second scheme which is designed to achieve such a significant reduction.

## 4.1  Outline

Recall that in the first scheme, we can broadcast each chunk to $n$ nodes in $\lceil \lg n \rceil$ steps using an overlay in which each node has $\mathcal{O}(\lg^2 n)$ children. Let $C(n)$ denote the overlay, and let us assume that $n$ is a power of two. $C(n)$ has a capability of forwarding each chunk to $n/2$ "outside" nodes in the $(\lg n)$th step of the broadcast of the chunk, if we allow that $n/2$ nodes in the overlay receive the chunk later than the $(\lg n)$th step. Let $f$ be a function such that $f(n) = \omega(1)$, $f(n) = o(n)$ and $f(n)$ is a power of two for any $n \geq 1$. The key idea of our second scheme is to use $C(f(n))$ as a basic component and to connect $n/f(n)$ copies of such component to form a $(f(n)/2)$-ary tree of depth $\frac{\lg n}{\lg f(n)-1} + \mathcal{O}(1)$. The out-degree of each node within each component is $\mathcal{O}(f(n)^2)$ by Theorem 2, and as will be described later, those components are connected so that each node has $\mathcal{O}(f(n))$ nodes as possible children in other components; i.e., the out-degree in the overall network is bounded by $\mathcal{O}(f(n)^2)$. Since the dissemination of chunk to a half of nodes in each component takes $\lg f(n)$ steps by assumption, even if the connection between two components in the tree consumes a constant number of steps, we can disseminate each chunk to at least $n/2$ nodes in $\lg n + o(\lg n)$ steps. Thus, if we could complete the broadcast to all nodes in the network by spending $o(\lg n)$ more steps, we will have the following theorem on the second scheme.

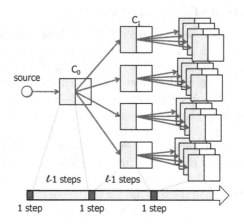

**Fig. 1.** Connection of basic components in the second scheme

**Theorem 3.** *The second scheme completes the broadcast of successive chunks in* $\lg n + o(\lg n)$ *steps using an overlay in which each node has* $\mathcal{O}((f(n))^2)$ *children for any* $f$ *such that* $f(n) = \omega(1)$ *and* $f(n) = o(n)$.

### 4.2   Basic Component

Let $\ell$ be a function from $\mathbb{N}$ to $\mathbb{N}$ such that $\ell(n) = \omega(1)$ and $\ell(n) = \mathcal{O}(\lg n)$, where $\mathbb{N}$ is the set of natural numbers. In the following, we denote $\ell(n)$ as $\ell$ if parameter $n$ is clear from the context. Consider $2^{\ell-1} + 1$ copies of component $C(2^{\ell})$, and call them as $C_0, C_1, \ldots, C_{2^{\ell}-1}$. We will connect those copies by a collection of links so that:

1. a node in $C_0$ receives a chunk from the source in the first step,
2. the received chunk is disseminated to $2^{\ell-1} - 1$ nodes in $C_0$ in the succeeding $\ell - 1$ steps (i.e., a half of nodes in $C_0$ receive the chunk in the first $\ell$ steps),
3. in the $(\ell+1)$st step, each of those $2^{\ell-1}$ nodes forwards the chunk to a node in a component so that exactly one node in $C_i$ receives the chunk for each $1 \le i \le 2^{\ell-1}$, and
4. in the succeeding $\ell - 1$ steps, the received chunk is disseminated to $2^{\ell-1} - 1$ nodes in $C_i$ for each $i$ (i.e., a half of nodes in $C_i$ receive the chunk in the next $\ell$ steps).

See Figure 1 for illustration. In the following, we say that $C_0$ is the parent component of $C_i$ and $C_i$ is a child component of $C_0$. Note that each child component successively receives chunks from the parent component, and $C_0$ simltaneously forwards chunks to $2^{\ell-1}$ succeeding nodes in a successive manner.

The dissemination of chunks in each component is conducted as in the first scheme except for the last one step (the design of links to be used in the last step will be described later). Thus, each node has at most $\mathcal{O}(\ell^2)$ children in each component. In each step later than the $(\ell-1)$st step of the broadcast of the

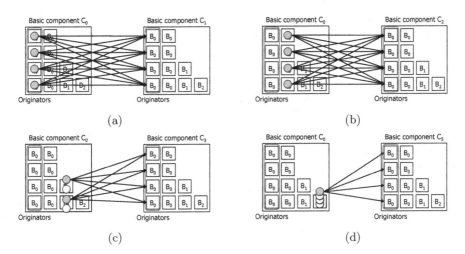

**Fig. 2.** Connection of nodes in adjacent blocks ($\ell = 4$)

first chunk, nodes in component $C_0$ dynamically organize $\ell$ threads of different sizes, i.e., two copies of 1-thread and one copy of $i$-thread for $2 \leq i \leq \ell - 1$. In addition, after the $\ell^{th}$ step of the broadcast of a chunk, all nodes contained in an $(\ell - 1)$-thread have successfully received the chunk. Any $(\ell - 1)$-thread organized during successive broadcasts consists of one left copy of $B_0$, one right copy of $B_0$ and exactly one copy of $B_i$ for each $1 \leq i \leq \ell - 2$. Thus, the forwarding of chunks from $C_0$ to $C_i$ is realized along a collection of static links connecting nodes in $C_0$ to nodes in the other components, in the following manner:

- For each $i$, component $C_i$ contains $\ell$ left copies of block $B_0$, which are used to receive chunks from the parent component. In the following, we call $\ell$ nodes in such copies of $B_0$ the "originators" of component $C_i$.
- Each node in a left copy of $B_0$ in $C_0$ is connected to the originators of component $C_1$; i.e., each originator of $C_1$ is connected with all originators of $C_0$ (Figure 2 (a)); each node in a right copy of $B_0$ in $C_0$ is connected to the originators of $C_2$ (Figure 2 (b)); two nodes in a copy of $B_1$ in $C_0$ are connected with the originators of $C_3$ and $C_4$, respectively (Figure 2 (c)); and four nodes in a copy of $B_2$ in $C_0$ are connected with the originators of $C_5, C_6, C_7$ and $C_8$, respectively, and so on (Figure 2 (d)).

Note that since each node in $C_0$ has $\ell$ children in the succeeding components, we can still bound the number of children of each node in $C_0$ by $\mathcal{O}(\ell^2)$.

### 4.3    $(2^{\ell-1})$-ary Tree

By repeating the above parent-children structure, we have a $(2^{\ell-1})$-ary tree in which the set of vertices corresponds to the set of components. Suppose that the given set of $n$ nodes is divided into $n/2^\ell$ components of size $2^\ell$ each. The minimum height of $(2^{\ell-1})$-ary tree consisting of $n/2^\ell$ vertices is calculated as follows. Let $T$ be a $(2^{\ell-1})$-ary tree consisting of infinitely many vertices. Since there are $2^{j(\ell-1)}$ vertices at depth $j$ in $T$, the number of vertices in a complete $(2^{\ell-1})$-ary tree of depth $h$ is given by

$$\sum_{j=0}^{h} 2^{j(\ell-1)} = \frac{2^{(h+1)(\ell-1)} - 1}{2^{\ell-1} - 1}.$$

Thus, the minimum depth of a $(2^{\ell-1})$-ary tree consisting of $n/2^\ell$ vertices is given by the smallest integer $h^*$ satisfying the following inequality:

$$\frac{2^{(h^*+1)(\ell-1)} - 1}{2^{\ell-1} - 1} \geq \frac{n}{2^\ell}.$$

By solving it, we have $h^* \geq \frac{\lg n}{\ell-1} + \mathcal{O}(1)$. Since each vertex in the tree associated with a component consumes $\ell$ steps for the broadcast of a chunk, the total number of steps required for completing a dissemination of a chunk to a "half" of nodes in $V$ (recall that by changing the receiver in the $(\ell+1)$st step of the broadcast in $C_0$ to the nodes in child components, a half of nodes in $C_0$ can not receive the chunk at that time) is

$$\ell \times \frac{\lg n}{\ell-1} + \mathcal{O}(\ell) = \left(1 + \frac{1}{\ell-1}\right) \lg n + \mathcal{O}(\ell) = \lg n + o(\lg n),$$

where the last equality is due to $\ell(n) = \omega(1)$ and $\ell(n) = \mathcal{O}(\lg n)$. The above analysis also indicates that by fixing $\ell$ to an appropriate constant, we have a scheme in which each node is connected with constant number of children. More concretely, if $\ell = 4$, the out-degree of each node is bounded by $\frac{(4-2)(4-1)}{2} + 2 + 4 = \frac{6}{2} + 6 = 9$ and the resulting broadcast time is $\left(1 + \frac{1}{4-1}\right) \lg n + \mathcal{O}(1) \leq 1.334 \lg n + \mathcal{O}(1)$. That is, the coefficient of $\lg n$ in the broadcast time is bounded by 1.334. The result of similar analysis for small $\ell$'s is summarized in Table 1.

### 4.4    Backward Propagation of Chunks

Suppose that we complete the dissemination of a chunk, say $c^*$, to a half of nodes in each non-leaf component and all nodes in each leaf component at time $t = \tau$ (recall that in each component which has no children in the $(2^{\ell-1})$-ary tree, we can complete the broadcast of $c^*$ in $\ell$ steps using the basic scheme described in Section 3). To complete the dissemination of $c^*$ to the remaining half of nodes in non-leaf nodes by spending $o(\lg n)$ more steps, we slightly modify the way of dissemination in each component as follows.

Let us assume $\ell \geq 2$, without loss of generality. Recall that in the basic scheme, each component contains two 1-threads, say $T_1^a$ and $T_1^b$. At first, we partition $T_1^a$ into two 0-threads and remove one 0-thread (consisting of one node) from the task of broadcasting within each component and the forwarding to a child component. More concretely, we statically separate nodes in each component to

**Table 1.** The performance of the second scheme, where the "broadcast time" represents the coefficient of $\lg n$ in the broadcast time

| $\ell$ | 4 | 5 | 6 | 7 |
|---|---|---|---|---|
| out-degree | 9 | 13 | 18 | 24 |
| broadcast time | 1.334 | 1.250 | 1.200 | 1.167 |

one excluded node and the remaining $2^\ell - 1$ nodes, and use the latter nodes for the broadcasting and the forwarding of chunks. Let us call the excluded node a preserved node. Note that by this modification, every component contains exactly one preserved node.

Let us focus on preserved nodes in leaf components. Those nodes receive chunk $c^*$ at time $t = \tau$, by assumption. Since each child component receives $c^*$ from a node in the parent component by postponing the forward of $c^*$ to a node, say $u^*$, in the parent component, and since the preserved node is exempted from the task of forwarding and broadcasting chunks, in the $(\tau + 1)$st step, the preserved node in the child component can send $c^*$ to node $u^*$ in the backward direction. Since such a backward transmission is conducted from all children of the parent component at the same time, in the $(\tau + 1)$st step, all of the remaining nodes in the parent component (including preserved node in the component) successfully receive chunk $c^*$. Such a backward propagation of chunk is repeated until it reaches the root component of the tree, and since the depth of the $(2^{\ell-1})$-ary tree is $(\lg n)/(\ell - 1) + \mathcal{O}(1)$, we can complete the broadcast to all nodes by spending $o(\lg n)$ additional steps. It is not difficult to show that the number of possible children of preserved nodes is bounded by $\mathcal{O}(\ell^2)$ since the parent of each component is statically determined. Hence Theorem 3 follows.

## 5 Concluding Remarks

This paper studies the problem of broadcasting several chunks in distributed networks under an assumption such that each node can upload a chunk to exactly one neighbor at a time and all chunks are successively given by the media server. Such a delivery of successive chunks commonly occurs in P2P video streaming systems such as SplitStream and CoolStreaming. We proposed two broadcast schemes. When the given system consists of $n$ nodes, the first scheme attains an optimum broadcast time of $\lceil \lg n \rceil$ steps using an overlay network in which each node has $\mathcal{O}(\lg^2 n)$ children, whereas the second scheme (significantly) reduces the number of children by (slightly) increasing the broadcast time. An important future work is to introduce the heterogeneity of nodes and links in the design and analysis of optimal broadcast schemes. Performance evaluation using extensive simulations is also an important issue to be addressed.

# References

1. Bianchi, G., Melazzi, N.B., Bracciale, L., Lo Piccolo, F., Salsano, S.: Streamline: An Optimal Distribution Algorithm for Peer-to-Peer Real-Time Streaming. IEEE Trans. on Parallel and Distributed Systems 21(6), 857–871 (2010)
2. Castro, M., Druschel, P., Kermarrec, A.-M., Nandi, A., Rowstron, A., Singh, A.: SplitStream: High-Bandwidth Content Distribution in Cooperative Environments. In: Proc. of ACM SOSP 2003, pp. 298–313 (2003)
3. Chen, M., Chiang, M., Chou, P., Li, J., Liu, S., Sengupta, S.: P2P Streaming Capacity: Survey and Recent Results. In: Proc. of 47th Annual Allerton Conference on Communication, Control, and Computing, pp. 378–387 (2009)
4. Fujita, S., Yamashita, M.: Optimal Group Gossiping in Hypercubes under a Circuit-Switching Model. SIAM J. Computing 25(5), 1045–1060 (1996)
5. Fujita, S.: A Fault Tolerant Broadcast Scheme in Star Graphs under the Single-Port Communication Model. IEEE Trans. on Computers 48(10), 1123–1126 (1999)
6. Guo, Y., Liang, C., Liu, Y.: AQCS: Adaptive Queue-Based Chunk Scheduling for P2P Live Streaming. In: Proc. of IFIP Networking, pp. 433–444 (2008)
7. Kumar, R., Liu, Y., Ross, K.W.: Stochastic Fluid Theory for P2P Streaming Systems. In: Proc. of INFOCOM 2007, pp. 919–927 (2007)
8. Liang, C., Guo, Y., Liu, Y.: Is Random Scheduling Sufficient in P2P Video Streaming? In: Proc. of 28th ICDCS, pp. 53–60 (2008)
9. Liu, Y.: On the Minimum Delay Peer-to-Peer Video Streaming: How Realtime Can It Be? In: Proc. of MULTIMEDIA 2007, pp. 127–136 (2007)
10. Liu, S., Zhang-Shen, R., Jiang, W., Rexford, J., Chiang, M.: Performance Bounds for Peer-Assisted Live Streaming. In: Proc. of ACM SIGMETRICS, pp. 313–324 (2008)
11. Liu, S., Chen, M., Sengupta, S., Chiang, M., Li, J., Chou, P.A.: P2P Streaming Capacity under Node Degree Bound. In: Proc. of ICDCS 2010, pp. 587–598 (2010)
12. Massoulie, L., Twig, A., Gkantsidis, C., Rodriguez, P.: Randomized Decentralized Broadcasting Algorithm. In: Proc. of IEEE Infocom, pp. 1073–1081 (2007)
13. Nguyen, T., Kolazhi, K., Kamath, R., Cheung, S., Tran, D.: Efficient Multimedia Distribution in Source Constraint Networks. IEEE Trans. on Multimedia 10(3), 532–537 (2008)
14. Sengupta, S., Liu, S., Chen, M., Chiang, M., Li, J., Chou, P.A.: Peer-to-Peer Streaming Capacity. IEEE Trans. on Information Theory 57(8), 5072–5087 (2011)
15. Xie, S., Li, B., Keung, Y., Zhang, X.: Coolstreaming: Design, Theory, and Practice. IEEE Trans. on Multimedia 9(8), 1661–1671 (2007)

# A    Proof of Remark 1

The number of nodes in the copies of blocks $B_1, \ldots, B_{m-2}$ is calculated as

$$\sum_{i=1}^{m-2} (m - i - 1)x_i = \sum_{j=1}^{m-2} \sum_{i=1}^{j} x_i = \sum_{j=1}^{m-2} (2x_j - 2)$$

where the last equality is due to $\sum_{i=0}^{j} x_i = 2x_j - 1$ and $x_0 = 1$. Thus, the amount is represented as

$$2 \sum_{j=1}^{m-2} x_j - 2(m-2) = 2(2x_{m-2} - 2) - 2(m-2) \ = \ 4 \times 2^{m-2} - 2m.$$

Since there are $2m$ copies of block $B_0$ consisting of one node, the remark follows.

## B     Proof of Theorem 2

Let $\nu(u, \tau)$ denote the maximum number of nodes which have a possibility of receiving a chunk from $u$ in the $\tau^{th}$ step of the broadcast of the chunk. In the following, we will evaluate the value of $\nu(u, \tau)$ for each $1 \le \tau \le m$. The case of $\tau = 1$ is obvious since $\nu(u, 1) = 1$ if $u$ is a node in the left copy of block $B_0$ which (directly) receives a chunk from the source and $\nu(u, 1) = 0$ otherwise. The evaluation for $2 \le \tau \le m-1$ is conducted as follows. Since there are $m-j-1$ copies of block $B_j$ for any $1 \le j \le m-2$, for each $1 \le i \le m-3$ and $i < j \le m-2$, a copy of block $B_i$ can have at most $m-j-1$ copies of block $B_j$ as the successor in the resulting broadcast trees. In addition, for any pair of copies of such $B_i$ and $B_j$, we can fix an injection from the set of nodes in the copy of $B_i$ to the set of nodes in the copy of $B_j$ so that all nodes in the copy of $B_j$ successfully receive the chunk, say $c^*$, in the corresponding step, since when $B_i$ sends $c^*$ to $B_j$, all blocks in the $j$-thread containing $B_j$ simultaneously send $c^*$ and the number of nodes in the $j$-thread equals to the number of nodes in $B_j$. In the $\tau^{th}$ step of the broadcast of $c^*$, merely nodes in $B_{\tau-1}$ receive $c^*$. Hence, for each $2 \le \tau \le m-1$, $\nu(u, \tau) \le m - \tau$ for any $u \in V$, i.e., $\sum_{\tau=1}^{m-1} \nu(u, \tau) \le 1 + \sum_{i=1}^{m-2} i$.

Finally, the value of $\nu(u, m)$ is evaluated as follows. Recall that in the $m^{th}$ step, nodes in the $(m-1)$-thread, which is organized in the $(m-1)$st step and disappear in the $m^{th}$ step, must send the received chunk, say $c^*$, to the remaining $|V|/2$ nodes. Consider the following assignment of $|V|/2$ nodes in the $(m-1)$-thread to the remaining $|V|/2$ nodes:

- Nodes in (copies of) blocks $B_0, B_0, B_1, B_2, \ldots, B_{m-3}$ in the $(m-1)$-thread, i.e., nodes which have received $c^*$ earlier than the $(m-1)$st step, are responsible to send $c^*$ to the nodes in the $(m-2)$-thread which is organized in the $(m-1)$st step for the broadcast of the next chunk. Since there is a one-to-one correspondence between such pair of node sets, we have $\nu(u, m) = 1$ for such $u$.
- Nodes in block $B_{m-2}$ in the $(m-1)$-thread are responsible to send $c^*$ to the remaining $|V|/4$ nodes, where the role of those nodes is fixed in advance such that a node in block $B_{m-2}$ is responsible to send $c^*$ to a node in block $B_j$ for some fixed $j$. Since there are at most $m$ copies of $B_j$ for any $j$, we have $\nu(u, m) \le m$ for such $u$.

Hence, we have $\sum_{t=1}^{m} \nu(u, t) \le \sum_{i=1}^{m-2} i + 2$ for node $u$ in a copy of block $B_j$ for some $0 \le j \le m-3$ and $\sum_{t=1}^{m} \nu(u, t) \le m$ for node $u$ in the copy of block $B_{m-2}$. Since $\sum_{i=1}^{m-2} i + 2 \ge m$ for any $m \ge 3$, the theorem follows.

# The $\mathcal{G}$-Packing with $t$-Overlap Problem

Jazmín Romero and Alejandro López-Ortiz

David R. Cheriton School of Computer Science, University of Waterloo, Canada

**Abstract.** We introduce the $k$-$\mathcal{G}$-Packing with $t$-Overlap problem to formalize the problem of finding communities in a network. In the $k$-$\mathcal{G}$-Packing with $t$-Overlap problem, we search for at least $k$ communities with possible overlap. In contrast with previous work where communities are disjoint, we regulate the overlap through a parameter $t$. Our focus is the parameterized complexity of the $k$-$\mathcal{G}$-Packing with $t$-Overlap problem. Here, we provide a new technique for this problem generalizing the crown decomposition technique [2]. Using our global rule, we achieve a kernel with size bounded by $2(rk - r)$ for the $k$-$\mathcal{G}$-Packing with $t$-Overlap problem when $t = r - 2$ and $\mathcal{G}$ is a clique of size $r$.

## 1 Introduction

Many complex systems that exist in real applications can be represented by networks, where each node is an entity, and each edge represents a relationship between two nodes [10]. A community is a part of the network in which the nodes are more highly interconnected to each other than to the rest [12]. To extract these communities is known as the *community discovering problem* [1]. There are approaches for this problem that determine separate communities [1,8]. However, most real networks are characterized by well-defined communities that share members with others [4,12]. Moreover, these approaches model a community as a fixed graph $G$, when in real applications there are different models for communities. To overcome these deficiencies, we introduce the $k$-$\mathcal{G}$-*Packing with t-Overlap problem* as a more realistic formalization of the community discovering problem. To the best of our knowledge, the $k$-$\mathcal{G}$-Packing with $t$-Overlap problem has not been studied before.

In the $k$-$\mathcal{G}$-Packing with $t$-Overlap problem, the goal is to find at least $k$ subgraphs (the communities) each isomorphic to a member of a family $\mathcal{G}$ of graphs (the community models) in a graph $H$ (the network) where each pair of subgraphs can overlap in at most $t$ vertices (the shared members).

A problem related to the $k$-$\mathcal{G}$-Packing with $t$-Overlap problem when $\mathcal{G}$ is the family of cliques is the *cluster editing problem*. This problem consists of modifying a graph by adding or deleting edges such that the modified graph is composed of a vertex-disjoint union of cliques. Some works have considering overlap in the cluster editing problem [3,5]. Fellows et al. [5] proposed the conditions of $s$-vertex overlap and $s$-edge overlap where each vertex and each edge, respectively, is contained in at most $s$ maximal cliques. This implies that instead of a member of a network belonging to only one community, it can belong to $s$

S.P. Pal and K. Sadakane (Eds.): WALCOM 2014, LNCS 8344, pp. 114–124, 2014.

communities. However, in practical applications, there is no restriction on the number of communities to which a member can belong, but instead it is expected that communities overlap and that the overlap should be low [7]. Thus, it is more realistic to impose an overlap condition directly on the communities as in the $k$-$\mathcal{G}$-Packing with $t$-Overlap problem. This problem is also a generalization of the well-studied $\mathcal{G}$-packing problem which consists of finding disjoint subgraphs in a graph.

The $k$-$\mathcal{G}$-Packing with $t$-Overlap problem is NP-complete. This follows since every instance of the $\mathcal{G}$-packing problem, which is NP-complete [9], is mapped to an instance of the $k$-$\mathcal{G}$-Packing with $t$-Overlap problem by making $t = 0$. Besides introducing the $k$-$\mathcal{G}$-Packing with $t$-Overlap problem, we study its parameterized complexity. Our goal is to obtain *problem kernels*; that is, reduced instances with size bounded by $f(k)$. In this paper, we introduce a global reduction rule, the *clique-crown decomposition rule*, based on a non-trivial generalization of the crown decomposition technique [2]. To the best of our knowledge, the crown decomposition technique has not been adapted to obtain kernels for problems that find subgraphs with arbitrary overlap. Using our clique-crown decomposition rule, we achieve a problem kernel of size $2(rk - r)$ for the $k$-$\mathcal{G}$-Packing with $t$-Overlap problem, when $\mathcal{G} = K_r$ and $t = r - 2$.

This paper is organized as follows. Notation and terminology is provided in Section 2. In Section 3, we introduce our global reduction rule, as well as a method to compute it. Section 4 shows how to apply our clique-crown reduction rule to obtain a problem kernel. Finally, conclusions are stated in Section 5.

## 2   Preliminaries

All graphs in this document are undirected, simple, and connected. For a graph $G$, $V(G)$ and $E(G)$ denote its sets of vertices and edges, respectively. $|V(G)|$ is the size of the graph. For a set of vertices $A \subseteq V(G)$, the neighborhood of $A$ is defined as $N(A) = \{v \notin A \mid (u, v) \in E(G) \text{ and } u \in A\}$. The subgraph induced by $A$ in $G$ is denoted as $G[A]$. For a set of cliques $C$, $|C|$ is the number of cliques in $C$ while $V(C) = \bigcup_{i=1}^{|C|} V(C_i)$ where $C_i \in C$.

A clique of $r$ vertices, i.e., $K_r$, is also called an $r$-clique. For any pair of vertex-disjoint cliques $A$ and $B$, $A$ *completes* $B$ (or viceversa) if $G[V(A) \cup V(B)]$, denoted $A \cdot B$, is a clique of $r$ vertices. That is, $A$ *completes* $B$ into the clique $A \cdot B$. If $G[V(A) \cup V(B)]$ does not have $r$ vertices then $A$ does not complete $B$ even if $G[V(A) \cup V(B)]$ is a clique.

Given a set of cliques $Q$, a clique $B$ is *minimal-completed* in $Q$ if there is at least one clique $A$ in $Q$ that completes $B$ and none subgraph (called a *subclique*) $A' \subset A$ is completed by a clique in $Q$. Also, $A$ is *minimal(t)-completed* in $Q$ if the size of any subclique $A' \subset A$ completed by a clique in $Q$ is at most $t$. The definitions are also applied when instead of having a set of cliques $Q$, we have a graph $R$. Figure 1 provides an example of these definitions.

Our clique-crown decomposition is a generalization of the crown decomposition technique. This technique was introduced by Chor et al. [2], and it has been adapted to obtain kernels for packing problems [6,11,14].

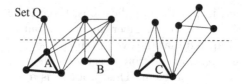

**Fig. 1.** For $r=4$ and $t = 1$, the cliques $B$ and $C$ are minimal-completed and minimal($t$)-completed, respectively. Clique $A$ is neither of both.

**Definition 1.** A *crown decomposition* $(H, C, R)$ in a graph $G$ is a partitioning of $V(G)$ into three sets $H$, $C$, and $R$ that have the following properties:

1. $C = C_m \cup C_u$ (the crown) is an independent set in $G$.
2. $H$ (the head) is a separator in $G$ such that there are no edges in $G$ between vertices belonging to $C$ and vertices belonging to $R$.
3. $R$ is the rest of the graph, i.e., $R = V(G) \backslash (C \cup H)$.
4. There is a perfect matching between $C_m$ and $H$.

Generally, vertices in $C_m$ and in $H$ are part of a desired solution while vertices in $C_u$ can be removed from $G$.

A crown decomposition can be computed in polynomial time for a graph $G$ given certain conditions.

**Lemma 1.** *[2] If a graph $G = (V, E)$ has an independent set of vertices $I \subseteq V(G)$ such that $|I| \geq |N(I)|$, then $G$ has a crown decomposition, where $C \subseteq I$ and $H \subseteq N(I)$, that can be found in time $O(|V(G)| + |E(G)|)$, given $I$ .*

## 3    Reduction Rules for the $k$-$K_r$-Packing with $t$-Overlap Problem

A parameterized problem is reduced to a *problem kernel*, if any instance can be reduced to a smaller instance such that: the reduction is in polynomial time, the size of the new instance is depending only on an input parameter, and the smaller instance has a solution if and only if the original instance has one.

Our goal is to reduce the $k$-$K_r$-Packing with $t$-Overlap problem to a problem kernel. The formal definition of our studied problem is as follows.

**Definition 2.** $k$-$K_r$-PACKING WITH $t$-OVERLAP PROBLEM
*Instance*: A graph $G = (V, E)$, and non-negative integers $k$ and $t$.
*Parameter*: $k$
*Question*: Does $G$ contain a set of $r$-cliques $\mathcal{K} = \{S_1, S_2, ..., S_l\}$ for $l \geq k$, where $|V(S_i) \cap V(S_j)| \leq t$, for any pair $S_i, S_j$?

We next apply a natural reduction rule to the input graph $G$.

**Reduction Rule 1.** Delete any vertex $v$ and any edge $e$ that are not included in a $K_r$.

To further reduce the graph $G$, we design the clique-crown reduction rule which is based on our proposed clique-crown decomposition.

## 3.1   The Clique-Crown Reduction Rule

In the clique-crown decomposition, we have cliques in both the head $H$ and the crown $C$, and each clique in $H$ is completed by at least one clique in $C$.

**Definition 3.** A *clique-crown decomposition* $(H, C, R)$ is a partition of $G$ that have the following properties:

1. $C = C_m \cup C_u$ (the crown) is a set of cliques in $G$ where each clique has size at most $r - (t + 1)$. Cliques in $C$ are denoted with letters $\alpha, \beta, \ldots$.
2. $H$ (the head) is a set of cliques in $G$ where each clique has size at least $t + 1$ and at most $r - 1$. Cliques in $H$ are denoted with letters $\mathbb{A}, \mathbb{B}, \ldots$. The head satisfies the following conditions.
   i. Each $\mathbb{A} \in H$ is minimal-completed by at least one clique in $C$.
   ii. Each pair of cliques $\mathbb{A}$ and $\mathbb{B}$ in $H$ shares at most $t$ vertices, i.e., $|V(\mathbb{A}) \cap V(\mathbb{B})| \le t$.
   iii. Each $\mathbb{A} \in H$ is minimal($t$)-completed by a clique in $R$, defined below.
3. $R$ is the rest of the graph, i.e., $R = G[V(G)\backslash(V(C) \cup V(H))]$.
4. The set of vertices of the cliques in $H$, $V(H)$, is a separator such that there are no edges in $G$ from $C$ to $R$.
5. There exists an injective function $f$ mapping each clique $\mathbb{A} \in H$ to a distinct clique $\alpha \in C_m$ such that $\alpha$ completes $\mathbb{A}$. In this way, $\mathbb{A} \cdot \alpha$ is an $r$-clique that we call *a mapped $r$-clique*. We impose the condition that any pair of mapped $r$-cliques shares at most $t$ vertices.

Figure 2 shows an example of a clique-crown decomposition. Cliques that belong to $H$ are highlighted with thicker lines.

To design the clique-crown reduction rule, we use an annotated version of the $k$-$K_r$-Packing with $t$-Overlap problem. In this annotated version, any $r$-clique of the solution overlaps in at most $t$ vertices with any clique from a set of cliques $Q$ given as part of the input.

**Definition 4.** ANNOTATED $k$-$K_r$-PACKING WITH $t$-OVERLAP PROBLEM
*Instance*: A graph $G = (V, E)$, a collection $Q$ of cliques from $G$ where any clique in $Q$ has size at least $t + 1$ and at most $r - 1$ and a non-negative integer $k$.
*Parameter*: $k$
*Question*: Does $G$ contain a set of $r$-cliques $\mathcal{K} = \{S_1, S_2, ..., S_l\}$ for $l \ge k$, where $|V(S_i) \cap V(S_j)| \le t$, for any pair $S_i, S_j$ and $|V(S) \cap V(C)| \le t$ for any $S \in \mathcal{K}$ and $C \in Q$?

**Reduction Rule 2.** *The Clique-Crown Reduction.* If $G$ admits a clique-crown decomposition $(H, C, R)$ then reduce $G$ as $G' = G[V(G)\backslash V(C)]$, and $k = k - |H|$. Make $H$ be the set $Q$ of the annotated $k$-$K_r$-Packing with $t$-Overlap problem.

The goal of the clique-crown reduction is to make the mapped $r$-cliques part of the solution and remove unnecessary vertices from $G$. As part of the correctness of Rule 2, we prove first that the vertices in $V(C_u)\backslash V(C_m)$ are not included in any $r$-clique of the solution.

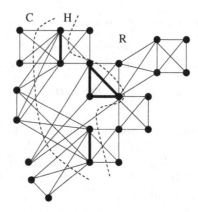

**Fig. 2.** Example of a clique-crown decomposition

**Lemma 2.** *The instance $(G, k)$ has a $k$-$K_r$-Packing with $t$-Overlap if and only if the instance $(G\backslash(V(C_u)\backslash V(C_m)), k)$ has a $k$-$K_r$-Packing with $t$-Overlap.*

*Proof.* Cliques in $C_u$ only complete cliques from the set $H$; otherwise $V(H)$ would not be a separator. However, every clique in $H$ is mapped to a unique completing clique in $C_m$ by the injective function. On the other hand, cliques in $H$ are minimal-completed in $C$ which implies that they cannot be partitioned in more cliques than $|H|$ that could be completed by cliques in $C_u$. Since cliques in $C$ are not necessarily vertex disjoint, then we can remove $V(C_u)\backslash V(C_m)$ from $G$.                                                                     □

We use the next observation for the proof of correctness of the following lemmas.

*Observation 1.* Any clique $\mathbb{A}' \subseteq \mathbb{A} \in H$, where $|V(\mathbb{A}')| \geq t + 1$, is an induced subgraph of at most one $r$-clique of any solution, since the $k$-$K_r$-Packing with $t$-Overlap problem allows overlap at most $t$.

**Lemma 3.** *If $G$ admits a clique-crown decomposition $(H, C, R)$, then the set of mapped $r$-cliques is an $|H|$-$K_r$-Packing with $t$-Overlap in $G$.*

*Proof.* Follows from Definition 3.

The input graph $G'$ for the annotated $k$-$K_r$-packing with $t$-Overlap problem is obtained by removing $C$ from $G$. Since a clique $\mathbb{A} \in H$ is already a subgraph of an $r$-clique from the solution, $\mathbb{A}$ cannot be included in another $r$-clique from the solution (Observation 1). Hence, we make the set $H$ to be the set $Q$ in the annotated $k$-$K_r$-packing with $t$-Overlap problem.

**Lemma 4.** *The instance $(G, k)$ has a $k$-$K_r$-Packing with $t$-Overlap if and only if the instance $(G', H, k - |H|)$ has an annotated $(k - |H|)$-$K_r$-packing with $t$-Overlap.*

*Proof.* Assume by contradiction that $G$ admits a clique-crown decomposition $(H, C, R)$ and has a $k$-$K_r$-Packing with $t$-Overlap, but $(G', H, k - |H|)$ does not have an annotated $(k-|H|)$-$K_r$-Packing with $t$-Overlap. By Observation 1, every $\mathbb{A} \in H$ is in at most one $r$-clique of the solution. Therefore, we cannot form more than $|H|$ $r$-cliques by completing each clique of $H$ with cliques in $R$ rather than with cliques in $C$. We could have more than $|H|$ $r$-cliques, if there is a clique $\mathbb{A} \in H$ that has at least two subcliques $\mathbb{A}', \mathbb{A}''$ each of size at least $t+1$ that are completed by some clique in $R$. A contradiction since $\mathbb{A}$ is minimal($t$)-completed in $R$.

Assume now that $(G', H, k - |H|)$ has an annotated $(k - |H|)$-$K_r$-Packing with $t$-Overlap, but $(G, k)$ does not have a $k$-$K_r$-Packing with $t$-Overlap. This would imply that the sets $H$ and $C$ form more than $|H|$ $r$-cliques which is a contradiction by Lemma 3. $\qquad\square$

### 3.2   Computing the Clique-Crown Decomposition

We next present a method to find a clique-crown decomposition in $G$ given two sets of cliques $O$ and $\mathsf{Cliques}(O)$. Every clique $\mathbb{A} \in \mathsf{Cliques}(O)$ is minimal-completed by at least one clique in $O$. Also, $\mathbb{A}$ is minimal($t$)-completed in the induced subgraph $G[V(G)\backslash(V(O) \cup V(\mathsf{Cliques}(O)))]$. Any pair of cliques in $\mathsf{Cliques}(O)$ shares at most $t$ vertices. Each clique in $O$ has size at most $r - (t+1)$. Since any clique in $\mathbb{A} \in \mathsf{Cliques}(O)$ is minimal-completed by a clique in $O$, then the size of $\mathbb{A}$ is at least $t + 1$ and at most $r - 1$. This method is a generalization of the method used to compute a crown-decomposition for the edge disjoint $K_3$-packing problem [14].

**Lemma 5.** *Any graph $G$ with a set $O$ of vertex-disjoint cliques where each clique has size at most $r - (t+1)$ and $|O| \geq |\mathsf{Cliques}(O)|$, has a clique-crown decomposition $(H, C, R)$ where $H \subseteq \mathsf{Cliques}(O)$, that can be found in $O(|V(G)|+|E(G)|)$ time given $O$ and $\mathsf{Cliques}(O)$.*

*Proof.* First, we construct a graph $G'$ from $G$ as follows. We initialize $V(G') = V(G)$ and $E(G') = E(G)$. We contract in $G'$ each clique $\alpha \in O$ into a single vertex $v_\alpha$, and we denote the set of contracted cliques as $O_{cont}$. After that for each clique $\mathbb{A} \in \mathsf{Cliques}(O)$, we add a vertex $v_\mathbb{A}$ to $V(G')$, i.e., *a representative vertex*; we denote as $Rep$ the set of all representative vertices. We say that $v_\alpha$ "completes" $v_\mathbb{A}$ if the clique $\alpha$ completes the clique $\mathbb{A}$. For every vertex $v_\alpha \in O_{cont}$ that completes $v_\mathbb{A}$, add $(v_\alpha, v_\mathbb{A})$ to $E(G')$. After that, add to $E(G')$ an edge from $v_\mathbb{A}$ to each vertex of $\mathbb{A}$. Finally, remove from $E(G')$ the edges from $v_\alpha$ to $\mathbb{A}$.

We next show that $G'$ has a crown decomposition $(H', C', R')$. In $G'$, the set of contracted cliques $O_{cont}$ is an independent set. By the construction of $G'$, we know that $N(O_{cont})$ is the set of representative vertices $Rep$. Since we introduced a representative vertex per clique in $\mathsf{Cliques}(O)$, then $|N(O_{cont})| = |\mathsf{Cliques}(O)|$. Thus, since $|O| \geq |\mathsf{Cliques}(O)|$ then $|O_{cont}| \geq |N(O_{cont})|$ in $G'$. By Lemma 1, $G'$ admits a crown decomposition $(H', C', R')$ that is computed in linear time, where $C' \subseteq O_{cont}$ and $H' \subseteq N(O_{cont}) = Rep$.

Now, we use the crown decomposition $(H', C', R')$ of $G'$ to construct the clique-crown decomposition $(H, C, R)$ of $G$ where $H \subseteq \mathsf{Cliques}(O)$.

1. For each vertex $v_\alpha \in C' \subseteq O_{cont}$, add the clique $\alpha$ to $C$. The size of each clique in $C$ is at most $r - (t+1)$. This follows because $\alpha \in O$ and each clique in $O$ has size at most $r - (t+1)$.
2. For each vertex $v_{\mathbb{A}} \in H'$, where $H' \subseteq Rep$, we assign to $H$ the clique that this vertex represents, i.e., $\mathbb{A}$. Since $H \subseteq \mathsf{Cliques}(O)$, then each clique in $H$ has size at least $t + 1$ and at most $r - 1$. Likewise, properties i-iii from Definition 3 follow.
3. $R = G[V(G) \backslash (C \cup H)]$
4. The set of vertices of the cliques in $H$, $V(H)$, is a separator. This follows since cliques in $C$ complete only cliques on $H$; thus, vertices in $V(C)$ are only adjacent to vertices in $V(H)$.
5. We make the perfect matching between $C'$ and $H'$ correspond to the injective function $f$ in the following way. For any matched edge $(v_\alpha, v_{\mathbb{A}})$ complete $\mathbb{A}$ with $\alpha$. For any pair of mapped $r$-cliques $\mathbb{A} \cdot \alpha$ and $\mathbb{B} \cdot \beta$ completed in this way, $|V(\mathbb{A} \cdot \alpha) \cap V(\mathbb{B} \cdot \beta)| \leq t$. This follows because $\alpha$ and $\beta$ are vertex-disjoint, and $|V(\mathbb{A}) \cap V(\mathbb{B})| \leq t$ by assumption in the set $\mathsf{Cliques}(O)$.

Thus, if $|O| \geq \mathsf{Cliques}(O)$ then $G$ admits a clique-crown decomposition. □

One method to obtain the sets of $O$ and $\mathsf{Cliques}(O)$ is to compute a maximal $K_r$-packing with $t$-Overlap $\mathcal{M}$ from $G$. $O$ will be the set of all cliques in $G[V(G) \backslash V(\mathcal{M})]$ such that each clique in $O$ completes at least one clique in $G[\mathcal{M}]$. $\mathsf{Cliques}(O)$ is therefore the set of cliques in $G[\mathcal{M}]$ completed by cliques in $O$. The overlap between an $r$-clique $\mathbb{A} \cdot \alpha$, where $\mathbb{A} \in \mathsf{Cliques}(O)$ and $\alpha \in O$, with some clique in $\mathcal{M}$ is at least $t + 1$. Therefore, the size of $\mathbb{A}$ is at least $t + 1$ and the size of $\alpha$ is at most $r - (t + 1)$. It has to be verified if the sets $O$ and $\mathsf{Cliques}(O)$ follow the properties of the crown and the head, respectively, from Definition 3.

## 4    A Kernel for the $k$-$K_r$-Packing with $(r - 2)$-Overlap Problem

Using the clique-crown reduction rule, we introduce an algorithm to obtain the kernel for the $k$-$K_r$-packing with $(r - 2)$-Overlap problem. First, we compute a maximal solution for the $k$-$K_r$-packing with $(r - 2)$-Overlap problem. Next, we show that the sets $O$ and $\mathsf{Cliques}(O)$ are composed of vertices that are outside and inside, respectively, of the maximal solution. Finding a maximal solution and then reducing the instance by a set of rules, in this case the clique-crown reduction rule, is known as the algorithmic version of the method of extremal structure [13].

$k$-$K_r$-PACKING WITH $(r-2)$-OVERLAP ALGORITHM
*Input*: A graph $G = (V, E)$ and a non-negative integer $k$.

1. Reduce $G$ by Reduction Rule 1.
2. Greedily, find a maximal $K_r$-Packing with $(r-2)$-Overlap $\mathcal{M}$ in $G$. If $|\mathcal{M}| \geq k$ then ACCEPT.
3. Let $O$ be $V(G)\backslash V(\mathcal{M})$ and Cliques($O$) be the set of cliques completed by vertices in $O$. If $|O| \geq |\text{Cliques}(O)|$, reduce $G$ by computing a clique-crown decomposition in $G$ using $O$ (Rule 2).

We next introduce a series of lemmas that characterize the sets $O$ and Cliques($O$).

**Claim 1.** $O = V(G)\backslash V(\mathcal{M})$ is an independent set.

*Proof.* Assume by contradiction that there exists an edge $(u, v)$ in $G[O]$. After applying Reduction Rule 1, each edge in the reduced graph is included in at least one $r$-clique; thus, $(u, v)$ belongs to at least one $r$-clique $S'$. $S'$ is not in $\mathcal{M}$; otherwise $u, v$ would not be in $O$. $S'$ is not in $O$; otherwise as $S'$ would be disjoint from $\mathcal{M}$, and hence, $S'$ could be added to $\mathcal{M}$, contradicting the maximality of $\mathcal{M}$. Thus, $S'$ should overlap with at least one $r$-clique $S \in \mathcal{M}$, for $S \neq S'$.

Since $u, v$ are both in $O$, the overlap with $S$ is at most $r - 2$, i.e., $|V(S) \cap V(S')| = r - 2$, but in this case, $S'$ could be added to $\mathcal{M}$ because the $k$-$K_r$-packing with $(r-2)$-Overlap problem allows overlap at most $r-2$, contradicting the maximality of $\mathcal{M}$.                                      □

**Claim 2.** Each clique $\mathbb{A} \in \text{Cliques}(O)$ is minimal-completed by at least one clique in $O$, and $\mathbb{A}$ is minimal($t$)-completed in $G\backslash(V(O) \cup V(\text{Cliques}(O)))$. Also, any pair of cliques in Cliques($O$) shares at most $t$ vertices.

*Proof.* Each clique in Cliques($O$) is completed by at least one clique in $O$. Since $O$ is an independent set, then the size of every clique in Cliques($O$) is $r-1$. Each clique $\mathbb{A} \in \text{Cliques}(O)$ is minimal-completed in $O$. Assume by contradiction that there is a subclique $\mathbb{A}' \subset \mathbb{A}$ of size $s < r - 1$ completed by a clique in $V(O)$. This would be possible if there would be a $K_{r-s}$ in $O$, but we already prove that $O$ is an independent set. Thus, there is no subclique of $\mathbb{A}$ completed by a clique in $C$. Any pair of cliques $\mathbb{A}, \mathbb{B} \in \text{Cliques}(O)$ share at most $t = r - 2$ vertices; this fact follows since $\mathbb{A} \neq \mathbb{B}$ and $|\mathbb{A}| = |\mathbb{B}| = r - 1$. Finally, the size of the biggest subclique of any $\mathbb{A} \in \text{Cliques}(O)$ is at most $t - 1$, and therefore, $\mathbb{A}$ is minimal($t$)-completed.                                      □

**Claim 3.** Each $K_{r-1}$ $T$ completed by any $u \in O$ is contained in an $r$-clique $S \in \mathcal{M}$.

*Proof.* $V(T) \cap V(O) = \emptyset$. Assume otherwise that there is a vertex $v \in V(T)$ contained in $O$. However, since $T \cdot u$ forms an $r$-clique this would imply that there is an edge $(u, v)$ in $O$ which it is a contradiction since $O$ is an independent set. Thus, $V(T) \subset V(\mathcal{M})$.

We claim that $|V(T) \cap V(S)| = r - 1$ for some $S \in \mathcal{M}$, i.e., $V(T) \subset V(S)$. Suppose otherwise that $|V(T) \cap V(S)| < r - 1$, for any $S \in \mathcal{M}$. Since $u \in O$, this would imply that $|V(T \cdot u) \cap V(S)| \leq r - 2$ for every $S \in \mathcal{M}$, and $T \cdot u$ could be added to $\mathcal{M}$ as the $k$-$K_r$-packing with $(r - 2)$-Overlap problem allows overlap at most $r - 2$, contradicting the assumption of maximality of $\mathcal{M}$.    □

**Claim 4.** The set of vertices $O$ completes at most $rk - r$ $K_{r-1}$'s.

*Proof.* By Claim 3, vertices in $O$ only complete $K_{r-1}$'s contained in $K_r$'s in $\mathcal{M}$. There are $r$ $K_{r-1}$'s in a $K_r$ and at most $k - 1$ $K_r$'s in $\mathcal{M}$; thus, there are at most $rk - r$ $K_{r-1}$'s that can be completed by vertices in $O$.    □

**Claim 5.** $|O| < rk - r$

*Proof.* In Step 3 of the $k$-$K_r$-packing with $(r - 2)$-Overlap algorithm, if $|O| \geq |\mathsf{Cliques}(O)|$, $O$ is reduced by the clique-crown reduction rule (Rule 2). Since $|\mathsf{Cliques}(O)| < rk - r$ then $|O| < rk - r$, after applying that rule.    □

**Lemma 6.** *If $|V(G)| > 2(rk - r)$ then the above algorithm will either find a $k$-$k$-$K_r$-Packing with $t$-Overlap, or it will reduce $G$.*

*Proof.* Assume by contradiction that $|V(G)| > 2(rk - r)$, but the algorithm neither finds a $k$-$K_r$-packing with $(r - 2)$-Overlap nor reduces the graph $G$. Any vertex $v \in V(G)$ that was not reduced by Rule 1 is in $V(\mathcal{M})$, or it is in $O = V(G) \backslash V(\mathcal{M})$; thus, $|V(G)| = |V(\mathcal{M})| + |O|$.

The size of $\mathcal{M}$ is at most $k - 1$; thus, $|V(\mathcal{M})|$ is at most $rk - r$, and by Claim 5 we know that an upper bound for $|O|$ is $rk - r$.

In this way, the size of the instance is at most $2(rk - r)$ which contradicts the assumption that $|V(G)| > 2(rk - r)$.    □

**Claim 6.** The $k$-$K_r$-packing with $(r - 2)$-Overlap problem admits a $2(rk - r)$ kernel which can be found in $O(n^r)$ time.

*Proof.* By Lemma 6, the reduced instance has size at most $2(rk - r)$. Rule 1 is computed in time $O(n^r)$, which is also the same time to compute the maximal solution $\mathcal{M}$ and $\mathsf{Cliques}(O)$. Lemma 5 shows that the clique-crown decomposition is computed in linear time given the set of cliques $O$ and $\mathsf{Cliques}(O)$. The set $O$ corresponds to the independent set $V(G) \backslash V(\mathcal{M})$ (Lemma 1), and the set $\mathsf{Cliques}(O)$ is the set of $K_{r-1}$'s completed by vertices in $O$. By Claim 3, all these $K_{r-1}$'s are contained in the $K_r$'s of $\mathcal{M}$. Moreover, in Rule 1, we already compute all $K_{r-1}$'s that a vertex completes. Thus, the time to obtain $\mathsf{Cliques}(O)$ is $O(n^r)$.    □

## 5   Conclusion

In this work, we have introduced the $k$-$\mathcal{G}$-Packing with $t$-Overlap problem to overcome the deficiencies of previous work on the community discovering problem. We have also introduced a generalized global reduction rule, the clique-crown decomposition, for the $k$-$\mathcal{G}$-Packing with $t$-Overlap problem, when

$\mathcal{G} = K_r$. Using our reduction rule, we achieved reductions to a kernel for this problem when $t = r - 2$. We emphasize that the clique-crown reduction rule can be extended to consider other families of graphs as well. For example, it would be interesting to consider community models less restrictive than cliques such as $s$-cliques, $s$-clubs, and $s$-plexes.

When computing the clique-crown decomposition in Section 5, it is assumed that cliques in $O$ are vertex disjoint and cliques in $\mathsf{Cliques}(O)$ overlap in at most $t + 1$. If these two conditions do not follow then a perfect matching would not guarantee that a pair of mapped $r$-cliques overlap in at most $t$ vertices. Therefore, it remains how to design a different injective function that satisfies this overlap condition.

**Acknowledgments.** We would like to thank to Naomi Nishimura for her invaluable comments and contribution to complete this paper.

# References

1. Balakrishman, H., Deo, N.: Detecting communities using bibliographic metrics. In: Proceedings of the IEEE International Conference on Granular Computing, pp. 293–298 (2006)
2. Chor, B., Fellows, M., Juedes, D.: Linear Kernels in Linear Time, or How to Save $k$ Colors in $O(n^2)$ Steps. In: Hromkovič, J., Nagl, M., Westfechtel, B. (eds.) WG 2004. LNCS, vol. 3353, pp. 257–269. Springer, Heidelberg (2004)
3. Damaschke, P.: Fixed-parameter tractable generalizations of cluster editing. In: Calamoneri, T., Finocchi, I., Italiano, G.F. (eds.) CIAC 2006. LNCS, vol. 3998, pp. 344–355. Springer, Heidelberg (2006)
4. Everett, M.G., Borgatti, S.P.: Analyzing clique overlap. Connections 21, 49–61 (1998)
5. Fellows, M.R., Guo, J., Komusiewicz, C., Niedermeier, R., Uhlmann, J.: Graph-based data clustering with overlaps. In: Ngo, H.Q. (ed.) COCOON 2009. LNCS, vol. 5609, pp. 516–526. Springer, Heidelberg (2009)
6. Fellows, M., Heggernes, P., Rosamond, F., Sloper, C., Telle, J.A.: Finding $k$ disjoint triangles in an arbitrary graph. In: Hromkovič, J., Nagl, M., Westfechtel, B. (eds.) WG 2004. LNCS, vol. 3353, pp. 235–244. Springer, Heidelberg (2004)
7. Freeman, L.: The sociological concept of "group": An empirical test of two models. American Journal of Sociology 98, 152–166 (1992)
8. Girvan, M., Newman, M.: Community Structure in Social and Biological Networks. Proceedings of the National Academy of Science of the United States of America, 7821–7826 (2002)
9. Kirkpatrick, D., Hell, P.: On the completeness of a generalized matching problem. In: Proceedings of the Tenth Annual ACM Symposium on Theory of Computing (STOC), pp. 240–245 (1978)
10. Luo, F., Wang, J., Promislow, E.: Exploring local community structures in large networks. In: Proceedings of the 2006 IEEE/WIC/ACM International Conference on Web Intelligence (WI), pp. 233–239 (2006)

11. Mathieson, L., Prieto, E., Shaw, P.: Packing Edge Disjoint Triangles: A Parameterized View. In: Downey, R.G., Fellows, M.R., Dehne, F. (eds.) IWPEC 2004. LNCS, vol. 3162, pp. 127–137. Springer, Heidelberg (2004)
12. Palla, G., Derenyi, I., Farkas, I., Vicsek, T.: Uncovering the overlapping community structure of complex networks in nature and society. Nature 435(7043), 814–818 (2005)
13. Prieto, E.: Systematic Kernelization in FPT Algorithm Design. Ph.D. thesis, The university of Newcastle (2005)
14. Prieto, E., Sloper, C.: Looking at the Stars. Theoretical Computer Science 351(3), 437–445 (2006)

# Minimax Regret Sink Location Problem in Dynamic Tree Networks with Uniform Capacity

Yuya Higashikawa[1], Mordecai J. Golin[2], and Naoki Katoh[1,*]

[1] Department of Architecture and Architectural Engineering,
Kyoto University, Japan
{as.higashikawa,naoki}@archi.kyoto-u.ac.jp
[2] Department of Computer Science and Engineering,
The Hong Kong University of Science and Technology, Hong Kong
golin@cs.ust.hk

**Abstract.** This paper addresses the minimax regret sink location problem in dynamic tree networks. In our model, a dynamic tree network consists of an undirected tree with positive edge lengths and uniform edge capacity, and the vertex supply which is nonnegative value is unknown but only the interval of supply is known. A particular realization of supply to each vertex is called a scenario. Under any scenario, the cost of a sink location $x$ is defined as the minimum time to complete the evacuation to $x$ for all supplies (evacuees), and the regret of $x$ is defined as the cost of $x$ minus the cost of the optimal sink location. Then, the problem is to find a sink location minimizing the maximum regret for all possible scenarios. We present an $O(n^2 \log^2 n)$ time algorithm for the minimax regret sink location problem in dynamic tree networks with uniform capacity, where $n$ is the number of vertices in the network. As a preliminary step for this result, we also address the minimum cost sink location problem in a dynamic tree networks under a fixed scenario and present an $O(n \log n)$ time algorithm, which improves upon the existing time bound of $O(n \log^2 n)$ by [11] if edges of a tree have uniform capacity.

**Keywords:** minimax regret, sink location, dynamic flow, evacuation planning.

## 1 Introduction

The Tohoku-Pacific Ocean Earthquake happened in Japan on March 11, 2011, and many people failed to evacuate and lost their lives due to severe attack by tsunamis. From the viewpoint of disaster prevention from city planning and evacuation planning, it has now become extremely important to establish effective evacuation planning systems against large scale disasters in Japan. In particular, arrangements of tsunami evacuation buildings in large Japanese cities near the coast has become an urgent issue. To determine appropriate tsunami evacuation buildings, we need to consider where evacuation buildings are assigned and how

---

* Supported by JSPS Grant-in-Aid for Scientific Research(A)(25240004).

S.P. Pal and K. Sadakane (Eds.): WALCOM 2014, LNCS 8344, pp. 125–137, 2014.

to partition a large area into small regions so that one evacuation building is designated in each region. This produces several theoretical issues to be considered. Among them, this paper focuses on the location problem of the evacuation building assuming that we fix the region such that all evacuees in the region are planned to evacuate to this building. An evaluation criterion of the building location is the time required to complete the evacuation. In order to represent the evacuation, we consider the *dynamic* setting in graph networks, which was first introduced by Ford et al. [8]. Under the dynamic setting, each edge of a given graph has capacity which limits the value of the flow into the edge at each time step. We call such networks under the dynamic setting *dynamic networks*.

This paper addresses *the minimax regret sink location problem in dynamic tree networks*. In our model, a dynamic tree network consists of an undirected tree with positive edge lengths and uniform edge capacity, and the vertex supply which is nonnegative value is unknown but only the interval of supply is known. Generally, the number of evacuees in an area (the initial supply at a vertex) may vary depending on the time (e.g., in an office area in a big city there are many people during the daytime on weekdays while there are much less people on weekends or during the night time). So, in order to take into account the uncertainty of the vertex supplies, we adopt the *maximum regret* for a particular sink location as another evaluation criterion assuming that we only know the interval of supply for each vertex. A particular realization (assignment of supply to each vertex) is called a *scenario*. Under any scenario, the *cost* of a sink location $x$ is defined as the minimum time to complete the evacuation to $x$ for all supplies, and the *regret* of $x$ is defined as the cost of $x$ minus the cost of the optimal sink location. Then, the problem can be understood as a 2-person Stackelberg game as follows. The first player picks a sink location $x$ and the second player chooses a scenario $s$ that maximizes the regret of $x$ under $s$. The objective of the first player is to choose $x$ that minimizes the maximum regret.

Several researchers have studied the minimax regret facility location problems [6, 10, 12, 13]. Especially, for tree networks, some efficient algorithms have been presented by [1–4, 7]. For dynamic networks, Cheng et al. [5] have studied the minimax regret sink location problem in dynamic path networks with uniform capacity and presented an $O(n \log^2 n)$ time algorithm. Recently, Wang [14] improved the time bound by [5] to $O(n \log n)$. In this paper, we extend the problem studied by [5, 14] from path networks to tree networks, that is, address the minimax regret sink location problem in dynamic tree networks with uniform capacity and present an $O(n^2 \log^2 n)$ time algorithm.

In order to develop the above mentioned algorithm, we consider the case where supply at each vertex is fixed to a given value. The problem is to find a sink location in a given tree which minimizes the time to complete the evacuation to the sink for all supplies under a fixed scenario, which is called *the minimum cost sink location problem in dynamic tree networks*. An algorithm for this problem can be used as a subroutine of the algorithm to solve the minimax regret sink location problem in dynamic tree networks. Mamada et al. [11] have studied the minimum cost sink location problem in dynamic tree networks with general capacity and presented an $O(n \log^2 n)$ time algorithm. In this paper, we present an $O(n \log n)$ time algorithm for the minimum cost sink location problem in dynamic tree networks with uniform capacity. Note that the paper by [11] assumed that a sink is located on a vertex while our paper assumes that a sink can be located at any point in the network.

This paper is the first one which studies the minimax regret sink location problem in dynamic tree networks and presents an efficient algorithm by developing new properties. We also study the minimum cost sink location problem in dynamic tree networks and present a whole new algorithm which improves upon the existing time bound by [11] if edges of a given tree have uniform capacity and gives the more clear proof of the time complexity.

# 2 Minimax Cost Sink Location Problem in Dynamic Tree Networks with Uniform Capacity

Let $T = (V, E)$ be an undirected tree with a vertex set $V$ and an edge set $E$. Let $\mathcal{N} = (T, l, w, c, \tau)$ be a dynamic network with the underlying graph being a tree $T$, where $l$ is a function that associates each edge $e \in E$ with positive length $l(e)$, $w$ is also a function that associates each vertex $v \in V$ with supply $w(v)$ (which takes a nonnegative integer) representing the number of evacuees at $v$, $c$ is a positive integer constant representing the capacity of each edge: the least upper bound for the number of the evacuees entering an edge per unit time, and $\tau$ is also a constant representing the time required for traversing the unit distance of each evacuee. We call such networks with tree structures *dynamic tree networks*.

## 2.1 Formula for the Minimum Completion Time of the Evacuation

In the following, for two integers $i$ and $j$, let $[i, j] = \{k \in \mathbb{Z} \mid i \leq k \leq j\}$. We first show a formula representing the minimum completion time for the evacuation in a dynamic tree network with uniform capacity. In the following, we also use a notation $T$ to denote a set of all points on edges in $E$ including all vertices in $V$. For two points $p, q \in T$, let $d(p, q)$ denote the distance between $p$ and $q$ in $T$. For a vertex $v \in V$, let $\delta(v)$ denote a set of vertices adjacent to $v$. For a sink location $x$ given at a point in $T$, let $\Theta(x)$ denote the minimum time required for all evacuees on $T$ to complete the evacuation to $x$. In this paper, we assume that for any vertex $v \in V$, any number of evacuees can reach $v$ in unit time step, any number of evacuees can stay at $v$, and if the sink is located at $v$, all evacuees on $v$ can finish their evacuation in no time. In the following discussion, we suppose that a sink location $x$ is given at a vertex. Then, let $T(x)$ be a rooted tree made from $T$ such that each edge has a natural orientation towards the root $x$. For any vertex $v \in V$, let $T(x, v)$ be a subtree of $T(x)$ rooted at $v$, and $\Theta(x, v)$ denote the minimum time required for all evacuees on $T(x, v)$ to complete the evacuation to $x$. Then, we clearly have

$$\Theta(x) = \max\{\Theta(x, u) \mid u \in \delta(x)\}. \tag{1}$$

Here, we need only to consider $\Theta(x, \hat{u})$ for $\hat{u} = \text{argmax}\{\Theta(x, u) \mid u \in \delta(x)\}$. Suppose that there are $n'$ vertices in $T(x, \hat{u})$ named $v_1(= \hat{u}), v_2, \ldots, v_{n'}$ such that $d(x, v_i) \leq d(x, v_{i+1})$ for $i \in [1, n'-1]$. Then, Kamiyama et al. [9] have observed that the value of $\Theta(x, \hat{u})$ does not change if $x$ and $v_i$ for $i \in [1, n']$ are relocated on a line with the same capacity so that $d(x, v_i)$ $i \in [1, n']$ remain the same (see Fig. 1), and $\Theta(x, \hat{u})$ can be represented as follows:

$$\Theta(x, \hat{u}) = \max_{j \in [1, n']} \left\{ d(x, v_j)\tau + \left\lceil \frac{\sum_{i \in [j, n']} w(v_i)}{c} \right\rceil - 1 \right\}. \tag{2}$$

For the completeness, we now see why this formulation holds. We first define a *group* as a set of evacuees who simultaneously reach $x$ from $\hat{u}$ and the *size* of a group as the number of evacuees in the group. Suppose that a group whose size is less than $c$ reaches $x$ at time $t'$. Then, we call a group which first reaches $x$ after $t'$ a *leading group* (see Fig. 2). We also call a group which first reaches $x$ after time 0 a leading group. Let $t_{\text{last}}$ denote the time when the last group reaches $x$ (i.e., the whole evacuation finishes at $t_{\text{last}}$). Suppose that a leading group reaches $x$ at time $t''$ and there is no leading group which reaches $x$ after $t''$ until $t_{\text{last}}$. Then, we call a leading group reaching $x$ at $t''$ the *last leading group* and a set of groups reaching $x$ from $t''$ to $t_{\text{last}}$ the *last cluster*. In order to derive $\Theta(x, \hat{u})$, we need only to observe the last cluster. Suppose that the last leading group is located at $v_l$ for some integer $l \in [1, n']$ at time 0. We notice that any leading group reaches $x$ without being blocked. Thus, the last leading group reaches $x$ at time $d(x, v_l)\tau$, and then, all groups except ones which belong to the last cluster have already reached $x$. If $d(x, v_l)\tau < t_{\text{last}}$, the size of a group reaching $x$ at each time $t \in [d(x, v_l)\tau, t_{\text{last}} - 1]$ is exactly $c$ because of definition of the last leading group. Therefore, $\Theta(x, \hat{u})$ can be represented as follows:

$$\Theta(x, \hat{u}) = d(x, v_l)\tau + \left\lceil \frac{\sum_{i \in [l, n']} w(v_i)}{c} \right\rceil - 1. \tag{3}$$

Note that this still holds for the case of $d(x, v_l)\tau = t_{\text{last}}$. We next see that the right hand of the formula (2) is the lower bound for $\Theta(x, \hat{u})$. For all evacuees located at $v_j, \ldots, v_{n'}$ with some integer $j \in [1, n']$, the time of $d(x, v_j)\tau + \lceil \sum_{i \in [j, n']} w(v_i)/c \rceil - 1$ is at least required to complete the evacuation to $x$, thus we have $\Theta(x, \hat{u}) \geq d(x, v_j)\tau + \lceil \sum_{i \in [j, n']} w(v_i)/c \rceil - 1$ for any integer $j \in [1, n']$. From the above discussion, we can derive the formula (2).

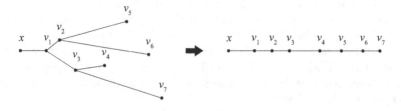

**Fig. 1.** Vertices of the tree can be relocated on a line with the same capacity

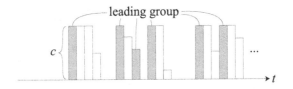

**Fig. 2.** The size of groups reaching $x$ from $\hat{u}$ for each time

## 2.2 Properties

In this section, we prove the two lemmas which are key to our presented algorithm. Let $x_{\mathrm{opt}}$ denote a point in $T$ which minimizes $\Theta(x)$. For two vertices $v, v' \in V$, let $V(v, v')$ denote the set of all vertices in $T(v, v')$ and $T(V')$ denote a subgraph induced by a vertex set $V' \subseteq V$.

**Lemma 1.** *Along a path from a leaf to another leaf in $T$, a function $\Theta(x)$ is unimodal in $x$.*

**Lemma 2.** *For a vertex $v \in V$, if $\hat{u} = \mathrm{argmax}\{\Theta(v, u) \mid u \in \delta(v)\}$ holds, there exists $x_{\mathrm{opt}} \in T(V(v, \hat{u}) \cup \{v\})$.*

In the proofs of two lemmas, we use the following notations. Let $P$ be a simple path in $T$ from a leaf to another leaf. Suppose that there are $k + 1$ vertices $v_0, v_1, \ldots, v_k$ in $P$ such that $v_0$ and $v_k$ are leaves and $d(v_0, v_i) \leq d(v_0, v_j)$ if $0 \leq i \leq j \leq k$. In the following, for a sink location $x \in P$, we abuse the notation $x$ to denote $d(v_0, x)$, and for a vertex $v_i \in P$ with $i \in [0, k]$, $v_i$ to denote $d(v_0, v_i)$. For a point $p \in P$, we call the direction to $v_0$ (resp. $v_k$) from $p$ the *left direction* (resp. *right direction*). For a sink location $x \in P$, let $\Theta_L(x; P)$ (resp. $\Theta_R(x; P)$) denote the minimum time required to complete the evacuation for all evacuees on $T$ who come to $x$ from the left direction (resp. right direction). For a vertex $v_i \in P$ with $i \in [1, k-1]$, let

$$\Theta_L^{+0}(v_i; P) = \lim_{\epsilon \to +0} \{\Theta_L(v_i + \epsilon; P)\}, \tag{4}$$

$$\Theta_R^{-0}(v_i; P) = \lim_{\epsilon \to +0} \{\Theta_R(v_i - \epsilon; P)\}. \tag{5}$$

We first show the following claim.

**Claim 1.** *Along a path $P$, a function $\Theta_L(x; P)$ is increasing in $x$ and a function $\Theta_R(x; P)$ is decreasing in $x$.*

*Proof.* By (2), (4) and (5), we can see the following three properties of $\Theta_L(x; P)$ and $\Theta_R(x; P)$ (see Fig. 3(a)): (i) for an open interval $(v_{i-1}, v_i)$ with $i \in [1, k]$, $\Theta_L(x; P)$ (resp. $\Theta_R(x; P)$) is linear in $x$ with slope $\tau$ (resp. $-\tau$), (ii) $\Theta_L(x; P)$ (resp. $\Theta_R(x; P)$) is left-continuous (resp. right-continuous) at $x = v_i$ for $i \in [1, k]$ (resp. $i \in [0, k-1]$), (iii) $\Theta_L(v_i; P) \leq \Theta_L^{+0}(v_i; P)$ and $\Theta_R^{-0}(v_i; P) \geq \Theta_R(v_i; P)$ hold at $v_i$ for $i \in [1, k-1]$. From these properties, $\Theta_L(x; P)$ (resp. $\Theta_R(x; P)$) is piecewise linear increasing (resp. decreasing) in $x$. $\qquad\square$

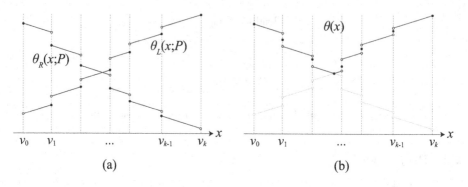

**Fig. 3.** Functions along $P$: **(a)** $\Theta_L(x; P)$, $\Theta_R(x; P)$ and **(b)** $\Theta(x)$

By Claim 1, there uniquely exists $x \in P$ minimizing $\max\{\Theta_L(x; P), \Theta_R(x; P)\}$, which we call $x_{\mathrm{opt}}(P)$ in the following. Then, we have the following claim.

**Claim 2.** (i) *For a vertex $v_i \in P$ such that $v_i \geq x_{\mathrm{opt}}(P)$, $\Theta_L(v_i; P) \leq \Theta(v_i) \leq \Theta_L^{+0}(v_i; P)$. (ii) For a vertex $v_i \in P$ such that $v_i \leq x_{\mathrm{opt}}(P)$, $\Theta_R^{-0}(v_i; P) \geq \Theta(v_i) \geq \Theta_R(v_i; P)$.*

*Proof.* Here, we prove only (i) ((ii) can be similarly proved). Let us look at a vertex $v_i \in P$ such that $v_i \geq x_{\mathrm{opt}}(P)$ (see Fig. 3(b)). By definition of $\Theta(v_i)$, we have $\Theta(v_i) \geq \Theta_L(v_i; P)$. Thus, in order to prove (i), we need only to show that

$$\Theta(v_i) \leq \Theta_L^{+0}(v_i; P). \tag{6}$$

By the condition of $v_i \geq x_{\mathrm{opt}}(P)$, $\Theta_L^{+0}(v_i; P) \geq \Theta_R(v_i; P)$ holds. Therefore, if $\Theta(v_i) = \Theta_R(v_i; P)$, (6) holds. If $\Theta(v_i) = \Theta_L(v_i; P)$, (6) also holds by (2). Otherwise, for a sink location $x = v_i$, an evacuee who lastly reaches $v_i$ arrives at $v_i$ through some adjacent vertex $u \in \delta(v_i)$ which is not on $P$. Suppose that we move the sink location from $x = v_i$ towards a point along $P$ with distance $\epsilon$ in the right direction (i.e., $x = v_i + \epsilon$) where $\epsilon$ is a sufficiently small positive number. Then, the last evacuee first reaches $v_i$ at time $\Theta(v_i)$, may be blocked there, and eventually reaches $x = v_i + \epsilon$, thus, he/she can reach $x = v_i + \epsilon$ after time $\Theta(v_i) + \epsilon\tau$, that is, $\Theta(v_i) + \epsilon\tau \leq \Theta_L(v_i + \epsilon; P)$ holds. By definition of (4), we obtain (6). □

**Proof of Lemma 1.** By Claims 1 and 2, we obtain that $\Theta(x)$ may possibly be discontinuous at $v_i$ for $i \in [1, k-1]$ but it is always unimodal in $x$ along $P$. □

**Proof of Lemma 2.** Let us consider a path $P$ from a leaf to another leaf through adjacent vertices $v$ and $\hat{u}$ where $\hat{u} = \mathrm{argmax}\{\Theta(v, u) \mid u \in \delta(v)\}$. Let us define the left direction in $P$ as the direction from $v$ to $\hat{u}$ and the right direction as the other direction. Suppose that there are $k+1$ vertices $v_0, v_1, \ldots, v_k$ in $P$, and $v = v_i$ and $\hat{u} = v_{i-1}$ with $i \in [1, k-1]$. Let $\Theta^{+0}(v_i; P) = \max\{\Theta_L^{+0}(v_i; P), \Theta_R(v_i; P)\}$. If we can show $\Theta(v_i) \leq \Theta^{+0}(v_i; P)$, there never exists $x_{\mathrm{opt}}$ in the right direction from $v_i$ along $P$ by Lemma 1. Then, this lemma can be proved by repeatedly applying the same discussion to all the other paths through $v$ and $\hat{u}$. By the assumption of $\hat{u} = \operatorname{argmax}\{\Theta(v, u) \mid u \in \delta(v)\}$, $\Theta_L(v_i; P) \geq \Theta_R(v_i; P)$ holds, and by (2), $\Theta_L(v_i; P) \leq \Theta_L^{+0}(v_i; P)$ holds. Thus, we have $\Theta_R(v_i; P) \leq \Theta_L^{+0}(v_i; P)$, which implies that

$$\Theta^{+0}(v_i; P) = \Theta_L^{+0}(v_i; P). \tag{7}$$

Then, by Lemma 1, we also have $v_i \geq x_{\mathrm{opt}}(P)$ where $x_{\mathrm{opt}}(P)$ is the unique point in $P$ minimizing $\max\{\Theta_L(x; P), \Theta_R(x; P)\}$. Thus, by Claim 2(i), $\Theta(v_i) \leq \Theta_L^{+0}(v_i; P)$ holds, and from this and (7), we derive $\Theta(v_i) \leq \Theta^{+0}(v_i; P)$.  □

## 2.3   Algorithm

In this section, we present an $O(n \log n)$ time algorithm for the minimum cost sink location problem in dynamic tree networks with uniform capacity, which we call BST (Binary Search in Tree).

First, we introduce the concept of *median*.

**Definition 1.** *For a set of vertices $U \subseteq V$ inducing a connected subgraph in $T$, a median of $U$ is a vertex $m \in U$ such that*

$$\max\{|V(m, u) \cap U| \mid u \in \delta(m)\} \leq \frac{|U|}{2}. \tag{8}$$

Note that for any set of vertices inducing a connected subgraph, there always exists a median.

Let us look at the first iteration by algorithm BST. Letting $U_1 = V$, the algorithm first finds a median $m_1$ of $U_1$ and computes $d(m_1, v)$ for every $v \in U_1$. Then, for each $u \in \delta(m_1)$, the algorithm computes $\Theta(m_1, u)$ in the following manner: the algorithm first creates a list $L(u)$ of all vertices $v \in U_1 \cap V(m_1, u)$ which are arranged in the nondecreasing order of $d(m_1, v)$, and after that, it computes $\Theta(m_1, u)$ by (2). The algorithm computes $u_1 = \operatorname{argmax}\{\Theta(m_1, u) \mid u \in \delta(m_1)\}$. It also sets $V_1 = U_1 \setminus (V(m_1, u_1) \cup \{m_1\})$ and merges lists $L(u)$ for $u \in \delta(m_1) \setminus \{u_1\}$ into a new list $L_1$. At the end of the first iteration, the algorithm sets $U_2 = U_1 \cap (V(m_1, u_1) \cup \{m_1\})$. Note that by Lemma 2, there exists $x_{\mathrm{opt}}$ in $T(U_2)$ and by Definition 1, $|U_2| \leq |U_1|/2 + 1$ holds. The algorithm iteratively performs the same procedure (see Fig. 4(a)(b)). Namely, at the $i$-th iteration, it finds a median $m_i$ of $U_i$, computes $u_i = \operatorname{argmax}\{\Theta(m_i, u) \mid u \in \delta(m_i)\}$, sets $V_i = U_i \setminus (V(m_i, u_i) \cup \{m_i\})$, creates a list $L_i$ of vertices $v \in V_i$ arranged in the nondecreasing order of $d(m_i, v)$ and also sets $U_{i+1} = U_i \cap (V(m_i, u_i) \cup \{m_i\})$. Since the algorithm reduces the size of the subgraph where $x_{\mathrm{opt}}$ exists roughly by half at each iteration, it halts after $l = O(\log |V|)$ iterations. At this point, it finds two vertices $m_l$ and $u_l \in U_l$ connected by an edge on which $x_{\mathrm{opt}}$ lies. Then, $x_{\mathrm{opt}}$ can be computed as follows. Let $x(t)$ denote a point dividing the edge $(m_l, u_l)$ with the ratio of $t$ to $1 - t$ for some $t$ $(0 \leq t \leq 1)$, and $\Theta(x(t), m_l)$ (resp. $\Theta(x(t), u_l)$) denote the minimum time required for all evacuees passing

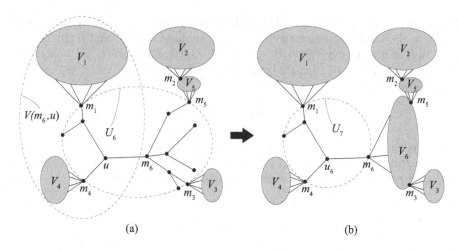

**Fig. 4.** Illustration of the $i$-th iteration: **(a)** $i = 6$ and **(b)** $i = 7$

through $m_l$ (resp. $u_l$) to complete the evacuation to $x(t)$. Then, $\Theta(x(t), m_l)$ and $\Theta(x(t), u_l)$ can be represented as follows:

$$\Theta(x(t), m_l) = \Theta(u_l, m_l) - (1 - t)d(m_l, u_l)\tau, \tag{9}$$
$$\Theta(x(t), u_l) = \Theta(m_l, u_l) - td(m_l, u_l)\tau. \tag{10}$$

If there exists $t$ such that $\Theta(x(t), m_l) = \Theta(x(t), u_l)$ and $0 \le t \le 1$, $x_{\mathrm{opt}} = x(t)$ holds by the unimodality of $\Theta(x)$. If the solution of $\Theta(x(t), m_l) = \Theta(x(t), u_l)$ satisfies $t < 0$, then $\Theta(m_l, u_l) < \Theta(u_l, m_l) - d(m_l, u_l)\tau$ holds, which implies $x_{\mathrm{opt}} = m_l$. Similarly, if $t > 1$, $x_{\mathrm{opt}} = u_l$ holds. Therefore, the algorithm can correctly output the optimal sink location $x_{\mathrm{opt}}$.

Now, let us analyze the time complexity of algorithm BST. We first show that the running time is $O(n \log^2 n)$ which will be improved to $O(n \log n)$ later, where $n = |V|$. In the following, suppose that $n = |V|$. Let us look at the running time for each iteration required by the algorithm. At the $i$-th iteration for $i \ge 2$, the algorithm finds a median $m_i$ of $U_i$ in $O(|U_i|)$ time, and computes $d(m_i, v)$ for all of $v \in V$ by depth-first search in $O(|V|)$ time. For each $u \in \delta(m_i)$, in order to compute $\Theta(m_i, u)$, the algorithm can create a list $L'(u)$ of vertices $v \in V(m_i, u)$ arranged in the nondecreasing order of $d(m_i, v)$ in $O(|V(m_i, u)| \log |V(m_i, u)|)$ by applying a simple merge sort. Thus, $u_i = \mathrm{argmax}\{\Theta(m_i, u) \mid u \in \delta(m_i)\}$ can be computed in $O(|V| \log |V| + |V|)$ time. Since the algorithm halts after $O(\log |V|)$ iterations as mentioned above, our problem can be solved in $O(|U_i| + |V| + |V| \log |V| + |V|) \times O(\log |V|) = O(n \log^2 n)$ time.

Next, we show that the running time required to create lists $L'(u)$ for $u \in \delta(m_i)$ can be improved from $O(n \log n)$ to $O(n + |U_i| \log |U_i|)$. We first show the following claim.

**Claim 3.** $|U_i| = O(\frac{n}{2^{i-1}})$ and $|V_i| = O(\frac{n}{2^{i-1}})$ hold for $i \ge 1$.

*Proof.* By definition of $U_i$, we can clearly see that $|U_i| = O(n/2^{i-1})$ holds. Remind that $V_i = U_i \setminus (V(m_i, u_i) \cup \{m_i\})$ and $|U_i \cap V(m_i, u_i)| = O(|U_i|/2)$, thus we have $|V_i| = O(n/2^{i-1})$.                                                                      □

The idea to improve the running time is to use the sorted lists $L_j$ with $j = 1, 2, \ldots, i - 1$. To see the idea to improve the running time, let us look at Fig. 4(a), and focus on a vertex $u \in \delta(m_6)$ in the figure. The computation of $L'(u)$ can be done in $O(n \log n)$ time if we know $d(m_6, v)$ for all $v \in V(m_6, u)$. But, since we know the $L_1$ and $L_4$, $L'(u)$ can be computed faster only by computing $d(m_6, v)$ for all $v \in U_6 \cap V(m_6, u)$. The idea is formalized as follows. In detail, for each $u \in \delta(m_i)$, it first creates a list $L(u)$ of vertices $v \in U_i \cap V(m_i, u)$ arranged in the nondecreasing order of $d(m_i, v)$, which takes $O(n' \log n')$ time where $n' = |U_i \cap V(m_i, u)|$. Thus, lists $L(u)$ for all $u \in \delta(m_i)$ can be created in $O(|U_i| \log |U_i|)$ time. For each $u \in \delta(m_i)$, the algorithm merges $L(u)$ and all lists $L_j$ with $V_j \subseteq V(m_i, u)$ into $L'(u)$ (at this point, all of the original lists are maintained since these will be used later). For this merging operation, if we apply a simple merge sort, it takes $O(|V(m_i, u)| \log |V(m_i, u)|)$ time, which does not improve the running time. Here, we notice that the $|L_j| = |V_j|$ for $j \in [1, i - 1]$. Instead, the algorithm basically takes the following two steps to create each list $L'(u)$ for $u \in \delta(m_i)$:

**[Step 1.]** For $L_j$ such that $V_j \subseteq V(m_i, u)$, choose $L_p = \operatorname{argmin}\{|L_j| \mid V_j \subseteq V(m_i, u)\}$ and merge each $L_j$ in the increasing order of size (i.e., the decreasing order of $j$) with $L_p$ one by one.
**[Step 2.]** Merge the list obtained at Step 1 and $L(u)$ into $L'(u)$.
For all $u \in \delta(m_i)$, it takes in $O(\sum_{j=1}^{i-1} jn/2^{j-1}) = O(n)$ time at Step 1, and thus, it takes $O(n + |U_i|) = O(n)$ time at Step 2. Recall that $L(u)$ for all $u \in \delta(m_i)$ can be created in $O(|U_i| \log |U_i|)$ time. Then, by Claim 3, it takes $O(n + |U_i| \log |U_i|) = O(n + (n/2^{i-1}) \log(n/2^{i-1}))$ time to create lists $L'(u)$ for all $u \in \delta(m_i)$.

**Lemma 3.** *The $i$-th iteration of algorithm* BST *takes* $O(n + \frac{n}{2^{i-1}} \log \frac{n}{2^{i-1}})$ *time.*

Recall that the algorithm halts after $O(\log n)$ iterations. Thus, by Lemma 3, it takes $O(n \log n + \sum\{(n/2^{i-1}) \log(n/2^{i-1}) \mid i \in [1, \log n]\}) = O(n \log n)$ time for the entire iterations. Therefore, we obtain the following theorem.

**Theorem 1.** *The minimum cost sink location problem in a dynamic tree network with uniform capacity can be solved in* $O(n \log n)$ *time.*


## 3   Minimax Regret Sink Location Problem in Dynamic Tree Networks with Uniform Capacity

Let $\mathcal{N} = (T, l, W, c, \tau)$ be a dynamic tree network with the underlying graph being a tree $T = (V, E)$, where $l$, $c$ and $\tau$ are the functions which are the same as those defined in Section 2, and $W$ is also a function that associates each vertex $v \in V$ with an interval of supply $W(v) = [w^-(v), w^+(v)]$, Let $\mathcal{S}$ denote the Cartesian product of all $W(v)$ for $v \in V$ (i.e., a set of scenarios):

$$\mathcal{S} = \prod_{v \in V} [w^-(v), w^+(v)]. \tag{11}$$

When a scenario $s \in \mathcal{S}$ is given, we use the notation $w^s(v)$ to denote the supply of a vertex $v \in V$ under the scenario $s$.

For a sink location $x$ given at a point in $T$ and a given scenario $s \in \mathcal{S}$, let $\Theta^s(x)$ denote the minimum time required for all evacuees on $T$ to complete the evacuation to $x$ under $s$. Suppose that a sink location $x$ is given at a vertex. For $u \in \delta(x)$, let $\Theta^s(x, u)$ denote the minimum time required for all evacuees on $T(x, u)$ to complete the evacuation to $x$. Then, we have

$$\Theta^s(x) = \max\{\Theta^s(x, u) \mid u \in \delta(x)\}. \tag{12}$$

For $\hat{u} = \operatorname{argmax}\{\Theta^s(x, u) \mid u \in \delta(x)\}$, we also have by (2)

$$\Theta^s(x, \hat{u}) = \max_{j \in [1, n']} \left\{ d(x, v_j)\tau + \left\lceil \frac{\sum_{i \in [j, n']} w^s(v_i)}{c} \right\rceil - 1 \right\}, \tag{13}$$

where $n'$ is the number of vertices in $T(x, \hat{u})$ and $v_1(= \hat{u}), v_2, \ldots, v_{n'}$ are vertices in $T(x, \hat{u})$ such that $d(x, v_i) \leq d(x, v_{i+1})$ for $1 \leq i \leq n' - 1$. For the ease of exposition, we assume that $c = 1$ (the case of $c > 1$ can be treated in essentially the same manner) and also omit the constant part (i.e., $-1$) from the formula in the following discussion, then we have

$$\Theta^s(x, \hat{u}) = \max_{j \in [1, n']} \left\{ d(x, v_j)\tau + \sum_{i \in [j, n']} w^s(v_i) \right\}, \tag{14}$$

Here, let $x_{\text{opt}}^s$ denote a point in $T$ which minimizes $\Theta^s(x)$ under a scenario $s \in (S)$. In the following, we use the notation $\Theta_{\text{opt}}^s$ for a scenario $s \in \mathcal{S}$ to denote $\Theta^s(x_{\text{opt}}^s)$. We now define the *regret* for $x$ under $s$ as

$$R^s(x) = \Theta^s(x) - \Theta_{\text{opt}}^s. \tag{15}$$

Moreover, we also define the *maximum regret* for $x$ as

$$R_{\max}(x) = \max\{R^s(x) \mid s \in \mathcal{S}\}. \tag{16}$$

If $\hat{s} = \operatorname{argmax}\{R^s(x) \mid s \in \mathcal{S}\}$, we call $\hat{s}$ the *worst case scenario* for a sink location $x$. The goal is to find a point $x^* \in T$, called the *minimax regret sink location*, which minimizes $R_{\max}(x)$ over $x \in T$, i.e., the objective is to

$$\text{minimize } \{R_{\max}(x) \mid x \in T\}. \tag{17}$$

## 3.1  Properties

First, we define a set of so-called *dominant scenarios* for a vertex $v \in V$ among which the worst case scenario exists when the sink is located at $v$. Suppose that $v_1$ is a vertex adjacent to $v$, $n'$ is the number of vertices in $T(v, v_1)$ and $v_2, \ldots, v_{n'}$ are vertices in $T(v, v_1)$ such that $d(v, v_i) \le d(v, v_{i+1})$ for $1 \le i \le n' - 1$. We now consider a scenario $s \in S$ such that $w^s(v_i) = w^+(v_i)$ for $v_i \in T(v, v_1)$ such that $l \le i \le n'$ with some $l \in [1, n']$ and $w^s(v') = w^-(v')$ for all the other vertices $v' \in V$. In the following, such a scenario is said to be *dominant* for $v$, and represented by $s(v, v_l)$. Then, let $S_d(v)$ denote the set of all dominant scenarios for $v$. Note that $S_d(v)$ consists of $n - 1$ scenarios. The following is a key lemma.

**Lemma 4.** *If a sink is located at a vertex $v \in V$, there exists a worst case scenario for $v$ which belongs to $S_d(v)$.*

Basically, Lemma 4 can be obtained from a lemma in [5], so we omit its proof. Here, we have the following claim by Lemma 1.

**Claim 4.** *For a scenario $s \in S$, a function $\Theta^s(x)$ is unimodal in $x$ when $x$ moves along a path from a leaf to another leaf in $T$.*

For a given scenario $s \in S$, by definition of (15) and Claim 4, a function $R^s(x)$ is unimodal in $x$ along a path from a leaf to another leaf in $T$. Thus, a function $R_{max}(x)$ is also unimodal in $x$ since it is the upper envelope of unimodal functions by (16).

**Lemma 5.** *Along a path from a leaf to another leaf in $T$, a function $R_{max}(x)$ is unimodal in $x$.*

We also have the following claim by Lemma 2.

**Claim 5.** *For a scenario $s \in S$ and a vertex $v \in V$, if $\hat{u} = \operatorname{argmax}\{\Theta^s(v, u) \mid u \in \delta(v)\}$ holds, there exists $x_{opt}^s \in T(V(v, \hat{u}) \cup \{v\})$.*

Here, suppose that $\hat{s} = \operatorname{argmax}\{R^s(v) \mid s \in S\}$ and $\hat{u} = \operatorname{argmax}\{\Theta^{\hat{s}}(v, u) \mid u \in \delta(v)\}$ hold for a vertex $v \in V$. We now show that there also exists the minimax regret sink location $x^*$ in $T(V(v, \hat{u}) \cup \{v\})$. Suppose otherwise: there exists $x^*$ in $T(v, u)$ or on an edge $(v, u)$ (not including endpoints) for some $u \in \delta(v)$ with $u \ne \hat{u}$. By Claim 5, there exists $x_{opt}^{\hat{s}}$ in $T(V(v, \hat{u}) \cup \{v\})$. Then, $\Theta^{\hat{s}}(x^*) > \Theta^{\hat{s}}(v)$ holds by Claim 4, thus $R^{\hat{s}}(x^*) > R^{\hat{s}}(v)$ also holds by (15). We have $R_{max}(x^*) \ge R^{\hat{s}}(x^*)$ by the maximality of $R_{max}(x^*)$ and $R^{\hat{s}}(v) = R_{max}(v)$ by definition of $\hat{s}$, thus $R_{max}(x^*) > R_{max}(v)$ holds, which contradicts the optimality of $x^*$. By the above discussion, we obtain the following lemma.

**Lemma 6.** *For a vertex $v \in V$, if $\hat{s} = \operatorname{argmax}\{R^s(v) \mid s \in S\}$ and $\hat{u} = \operatorname{argmax}\{\Theta^{\hat{s}}(v, u) \mid u \in \delta(v)\}$ hold, there exists the minimax regret sink location $x^* \in T(V(v, \hat{u}) \cup \{v\})$.*

## 3.2 Algorithm

In this section, we present an $O(n^2 \log^2 n)$ time algorithm that computes $x^* \in T$ which minimizes a function $R_{\max}(x)$.

We first show how to compute $R_{\max}(v)$ for a given vertex $v \in V$. Given a dominant scenario $s \in \mathcal{S}_d(v)$, $\Theta^s(v)$ can be computed in $O(n \log n)$ time, and by Theorem 1, $\Theta^s_{opt}$ can be computed in $O(n \log n)$ time. Thus by (15), $R^s(v)$ can be computed in $O(n \log n)$ time. By Lemma 4, we need only to consider $n-1$ dominant scenarios for $v$, thus, $R_{\max}(v)$ can be computed by (16) in $O(n^2 \log n)$ time.

**Lemma 7.** *For a vertex $v \in V$, $R_{\max}(v)$ can be computed in $O(n^2 \log n)$ time.*

In order to find the minimax regret sink location $x^* \in T$, we apply an algorithm similar to the one presented at Section 2.3. The algorithm maintains a vertex set $U_i \subseteq V$ which induces a connected subgraph of $T$ including $x^*$. At the beginning of the procedure, the algorithm sets $U_1 = V$, and at $i$-th iteration, it finds a median $m_i$ of $U_i$, computes $R_{\max}(m_i)$ in the above mentioned manner, and sets $U_{i+1} = U_i \cap (V(m_i, u_i) \cup \{v\})$ where $u_i = \text{argmax}\{\Theta^{\hat{s}}(m_i, u) \mid u \in \delta(m_i)\}$ and $\hat{s} = \text{argmax}\{R^s(m_i) \mid s \in \mathcal{S}_d(m_i)\}$. Note that, by Lemma 6, $T(U_{i+1})$ contains $x^*$ if $T(U_i)$ includes $x^*$. The algorithm iteratively performs the same procedure until $|U_l|$ becomes two where $l = O(\log n)$. Suppose that there eventually remain two vertices $m_l$ and $u_l \in U_l$. Then, the algorithm has already known that

$$R_{\max}(m_l) = \Theta^{\hat{s}_1}(m_l) - \Theta^{\hat{s}_1}_{opt}, \tag{18}$$

$$R_{\max}(u_l) = \Theta^{\hat{s}_2}(u_l) - \Theta^{\hat{s}_2}_{opt}, \tag{19}$$

$$m_l = \text{argmax}\{\Theta^{\hat{s}_2}(u_l, u) \mid u \in \delta(u_l)\}, \tag{20}$$

$$u_l = \text{argmax}\{\Theta^{\hat{s}_1}(m_l, u) \mid u \in \delta(m_l)\}, \tag{21}$$

where $\hat{s}_1$ and $\hat{s}_2$ are worst case scenarios for $m_l$ and $u_l$, respectively. Let $x(t)$ denote a point dividing the edge $(m_l, u_l)$ with the ratio of $t$ to $1-t$ for some $t$ ($0 \le t \le 1$). If there exists $t$ such that $R_{\max}(m_l) - t d(m_l, u_l)\tau = R_{\max}(u_l) - (1-t)d(m_l, u_l)\tau$ and $0 \le t \le 1$, $x^* = x(t)$ holds by the unimodality of $R_{\max}(x)$. If the solution satisfies $t < 0$, then $R_{\max}(m_l) < R_{\max}(u_l) - d(m_l, u_l)\tau$ holds, which implies $x^* = m_l$. Similarly, if $t > 1$, $x^* = u_l$ holds. As above, the algorithm correctly outputs the minimax regret sink location $x^*$ after $O(\log n)$ iterations. Thus, by Lemma 7, we obtain the following theorem.

**Theorem 2.** *The minimax regret sink location problem in a dynamic tree network with uniform capacity can be solved in $O(n^2 \log^2 n)$ time.*

## 4    Conclusion

In this paper, we develop an $O(n^2 \log^2 n)$ time algorithm for the minimax regret sink location problem in dynamic tree networks with uniform capacity. We also develop an $O(n \log n)$ time algorithm for the minimum cost sink location problem in dynamic tree networks with uniform capacity.

On the other hand, we leave as an open problem to extend the solvable networks for the minimax regret sink location problem to dynamic tree networks with general capacities. Indeed, under a fixed scenario, the algorithm by [11] can solve the minimum cost sink location problem in dynamic tree networks with general capacity, we cannot simply apply this as a subroutine to solve the minimax regret sink location problem. For example, if Lemmas 4 and 6 hold in a dynamic tree network with general capacities, we can expect that an $O(n^2 \log^3 n)$ time algorithm will be achieved.

# References

1. Averbakh, I., Berman, O.: Algorithms for the robust 1-center problem on a tree. European Journal of Operational Research 123(2), 292–302 (2000)
2. Bhattacharya, B., Kameda, T.: A linear time algorithm for computing minmax regret 1-median on a tree. In: Gudmundsson, J., Mestre, J., Viglas, T. (eds.) COCOON 2012. LNCS, vol. 7434, pp. 1–12. Springer, Heidelberg (2012)
3. Brodal, G.S., Georgiadis, L., Katriel, I.: An $O(n \log n)$ version of the Averbakh-Berman algorithm for the robust median of a tree. Operations Research Letters 36(1), 14–18 (2008)
4. Chen, B., Lin, C.: Minmax-regret robust 1-median location on a tree. Networks 31(2), 93–103 (1998)
5. Cheng, S.-W., Higashikawa, Y., Katoh, N., Ni, G., Su, B., Xu, Y.: Minimax regret 1-sink location problems in dynamic path networks. In: Chan, T.-H.H., Lau, L.C., Trevisan, L. (eds.) TAMC 2013. LNCS, vol. 7876, pp. 121–132. Springer, Heidelberg (2013)
6. Conde, E.: Minimax regret location-allocation problem on a network under uncertainty. European Journal of Operational Research 179(3), 1025–1039 (2007)
7. Conde, E.: A note on the minmax regret centdian location on trees. Operations Research Letters 36(2), 271–275 (2008)
8. Ford Jr., L.R., Fulkerson, D.R.: Constructing maximal dynamic flows from static flows. Operations Research 6, 419–433 (1958)
9. Kamiyama, N., Katoh, N., Takizawa, A.: An efficient algorithm for evacuation problem in dynamic network flows with uniform arc capacity. IEICE Transactions 89-D(8), 2372–2379 (2006)
10. Kouvelis, P., Yu, G.: Robust discrete optimization and its applications. Kluwer Academic Publishers, Dordrecht (1997)
11. Mamada, S., Uno, T., Makino, K., Fujishige, S.: An $O(n \log^2 n)$ algorithm for the optimal sink location problem in dynamic tree networks. Discrete Applied Mathematics 154(16), 2387–2401 (2006)
12. Ogryczak, W.: Conditional median as a robust solution concept for uncapacitated location problems. TOP 18(1), 271–285 (2010)
13. Puerto, J., Rodríguez-Chía, A.M., Tamir, A.: Minimax regret single-facility ordered median location problems on networks. INFORMS Journal on Computing 21(1), 77–87 (2009)
14. Wang, H.: Minmax regret 1-facility location on uncertain path networks. In: Cai, L., Cheng, S.-W., Lam, T.-W. (eds.) ISAAC 2013. LNCS, vol. 8283, pp. 733–743. Springer, Heidelberg (2013)

# On a Class of Covering Problems with Variable Capacities in Wireless Networks

Selim Akl[1], Robert Benkoczi[2], Daya Ram Gaur[2], Hossam Hassanein[1], Shahadat Hossain[2], and Mark Thom[2]

[1] Queen's University
{akl,hossam}@cs.queensu.ca
[2] University of Lethbridge
{robert.benkoczi,shahadat.hossain,mark.thom2}@uleth.ca, gaur@cs.uleth.ca

**Abstract.** We consider the problem of allocating clients to base stations in wireless networks. Two design decisions are the location of the base stations, and the power levels of the base stations. We model the interference due to the increased power usage resulting in greater serving radius, as capacities that are non-increasing with respect to the covering radius. We consider three models. In the first model the location of the base stations and the clients are fixed, and the problem is to determine the serving radius for each base station so as to serve a set of clients with maximum total profit subject to the capacity constraints of the base stations. In the second model, each client has an associated demand in addition to its profit. A fixed number of facilities have to be opened from a candidate set of locations. The goal is to serve clients so as to maximize the profit subject to the capacity constraints. In the third model the location and the serving radius of the base stations are to be determined. There are costs associated with opening the base stations, and the goal is to open a set of base stations of minimum total cost so as to serve the entire client demand subject to the capacity constraints at the base stations. We show that for the first model the problem is NP-complete even when there are only two choices for the serving radius, and the capacities are $1, 2$. For the second model we give a $1/2 - \epsilon$ approximation algorithm. For the third model we give a column generation procedure for solving the standard linear programming model, and a randomized rounding procedure. We establish the efficacy of the column generation based rounding scheme on randomly generated instances.

## 1 Introduction

Given a set of client locations the covering facility location problem is to determine optimal locations for facilities to serve all the clients. Facilities can serve clients within a prescribed radius only. A client is *covered* or served by a facility if it is within the covering range of the facility, otherwise it is not covered. This "all or nothing" covering model is too restrictive for many practical applications and several relaxations of the model have been proposed and studied in the last decade [3]. One relaxation is the *gradual cover* model, where the degree with

S.P. Pal and K. Sadakane (Eds.): WALCOM 2014, LNCS 8344, pp. 138–150, 2014.
© Springer International Publishing Switzerland 2014

which a customer is served decreases as its distance to the facility increases. In the *cooperative cover* model several facilities contribute to serving a customer. In the *variable covering radius* model a covering range is assigned to facilities but the cost of opening the facility increases with its range.

We introduce a new family of covering problems, Covering with Variable Capacities (*CVC*), which addresses the client coverage problem in the presence of interference in wireless networks. Facilities correspond to wireless base stations employing omni-directional antennas and clients represent service subscribers. We assume that the location of the clients is given. Associated with the clients may be demands (bandwidth requirements) and profits (revenue). The facilities can be opened at any location with a covering range. The range corresponds to the power consumption by the base station. The capacity of a facility to serve the clients is a non-increasing function of the serving radius. A client is covered by a single facility if it is within the range assigned to the facility. The total demand of the clients served by the facility is no larger than the available capacity for the facility. In our model, every facility has a variable covering range. The range can only be increased at the expense of the capacity, as increasing the power of the radio transmitter causes more interference in the network [4,8,10]. Radwan and Hassanein [14] show that there are significant savings in resource utilization for wCDMA networks when the range of the base stations is appropriately chosen. Tam *et al.* [16] describe a cellular network that exploits this phenomenon.

In this paper, we extend the work of Tam *et al.* [16]. Arguably our models and solutions have immediate applications in the mobile telephony industry. In addition we extend the theory of facility location problems by studying a new class of location problems in which the facilities have variable capacities, and the designer not only has to choose a location for the facility but also the capacity at which the facility should operate.

We study three variants for *CVC* that are typically investigated for covering facility location problems: set-cover *CVC* where the entire set of customers must be covered by a set of facilities of minimum cost, maximum *CVC* where a set of customers with maximum total profit must be covered by a fixed number of facilities, and *CVC* with fixed facilities (or simply *CVC*) where the location of the facilities is given and the objective is to maximize the total profit of the covered customers. We define the problems formally below. We use index notation to refer to customers and facilities, and so $u_i$ for some index $i$ refers to a customer and $a_j$ for some index $j$ refers to a facility.

*Problem 1 (*CVC*)*

Input:
- A set $\mathcal{U} = \{u_i : i \in \mathcal{I}\}$ of customers where $\mathcal{I}$ is the index set of customers.
- For each customer $u_i$, a non-negative demand or size $s_i$ and profit $p_i$.
- A set $\mathcal{A} = \{a_j : j \in \mathcal{J}\}$ of open facilities, where $\mathcal{J}$ is the index set of facilities.
- For each facility $a_j$, a set $R_j$ of allowed ranges and for each $r \in R_j$ a corresponding capacity $c_{jr}$. We denote by $N_{jr}$ the set of customers within the covering range $r$ of facility $a_j$.

Output:
- For each facility $a_j$, a range $r_j \in R_j$.
- For each facility $a_j$, a subset of clients indexed by $\mathcal{I}_j \subseteq \mathcal{I}$ that are served exclusively by $a_j$ so that $\sum_{i \in \mathcal{I}_j} s_i \leq c_{jr_j}$ and $u_i \in N_{jr_j}$ for all $i \in \mathcal{I}_j$.

Objective:
- To maximize the total profit of clients served, $\max \sum\limits_{i \in \bigcup_{j \in \mathcal{J}} I_j} p_i$.

### Problem 2 (Maximum CVC)

Input:
- Same as for Problem 1 with the observation that the set of facilities represents *candidate* facility locations.
- A positive integer $k$.

Output:
- $k$ facilities are to be opened, indexed by $\mathcal{J}^* \subseteq \mathcal{J}$ where $|\mathcal{J}^*| = k$.
- For each open facility $a_j$ for $j \in \mathcal{J}^*$, a range $r_j \in R_j$.
- For each open facility $a_j$ for $j \in \mathcal{J}^*$, a subset of clients indexed by $\mathcal{I}_j \subseteq \mathcal{I}$ that are served exclusively by $a_j$ so that $\sum_{i \in \mathcal{I}_j} s_i \leq c_{jr_j}$ and $u_i \in N_{jr_j}$ for all $i \in \mathcal{I}_j$.

Objective:
- To maximize the total profit of clients served, $\max \sum\limits_{i \in \bigcup_{j \in \mathcal{J}^*} I_j} p_i$.

### Problem 3 (Set cover CVC)

Input:
- Same as for Problem 1 with the observation that the set of facilities represents *candidate* facility locations. The model can be augmented to handle arbitrary costs to open facilities.

Output:
- A subset of facilities to be opened, indexed by $\mathcal{J}^* \subseteq \mathcal{J}$.
- For each open facility $a_j$ for $j \in \mathcal{J}^*$, a range $r_j \in R_j$.
- For each open facility $a_j$ for $j \in \mathcal{J}^*$, a subset of clients indexed by $\mathcal{I}_j \subseteq \mathcal{I}$ that are served exclusively by $a_j$ so that $\sum_{i \in \mathcal{I}_j} s_i \leq c_{jr_j}$, $u_i \in N_{jr_j}$ for all $i \in \mathcal{I}_j$, and all clients are served $\bigcup_{j \in \mathcal{J}^*} \mathcal{I}_j = \mathcal{I}$.

Objective:
- To minimize the number of open facilities, $\min |\mathcal{J}^*|$.

The special case when customers have unit demand and profit, $s_i = p_i = 1$ for all $i \in \mathcal{I}$, is called the uniform version of the corresponding *CVC* problem.

*CVC* generalizes the classical capacitated covering. The latter corresponds to a *CVC* instance where the set $R_j$ of covering ranges for each facility has cardinality one. Using a set to model the dependency between capacity and covering range defines, in fact, a step function for capacity. This model is sufficiently general for applications such as wireless networks where measurements about the performance of the wireless channel are discrete. However, *CVC* can be studied with other capacity functions as well. Our results indicate that both the capacity function and the metric space where customers and facilities are located, are significant. For example, $CVC$ with fixed location of facilities is NP-hard even when the capacity function is a step function with two steps. In contrast, the corresponding classical covering with fixed facilities can be solved in polynomial time using flows.

Problem 1 is also related to the separable assignment problem (SAP). Given a set of bins and a set of items, a value $f_{ij}$ for placing item $j$ in bin $i$, and for each bin $i$, a family of sets of items $\mathcal{I}_i$ that fit in bin $i$, the SAP problem is to pack items into bins so as to maximize the total value of packed items. Note that only subsets of the sets in $\mathcal{I}_i$ can be packed into bin $i$. SAP was studied by Fleisher *et al.* [7] who gave two approximation algorithms; an LP-rounding based $1 - 1/e$ approximation algorithm, and a local search based $1/2$ approximation algorithm. Consider an instance of the $CVC$ problem. Construct an instance of SAP as follows: each facility corresponds to a bin, for a given range $r$ and location $i$, we know the capacity $c_r$ of the facility. Define $\mathcal{I}_{i,r}$ to be the set of all valid subsets of total size at most $c_r$ for candidate location $i$ of the facility, let $\mathcal{I}_i = \bigcup_{r \in R} \mathcal{I}_{i,r}$. This reduction implies the existence of a $1 - 1/e$, LP rounding based approximation algorithm for the $CVC$ problem. It should be noted that the size of the resulting LP is exponential, however it can be solved in polynomial time as there exists a separation oracle [7].

The separable assignment problem can be reduced to the problem of maximization of a sub-modular function over matroids (SFM); [7] attributes this reduction to Chekuri. Fisher *et al.* [6] describe two $1/2$ approximation algorithms for the SFM problem; one greedy and the other local search based. We give a simple $1/2$ approximation algorithm for the MAX $CVC$ problem. The bound in Theorem 3 for the greedy algorithm also follows from Fisher *et al.* [6]. Following the reduction (SAP $\propto$ SFM) in [7] the MAX $CVC$ problem can also be reduced to SFM. The details of the reduction are in the full version of the paper. In light of the reductions above the results of Vondrak [17] and Fleischer *et al.* [7] imply an $1 - 1/e$ approximation algorithm for the MAX $CVC$ problem.

**Our Contributions:** We introduce a new class of covering problems with variable capacities that arise in wireless networks. We show that $CVC$ with fixed facilities, uniform clients (the profit and demand equals one for every client) and a capacity function with two ranges serving either two clients or one, is NP-complete. In contrast, under the same conditions, the classical capacitated covering problem where facilities have a fixed capacity and a fixed covering range is equivalent to the maximum flow problem and can be solved optimally in polynomial time [15]. Three natural integer programming formulations for the $CVC$ problem with fixed facilities are possible but they have a large integrality gap. The formulations can be extended to handle both the set cover and maximization versions. Unfortunately, the formulations are too large to be solved in practice even if we relax the integrality constraints. We give strong evidence that a set cover based linear programming formulation that is specific for the set cover $CVC$ combined with a simple rounding procedure is very effective at finding approximate solutions. We then give a greedy $\frac{1}{2} - \epsilon$ approximation algorithm with simple analysis for the maximum $CVC$ where the clients have arbitrary demands and profits and there are no assumptions on the capacity function. The running time of the greedy algorithm can be significantly reduced for the version of the problem with unitary demands and profits. Finally, we mention that all three types of uniform *CVC* can be solved in polynomial time when the clients and the facilities are located on a line. The proofs are provided in the full version of the paper.

## 2    Complexity

We note that the *CVC* problem with only one fixed facility, clients with arbitrary demands and profits, and a constant capacity (single range) function is NP-complete as it is equivalent to the Knapsack problem. Hence, it is natural to examine the case of uniform demands and profits with variable capacities for fixed facilities. The following theorem establishes the complexity of the problem. The proof is omitted.

**Theorem 1.** *The uniform* CVC *problem with fixed facilities (Problem 1) is NP-complete even when facilities are identical and use two ranges with capacities in the set* $\{1, 2\}$.

## 3    Algorithms for *CVC* Problems

We investigated three natural compact integer programming formulations for *CVC* with fixed facilities. We have shown that they exhibit a rather large integrality gap. The formulations can be adapted to handle maximum and set cover *CVC* but they require a quadratic number of constraints and therefore they are impractical. Details about the compact formulations are omitted in this paper.

In Section 3.1 we provide a column generation formulation for the set cover *CVC* that runs well in practice. When combined with a simple rounding scheme, our method solves uniform set cover geometric[1] *CVC* problems with a performance ratio close to one. In the following section we describe a simple approximation algorithm for the maximum uniform *CVC* problem with a performance ratio of $\frac{1}{2} - \epsilon$. The algorithm can be extended to maximum *CVC* instances that are not uniform, but at the expense of the approximation factor.

On the positive side, we argue that both maximum and set cover *CVC* with uniform demands and profits can be solved optimally in polynomial time if the clients and the facilities are located on a line. Due to space constraints, details of the latter algorithms are included in the full version of this paper.

### 3.1    Column Generation and Rounding for Set Cover *CVC*

In the compact integer programming formulations for *CVC*, the number of constraints are proportional to the product of the number of customers and the

---

[1] Customers and facilities are points in the plane and distances are Euclidean.

number of candidate facility locations. An alternative formulation where the number of constraints is equal to the number of customers, transforms the set cover $CVC$ problem into a *weighted set cover* problem.

In the weighted set cover, a set $U$ of elements called the universe is given along with a family $\mathcal{S}$ of subsets of $U$. Each subset $S \in \mathcal{S}$ has a known weight $w_S$. The goal is to find a minimum weight cover, *i.e.* $\mathcal{F} \subseteq \mathcal{S}, \bigcup_{S \in \mathcal{F}} S = U$ that minimizes $\sum_{S \in \mathcal{F}} w_S$.

We now describe how to transform an instance of the set cover $CVC$ to an instance of the weighted set cover. We use the notation introduced in problems 1-3 and we remind the reader that any feasible allocation of customers $\mathcal{I}_j$ to facility $a_j$ that is assigned a range $r$ must satisfy the following three conditions.

$$j \in \mathcal{J}^* \tag{1}$$

$$\sum_{i \in \mathcal{I}_j} s_i \leq c_{ar}, \tag{2}$$

$$u_i \in N_{jr}, \quad \text{for all } i \in \mathcal{I}_j \tag{3}$$

Constraint (1) states that customers can only be assigned to open facilities. Inequality (2) is the capacity constraint and inequality (3) is the coverage constraint.

We can construct an equivalent weighted set cover as follows. The universe $U$ corresponds to the set of customers. Set $\mathcal{S}$ corresponds to the set of all feasible assignments of customers to candidate facility $a_j$ for a covering range $r \in R_j$ that satisfies constraints (2)-(3), for all choices of $a$ and $r$. The size of set $\mathcal{S}$ is exponential in the number of customers and facilities, however the linear programming (LP) relaxation of the weighted set cover can be solved by column generation.

**Column Generation.** Consider the LP relaxation of the weighted set cover problem,

$$\min \sum_{S \in \mathcal{S}} w_S x_S, \tag{4}$$

$$\sum_{S, u \in S} x_S \geq 1, \quad \forall u \in U, x_S \geq 0. \tag{5}$$

By convention, $x_S = 1$ if subset $S$ is chosen in the cover, otherwise $x_S = 0$. Problem (4) is called the master problem. It is much simpler than the compact formulations for $CVC$ since the coverage and capacity constraints are implicit in the structure of the columns of the constraint matrix.

As hinted in the previous paragraph, the master problem is solved iteratively. In each iteration, only a small set of variables $x_S$ are explicit. Once an optimal fractional solution to the master problem (4) is computed, a column generation procedure attempts to compute a new set $\mathcal{I}_{j'}$ of customers that are served by some facility $a_{j'}$ at some covering range $r \in R_{j'}$ subject to capacity and coverage constraints and that minimizes the reduced cost $\rho_{j'} = 1 - \sum_{i \in \mathcal{I}_{j'}} y_i$, where $y_i$ is the optimal dual variable of the master problem (4) corresponding to customer $i$.

If there is no column with negative reduced cost then the optimal fractional solution to the master problem is also an optimal fractional solution for the set cover $CVC$ and the iterative procedure stops. Otherwise, the new column corresponding to set $\mathcal{I}_{j'}$ is added to the constraint matrix of the master problem and the entire procedure is repeated.

We now describe the sub-problem, the procedure for generating the column with negative reduced cost. We iterate over every candidate facility location $a_j$ and covering range $r \in R_j$. To minimize $\rho_j$, we seek a subset of customers reachable from facility $a_j$ with range $r$ that maximizes the sum of the dual variables subject to the capacity constraint. This is a knapsack problem with the set of items $i \in N_{jr}$, profit $y_i$ and knapsack capacity $c_{jr}$. In particular, when the demand is equal to one for all customers, this knapsack problem can be solved by a greedy procedure.

**Rounding.** At the end of the column generation phase, we have an optimal fractional solution to the set cover $CVC$. If the solution is not integral, we round the fractional solution to obtain an integral solution. The solution may not be optimal, and the cost of the linear programming solution is a lower bound on the optimal solution which gives us a measure of the quality of the solution of the rounding step.

The rounding proceeds in two phases. First, we obtain a partial integral set cover solution by rounding with probabilities given by the value of the fractional optimal solution. The outcome of this rounding step might not be feasible. We repeat the rounding process until a feasible integral cover is obtained. In phase two, we obtain a minimal set cover from our solution by removing some facilities. The two phases are as in Algorithm 1. An average of 20 rounding experiments is used to determine the performance ratio for a single instance.

**Experiments.** To evaluate the effectiveness of the column generation and the randomized rounding procedure on the set cover $CVC$, we generated close to 500 geometric instances consisting of $n$ points placed uniformly at random in the unit square representing both candidate facility locations and customers. The number of points $n$ varies between 10 and 500 in unit increments. Each candidate facility is assigned 5 covering ranges in the interval $(0, 1)$ and five capacities from set $\{1, \ldots, 5\}$ uniformly at random in such a way that the capacity is a non-increasing function of the covering range. We use a compact formulation similar to the first integer programming formulation. The column generation, the rounding, and the compact integer and linear formulations were coded in Octave and the GNU LP solver was used.

The results are depicted in Fig. 1. The $x$ axis represents the number of customers in the problem instances solved. Figure 1.a shows the performance ratio, *i.e.* the ratio between the cost of the cover obtained by column generation and

---

**Algorithm 1.** Rounding

---

Let $\mathcal{S}$ be the set of generated columns.
Let $x_S^*$ for $S \in \mathcal{S}$ be the optimal LP solution.
$F \leftarrow \emptyset$.
**repeat**
  **for all** $S \in \mathcal{S}$ **do**
    $x_S \leftarrow 1$ with probability $x_S^*$, otherwise $x_S \leftarrow 0$.
    $F \leftarrow F \cup \{S : x_S = 1\}$
  **end for**
**until** $F$ is a set cover
**for all** $S \in F$ in random order **do**
  **if** $F \setminus \{S\}$ is a set cover **then**
    $F \leftarrow F \setminus \{S\}$
  **end if**
**end for**
Return $F$

---

rounding and the cost of the optimal solution to the LP relaxation. The experiments include the integrality gap for the compact formulation on the instances for which the integer program reported an optimal solution. From Fig. 1.b we see clearly that the compact formulations cannot handle instances beyond 90 clients because of thrashing. In contrast, the running time for the column generation with rounding seems to scale well with the size of the problem instance, a fact also supported by Fig. 1.c that shows an almost linear dependency of the number of generated columns with the problem size. Finally, Fig. 1.d shows that the average number of rounding iterations is approximately equal to 1.4 and it seems to be independent of the size of the problem. The average was calculated over twenty rounding experiments.

## 3.2   A Greedy Algorithm for Maximum *CVC*

We describe our results for the *CVC* with fixed facilities and show at the end that the algorithm can handle maximum *CVC* problems with the same approximation factor.

The greedy algorithm relies on repeatedly solving instances of the Knapsack problem [12]. We denote by $K(\alpha)$ an $\alpha$-approximation algorithm for the knapsack problem, and by $V(\alpha)$ the greedy algorithm for *CVC* with fixed facilities that uses $K(\alpha)$ as a subroutine. By convention, we use an approximation factor less than one for maximization problems, $\alpha < 1$. The crucial idea is to examine facilities in an arbitrary but fixed order and, for each facility, to decompose the *CVC* instance into several independent knapsack instances, one for each range and capacity. The solution to the *CVC* instance is then constructed greedily, by selecting the maximum profit knapsack solutions (see Algorithm 2). The algorithm outputs the total profit of clients covered, the set of clients covered by each facility, and the range for each facility.

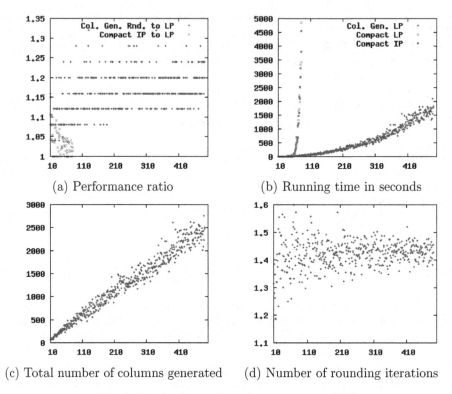

(a) Performance ratio    (b) Running time in seconds

(c) Total number of columns generated    (d) Number of rounding iterations

**Fig. 1.** Results for random planar set cover $CVC$

**Theorem 2.** *Algorithm* $V(\alpha)$ *computes an* $\frac{\alpha}{\alpha+1}$ *approximate solution to the* CVC *problem with fixed facilities.*

Intuitively, one expects that the knapsack subroutine in Algorithm $V(\alpha)$ would not be much worse than the optimal assignment of clients to facilities in the $CVC$ instance. The only reason why there might be a significant difference between the total profit packed by Algorithm $K(\alpha)$ and the optimal solution is that Algorithm $K(\alpha)$ runs on an instance consisting of those objects not already covered by previous iterations in $V(\alpha)$, and the optimal $CVC$ uses too many of these missing objects. However, the "missing" objects are by definition covered by $V(\alpha)$ as well, only they are covered by different facilities. Next we formalize this intuition.

**Proof of Theorem 2:**    Let $a_j$ be the facility examined by the algorithm in the $j^{th}$ iteration. Let $Q_j$ be the set of clients assigned to $a_j$ in the optimal $CVC$ solution but not available to Algorithm $V(\alpha)$ because they were covered by $V(\alpha)$ in previous iterations $1, \ldots, j-1$. Let $OPT_j$ be the set of clients assigned to $a_j$ in the optimal solution. Let $A_j = OPT_j \setminus Q_j$, be the set of clients in the optimal solution that are available to the knapsack subroutine of Algorithm $V(\alpha)$.

---

**Algorithm 2.** $V(\alpha)$:

---

let $U$ be the set of all available clients
**for all** $j \in \mathcal{J}$, consider facility $a_j$ **do**
    **for all** ranges $r \in R_j$ **do**
        use Algorithm $K(\alpha)$ to solve the knapsack instance with capacity $c_{jr}$ and
        set of items $N_{jr}$.
        let $p_{jr}$ be the profit returned by Algorithm $K(\alpha)$, and let $\mathcal{I}_{jr}$ be the set
        of packed clients.
    **end for**
    let $p_j \leftarrow \max_{r \in R_j} p_{jr}$ ($p_j$ is the solution with largest profit)
    let $\mathcal{I}_j$ be the set of clients packed by the most profitable solution
    let $r_j$ be the covering range for the most profitable solution.
    let $\mathcal{I} \leftarrow \mathcal{I} \setminus \mathcal{I}_j$.
**end for**
return $\sum_{j=1}^{f} p_j$

---

If $X$ is a subset of clients, then $p_X$ denotes the total profit for the clients in $X$. Term $p_{K(\alpha,r,j)}$ represents the total profit for the solution returned by Algorithm $K(\alpha)$ on the knapsack instance used by Algorithm $V(\alpha)$ with facility $a_j$ and range $r$.

Let $r_j$ be the range chosen by Algorithm $V(\alpha)$ for facility $a_j$. Because $r_j$ was chosen greedily (with maximum profit) by Algorithm $K(\alpha)$, and the clients in $A_j$ were available in the $j^{th}$ iteration,

$$p_{K(\alpha,r_j,j)} \geq \alpha p_{A_j}.$$

Summing up over all facilities $a_j$, we obtain

$$\sum_{j=1}^{f} p_{A_j} \leq \frac{1}{\alpha} \sum_{j=1}^{f} p_{K(\alpha,r_j,j)} = \frac{V}{\alpha},$$

where $V$ is the total profit returned by Algorithm $V(\alpha)$. Since set $Q_j$ represents the clients covered by both the optimal $CVC$ solution and the solution returned by Algorithm $V(\alpha)$, we also have,

$$\sum_{j=1}^{f} p_{Q_j} \leq V.$$

Summing the last two inequalities we obtain immediately

$$OPT \leq V\left(1 + \frac{1}{\alpha}\right),$$

and the theorem follows. $\qquad\qquad\qquad\qquad\qquad\qquad\qquad\qquad\qquad\qquad$ $\square$

**Fig. 2.** Tight example for the $\frac{1}{2}$-factor approximation algorithm for uniform $CVC$ with fixed facilities

A consequence of Theorem 2 is that the greedy algorithm for the uniform $CVC$ problem admits a $\frac{1}{2}$ approximation factor. For objects with unit sizes and profits, the knapsack problem has trivial optimal solutions, and therefore $\alpha = 1$.

**Corollary 1.** *The greedy algorithm for the uniform* CVC *problem with fixed facilities has an approximation factor of $\frac{1}{2}$.*

Next we show that the bound of $\frac{1}{2}$ is tight for the uniform $CVC$ problem. Consider the following instance on the real line (Fig. 2): the two facilities are at points 1 and 3, and the two clients are on points 0 and 2. Each facility has only one range (radius 1) with capacity 1. If the greedy approximation algorithm chooses the facilities in the order $(1, 3)$ and the client on point 2 is assigned to the facility on point 1, then the facility at point 3 cannot cover the client at point 0. In the optimal solution, the facility at point 1 (3) covers the client at point 0 (2).

*Remark:* Note that the tight example also tells us that a modification of the approximation algorithm in which we use the facility that covers the most number of clients (greedily) in each iteration also has performance ratio 2.

Next we analyze the running time for the approximation algorithm. For each facility there are at most $n$ knapsack instances. If we use the FPTAS for Knapsack due to Ibarra and Kim [11] with running time $O(\frac{n^3}{\epsilon})$ to obtain a $\alpha = 1 - \epsilon$ approximate solution to the Knapsack instance, then the total running time is $O(f\frac{n^4}{\epsilon})$ where $n$ is the number of clients and $f$ is the number of facilities. For the uniform $CVC$ problem the running time can be further reduced to $O(fn \log n)$. In this case, the best knapsack solution for each iteration can be obtained by sorting all clients in non-decreasing order of their distance to the facility (time complexity $O(n \log n)$) and finding the maximum range $r_j$ for which the capacity of the facility is larger than the number of clients covered (time complexity $O(n)$). The same running time complexity can be obtained for the general $CVC$ problem if we choose $\alpha = \frac{1}{2}$, that is, if we use the greedy algorithm for the knapsack problem [12]. In each iteration, we can find the profit of the knapsack sub-problem by answering an orthogonal two dimensional range search query [2]. The set of points over which the range query is invoked corresponds to the set of clients in the $CVC$ instance and the coordinates of the points are defined by profit density and distance to the facility. The range search query is unbounded and can be implemented in $O(\log n)$ time. The data structures required can be initialized in $O(n \log n)$ time per facility.

*Maximum* CVC: Algorithm 2 can be extended to handle maximum $CVC$ as follows. The algorithm consists of $k$ iterations, where $k$ is the number of facilities to be opened. In each iteration, we solve a knapsack instance for all remaining candidate facilities and all available covering ranges. We open the facility corresponding to the most profitable knapsack solution found. With this change, the conditions necessary for the proof of Theorem 2 are satisfied and we can claim the following theorem.

**Theorem 3.** *There exists a greedy algorithm that computes an $\frac{\alpha}{\alpha+1}$ approximate solution to the maximum CVC problem.*

**Proof:** Details omitted.                                                          □

## 4  Conclusion

In this paper, we introduce a new class of covering problems called "Covering with Variable Capacities" which generalizes the well known covering and capacitated facility location problems. We define three classes of problems, set cover *CVC*, maximum *CVC*, and *CVC* with fixed facilities. Our model is inspired by an application in wireless cellular networks, but is appropriate for other applications in facility location as well. We show that *CVC* with fixed facilities and uniform demands is NP-hard. In contrast, under the same conditions, the capacitated facility location problem can be solved in polynomial time using flows. We exhibit large integrality gaps for three different integer programming formulations for *CVC* with fixed facilities.

Fast approximation algorithms are of particular interest for the telecommunication industry. We give evidence that random instances of the set cover *CVC* have a small integrality gap on average and we propose an efficient column generation algorithm followed by randomized rounding.

We describe a simple greedy algorithm for the maximum *CVC* that achieves a performance bound of $\frac{1}{2} - \epsilon$. The running time of the algorithm is determined by the knapsack sub-problem and can be quite expensive if we want to solve knapsack very precisely using the FPTAS [11]. However, for a slightly worse performance bound of $\frac{1}{3}$, we can solve the general instance *CVC* problem efficiently, in $O(mn \log n)$ time, where $m$ is the number of facilities and $n$ is the number of clients. In practice, $m \ll n$. For the uniform *CVC* instance, knapsack can be solved optimally, and the greedy algorithm achieves a performance bound of $\frac{1}{2}$ with the same time complexity.

Finally, we mention that all three types of *CVC* problems when clients and facilities are constrained to lie on a line can be solved in polynomial time using a greedy algorithm for set cover *CVC* and dynamic programming for maximum *CVC*. Details of this result will be presented in the full version of the paper.

*Open Problems:* It would be interesting to develop a fast $1 - 1/e$ approximation algorithm for the maximum *CVC* problem that does not use the connection with the separable assignment problem and constant factor approximations for the set cover version. Following our results on *CVC* on the line, it would be interesting to examine other network topologies. Finally, *CVC* can be investigated with capacity functions other than the step function, for example linear continuous functions.

**Acknowledgements.** Selim Akl, Robert Benkoczi, Hossam Hassanein, Shahadat Hossain and Mark Thom acknowledge the funding support from NSERC. Selim Akl also aknowledges the funding support from Queen's University.

# References

1. Aardal, K.: Capacitated facility location: Separation algorithms and computational experience. Math. Program. 81, 149–175 (1998)
2. de Berg, M., Cheong, O., van Kreveld, M., Overmars, M.: Computational Geometry: Algorithms and Applications, ch. 5, 3rd edn. Springer (2008) ISBN: 978-3-540-77973-5
3. Berman, O., Drezner, Z., Krass, D.: Generalized coverage: New developments in covering location models. Computers & Operations Research 37(10), 1675–1687 (2010)
4. Catrein, D., Imhof, L.A., Mathar, R.: Power control, capacity, and duality of uplink and downlink in cellular CDMA systems. IEEE Transactions on Communications 52(10), 1777–1785 (2004)
5. Chudak, F.A., Williamson, D.P.: Improved approximation algorithms for capacitated facility location problems. Math. Program. 102(2), 207–222 (2005)
6. Fisher, M.L., Nemhauser, G.L., Wolsey, L.A.: An analysis of the approximations for maximizing submodular set functions II. Mathematical Programming Study 8, 73–87 (1978)
7. Fleischer, L., Goemans, M.X., Mirrokni, V.S., Sviridenko, M.: Tight approximation algorithms for maximum general assignment problem. In: SODA, pp. 611–620 (2006)
8. Hanly, S.: Congestion measures in DS-CDMA networks. IEEE Transactions on Communications 47(3), 426–437 (1999)
9. Hochbaum, D.S., Maass, W.: Approximation schemes for covering and packing problems in image processing and VLSI. J. ACM 32(1), 130–136 (1985)
10. Holma, H., Toskala, A.: WCDMA for UMTS: HSPA Evolution and LTE, 4th edn. Wiley (2007) ISBN: 978-0470319338
11. Ibarra, O.H., Kim, C.E.: Fast approximation algorithms for the knapsack and sum of subsets problems. J. ACM 22, 463–468 (1975)
12. Martello, S., Toth, P.: Knapsack Problems: Algorithms and Computer Implementations, Revised edn. Wiley (1990) ISBN: 978-0471924203
13. Mulvey, J.M., Beck, M.P.: Solving capacitated clustering problems. European Journal of Operational Research 18(3), 339–348 (1984)
14. Radwan, A., Hassanein, H.: Capacity enhancement in CDMA cellular networks using multi-hop communication. In: Proceedings of the 11th IEEE Symposium on Computers and Communications, June 26-29, pp. 832–837. IEEE (2006)
15. Schrijver, A. Combinatorial Optimization, first ed., vol. A, part II. Springer, ch. 21, pp. 337–377. ISBN: 978-3540443896 (2003)
16. TAM, Y. H., HASSANEIN, H. S., AKL, S. G., AND BENKOCZI, R. Optimal multi-hop cellular architecture for wireless communications. In *Proceedings of the 31st IEEE Conference on Local Computer Networks* (Nov. 2006), pp. 738–745. **Optimal multi-hop cellular architecture for wireless communications. In *Proceedings of the 31st Conference on Local Computer Networks* (pp**
17. VONDRAK, J. Optimal approximation for the submodular welfare problem in value oracle model. STOC, (2008), 67–74. **Optimal approximation for the submodular welfare problem in value oracle model. STOC**

# Algorithm and Hardness Results
# for Outer-connected Dominating Set in Graphs

B.S. Panda and Arti Pandey

Department of Mathematics
Indian Institute of Technology Delhi
Hauz Khas, New Delhi 110016, India
{bspanda,artipandey}@maths.iitd.ac.in

**Abstract.** A set $D \subseteq V$ of a graph $G = (V, E)$ is called an *outer-connected dominating set* of $G$ if for all $v \in V$, $|N_G[v] \cap D| \geq 1$, and the induced subgraph of $G$ on $V \setminus D$ is connected. The MINIMUM OUTER-CONNECTED DOMINATION problem is to find an outer-connected dominating set of minimum cardinality of the input graph $G$. Given a positive integer $k$ and a graph $G = (V, E)$, the OUTER-CONNECTED DOMINATION DECISION problem is to decide whether $G$ has an outer-connected dominating set of cardinality at most $k$. The OUTER-CONNECTED DOMINATION DECISION problem is known to be NP-complete for bipartite graphs. In this paper, we strengthen this NP-completeness result by showing that the OUTER-CONNECTED DOMINATION DECISION problem remains NP-complete for perfect elimination bipartite graphs. On the positive side, we propose a linear time algorithm for computing a minimum outer-connected dominating set of a chain graph, a subclass of bipartite graphs. We propose a $\Delta(G)$-approximation algorithm for the MINIMUM OUTER-CONNECTED DOMINATION problem, where $\Delta(G)$ is the maximum degree of $G$. On the negative side, we prove that the MINIMUM OUTER-CONNECTED DOMINATION problem cannot be approximated within a factor of $(1 - \varepsilon) \ln |V|$ for any $\varepsilon > 0$, unless NP $\subseteq$ DTIME($|V|^{O(\log \log |V|)}$). We also show that the MINIMUM OUTER-CONNECTED DOMINATION problem is APX-complete for graphs with bounded degree 4 and for bipartite graphs with bounded degree 7.

**Keywords:** Domination, outer-connected domination, NP-completeness, APX-completeness.

# 1 Introduction

A vertex $v$ of a graph $G = (V, E)$ is said to *dominate* a vertex $w$ if either $v = w$ or $vw \in E$. A set of vertices $D$ is a *dominating set* of $G$ if every vertex of $G$ is dominated by at least one member of $D$. The *domination number* of a graph $G$, denoted by $\gamma(G)$, is the cardinality of a minimum dominating set of $G$. The MINIMUM DOMINATION problem is to find a dominating set of minimum cardinality of the input graph $G$. The concept of domination and its variations are widely studied as can be seen in [1, 2].

S.P. Pal and K. Sadakane (Eds.): WALCOM 2014, LNCS 8344, pp. 151–162, 2014.

For a set $S \subseteq V$ of the graph $G = (V, E)$, the subgraph of $G$ induced by $S$ is defined as $G[S] = (S, E_S)$, where $E_S = \{xy \in E | x, y \in S\}$. A set $D \subseteq V$ of a graph $G = (V, E)$ is called an *outer-connected dominating set* of $G$ if $D$ is a dominating set of $G$ and $G[V \setminus D]$ is connected. The *outer-connected domination number* of a graph $G$, denoted by $\gamma_c(G)$, is the cardinality of a minimum outer-connected dominating set of $G$. The concept of outer-connected domination number was introduced by Cyman [3] and further studied by others (see [4, 5, 6]). This problem has possible applications in computer networks. Consider a client-server architecture based network in which any client must be able to communicate to one of the servers. Since overloading of severs is a bottleneck in such a network, every client must be able to communicate to another client directly (without interrupting any of the server). A smallest group of servers with these properties is a minimum outer-connected dominating set for the graph representing the computer network.

The MINIMUM OUTER-CONNECTED DOMINATION (MOCD) problem is to find an outer-connected dominating set of minimum cardinality of the input graph $G$. Given a positive integer $k$ and a graph $G = (V, E)$, the OUTER-CONNECTED DOMINATION DECISION (OCDD) problem is to decide whether $G$ has an outer-connected dominating set of cardinality at most $k$. MINIMUM OUTER-CONNECTED DOMINATION problem is studied for some subclasses of chordal graphs (doubly chordal graphs, undirected path graphs and proper interval graphs) [5].

In this paper, we study the algorithmic aspect of the MINIMUM OUTER-CONNECTED DOMINATION problem. The contributions made in this paper are summarized below.

1. We strengthen the NP-completeness result of the OCDD problem by showing that this problem remains NP-complete for perfect elimination bipartite graphs. On the positive side, we propose a linear time algorithm for computing a minimum outer-connected dominating set of a chain graph.
2. We propose a $\Delta(G)$-approximation algorithm for the MOCD problem, where $\Delta(G)$ is the maximum degree of $G$. On the negative side, we prove that the MOCD problem cannot be approximated within a factor of $(1 - \varepsilon) \ln |V|$ for any $\varepsilon > 0$, unless NP $\subseteq$ DTIME($|V|^{O(\log \log |V|)}$).
3. We show that the MOCD problem is APX-complete for graphs with bounded degree 4 and for bipartite graphs with bounded degree 7.

## 2    Preliminaries

For a graph $G = (V, E)$, the sets $N_G(v) = \{u \in V(G) | uv \in E\}$ and $N_G[v] = N_G(v) \cup \{v\}$ denote the *open neighborhood* and *closed neighborhood* of a vertex $v$, respectively. For a connected graph $G$, a vertex $v$ is a cut vertex if $G \setminus \{v\}$ is disconnected. The *degree* of a vertex $v$ is $|N_G(v)|$ and is denoted by $d_G(v)$. If $d_G(v) = 1$, then $v$ is called a *pendant vertex*. For $S \subseteq V$, let $G[S]$ denote the subgraph induced by $G$ on $S$. A graph $G = (V, E)$ is said to be *bipartite* if

$V(G)$ can be partitioned into two disjoint sets $X$ and $Y$ such that every edge of $G$ joins a vertex in $X$ to another vertex in $Y$. A partition $(X, Y)$ of $V$ is called a *bipartition*. A bipartite graph with bipartition $(X, Y)$ of $V$ is denoted by $G = (X, Y, E)$. Let $n$ and $m$ denote the number of vertices and number of edges of $G$, respectively. A graph $H = (V', E')$ is a *spanning subgraph* of $G = (V, E)$ if $V' = V$ and $E' \subseteq E$. A connected acyclic spanning subgraph of $G$ is a *spanning tree* of $G$. A tree with exactly one non-pendant vertex is a *star* and a tree with exactly two non-pendant vertices is called a *bi-star*.

In the rest of the paper, by a graph we mean a connected graph with at least two vertices unless otherwise mentioned specifically. The following observations regarding outer-connected dominating set of a graph are straightforward and hence the proofs are omitted.

**Observation 1.** *(a) If $v$ is a pendant vertex of $G = (V, E)$, then either $v \in D$ or $D = V \setminus \{v\}$ for every outer-connected dominating set $D$ of $G$.*
*(b) Let $G = (V, E)$ be a connected graph having at least three vertices. Then there is a minimum outer-connected dominating set of $G$ containing all the pendant vertices of $G$.*
*(c) Every outer-connected dominating set $D$ of cardinality at most $n - 2$ of a graph $G = (V, E)$ having $n$ vertices contains all the pendant vertices of $G$.*

## 3   NP-completeness Proof for Perfect Elimination Bipartite Graphs

Let $G = (X, Y, E)$ be a bipartite graph. Then $uv \in E$ is a *bisimplicial edge* if $N_G(u) \cup N_G(v)$ induces a complete bipartite subgraph in $G$. Let $(e_1, e_2, \ldots, e_k)$ be an ordering of pairwise non-adjacent edges (no two edges have a common end vertex) of $G$ (not necessarily all edges of $E$). Let $S_i$ be the set of endpoints of edges $e_1, e_2, \ldots, e_i$ and let $S_0 = \emptyset$. Ordering $(e_1, e_2, \ldots, e_k)$ is a *perfect edge elimination ordering* for $G$ if $G[(X \cup Y) \setminus S_k]$ has no edge and each edge $e_i$ is bisimplicial in the remaining induced subgraph $G[(X \cup Y) \setminus S_{i-1}]$. $G$ is a *perfect elimination bipartite graph* if $G$ admits a perfect edge elimination ordering. The class of perfect elimination bipartite graphs was introduced by Golumbic and Goss [7].

To show the NP-completeness of the OCDD problem, we need to use a well known NP-complete problem, called VERTEX COVER DECISION problem [8]. A set $S \subseteq V$ of a graph $G = (V, E)$ is called a *vertex cover* of $G$ if for every edge $uv \in E$, either $u \in S$ or $v \in S$.
VERTEX COVER DECISION problem
**INSTANCE:** A graph $G = (V, E)$ and a positive integer $k$.
**QUESTION:** Does $G$ have a vertex cover of cardinality at most k?

We are now ready to prove the following theorem:

**Theorem 2.** *OCDD problem is NP-complete for perfect elimination bipartite graphs.*

*Proof.* Given a perfect elimination bipartite graph $G = (V, E)$, a positive integer $k$ and an arbitrary subset $D$ of $V$, we can check in polynomial time whether $|D| \leq k$ and $D$ is an outer-connected dominating set of $G$. Hence the OCDD problem for perfect elimination bipartite graphs is in NP. To show the hardness, we provide the polynomial time reduction from VERTEX COVER DECISION problem in general graphs to the OCDD problem in perfect elimination bipartite graphs.

Given a graph $G = (V, E)$, construct the graph $G' = (V', E')$ as follows: If $V = \{v_1, v_2, \ldots, v_n\}$ and $E = \{e_1, e_2, \ldots, e_m\}$, define $V' = \{v_i, x_i, y_i, w_i \mid 1 \leq i \leq n\} \cup \{e_i', g_i, h_i : 1 \leq i \leq m\}$ and $E' = \{v_i w_i, v_i x_i, x_i y_i \mid 1 \leq i \leq n\} \cup \{e_i' g_i, g_i h_i \mid 1 \leq i \leq m\} \cup \{e_i' v_j, e_i' v_k, g_i x_j, g_i x_k \mid 1 \leq i \leq m, v_j$ and $v_k$ are endpoints of edge $e_i\}$.

The graph $G = (V, E)$, where $V = \{v_1, v_2, v_3\}$ and $E = \{e_1 = v_1 v_2, e_2 = v_2 v_3, e_3 = v_3 v_1\}$ and the associated graph $G'$ are shown in Fig. 1 to illustrate the above construction.

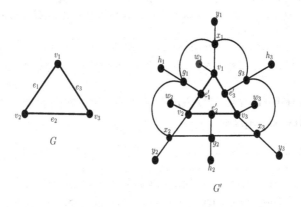

**Fig. 1.** An illustration to the construction of $G'$ from $G$

Clearly $G'$ is a perfect elimination bipartite graph and $(x_1 y_1, x_2 y_2, \ldots, x_n y_n, v_1 w_1, v_2 w_2, \ldots, v_n w_n, g_1 h_1, g_2 h_2, \ldots, g_m h_m)$ is perfect edge elimination ordering for $G'$.

*Claim.* $G$ has a vertex cover of size $k$ if and only if $G'$ has an outer-connected dominating set of size at most $2n + m + k$.

*Proof.* Let us first assume that $G$ has a vertex cover say $V_c$ of size $k$. Then $V_c \cup \{w_i, y_i \mid 1 \leq i \leq n\} \cup \{h_i \mid 1 \leq i \leq m\}$ is an outer-connected dominating set of $G'$ of size $2n + m + k$.

Conversely suppose that $D$ is an outer-connected dominating set of $G'$ of size $2n + m + k$. Define $S = \{h_i \mid 1 \leq i \leq m\} \cup \{w_i, y_i \mid 1 \leq i \leq n\}$ and $E' = \{e_i' \mid 1 \leq i \leq m\}$. By using Observation 1(c), all the pendant vertices must belong to $D$, hence $S \subseteq D$. But $S$ does not dominate the vertices of $E'$. Define $S' = D \setminus S$. Hence all the vertices of $E'$ are dominated using $S'$. Now to dominate $e_i'$, either $e_i' \in S'$ or $g_i \in S'$ or some $v_j \in S'$. If $e_i' \in S'$ or $g_i \in S'$, we remove it from $S'$ and add $v_j$ (i.e. adjacent to $e_i'$) in $S'$. Do this for all $i$ between 1 to $m$. Define $V_t = V \cap S'$. Note that $V_t$ is a vertex cover of $G$ and $|V_t| \leq k$. This proves our claim.                                                                  □

Hence our theorem is proved.                                                                  □

## 4    Chain Graphs

We have already seen that the OCDD problem is NP-complete even for perfect elimination bipartite graphs. In this section, we show that the problem of computing a minimum outer-connected dominating set of a chain graph can be solved in polynomial time.

A bipartite graph $G = (X, Y, E)$ is called a chain graph if the neighborhoods of the vertices of $X$ form a chain, that is, the vertices of $X$ can be linearly ordered, say $x_1, x_2, \ldots, x_p$, such that $N_G(x_1) \subseteq N_G(x_2) \subseteq \ldots \subseteq N_G(x_p)$. If $G = (X, Y, E)$ is a chain graph, then the neighborhoods of the vertices of $Y$ also form a chain [9]. An ordering $\alpha = (x_1, x_2, \ldots, x_p, y_1, y_2, \ldots, y_q)$ of $X \cup Y$ is called a chain ordering if $N_G(x_1) \subseteq N_G(x_2) \subseteq \cdots \subseteq N_G(x_p)$ and $N_G(y_1) \supseteq N_G(y_2) \supseteq \cdots \supseteq N_G(y_q)$. It is well known that every chain graph admits a chain ordering [9, 10].

**Theorem 3.** *Let $G = (X, Y, E)$ be a connected chain graph and $\alpha = (x_1, x_2, \ldots, x_p, y_1, y_2, \ldots, y_q)$ is chain ordering of $X \cup Y$. Then $r - 1 \leq \gamma_c(G) \leq r + 2$, where $r$ is the number of pendant vertices of $G$. Furthermore, the following are true.*

*(a) $\gamma_c(G) = r - 1$ if and only if $G = K_2$.*
*(b) $\gamma_c(G) = r$ if and only if $G$ is a star or bi-star of order greater than 2.*
*(c) Let $P$ denotes the set of all pendant vertices of $G$ and $P_A$ denotes the set of vertices adjacent to the vertices of $P$. Then $\gamma_c(G) = r + 1$ if and only if $G' = G[V \setminus (P \cup P_A)]$ is a star.*
*(d) If $G$ is a graph other than the graphs described in the above statements then $\gamma_c(G) = r + 2$.*

*Proof.* Suppose $D$ is a minimum outer-connected dominating set of $G$. Then $|D| = \gamma_c(G)$. Now by using Observation 1(a), either $D$ contains all the pendant vertices of $G$ or $D = V \setminus \{v\}$, where $v$ is some pendant vertex. So either $\gamma_c(G) \geq r$ or $\gamma_c(G) = n - 1 \geq r - 1$. Hence $\gamma_c(G) \geq r - 1$.

Let $P$ denotes the set of pendant vertices of $G$. Now $D = P \cup \{x_p, y_1\}$ is an outer-connected dominating set of $G$. Hence $\gamma_c(G) \leq r + 2$.

(a) If $G = K_2$. Then $r = 2$ and $\gamma_c(G) = 1$ and hence $\gamma_c(G) = r - 1$. Conversely suppose that $\gamma_c(G) = r - 1$ and $D$ be a minimum outer-connected dominating set of $G$. This implies that $D$ does not contain at least one pendant vertex. Then by using Observation 1(a), $D$ contains all the vertices of $G$ other than one pendant vertex and hence $|D| = n - 1$. This implies that $r - 1 = n - 1$ and hence $r = n$. So all the vertices of $G$ are pendant vertices. $K_2$ is the only such graph. Hence $G = K_2$.

(b) If $G$ is a star or a bi-star having at least 3 vertices, then clearly $\gamma_c(G) = r$.

Conversely suppose $\gamma_c(G) = r$. Let $D$ be a minimum outer-connected dominating set of $G$ and $P$ be the set of all pendant vertices of $G$. Since $|D| = r$, either $D = P$ or $r = n - 1$. This implies that either all the vertices are adjacent to one of the pendant vertices or all the vertices other than one are pendant vertices. Hence $G$ is a star or bi-star of order greater than 2.

(c) First suppose that $G' = G[V \setminus (P \cup P_A)]$ is a star. Note that $P$ is not a dominating set. Since by Observation 1(b), $P$ is properly contained in some minimum outer-connected dominating set of $G$, say $D$, $\gamma_c(G) \geq r + 1$. Let $u$ be the star center of $G'$. Then $D = P \cup \{u\}$ dominates all the vertices of $G$. Now the vertex adjacent to the pendant vertices in $X$, say $v$, is adjacent to all the vertices of $X$ and the vertex adjacent to the pendant vertices in $Y$, say $w$, is adjacent to all the vertices of $Y$. Also $v$ and $w$ both are not taken in $D$. Hence $G[V \setminus D]$ is connected. So $D$ is an outer-connected dominating set of $G$. So $\gamma_c(G) = r + 1$.

Conversely suppose that $\gamma_c(G) = r + 1$. By Observation 1(b), there is a minimum outer-connected dominating set, say $D$, of $G$ such that $P \subset D$. Now the vertices of $V \setminus (P \cup P_A)$ are dominated using only one vertex. This implies that $G[V \setminus (P \cup P_A)]$ is a star as it is a bipartite graph.

(d) Proof directly follows from above statements.                                    □

A chain ordering of a chain graph $G = (X, Y, E)$ can be computed in linear time [11]. The set $P$ of pendant vertices of $G$ can be computed in $O(n+m)$ time. If $|V(G)| = 2$, then take $D = \{v\}$, $v \in V(G)$. It can be checked in $O(n + m)$ whether $G$ is a star or a bi-star. In that case, take $D = P$. If $G' = G[V \setminus (P \cup P_A)]$, where $P(A)$ be the set of vertices adjacent to a vertex in $P$, is a star with star-center $v$, then take $D = P \cup \{v\}$, otherwise take $D = P \cup \{y_1, x_p\}$. By Theorem 3, $D$ is a minimum outer-connected dominating set of $G$. So we have the following theorem.

**Theorem 4.** *A minimum outer-connected dominating set of a chain graph can be computed in $O(n + m)$ time.*

## 5    Approximation Algorithm and Hardness of Approximation

Let $G = (V, E)$ be any graph. Let $D$ be any minimum outer-connected dominating set of $G$. Now $V = \cup_{v \in D} N_G[v]$. So, $n = |V| = |\cup_{v \in D} N_G[v]| \leq \sum_{v \in D} |N_G[v]| \leq \sum_{v \in D} (d_G(v) + 1) \leq \sum_{v \in D} (\Delta(G) + 1) \leq (\Delta(G) + 1)|D|$. So, $|D| \geq \lfloor \frac{n}{\Delta(G)+1} \rfloor$. So, we have the following result.

**Lemma 1.** *For any graph $G$ of order $n$ with maximum degree $\Delta(G)$,*

$$\gamma_c(G) \geq \lfloor (\frac{n}{\Delta(G) + 1}) \rfloor.$$

So, for a graph $G = (V, E)$, $D = V(G)$ is an outer-connected dominating set such that $|D| \leq (\Delta(G) + 1)OPT$, where $OPT$ is the cardinality of a minimum outer-connected dominating set of $G$. So we have the following theorem.

**Theorem 5.** *The MOCD problem in any graph* $G = (V, E)$ *with maximum degree* $\Delta(G)$ *can be approximated with an approximation ratio of* $\Delta(G) + 1$.

The following approximation hardness result of the MINIMUM DOMINATION problem will be used to establish an approximation hardness result of the MOCD problem.

**Theorem 6.** *[12]* MINIMUM DOMINATION *problem can not be approximated within a factor of* $(1 - \varepsilon) \ln |V|$ *in polynomial time for any constant* $\varepsilon > 0$ *unless* $NP \subseteq DTIME( |V|^{O(\log \log |V|)})$.

Now we are ready to prove an approximation hardness result for the MOCD problem.

**Theorem 7.** *MOCD problem for a graph* $G = (V, E)$ *can not be approximated within a factor of* $(1 - \varepsilon) \ln |V|$ *in polynomial time for any constant* $\varepsilon > 0$ *unless* $NP \subseteq DTIME(|V|^{O(\log \log |V|)})$.

*Proof.* We propose an approximation preserving reduction from the MINIMUM DOMINATION problem to the MOCD problem. This together with the non-approximability bound of the MINIMUM DOMINATION problem stated in Theorem 6 will provide the desired result.

Let us first describe the reduction from the MINIMUM DOMINATION problem to the MOCD problem. Given a graph $G = (V, E)$, where $V = \{v_1, v_2, \ldots, v_n\}$ construct a graph $G' = (V', E')$ as follows:
$V(G') = V(G) \cup \{w_1, w_2, \ldots, w_n\} \cup \{z\}$, and $E(G') = E(G) \cup \{v_i w_i | 1 \leq i \leq n\} \cup \{w_i w_j | 1 \leq i < j \leq n\} \cup \{z w_i | 1 \leq i \leq n\}$.

The graph $G = (V, E)$, where $V = \{v_1, v_2, v_3\}$ and $E = \{v_1 v_2, v_2 v_3\}$ and the associated graph $G'$ are shown in Fig. 2 to illustrate the above construction.

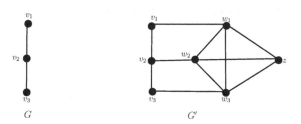

**Fig. 2.** An illustration to the construction of $G'$ from $G$

It is easy to see that if $D^*$ is a minimum dominating set of $G$, then $D^* \cup \{z\}$ is a an outer-connected dominating set of $G'$.

Now assume that the minimum outer-connected dominating set can be approximated within a ratio of $\alpha$, where $\alpha = (1 - \varepsilon) \ln |V|$ for some (fixed) $\varepsilon > 0$, by using some algorithm, say algorithm $A$, that runs in polynomial time. Let $l$ be a fixed positive integer. Consider the following algorithm:

**Algorithm B**

Input: A graph $G = (V, E)$

1. If a minimum dominating set $D$ of cardinality $< l$ exists, construct it Else:
2. Construct $G'$ as above.
3. Compute outer-connected dominating set $D_o$ in $G'$ using algorithm $A$.
4. Compute $D$ by following procedure
5. Define $D = D_o \bigcap V$
6. For each $w_i$, if $w_i \in D_o$ then $D = D \cup v_i$
6. Output $D$

This algorithm runs in polynomial time since algorithm A is a polynomial time algorithm and step 1 runs in polynomial time as $l$ is a constant. Note that if $D$ is a minimum dominating set of cardinality at most $l$, then it is optimal. In the following we will analyze the case where $D$ is not a minimum dominating set of cardinality at most $l$.

Let $D_o^*$ be a minimum outer-connected dominating set, then $|D_o^*| \geq l$. Given the graph $G = (V, E)$ algorithm $B$ computes a dominating set $D$ of cardinality $|D| \leq |D_o| \leq \alpha |D_o^*| \leq \alpha (1 + |D^*|) = \alpha (1 + 1/|D^*|) |D^*| \leq \alpha (1 + 1/l) |D^*|$

Hence Algorithm $B$ approximates minimum dominating set within ratio $\alpha (1 + 1/l)$. Since $\alpha = (1 - \varepsilon) \ln |V|$ for some (fixed) $\varepsilon > 0$, for some positive integer $l$ such that $1/l < \varepsilon/2$, algorithm $B$ approximates minimum dominating set within ratio

$\alpha (1 + 1/l) < (1 - \varepsilon)(1 + \varepsilon/2) \ln(|V|) = (1 - \varepsilon') \ln(|V|)$ for $\varepsilon' = \varepsilon/2 + \varepsilon^2/2$.

By Theorem 6, if the MINIMUM DOMINATION problem can be approximated within a ratio of $(1 - \varepsilon') \ln(|V|)$, then NP $\subseteq$ DTIME($|V|^{O(\log \log |V|)}$). It follows that if the MINIMUM OUTER-CONNECTED DOMINATION problem can be approximated within a ratio of $(1 - \varepsilon) \ln(|V|)$ then NP $\subseteq$ DTIME($|V|^{O(\log \log |V|)}$).

Since $\ln |V| \approx \ln(2|V| + 1)$ for sufficiently large values of $|V|$, for a graph $G' = (V', E')$, where $|V'| = 2|V| + 1$, MINIMUM OUTER-CONNECTED DOMINATION problem cannot be approximated within a ratio of $(1 - \varepsilon) \ln |V'|$ unless NP $\subseteq$ DTIME($|V'|^{O(\log \log |V'|)}$). $\qquad \square$

# 6    APX-completeness

In this section, we show that the MOCD problem is APX-complete for bounded degree graphs.

Since $\Delta(G) \leq k$ for some integer constant $k$, the following corollary follows from Theorem 5.

**Corollary 1.** *MOCD problem for bounded degree graphs is in APX.*

Next we prove that the MOCD problem for bounded degree graphs is APX-hard. To this end, we need the concept of a very popular reduction, known as L-reduction.

**Definition 1.** *Given two NP optimization problems $F$ and $G$ and a polynomial time transformation $f$ from instances of $F$ to instances of $G$, we say that $f$ is an*

*L-reduction if there are positive constants $\alpha$ and $\beta$ such that for every instance $x$ of $F$*

1. $opt_G(f(x)) \leq \alpha \cdot opt_F(x)$.
2. *for every feasible solution $y$ of $f(x)$ with objective value $m_G(f(x), y) = c_2$ we can in polynomial time find a solution $y'$ of $x$ with $m_F(x, y') = c_1$ such that $|opt_F(x) - c_1| \leq \beta|opt_G(f(x)) - c_2|$.*

*To show the APX-completeness of a problem $\Pi \in APX$, it is enough to show that there is an L-reduction from some APX-complete problem to $\Pi$.*

Now, we are ready to proof the following result.

**Theorem 8.** *MOCD problem is APX-complete for graphs with maximum degree 4.*

*Proof.* By Corollary 1, MOCD problem for bounded degree graphs is in APX. MINIMUM DOMINATION problem is known to be APX-hard for general graphs with maximum degree 3 [13]. We describe an L-reduction $f$ from instances of the MINIMUM DOMINATION PROBLEM for graphs with maximum degree 3 to the instances of the MOCD problem. Given a graph $G = (V, E)$ of maximum degree 3, we construct a graph $G' = (V', E')$ as follows. Let $V = \{v_1, v_2, \ldots, v_n\}$. Let $V' = V \cup \{z_1, z_2, \ldots, z_n\} \cup \{y_1, y_2, \ldots, y_n\}$ and $E' = E \cup \{v_i y_i, y_i z_i | 1 \leq i \leq n\} \cup \{y_i y_{i+1} | 1 \leq i \leq n-1\}$. Note that the maximum degree of $G'$ is 4. Now let us first prove the following claim:

*Claim.* If $D^*$ is a minimum cardinality dominating set of $G$, then the cardinality of minimum outer-connected dominating set, say $D_o^*$, in $G'$ is $|D^*| + n$, where $n = |V|$.

*Proof.* Suppose $D^*$ is a minimum cardinality dominating set of $G$, then $D^* \cup \{z_i \mid 1 \leq i \leq n\}$ is an outer-connected dominating set of cardinality $|D^*| + n$. Hence the cardinality of a minimum outer-connected dominating set, say $D_o^*$ is less than or equal to $|D^*| + n$, that is, $|D_o^*| \leq |D^*| + n$.

Next suppose that $D_o^*$ is a minimum cardinality outer- connected dominating set of $G'$. Define $D_o = V \cup \{z_i \mid 1 \leq i \leq n\}$. Then $D_o$ is an outer-connected dominating set of cardinality $2n$. Hence $|D_o^*| \leq 2n$. So by Observation 1(c), all the pendant vertices of $G'$ must belong to $D_o^*$. Hence $z_i$ must belong to $D_o^*$ for all $i, 1 \leq i \leq n$. Let $D' = D_o^* \setminus \{z_i \mid 1 \leq i \leq n\}$. Let $S = \{y_1, \ldots, y_n\} \cap D'$. Let $D'' = (D' \setminus S) \cup \{v_i | y_i \in S\}$. Then $D''$ is a dominating set of $G$ and cardinality of $D''$ is less than or equal to $|D_o^*| - n$. Hence if $D^*$ is minimum dominating set then $|D^*| \leq |D_o^*| - n$. So $|D_o^*| \geq |D^*| + n$.

This completes the proof of the claim.                                   $\square$

Let $D^*$ and $D_o^*$ be a minimum dominating set and a minimum outer-connected dominating set of $G$ and $G'$, respectively. Since $G$ is of bounded degree 3, by Lemma 1, $|D^*| \geq n/4$. Hence $|D_o^*| = |D^*| + n \leq |D^*| + 4|D^*|$ i.e. $|D_o^*| \leq 5|D^*|$. Now consider any outer-connected dominating set $D_o$ of $G'$, then we have the following two cases:

**Case 1:** $z_i$ belong to $D_o$ for all $i, 1 \leq i \leq n$.
Here $y_i$ may or may not belong to $D_o$. Let $|D_o \cap \{y_1, y_2, \ldots, y_n\}| = r$ and
$|D_o \cap V(G)| = k$. Then $|D_o| = n + r + k$. Now we try to find a dominating set
$D$ of $G$. First include those $k$ vertices of $V$ in $D$, which also belong to $D_o$. If
$y_i \in D_o$ but $v_i \notin D_o$, then include $v_i$ in $D$. Suppose this happens for $k'$ values
of $i$, where $k' \leq r$. Then $D$ is a dominating set of $G$ and $|D| = k' + k$. Now
$|D_o| - |D_o^*| = (n + r + k) - |D_o^*| = r + k - |D^*| \geq k' + k - |D^*| = |D| - |D^*|$ (as
$|D_o^*| = |D^*| + n$). This implies $|D| - |D^*| \leq |D_o| - |D_o^*|$ in this case.
**Case 2:** At least one of the $z_i$ does not belong to $D_o$ for some $i$, where $1 \leq i \leq n$.
In this case all the vertices except this particular $z_i$ belong to $D_o$. Hence $|D_o| =
3n - 1$. Now take $D = D_o \cap V$. Then $D$ is a dominating set of $G$ and $|D| = n$.
Then $|D_o| - |D_o^*| = (3n-1) - (|D^*| + n) = (2n-1) - |D^*| \geq n - |D^*| = |D| - |D^*|$.
This implies $|D| - |D^*| \leq |D_o| - |D_o^*|$ in this case.

Hence $|D| - |D^*| \leq |D_o| - |D_o^*|$ in both the cases and we have shown that $f$
is an $L$-reduction with $\alpha = 5$ and $\beta = 1$.

Thus, the MOCD problem in graphs of bounded degree 4 is APX-complete.

□

Next we prove the APX-completeness of the MOCD problem for bipartite
graphs of bounded degree. A set $S \subseteq V$ of a graph $G = (V, E)$ is a *total dominating set* if $N_G(v) \cap S \neq \emptyset$ for all $v \in V$. The MINIMUM TOTAL DOMINATION
problem is to find a total dominating set of minimum cardinality of the input
graph $G$. MINIMUM TOTAL DOMINATION problem is known to be APX-complete
for bipartite graphs with maximum degree 3 [14].

**Theorem 9.** *MOCD problem is APX-complete for bipartite graphs with maximum degree 7.*

*Proof.* By Corollary 1, the MOCD problem for bounded degree bipartite graphs
is in APX. We describe an L-reduction $f$ from instances of the MINIMUM TOTAL
DOMINATION problem for bipartite graphs with maximum degree 3 to the instances of the MOCD problem for bipartite graphs of maximum degree 7. Given
a bipartite graph $G = (V, E)$ of maximum degree 3 construct a graph $G' =
(V', E')$ as follows. Let $V(G) = \{v_1, v_2, \ldots, v_n\}$. Let $V' = V \cup \{w_1, w_2, \ldots, w_n\} \cup
\{z_1, z_2, \ldots, z_n\} \cup \{y_1, y_2, \ldots, y_n\}$. Construct a spanning tree $T = (V, E_1)$ of $G$.
Let $E_R = \{w_i w_j | v_i v_j \in E_1, 1 \leq i < j \leq n\}$. Let $E^i = \{w_i v_j | v_j \in N_G(v_i)\}$. Let
$E' = E \cup E_R \cup \{w_i z_i, z_i y_i, 1 \leq i \leq n\} \cup (\cup_{i=1}^n E^i)$.

Clearly $G'$ is a bipartite graph of maximum degree 7. The graph $G = (V, E)$,
where $V = \{v_1, v_2, v_3, v_4\}$ and $E = \{v_1 v_2, v_2 v_3, v_3 v_4, v_4 v_1\}$ and the associated
graph $G'$ are shown in Fig. 3 to illustrate the above construction.

Let us first prove the following claim:

*Claim.* If $D_T^*$ is a minimum total dominating set of $G$ and $D_o^*$ is a minimum
outer-connected dominating set of $G'$, then $|D_o^*| = |D_T^*| + n$.

*Proof.* Clearly $D_T^* \cup \{y_1, y_2, \ldots, y_n\}$ is an outer-connected dominating set. Hence
$|D_o^*| \leq |D_T^*| + n$.

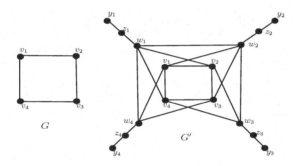

**Fig. 3.** An illustration to the construction of $G'$ from $G$

Now we construct a total dominating set of $G$ of cardinality at most $|D_o^*| - n$ from the minimum outer-connected dominating set $D_o^*$ of $G'$ as follows.

The minimum outer-connected dominating set $D_o^*$ of $G'$ will necessarily contain all the $y_i, 1 \leq i \leq n$. Given $D_o^*$, we construct an outer-connected dominating set $D_o^{**}$ such that $|D_o^*| = |D_o^{**}|$ and $D_o^{**} \cap \{w_1, w_2, \ldots, w_n\} = \emptyset$, as follows:

For each $i$, $1 \leq i \leq n$, if $w_i \in D_o^*$, then replace $w_i$ with $v_i$.

Let us call the resultant set $D_o^{**}$. Define $D' = D_o^{**} \setminus \{y_1, y_2, \ldots, y_n\}$. Now to dominate $w_i$, either $z_i$ belongs to $D'$ or some neighbor $v_j$ of $w_i$ belongs to $D'$. If $z_i$ belongs to $D'$, then remove it from $D'$ and add some neighbor $v_j$ of $w_i$ in $D'$. Then $D'$ is a total dominating set of $G$ and $|D'| \leq |D_o^{**}| - n = |D_o^*| - n$. Hence $|D_T^*| \leq |D_o^*| - n$. This proves our claim.                                    □

Since maximum degree of $G$ is 3, for any total dominating set $D_T$ of $G$, $|D_T| \geq n/3$. So $|D_T^*| \geq n/3$. Hence $|D_o^*| = |D_T^*| + n \leq |D_T^*| + 3|D_T^*|$. So $|D_o^*| \leq 4|D_T^*|$.

Now consider any outer-connected dominating set $D_o$, then we have following two cases:

**Case 1:** $y_i$ belong to $D_o$ for all $i, 1 \leq i \leq n$.

Define the sets $W = \{w_1, w_2, \ldots, w_n\}$ and $Z = \{z_1, z_2, \ldots, z_n\}$. Now we construct an outer-connected dominating set $D_o'$ from $D_o$, by replacing $w_i$ with $v_i$, whenever $w_i \in D_o, 1 \leq i \leq n$. Note that $\{y_1, y_2, \ldots, y_n\} \subseteq D_o'$ and $D_o' \cap W = \emptyset$.

So $D_o'$ is an outer-connected dominating set of same or lesser cardinality than that of $D_o$. Now suppose $|D_o' \cap Z| = r$ and $|D_o' \cap V| = k$, then $|D_o'| = n + r + k$.

Since for each $i$, $1 \leq i \leq n$, $N_{G'}(w_i) \cap V = N_G(v_i)$, $D_o' \cap V$ is a total dominating set of $G$ whenever $(N_{G'}(w_i) \cap V) \cap D_o' = N_G(v_i) \cap D_o' \neq \emptyset$ for all $i, 1 \leq i \leq n$. If not so, then suppose there exist a set of vertices $S \subseteq V$ such that for every vertex $v_j \in S$, $N_G(v_j) \cap D_o' = \emptyset$, that is, $(N_{G'}(w_j) \cap V) \cap D_o' = \emptyset$. Now since $N_{G'}(w_j) \subseteq V \cup W \cup \{z_j\}$, $z_j$ must belong to $D_o'$, as $N_G(v_j) \cap D_o' = \emptyset$. Now update $D_o'$ as $D_o' = (D_o' \setminus \{z_j\}) \cup \{v_k\}$, where $v_k \in N_G(v_j)$. Do this for all the vertices in $S$. Now define $D_T = D_o' \cap V$. Then $D_T$ is a total dominating set of $G$ and $|D_T| = k + k_1$, where $k_1 \leq r$. Now, $|D_T| - |D_T^*| = k + k_1 - |D_T^*| \leq n + r + k - (|D_T^*| + n) = |D_o'| - |D_o^*| \leq |D_o| - |D_o^*|$. This implies $|D_T| - |D_T^*| \leq |D_o| - |D_o^*|$ in this case.

**Case 2:** At least one $y_i$ does not belong to $D_o$ for some $i$, $1 \leq i \leq n$.
In this case all the vertices except this particular $y_i$ belong to $D_o$. Hence $|D_o| = 4n - 1$. Now take $D_T = D_o \cap V$. Then $D_T$ is a total dominating set of $G$ and $|D_T| = n$. Then $|D_o| - |D_o^*| = (4n - 1) - (|D_T^*| + n) = (3n - 1) - |D_T^*| \geq n - |D_T^*| = |D_T| - |D_T^*|$. This implies $|D_T| - |D_T^*| \leq |D_o| - |D_o^*|$ in this case.

Hence $|D_T| - |D_T^*| \leq |D_o| - |D_o^*|$ in both the cases and we have shown that $f$ is an $L$-reduction with $\alpha = 4$ and $\beta = 1$. $\qquad\square$

# References

[1] Haynes, T.W., Hedetniemi, S.T., Slater, P.J.: Domination in graphs: Advanced topics, vol. 209. Marcel Dekker Inc., New York (1998)
[2] Haynes, T.W., Hedetniemi, S.T., Slater, P.J.: Fundamentals of Domination in Graphs, vol. 208. Marcel Dekker Inc., New York (1998)
[3] Cyman, J.: The outer-connected domination number of a graph. Australas J. Comb. 38, 35–46 (2007)
[4] Akhbari, M.H., Hasni, R., Favaron, O., Karami, H., Sheikholeslami, S.M.: On the outer-connected domination in graphs. J. Comb. Optim. (2012), doi:10.1007/s10878-011-9427-x
[5] Keil, J.M., Pradhan, D.: Computing a minimum outer-connected dominating set for the class of chordal graphs. Inform. Process. Lett. 113, 552–561 (2013)
[6] Jiang, H., Shan, E.: Outer-connected domination number in graph. Utilitas Math. 81, 265–274 (2010)
[7] Golumbic, M.C., Gauss, C.F.: Perfect elimination and chordal bipartite graphs. J. Graph Theory 2, 155–163 (1978)
[8] Garey, M.R., Johnson, D.S.: Computers and Interactability: a guide to the theory of NP-completeness. W.H. Freeman and Co., San Francisco (1979)
[9] Yannakakis, M.: Node- and edge-deletion NP-complete problems. In: Conference Record of the Tenth Annual ACM Symposium on Theory of Computing, San Diego, Calif., pp. 253–264. ACM, New York (1978)
[10] Kloks, T., Kratsch, D., Müller, H.: Bandwidth of chain graphs. Inform. Process. Lett. 68(6), 313–315 (1998)
[11] Uehara, R., Uno, Y.: Efficient algorithms for the longest path problem. In: Fleischer, R., Trippen, G. (eds.) ISAAC 2004. LNCS, vol. 3341, pp. 871–883. Springer, Heidelberg (2004)
[12] Chlebík, M., Chlebíková, J.: Approximation hardness of dominating set problems in bounded degree graphs. Inform. and Comput. 206(11), 1264–1275 (2008)
[13] Alimonti, P., Kann, V.: Hardness of approximating problems on cubic graphs. Theoretical Computer Science 237, 123–134 (2000)
[14] Pradhan, D.: Algorithmic aspects of k-tuple total domination in graphs. Inform. Process. Lett. 112, 816–822 (2012)

# Some Results on Point Visibility Graphs

Subir Kumar Ghosh and Bodhayan Roy

School of Technology and Computer Science,
Tata Institute of Fundamental Research, Mumbai 400005, India
{ghosh,bodhayan}@tifr.res.in

**Abstract** In this paper, we present two necessary conditions for recognizing point visibility graphs. We show that this recognition problem lies in PSPACE. We state new properties of point visibility graphs along with some known properties that are important in understanding point visibility graphs. For planar point visibility graphs, we present a complete characterization which leads to a linear time recognition and reconstruction algorithm.

## 1   Introduction

The visibility graph is a fundamental structure studied in the field of computational geometry and geometric graph theory [4, 8]. Some of the early applications of visibility graphs included computing Euclidean shortest paths in the presence of obstacles [12] and decomposing two-dimensional shapes into clusters [16]. Here, we consider problems from visibility graph theory.

Let $P = \{p_1, p_2, ..., p_n\}$ be a set of points in the plane (see Fig. 1). We say that two points $p_i$ and $p_j$ of $P$ are *mutually visible* if the line segment $p_ip_j$ does not contain or pass through any other point of $P$. In other words, $p_i$ and $p_j$ are visible if $P \cap p_ip_j = \{p_i, p_j\}$. If two vertices are not visible, they are called an *invisible pair*. For example, in Fig. 1(c), $p_1$ and $p_5$ form a visible pair whereas $p_1$ and $p_3$ form an invisible pair. If a point $p_k \in P$ lies on the segment $p_ip_j$ connecting two points $p_i$ and $p_j$ in $P$, we say that $p_k$ blocks the visibility between $p_i$ and $p_j$, and $p_k$ is called a *blocker* in $P$. For example in Fig. 1(c), $p_5$ blocks the visibility between $p_1$ and $p_3$ as $p_5$ lies on the segment $p_1p_3$. The *visibility graph* (also called the *point visibility graph* (PVG)) $G$ of $P$ is defined by associating a vertex $v_i$ with each point $p_i$ of $P$ such that $(v_i, v_j)$ is an undirected edge of $G$ if and only if $p_i$ and $p_j$ are mutually visible (see Fig. 1(a)). Observe that if no three points of $P$ are collinear, then $G$ is a complete graph as each pair of points in $P$ is visible since there is no blocker in $P$. Sometimes the visibility graph is drawn directly on the point set, as shown in Figs. 1(b) and 1(c), which is referred to as a *visibility embedding* of $G$.

Given a point set $P$, the visibility graph $G$ of $P$ can be computed as follows. For each point $p_i$ of $P$, the points of $P$ are sorted in angular order around $p_i$. If two points $p_j$ and $p_k$ are consecutive in the sorted order, check whether $p_i$, $p_j$ and $p_k$ are collinear points. By traversing the sorted order, all points of $P$, that are not visible from $p_i$, can be identified in $O(n \log n)$ time. Hence, $G$ can be

S.P. Pal and K. Sadakane (Eds.): WALCOM 2014, LNCS 8344, pp. 163–175, 2014.

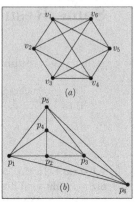

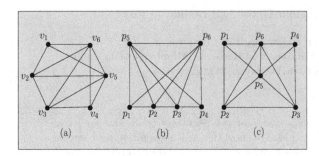

**Fig. 1.** (a) A point visibility graph with $(v_1, v_2, v_3, v_4)$ as a CSP. (b) A visibility embedding of the point visibility graph where $(p_1, p_2, p_3, p_4)$ is a GSP. (c) A visibility embedding of the point visibility graph where $(p_1, p_2, p_3, p_4)$ is not a GSP.

**Fig. 2.** (a) A planar graph $G$. (b) A planar visibility embedding of $G$.

computed from $P$ in $O(n^2 \log n)$ time. Using the result of Chazelle et al. [3] or Edelsbrunner et al. [6], the time complexity of the algorithm can be improved to $O(n^2)$ by computing sorted angular orders for all points together in $O(n^2)$ time.

Consider the opposite problem of determining if there is a set of points $P$ whose visibility graph is the given graph $G$. This problem is called the visibility graph *recognition* problem. Identifying the set of properties satisfied by all visibility graphs is called the visibility graph *characterization* problem. The problem of actually drawing one such set of points $P$ whose visibility graph is the given graph $G$, is called the visibility graph *reconstruction* problem.

Here we consider the recognition problem: Given a graph $G$ in adjacency matrix form, determine whether $G$ is the visibility graph of a set of points $P$ in the plane [9]. In Sect. 2, we present two necessary conditions for this recognition problem In the same section, we establish new properties of point visibility graphs, and in addition, we state some known properties with proofs that are important in understanding point visibility graphs. Though the first necessary condition can be tested in $O(n^3)$ time, it is not clear whether the second necessary condition can be tested in polynomial time. On the other hand, we show in Sect. 3 that the recognition problem lies in PSPACE.

If a given graph $G$ is planar, there can be three cases: (i) $G$ has a planar visibility embedding (Fig. 2), (ii) $G$ admits a visibility embedding, but no visibility embedding of $G$ is planar (Fig. 3), and (iii) $G$ does not have any visibility embedding (Fig. 4). Case (i) has been characterized by Eppstein [5] by presenting four infinite families of $G$ and one particular graph. In order to characterize graphs in Case (i) and Case (ii), we show that two infinite families and five particular graphs are

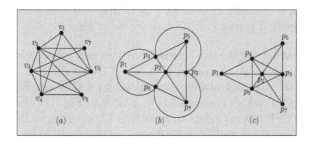
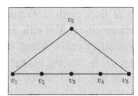

**Fig. 3.** (a) A planar graph $G$. (b) A planar embedding of $G$. (c) A non-planar visibility embedding of $G$.

**Fig. 4.** A planar graph $G$ that does not admit a visibility embedding

required in addition to graphs for Case (i). Using this characterization, we present an $O(n)$ algorithm for recognizing and reconstructing $G$ in Sect. 4. Note that this algorithm does not require any prior embedding of $G$. Finally, we conclude the paper with a few remarks.

## 2    Properties of Point Visibility Graphs

Consider a subset $S$ of vertices of $G$ such that their corresponding points C in a visibility embedding $\xi$ of $G$ are collinear. The path formed by the points of C is called a *geometric straight path* (GSP). For example, the path $(p_1, p_2, p_3, p_4)$ in Fig. 1(b) is a GSP as the points $p_1$, $p_2$, $p_3$ and $p_4$ are collinear. Note that there may be another visibility embedding $\xi$ of $G$ as shown in Fig. 1(c), where points $p_1$, $p_2$, $p_3$ and $p_4$ are not collinear. So, the points forming a GSP in $\xi$ may not form a GSP in every visibility embedding of $G$. If a GSP is a maximal set of collinear points, it is called a *maximal geometric straight path* (max GSP). A GSP of $k$ collinear points is denoted as $k$-*GSP*. In the following, we state some properties of PVGs and present two necessary conditions for recognizing $G$.

**Lemma 1.** *If $G$ is a PVG but not a path, then for any GSP in any visibility embedding of $G$, there is a point visible from all the points of the GSP[11].*

*Proof.* For every GSP, there exists a point $p_i$ whose perpendicular distance to the line containing the GSP is the smallest. So, all points of the GSP are visible from $p_i$. ☐

**Lemma 2.** *If $G$ admits a visibility embedding $\xi$ having a $k$-GSP, then the number of edges in $G$ is at least $(k-1) + k(n-k)$.*

*Proof.* Let $p_i$ and $p_j$ be two points of $\xi$ such that $p_i$ is a point of the $k$-GSP and $p_j$ is not. Consider the segment $p_i p_j$. If $p_i$ and $p_j$ are mutually visible, then $(v_i, v_j)$ is an edge in $G$. Otherwise, there exists a blocker $p_k$ on $p_i p_j$ such that $(v_j, v_k)$ is an edge in $G$. So, $p_j$ has an edge in the direction towards $p_i$. Therefore, for every such pair $p_i$ and $p_j$, there is an edge in $G$. So, $(n-k)k$ such pairs in $\xi$ correspond to $(n-k)k$ edges in $G$. Moreover, there are $(k-1)$ edges in $G$ corresponding to the $k$-GSP. Hence, $G$ has at least $(k-1) + k(n-k)$ edges. ☐

**Corollary 1.** *If a point $p_i$ in a visibility embedding $\xi$ of $G$ does not belong to a $k$-GSP in $\xi$, then its corresponding vertex $v_i$ in $G$ has degree at least $k$.*

Let $H$ be a path in $G$ such that no edges exist between any two non-consecutive vertices in $H$. We call $H$ a *combinatorial straight path* (*CSP*). Observe that in a visibility embedding of $G$, $H$ may not always correspond to a GSP. In Fig. 1(a), $H = (v_1, v_2, v_3, v_4)$ is a CSP which corresponds to a GSP in Fig. 1(b) but not in Fig. 1(c). Note that a CSP always refers to a path in $G$, whereas a GSP refers to a path in a visibility embedding of $G$. A CSP that is a maximal path, is called a *maximal combinatorial straight path* (*max CSP*). A CSP of $k$-vertices is denoted as $k$-*CSP*.

**Lemma 3.** *$G$ is a PVG and bipartite if and only if the entire $G$ is a CSP.*

*Proof.* If the entire $G$ can be embedded as a GSP, then alternating points in the GSP form the bipartition and the lemma holds. Otherwise, there exists at least one max GSP which does not contain all the points. By Lemma 1, there exists one point $p_i$ adjacent to all points of the GSP. So, $p_i$ must belong to one partition and all points of the GSP (having edges) belong to the other partition. Hence, $G$ cannot be a bipartite graph, a contradiction. The other direction of the proof is trivial.    □

**Corollary 2.** *$G$ is a PVG and triangle-free if and only if the entire $G$ is a CSP.*

**Lemma 4.** *If $G$ is a PVG, then the size of the maximum clique in $G$ is bounded by twice the minimum degree of $G$, and the bound is tight.*

*Proof.* In a visibility embedding of $G$, draw rays from a point $p_i$ of minimum degree through every visible point of $p_i$. Observe that any ray may contain several points not visible from $p_i$. Since any clique can have at most two points from the same ray, the size of the clique is at most twice the number of rays, which gives twice the minimum degree of $G$.    □

**Lemma 5.** *If $G$ is a PVG and it has more than one max CSP, then the diameter of $G$ is 2 [11].*

*Proof.* If two vertices $v_i$ and $v_j$ are not adjacent in $G$, then they belong to a CSP L of length at least two. By Lemma 1, there must be some vertex $v_k$ that is adjacent to every vertex in L. $(v_i, v_k, v_j)$ is the required path of length 2. Therefore, the diameter of $G$ cannot be more than two.    □

**Corollary 3.** *If $G$ is a PVG but not a path, then the BFS tree of $G$ rooted at any vertex $v_i$ of $G$ has at most three levels consisting of $v_i$ in the first level, the neighbours of $v_i$ in $G$ in the second level, and the rest of the vertices of $G$ in the third level.*

**Lemma 6.** *If $G$ is a PVG but not a path, then the subgraph induced by the neighbours of any vertex $v_i$, excluding $v_i$, is connected.*

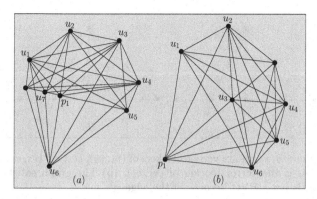

**Fig. 5.** (a) The points $(u_1, u_2, ..., u_7, u_1)$ are visible from an internal point $p_1$. (b) The points $(u_1, u_2, ..., u_6)$ are visible from a convex hull point $p_1$.

*Proof.* Consider a visibility embedding of $G$ where $G$ is not a path. Let $(u_1, u_2, ..., u_k, u_1)$ be the visible points of $p_i$ in clockwise angular order. If $p_i$ is not a convex hull point, then $(u_1, u_2), (u_2, u_3), ..., (u_{k-1}, u_k), (u_k, u_1)$ are visible pairs (Fig. 5(a)). If $p_i$, $u_1$ and $u_k$ are convex hull points, then $(u_1, u_2), (u_2, u_3), ..., (u_{k-1}, u_k)$ are visible pairs (Fig. 5(b)). Since there exists a path between every pair of points in $(u_1, u_2, ..., u_k, u_1)$, the subgraph induced by the neighbours of $v_i$ is connected.    □

**Necessary Condition 1** *If $G$ is not a CSP, then the BFS tree of $G$ rooted at any vertex can have at most three levels, and the induced subgraph formed by the vertices in the second level must be connected.*

*Proof.* Follows from Corollary 3 and Lemma 6.    □

As defined for point sets, if two vertices $v_i$ and $v_j$ of $G$ are adjacent (or, not adjacent) in $G$, $(v_i, v_j)$ is referred to as a *visible pair* (respectively, *invisible pair*) of $G$. Let $(v_1, v_2, ..., v_k)$ be a path in $G$ such that no two non-consecutive vertices are connected by an edge in $G$ (Fig. 6(a)). For any vertex $v_j$, $2 \leq j \leq k - 1$, $v_j$ is called a *vertex-blocker* of $(v_{j-1}, v_{j+1})$ as $(v_{j-1}, v_{j+1})$ is not an edge in $G$ and both $(v_{j-1}, v_j)$ and $(v_j, v_{j+1})$ are edges in $G$. In the same way, consecutive vertex-blockers on such a path are also called *vertex-blockers*. For example, $v_m * v_{m+1}$ is a vertex-blocker of $(v_{m-1}, v_{m+2})$ for $2 \leq m \leq k - 2$. Note that $*$ represents concatenation of consecutive vertex-blockers.

Consider the graph in Fig. 6(b). Though $G$ satisfies Necessary Condition 1, it is not a PVG because it does not admit a visibility embedding. It can be seen that this graph without the edge $(v_2, v_4)$ admits a visibility embedding (see Fig. 6(a)), where $(v_1, v_2, v_3, v_4, v_5)$ forms a GSP. However, $(v_2, v_4)$ demands visibility between two non-consecutive collinear blockers which cannot be realized in any visibility embedding.

**Necessary Condition 2** *There exists an assignment of vertex-blockers to invisible pairs in $G$ such that:*

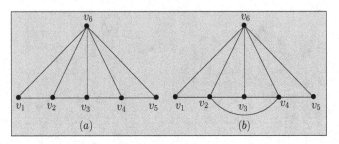

**Fig. 6.** (a) Vertices $v_2$, $v_3$, $v_4$ are vertex-blockers of $(v_1, v_3)$, $(v_3, v_4)$ $(v_3, v_5)$ respectively. Also, $v_2 * v_3 * v_4$ is the vertex-blocker of $(v_1, v_5)$. (b) The graph satisfies Necessary Condition 1 but is not a PVG because of the edge $(v_2, v_4)$.

1. *Every invisible pair is assigned one vertex-blocker.*
2. *If two invisible pairs in $G$ sharing a vertex $v_i$ (say, $(v_i, v_j)$ and $(v_i, v_k)$ ), and their assigned vertex-blockers are not disjoint, then all vertices in the two assigned vertex-blockers along with vertices $v_i$, $v_j$ and $v_k$ must be a CSP in $G$.*
3. *If two invisible paris in $G$ are sharing a vertex $v_i$ (say, $(v_i, v_j)$ and $(v_i, v_k)$), and $v_k$ is assigned as a vertex blocker to $(v_i, v_j)$, then $v_j$ is not assigned as a vertex blocker to $(v_i, v_k)$.*

*Proof.* In a visibility embedding of $G$, every segment connecting two points, that are not mutually visible, must pass through another point or a set of collinear points, and they correspond to vertex-blockers in $G$.

Since $(v_i, v_j)$ and $(v_i, v_k)$ are invisible pairs, the segments $(p_i, p_j)$ and $(p_i, p_k)$ must contain points. If there exists a point $p_m$ on both $p_i p_j$ and $p_i p_k$, then points $p_i$, $p_m$, $p_j$, $p_k$ must be collinear. So, $v_i$, $v_m$, $v_j$ and $v_k$ must belong to a CSP.

Since $(v_i, v_j)$ and $(v_i, v_k)$ are invisible pairs, the segments $(p_i, p_j)$ and $(p_i, p_k)$ must contain points. If the point $p_k$ lies on $p_i p_j$, then $p_j$ cannot lie on $p_i p_k$, because it contradicts the order of points on a line. □

**Lemma 7.** *If the size of the longest GSP in some visibility embedding of a graph $G$ with $n$ vertices is $k$, then the degree of each vertex in $G$ is at least $\lceil \frac{n-1}{k-1} \rceil$ [14, 13, 15].*

*Proof.* For any point $p_i$ in a visibility embedding of $G$, the degree of $p_i$ is the number of points visible from $p_i$ which are in angular order around $p_i$. Since the longest GSP is of size k, a ray from $p_i$ through any visible point of $p_i$ can contain at most $k - 1$ points excluding $p_i$. So there must be at least $\lceil \frac{n-1}{k-1} \rceil$ such rays, which gives the degree of $p_i$. □

**Theorem 1.** *If $G$ is a PVG but not a path, then $G$ has a Hamiltonian cycle.*

*Proof.* Let $H_1, H_2, ..., H_k$ be the convex layers of points in a visibility embedding of $G$, where $H_1$ and $H_k$ are the outermost and innermost layers respectively. Let $p_i p_j$ be an edge of $H_1$, where $p_j$ is the next clockwise point of $p_i$ on $H_1$ (Fig.

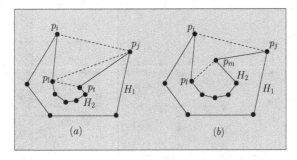

**Fig. 7.** (a) The left tangents of $p_i$ and $p_j$ meet $H_2$ at the same point $p_l$. (b) The left tangents of $p_i$ and $p_j$ meet $H_2$ at points $p_l$ and $p_m$ of the same edge.

7(a)). Draw the left tangent of $p_i$ to $H_2$ meeting $H_2$ at a point $p_l$ such that the entire $H_2$ is to the left of the ray starting from $p_i$ through $p_l$. Similarly, draw the left tangent from $p_j$ to $H_2$ meeting $H_2$ at a point $p_m$. If $p_l = p_m$ then take the next clockwise point of $p_l$ in $H_2$ and call it $p_t$. Remove the edges $p_i p_j$ and $p_l p_t$, and add the edges $p_i p_l$ and $p_j p_t$ (Fig. 7(a)). Consider the other situation where $p_l \neq p_m$. If $p_l p_m$ is an edge, then remove the edges $p_i p_j$ and $p_l p_m$, and add the edges $p_i p_l$ and $p_j p_m$ (Fig. 7(b)). If $p_l p_m$ is not an edge of $H_2$, take the next counterclockwise point of $p_m$ on $H_2$ and call it $p_q$. Remove the edges $p_i p_j$ and $p_q p_m$, and add the edges $p_i p_q$ and $p_j p_m$ (Fig. 8(a)).

Thus, $H_1$ and $H_2$ are connected forming a cycle $C_{1,2}$. Without the loss of generality, we assume that $p_m \in H_2$ is the next counter-clockwise point of $p_j$ in $C_{1,2}$ (Fig. 8(b)). Starting from $p_m$, repeat the same construction to connect $C_{1,2}$ with $H_3$ forming $C_{1,3}$. Repeat till all layers are connected to form a Hamiltonian cycle $C_{1,k}$. Note that if $H_k$ is just a path (Fig. 8(b)), it can be connected trivially to form $C_{1,k}$.                                                                    □

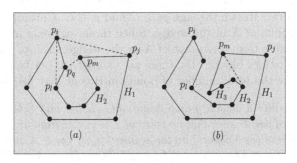

**Fig. 8.** (a) The left tangents of $p_i$ and $p_j$ meet $H_2$ points $p_l$ and $p_m$ of different edges. (b) The innermost convex layer is a path which is connected to $C_{1,2}$.

**Corollary 4.** *Given $G$ and a visibility embedding of $G$, a Hamiltonian cycle in $G$ can be constructed in linear time.*

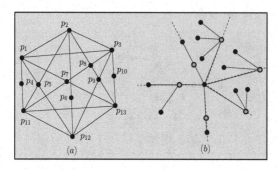

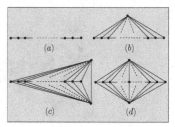

**Fig. 9.** (a) A PVG with A = $\{p_1, p_2, p_3\}$, B=$\{p_4,$ $p_5, p_6, p_7, p_8, p_9, p_{10}\}$ and C=$\{p_{11}, p_{12}, p_{13}\}$. (b) Points of A and C connected by edges representing blockers.

**Fig. 10.** These four infinite families admit planar visibility embedding (Eppstein [5])

*Proof.* This is because the combinatorial representation of G contains all its edges, and hence the gift-wrapping algorithm for finding the convex layers of a point set becomes linear in the input size.  □

**Lemma 8.** *Consider a visibility embedding of G. Let A, B and C be three nonempty, disjoint sets of points in it such that $\forall p_i \in A$ and $\forall p_j \in C$, the GSP between $p_i$ and $p_j$ contains at least one point from B, and no other point from A or C (Fig. 9(a)). Then $|B| \geq |A| + |C| - 1$ [14, 13, 15].*

*Proof.* Draw rays from a point $p_i \in A$ through every point of C (Fig. 9(b)). These rays partition the plane into $|C|$ wedges. Since points of C are not visible from $p_i$, there is at least one blocker lying on each ray between $p_i$ and the point of C on the ray. So, there are at least $|C|$ number of such blockers. Consider the remaining $|A - 1|$ points of A lying in different wedges. Consider a wedge bounded by two rays drawn through $p_k \in C$ and $p_l \in C$. Consider the segments from $p_k$ to all points of A in the wedge. Since these segments meet only at $p_k$, and $p_k$ is not visible from any point of A in the wedge, each of these segments must contain a distinct blocker. So, there are at least $|A| - 1$ blockers in all the wedges. Therefore the total number of points in B is at least $|A| + |C| - 1$.  □

**Lemma 9.** *Consider a visibility embedding of G. Let A and C be two nonempty and disjoint sets of points such that no point of A is visible from any point of C. Let B be the set of points (or blockers) on the segment $p_i p_j$, $\forall p_i \in A$ and $\forall p_j \in C$, and blockers in B are allowed to be points of A or C. Then $|B| \geq |A| + |C| - 1$ [15].*

*Proof.* Draw rays from a point $p_i \in A$ through every point of C. These rays partition the plane into at most $|C|$ wedges. Consider a wedge bounded by two rays drawn through $p_k \in C$ and $p_l \in C$. Since these rays may contain other points of A and C, all points between $p_i$ and the farthest point from $p_i$ on a ray, are blockers in B. Observe that all these blockers except one may be from A or C. Thus, excluding $p_i$, B has at least as many points as from A and C on the ray.

Consider the points of A inside the wedge. Draw segments from $p_k$ to all points of A in the wedge. Since these segments may contain multiple points from A, all points on a segment between $p_k$ and the farthest point from $p_k$ are blockers in B. All these points except one may be from A. Thus, B has at least as many points as from A inside the wedge. Therefore the total number of points in B is at least $|A| + |C| - 1$.                                                        □

## 3  Computational Complexity of the Recognition Problem

In this section we show that the recognition problem for a PVG lies in PSPACE. Our technique in the proof follows a similar technique used by Everett [7] for showing that the recognition problem for polygonal visibility is in PSPACE. We start with the following theorem of Canny [2].

**Theorem 2.** *Any sentence in the existential theory of the reals can be decided in PSPACE.*

A sentence in the first order theory of the reals is a formula of the form :

$$\exists x_1 \exists x_2 ... \exists x_n P(x_1, x_2, ..., x_n)$$

where the $x_i's$ are variables ranging over the real numbers and where $P(x_1, x_2, ..., x_n)$ is a predicate built up from $\neg$, $\wedge$, $\vee$, $=$, $<$, $>$ , $+$, $\times$, 0, 1 and -1 in the usual way.

**Theorem 3.** *The recognition problem for point visibility graphs lies in PSPACE.*

*Proof.* Given a graph $G(V, E)$, we construct a formula in the existential theory of the reals polynomial in size of $G$ which is true if and only if $G$ is a point visibility graph.

Suppose $(v_i, v_j) \notin E$. This means that if $G$ admits a visibility embedding, then there must be a blocker (say, $p_k$) on the segment joining $p_i$ and $p_j$. Let the coordinates of the points $p_i$, $p_j$ and $p_k$ be $(x_i, y_i)$, $(x_j, y_j)$ and $(x_k, y_k)$ respectively. So we have :

$$\exists t \in \mathbb{R} \Big( \big(0 < t\big) \wedge \big(t < 1\big) \wedge \big((x_k - x_i) = t \times (x_j - x_i)\big) \wedge \big((y_k - y_i) = t \times (y_j - y_i)\big) \Big)$$

Now suppose $(v_i, v_j) \in E$. This means that if $G$ admits a visibility embedding, no point in $P$ lies on the segment connecting $p_i$ and $p_j$ to ensure visibility. So, (i) either $p_k$ forms a triangle with $p_i$ and $p_j$ or (ii) $p_k$ lies on the line passing through $p_i$ and $p_j$ but not between $p_i$ and $p_j$. Determinants of non-collinear points is non-zero. So we have :

$$\exists t \in \mathbb{R} \Big( \big(det(x_i, x_j, x_k, y_i, y_j, y_k) > 0\big) \vee \big(det(x_i, x_j, x_k, y_i, y_j, y_k) < 0\big) \Big) \vee \Big( (t > 1) \vee \big(t < -1\big) \wedge \big((x_k - x_i) = t \times (x_j - x_i)\big) \wedge \big((y_k - y_i) = t \times (y_j - y_i)\big) \Big)$$

For each triple $(v_i, v_j, v_k)$ of vertices in $V$, we add a $t = t_{i,j,k}$ to the existential part of the formula and the corresponding portion to the predicate. So the formula becomes:

$$\exists x_1 \exists y_1 ... \exists x_n \exists y_n \exists t_{1,2,3} .... \exists t_{n-2,n-1,n} \ P(x_1, y_1, ..., x_n, y_n, t_{1,2,3}, ..., t_{n-2,n-1,n})$$

which is of size $O(n^3)$. This proves our theorem.                    □

## 4    Planar Point Visibility Graphs

In this section, we present a characterization, recognition and reconstruction of planar point visibility graphs. Let $G$ be a given planar graph. We know that the planarity of $G$ can be tested in linear time [1]. If $G$ is planar, a straight line embedding of $G$ can also be constructed in linear time. However, this embedding may not satisfy the required visibility constraints, and therefore, it cannot be a visibility embedding. We know that collinear points play a crucial role in a visibility embedding of $G$. It is, therefore, important to identify points belonging to a GSP of maximum length. Using this approach, we construct a visibility embedding of a given planar graph $G$, if it exists. We have the following lemmas on visibility embeddings of $G$.

**Lemma 10.** *Assume that $G$ admits a visibility embedding $\xi$. If $\xi$ has at least one $k$-GSP for $k \geq 4$, then the number of vertices in $G$ is at most*

$$k + \left\lfloor \frac{2k - 5}{k - 3} \right\rfloor$$

*Proof.* By Lemma 2, $G$ can have at least $(k - 1) + (n - k)k$ edges. By applying Euler's criterion for planar graphs, we have the following inequality on the number of permissible edges of $G$.

$$
\begin{aligned}
& (k - 1) + (n - k)k \leq 3(n) - 6 \\
\Rightarrow \; & (k - 1) + (n - k)k \leq 3(k + n - k) - 6 \\
\Rightarrow \; & (k - 1) + (n - k)k \leq 3k + 3(n - k) - 6 \\
\Rightarrow \; & (n - k)(k - 3) \quad\;\; \leq 2k - 5 \\
\Rightarrow \; & (n - k) \qquad\qquad\;\; \leq \frac{2k - 5}{k - 3}
\end{aligned}
$$

$$\tag{1}$$

Since $(n - k)$ must be an integer, we have

$$(n - k) \leq \left\lfloor \frac{2k - 5}{k - 3} \right\rfloor$$

$$\Rightarrow n \quad\; \leq k + \left\lfloor \frac{2k - 5}{k - 3} \right\rfloor \tag{2}$$

□

**Corollary 5.** *There are six infinite families of planar graphs $G$ that admit a visibility embedding $\xi$ with a $k$-GSP for $k \geq 5$ (Figs. 10 and 11).*

*Proof.* For $k \geq 5, n \leq k + 2$. There can be only six infinite families of graphs having at most two points outside a maximum size GSP in $\xi$ (denoted as $l$) as follows.

1. There is no point lying outside $l$ in $\xi$ (see Fig. 10(a)).
2. There is only one point lying outside $l$ in $\xi$ that is adjacent to all points in $l$ (see Fig. 10(b)).

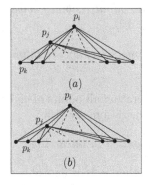

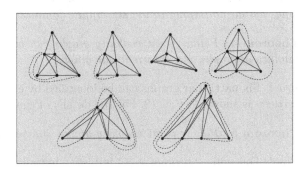

**Fig. 11.** These two infinite families do not admit planar visibility embedding

**Fig. 12.** Six planar PVGs that do not belong to any of the six families. Dotted lines show how the edge-crossings in the visibility embedding can be avoided in a planar embedding.

3. There are two points lying outside $l$ in $\xi$ that are adjacent to all other points in $\xi$ (see Fig. 10(c)).
4. There are two points lying outside $l$ in $\xi$ that are not adjacent to each other but adjacent to all points of $l$ in $\xi$ (see Fig. 10(d)).
5. There are two points $p_i$ and $p_j$ lying outside $l$ in $\xi$ such that $p_i$ and $p_j$ are adjacent to all other points in $\xi$ except an endpoint $p_k$ of $l$ as $p_j$ is a blocker on $p_i p_k$ (see Fig. 11(a)).
6. Same as the previous case, except $p_k$ is now an intermediate point of $l$ in $\xi$ (see Fig. 11(b)).

$\square$

Let us identify those graphs that do not belong to these six infinite families. We show in the following that such graphs can have a maximum of eight vertices.

**Lemma 11.** *Assume that $G$ admits a visibility embedding $\xi$. If $\xi$ has at least one 4-GSP, then the number of vertices in $G$ is at most seven.*

*Proof.* Putting $k = 4$ in the formula of Lemma 10, we get $n \leq 7$.     $\square$

**Lemma 12.** *Assume that $G$ admits a visibility embedding $\xi$. If $G$ has at least one 3-CSP but no 4-CSP, then $G$ has at most eight vertices.*

*Proof.* Since $G$ has no 4-CSP, and $G$ is not a clique, there is a 3-GSP in $\xi$. Starting from the 3-GSP, points are added one at a time to construct $\xi$. Since no subsequent point can be added on the line passing through points of the 3-GSP to prevent forming a 4-GSP, adding the fourth and fifth points gives at least three edges each in $\xi$. As $\xi$ does not have a 4-CSP, there can be at most one blocker between an invisible pair of points in $\xi$. So, for the subsequent points, at least $\lceil \frac{i-1}{2} \rceil$ edges are added for the $ith$ point. Since $G$ is planar, by Euler's condition we must have: $8 + \sum_{i=6}^{n} \lceil \frac{i-1}{2} \rceil \leq 3n - 6$. This inequality is valid only up to $n = 8$.     $\square$

**Lemma 13.** *There are six distinct planar graphs G that admit visibility embeddings but do not belong to the six infinite families (Fig. 12).*

**Theorem 4.** *Planar point visibility graphs can be characterized by six infinite families of graphs and six particular graphs.*

*Proof.* Six particular graphs can be identified by enumerating all points of eight vertices as shown in Fig. 12. For the details of the enumeration, see [10].    □

**Theorem 5.** *Planar point visibility graphs can be recognized in linear time.*

*Proof.* Following Theorem 4, $G$ is tested initially whether it is isomorphic to any of the six particular graphs for $n \leq 8$. Then, the maximum CSP is identified before its adjacency is tested with the remaining one or two vertices of $G$. The entire testing can be carried out in linear time.    □

**Corollary 6.** *Planar point visibility graphs can be reconstructed in linear time.*

*Proof.* Theorem 5 gives the relative positions and collinearity of points in the visibility embedding of $G$. Since each point can be drawn with integer coordinates of size $O(log n)$ bits, $G$ can be reconstructed in linear time.    □

## 5    Concluding Remarks

We have given two necessary conditions for recognizing point visibility graphs (which is still an open problem). Though the first necessary condition can be tested in $O(n^3)$ time, it is not clear how vertex-blockers can be assigned to every invisible pair in $G$ in polynomial time satisfying the second necessary condition. Observe that these assignments in a visibility embedding give the ordering of collinear points along any ray starting from any point through its visible points. These rays together form an arrangement of rays in the plane. It is open whether such an arrangement can be constructed satisfying assigned vertex-blockers.

Let us consider the complexity issues of the problems of Vertex Cover, Independent Set and Maximum Clique in a point visibility graph. Let $G$ be a graph of $n$ vertices, not necessarily a PVG. We construct another graph $G'$ such that (i) $G$ is an induced subgraph of $G'$, and (ii) $G'$ is a PVG. Let $C$ be a convex polygon drawn along with all its diagonals, where every vertex $v_i$ of $G$ corresponds to a vertex $p_i$ of $C$. For every edge $(v_i, v_j) \notin G$, introduce a blocker $p_t$ on the edge $(p_i, p_j)$ such that $p_t$ is visible to all points of $C$ and all blockers added so far. Add edges from $p_t$ to all vertices of $C$ and blockers in $C$. The graph corresponding to this embedding is called $G'$. So, $G'$ and its embedding can be constructed in polynomial time. Let the sizes of the minimum vertex cover, maximum independent set and maximum clique in $G$ be $k_1$, $k_2$ and $k_3$ respectively. If $x$ is the number of blockers added to $C$, then the sizes of the minimum vertex cover, maximum independent set and maximum clique in $G'$ are $k_1 + x$, $k_2$ and $k_3 + x$ respectively. Hence, the problems remain NP-Hard.

**Theorem 6.** *The problems of Vertex Cover, Independent Set and Maximum Clique remain NP-hard on point visibility graphs.*

**Acknowledgements.** The preliminary version of a part of this work was submitted in May, 2011 as a Graduate School Project Report of Tata Institute of Fundamental Research [15]. The authors would like to thank Sudebkumar Prasant Pal for his helpful comments during the preparation of the manuscript.

# References

[1] Boyer, J.M., Myrvold, W.J.: On the Cutting Edge: Simplified O(n) Planarity by Edge Addition. Journal of Graph Algorithms and Applications 8(3), 241–273 (2004)

[2] Canny, J.: Some algebraic and geometric computations in PSPACE. In: Proceedings of the 20th Annual ACM Symposium on Theory of Computing, pp. 460–467 (1988)

[3] Chazelle, B., Guibas, L.J., Lee, D.T.: The power of geometric duality. BIT 25, 76–90 (1985)

[4] de Berg, M., Cheong, O., Kreveld, M., Overmars, M.: Computational Geometry, Algorithms and Applications, 3rd edn. Springer (2008)

[5] Dujmovic, V., Eppstein, D., Suderman, M., Wood, D.R.: Drawings of planar graphs with few slopes and segments. In: Computational Geometry Theory and Applications, pp. 194–212 (2007)

[6] Edelsbrunner, H., O'Rourke, J., Seidel, R.: Constructing arrangements of lines and hyperplanes with applications. SIAM Journal on Computing 15, 341–363 (1986)

[7] Everett, H.: Visibility Graph Recognition. Ph. D. Thesis, University of Toronto, Toronto (January 1990)

[8] Ghosh, S.K.: Visibility Algorithms in the Plane. Cambridge University Press (2007)

[9] Ghosh, S.K., Goswami, P.P.: Unsolved problems in visibility graphs of points, segments and polygons. ACM Computing Surveys 46(2), 22:1–22:29 (2014)

[10] Ghosh, S.K., Roy, B.: Some results on point visibility graphs. arXiv:1209.2308 (September 2012)

[11] Kára, J., Pór, A., Wood, D.R.: On the Chromatic Number of the Visibility Graph of a Set of Points in the Plane. Discrete & Computational Geometry 34(3), 497–506 (2005)

[12] Lozano-Perez, T., Wesley, M.A.: An algorithm for planning collision-free paths among polyhedral obstacles. Communications of ACM 22, 560–570 (1979)

[13] Payne, M.S., Pór, A., Valtr, P., Wood, D.R.: On the connectivity of visibility graphs. Discrete & Computational Geometry 48(3), 669–681 (2012)

[14] Payne, M.S., Pór, A., Valtr, P., Wood, D.R.: On the connectivity of visibility graphs. arXiv:1106.3622v1 (June 2011)

[15] Roy, B.: Recognizing point visibility graphs. Graduate School Project Report, Tata Institute of Fundamental Research (May 2011)

[16] Shapiro, L.G., Haralick, R.M.: Decomposition of two-dimensional shape by graph-theoretic clustering. IEEE Transactions on Pattern Analysis and Machine Intelligence PAMI-1, 10–19 (1979)

# Some Extensions
# of the Bottleneck Paths Problem*

Tong-Wook Shinn and Tadao Takaoka

Department of Computer Science and Software Engineering
University of Canterbury
Christchurch, New Zealand

**Abstract.** We extend the well known bottleneck paths problem in two directions for directed unweighted graphs with positive real edge capacities. Firstly we narrow the problem domain and compute the bottleneck of the entire network in $O(m \log n)$ time, where $m$ and $n$ are the number of edges and vertices in the graph, respectively. Secondly we enlarge the domain and compute the shortest paths for all possible bottleneck amounts. We present a combinatorial algorithm to solve the Single Source Shortest Paths for All Flows (SSSP-AF) problem in $O(mn)$ worst case time, followed by an algorithm to solve the All Pairs Shortest Paths for All Flows (APSP-AF) problem in $O(\sqrt{t}n^{(\omega+9)/4})$ time, where $t$ is the number of distinct edge capacities and $O(n^\omega)$ is the time taken to multiply two $n$-by-$n$ matrices over a ring. We also discuss practical applications for these new problems.

## 1 Introduction

The bottleneck (capacity) of a path is the minimum capacity of all edges on the path. Thus the bottleneck is the maximum flow that can be pushed through this path. The bottleneck of a pair of vertices $(i, j)$ is the maximum of all bottleneck values of all paths from $i$ to $j$. The bottleneck paths problems are important in various areas, such as logistics and computer networks. In this paper we consider two extensions to this well known problem on directed unweighted (unit edge cost) graphs with positive real edge capacities.

The bottleneck of the (entire) network is the minimum bottleneck out of all bottlenecks for all pairs $(i, j)$. We refer to the problem of finding the bottleneck of the entire network as the Graph Bottleneck (GB) problem. In this paper we introduce a simple algorithm based on binary search to show that we can solve the GB problem faster than solving the All Pairs Bottleneck Paths (APBP) problem. The method is based on determining the strongly connected components and the time complexity of the algorithm is $O(m \log n)$, where $m$ is the number of edges and $n$ is the number of vertices in the graph. This algorithm is simple but effective, and provides a good starting point for this paper.

* This research was supported by the EU/NZ Joint Project, Optimization and its Applications in Learning and Industry (OptALI).

S.P. Pal and K. Sadakane (Eds.): WALCOM 2014, LNCS 8344, pp. 176–187, 2014.

Consider the shortest path from vertex $u$ to vertex $v$ that can push a flow of amount up to $f$. If the flow demand from $u$ to $v$ is less than $f$, however, there may be a shorter route, which is useful if one wishes to minimize the distance for a given amount of flow. Thus we compute the shortest path for each possible bottleneck (flow) value. We call this problem Shortest Paths for All Flows (SP-AF). We present a non-trivial $O(mn)$ algorithm to solve the Single Source Shortest Paths for All Flows (SSSP-AF) problem, that is, computing the shortest paths for all flows from one source vertex to all other vertices in the graph.

Naturally, we move onto the All Pairs Shortest Paths for All Flows (APSP-AF) problem, where we compute the shortest paths for all flows for all pairs of vertices in the graph. Note that this new problem is different from the All Pairs Bottleneck Shortest Paths (APBSP) problem [15], which is to compute the bottlenecks for the shortest paths of all pairs. Applying our algorithm for SSSP-AF $n$ times gives us $O(mn^2)$. If the graph is dense, however, $m = O(n^2)$, and the time complexity becomes $O(n^4)$. We can utilize faster matrix multiplication over a ring to achieve a sub-quartic time bound for dense graphs. We present an algorithm that runs in $O(\sqrt{t}n^{(\omega+9)/4})$ time, where $t$ is the number of distinct edge capacities and $\omega < 2.373$ [16].

The algorithms are presented in the order of increasing complexity. Section 3 details the algorithm for solving the GB problem. In Section 4 and Section 5 we present the algorithms for solving the SSSP-AF and APSP-AF problems, respectively. Finally, in Section 6, we describe some practical applications of the SP-AF problem in computer networking before concluding the paper.

## 2   Preliminaries

Let $G = \{V, E\}$ be a strongly connected directed unweighted graph with edge capacities of positive real numbers. Let $n = |V|$ and $m = |E|$. Vertices (or nodes) are given by integers such that $\{1, 2, 3, ..., n\} \in V$. Let $(i, j) \in E$ denote the edge from vertex $i$ to vertex $j$. Let $cap(i, j)$ denote the capacity of the edge $(i, j)$. Let $t$ be the number of distinct $cap(i, j)$ values.

We call $C = \{c_{ij}\}$ the capacity matrix, where $c_{ij}$ represents a capacity from $i$ to $j$. Let $C^\ell = \{c_{ij}^\ell\}$, where $c_{ij}^\ell$ is defined to be the maximum bottleneck out of all paths of lengths up to $\ell$ from $i$ to $j$. Clearly $c_{ij}^1 = cap(i, j)$ if $(i, j) \in E$, and 0 otherwise. Let $c_{ij}^*$ be the maximum bottleneck for all paths from $i$ to $j$. We call $C^* = \{c_{ij}^*\}$ the closure of $C$ and also refer to it as the bottleneck matrix. The problem of computing $C^*$ is formally known as the All Pairs Bottleneck Paths (APBP) problem. For graphs with unit edge costs, the APBP problem is well studied in [15] and [5]. The complexities of algorithms given by the two papers are $\tilde{O}(n^{2+\omega/3}) = \tilde{O}(n^{2.791})$ and $\tilde{O}(n^{(\omega+3)/2}) = \tilde{O}(n^{2.687})$, respectively.

Let $Q = A \star B$ denote the $(max, min)$-product of capacity matrices $A = \{a_{ij}\}$ and $B = \{b_{ij}\}$, where $Q = \{q_{ij}\}$ is given by:

$$q_{ij} = \max_{k=1}^{n}\{\min\{a_{ik}, b_{kj}\}\}$$

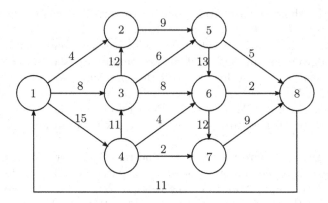

**Fig. 1.** An example of a directed unweighted graph with $n = 8$, $m = 16$ and $t = 10$. Capacities are shown beside each edge.

Note that if all elements in $A$ and $B$ are either 0 or 1, this becomes Boolean matrix multiplication. If we interpret "max" as addition and "min" as multiplication, the set of non-negative numbers forms a closed semi-ring. Similarly, the set of matrices where the product is defined as the $(max, min)$-product and the sum is defined as a component-wise "max" operation also forms a closed semi-ring. Then the bottleneck matrix is given by the closure of the capacity matrix, where the closure of matrix $A$ is defined by:

$$A^* = I + A + A^2 + A^3 + ...$$

and $I$ is the identity matrix with diagonal elements of $\infty$ and non-diagonal elements of 0. Although $A^*$ is defined by an infinite series we can stop at $n - 1$. The computational complexity of computing $A^*$ is asymptotically equal to that of the matrix product in the more general setting of closed semi-ring [1].

Similarly to the capacity matrix, we can define the distance matrix, where each element represents the distance from $i$ to $j$. The problem of computing the closure of the distance matrix is formally known as the All Pairs Shortest Paths (APSP) problem. Zwick achieved $\tilde{O}(n^{2.575})$ time for solving APSP on directed graphs with unit edge costs [17], which has recently been improved to $\tilde{O}(n^{2.53})$ thanks to Le Gall's new algorithm for rectangular matrix multiplication [7].

Let $Q = A * B$ denote the $(min, +)$-product, or the distance product, of distance matrices $A$ and $B$, where $Q = \{q_{ij}\}$ is given by:

$$q_{ij} = \min_{k=1}^{n} \{a_{ik} + b_{kj}\}$$

## 3   The Graph Bottleneck Problem

Let $\Theta$ be the bottleneck value of the entire network, that is, the solution to the Graph Bottleneck (GB) problem. Let the capacity matrix C be defined by

$c_{ij} = cap(i,j)$. One straightforward method to compute $\Theta$ would be to compute $C^*$ and find the minimum among them. We can solve the problem more efficiently by a simple but effective binary search as shown in Algorithm 1.

We begin by assuming that the edge capacities are integers bounded by $c$. Let the threshold value $h$ be initialized to $c/2$. Let $G' = \{V, E'\}$ such that $E'$ only contains edges that have capacities greater than or equal to $h$. Clearly $G'$ is strongly connected *iff* $\Theta \geq h$. We repeatedly halve the possible range $[\alpha, \beta]$ for $\Theta$ by adjusting the threshold, $h$, through binary search.

---

**Algorithm 1.** Solve the GB problem in $O(m \log n)$ time

---

1: $\alpha \leftarrow 0$
2: $\beta \leftarrow c$
3: **while** $\beta - \alpha > 0$ **do**
4:     $h \leftarrow (\alpha + \beta)/2$
5:     $G' \leftarrow$ Remove all $(i,j)$ from $G$ such that $cap(i,j) < h$
6:     **if** $G'$ is strongly connected **then**
7:         $\alpha \leftarrow h$
8:     **else**
9:         $\beta \leftarrow h$
10: $\Theta \leftarrow \alpha$

---

Obviously the iteration over the while loop is performed $O(\log c)$ times. We use the $O(m)$ algorithm by Tarjan [14] to compute the strongly connected components in line 6. Thus the total time becomes $O(m \log c)$. If $c$ is large, say $O(2^n)$, the algorithm is not very efficient, taking $O(n)$ halvings of the possible ranges of $\Theta$. In this case, we sort edges in ascending order. Since there are at most $m$ possible values of capacities, doing binary search over the sorted edges gives us $O(m \log m) = O(m \log n)$. Obviously this method also works for edge capacities of real numbers.

*Example 1.* The value of $\Theta$ for the graph in Figure 1 is 9, which is the capacity of edges $(2,5)$ and $(7,8)$.

## 4 The Single Source Shortest Paths for All Flows Problem

From a source vertex $s$ to all other vertices $v \in V$, we want to find the shortest paths for each flow value. The shortest path from $s$ to $v$ for a given flow value $f$ allows us to push flows up to $f$ as quickly as possible. For some $f' < f$, however, there may be a shorter path. Thus if we find the shortest path for all possible flows, we can respond to queries of flow demands from $s$ to $v$ with the shortest paths that can accommodate the flows. Let $t$ be the number of distinct edge capacities. If all edge capacities are distinct, then $t = m$. We refer to the distinct edge capacity values as maximal flows.

A straightforward method of solving the SSSP-AF problem is to solve the SSSP problem for each maximal flow value $f$, that is, we repeatedly solve SSSP using only $(u, v) \in E$ such that $cap(u, v) \geq f$, for all $f$. SSSP can be solved by a simple breadth-first-search (BFS) on graphs with unit edge costs, hence this method takes $O(tm)$ time, which is $O(m^2)$ in the worst case. Each BFS will result in a shortest path spanning tree (SPT) with $s$ as the root. Explicit paths can be retrieved by traversing up the SPTs.

One may be led to think that SSSP-AF can be solved with a simple decremental algorithm, that is, repeatedly removing edges in decreasing order of capacity, and checking for connectivity of vertices. This method, however, gives incorrect results because edges with larger capacities may later be required to provide shorter paths for smaller flows. The SP-AF problem requires solving the shortest paths problem and the bottleneck paths problem at the same time. This is not a trivial matter, as operations required to solve the two problems generally take us in opposite directions; maximizing bottlenecks comes at the cost of increased distances and minimizing distances comes at the expense of decreased bottlenecks.

We have achieved $O(mn)$ worst case time for solving SSSP-AF by fully exploiting the fact that all edges have unit costs, as shown in Algorithm 2. Let $B[v]$ be the largest bottleneck value currently known from $s$ to vertex $v$. Let $L[v]$ be the current shortest possible distance from $s$ to $v$. Let $T$ represent the SPT. $T$ can be considered to be a persistent data structure, that is, we do not compute $T$ from scratch for each maximal flow value. Let $Q[i]$ be a set of vertices that may be added to $T$ at distance $i$, such that $1 \leq i \leq n - 1$, i.e. one set for each possible path length from $s$.

**Theorem 1.** *Algorithm 2 correctly solves the SSSP-AF problem in $O(mn)$ time bound.*

*Proof.* We iterate through each maximal flow $f$ in increasing order. At each iteration, all $v \in V$ such that $B[v] < f$ is cut from $T$ and added to $Q[L[v] + 1]$. In other words, all paths that cannot accommodate the current flow value of $f$ are discarded. We then attempt to add all pruned vertices back to $T$ at the shortest possible distance, represented by $L[v]$. If it is possible to add $v$ at distance $L[v]$ from the source (line 14), we maximize the bottleneck value $B[v]$ by ensuring that the parent node in $T$ is the one that can give us the maximum $B[v]$ (line 15). Thus for each maximal flow value $f$, we are effectively solving the SSSP problem by making incremental updates to $T$. Explicit paths can be retrieved simply by traversing up the SPT.

We perform lifetime analysis to determine the worst case time complexity. Each vertex $v$ can be cut from $T$ and be re-added to $T$ $O(n)$ times, once per each possible distance from $s$. Cutting/adding $v$ from/to $T$ takes $O(1)$ time, achieved by setting the parent of $v$ to *null* or $u$, respectively. Thus the total time taken for all operations involving $T$ is $O(n^2)$. $Q$ can be implemented with a simple linked list structure. Therefore the total time complexity of all operations involving $Q$ is also $O(n^2)$. Finally we analyse the time complexity of edge inspections. $O(n)$ vertices can be observed at each possible distance from $s$. At each distance, for

---

**Algorithm 2.** Solve single source shortest paths for all flows problem

---
1: **for** $i \leftarrow 1$ to $n$ **do**
2:     $B[i] \leftarrow 0, L[i] \leftarrow 0$
3: $B[s] \leftarrow \infty, T \leftarrow s$ /* $T$ is for $SPT$, initially only root $s$ */
4: **for all** maximal flows $f$ in increasing order **do**
5:     **for all** $v \in V$ such that $B[v] < f$ **do**
6:         **if** $v$ is in $T$ **then**
7:             Cut $v$ from $T$
8:         $L[v] \leftarrow L[v] + 1$
9:         Push $v$ to $Q[L[v]]$ /* $v$ to be processed later */
10:     **for** $\ell \leftarrow 1$ to $n - 1$ **do**
11:         **while** $Q[\ell]$ is not empty **do**
12:             Pop $v$ from $Q[\ell]$
13:             **for all** $(u, v) \in E$ **do**
14:                 **if** $L[u] = L[v] - 1$ **then**
15:                     **if** MIN$(cap(u, v), B[u]) > B[v]$ **then**
16:                         $B[v] \leftarrow$ MIN$(cap(u, v), B[u])$ /* $B[v]$ increased */
17:                         Add $v$ to $T$ with $u$ as the parent
18:             **if** $v$ is not in $T$ **then**
19:                 $L[v] \leftarrow L[v] + 1$
20:                 Push $v$ to $Q[L[v]]$ /* $v$ to be processed later */

---

each vertex, we inspect all incoming edges (line 13). Therefore the time taken for edge inspections is $O(m)$ for each possible distance from $s$, resulting in $O(mn)$ total time complexity, which subsequently becomes the total worst case time complexity of the algorithm.                                                                    □

Note that the worst case time complexity of Algorithm 2 is $O(mn)$ regardless of the value for the number of maximal flows, $t$. For $n < t$, Algorithm 2 is faster than the straightforward method of repeatedly solving the SSSP problem $t$ times.

*Example 2.* Figure 2 shows Algorithm 2 being used to solve the SSSP-AF problem on the example graph shown in Figure 1, with $s = 1$. At iteration $f = 4$, the edge $(4, 7)$ is cut, because the path to vertex 7 cannot push flow $f = 4$. The next shortest possible distance for vertex 7 is 3, with the new bottleneck value of 8. That is, $L[7]$ is increased from 2 to 3 and $B[7]$ is increased from 2 to 8.

## 5   All Pairs Shortest Paths for All Flows Problem

We firstly focus on solving the All Pairs Shortest Distances for All Flows (APSD-AF) problem, then show that the APSP-AF problem can be solved with an additional polylog factor with a minor modification to the algorithm. For each pair of vertices $(i, j)$ for each maximal flow, we want to compute the shortest distances. Thus our aim here is to obtain sets of $(d, f)$ pairs for all $(i, j)$, where

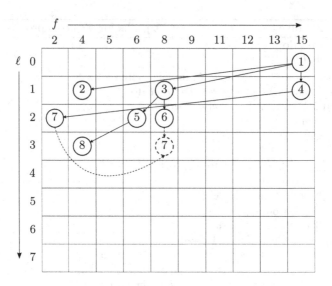

**Fig. 2.** Changes to $T$ at iteration $f = 4$

$f$ is the maximum flow that can be pushed through the shortest path whose length (distance) is $d < n$.

Let $C = \{c_{ij}\}$ be the capacity matrix and let $D^f = \{d_{ij}^f\}$ be the approximate distance matrix for paths that can accommodate flows up to $f$. A more detailed description of $D^f$ follows shortly. Let $S = \{s_{ij}\}$ be a matrix such that $s_{ij}$ is a set of $(d, f)$ pairs as described above, from vertex $i$ to vertex $j$. Let both $(d, f)$ and $(d', f')$ be in $s_{ij}$ such that $d < d'$. We keep $(d', f')$ *iff* $f < f'$ i.e. a longer path is only relevant if it can accommodate a greater flow. If $d = d'$, we keep the the pair that can accomodate the greater flow. Thus each $s_{ij}$ has at most $n - 1$ pairs of $(d, f)$. We assume the pairs are sorted in ascending order of $d$. We make an interesting observation here that the set of first pairs for all $s_{ij}$ is the solution to the All Pairs Bottleneck Shortest Distances (APBSD) problem, and the set of last pairs for $s_{ij}$ is the solution to the All Pairs Bottleneck Distances (APBD) problem. For APSD-AF, all relevant $(d, f)$ pairs for all $s_{ij}$ are computed.

*Example 3.* Solving APSD-AF on the graph given in Figure 1 results in a set of four $(d, f)$ pairs from vertex 4 to vertex 7, that is, $s_{47} = \{(1, 2), (2, 4), (3, 8), (5, 9)\}$.

Algorithm 3 is largely based on the method given by Alon, Galil and Margalit in [2], which is commonly used to solve various all pairs path problems [8,12,17,15]. This method has been reviewed in [12] and we use the same set of terminologies as the review. The algorithm consists of two phases; the *acceleration* phase and the *cruising* phase. Simply speaking, we run the algorithm by Alon *et al.* for all $f$ in parallel with a modified acceleration phase.

**Lemma 1.** *Algorithm 3 correctly solves the APSD-AF problem.*

---

**Algorithm 3.** Solve the APSD-AF problem

---

/* Initialization for the acceleration phase */
1: $C^0 \leftarrow I$
2: **for** $i \leftarrow 1$ to $n$; $j \leftarrow 1$ to $n$ **do**
3:     $s_{ij} \leftarrow \phi$ /* $\phi$ is empty */
    /* Acceleration phase */
4: **for** $\ell \leftarrow 1$ to $r$ **do**
5:     $C^\ell \leftarrow C^{\ell-1} \star C$
6:     **for** $i \leftarrow 1$ to $n$; $j \leftarrow 1$ to $n$ such that $i \neq j$ **do**
7:         $f \leftarrow c_{ij}^\ell$
8:         **if** $f > c_{ij}^{\ell-1}$ **then**
9:             $s_{ij} \leftarrow s_{ij}||(\ell, f)$ /* append $(\ell, f)$ to $s_{ij}$ */
    /* Initialization for the cruising phase */
10: **for** $i \leftarrow 1$ to $n$; $j \leftarrow 1$ to $n$ such that $i \neq j$ **do**
11:     **for all** $x$ in $s_{ij}$ **do**
12:         **if** $x \neq \phi$ **then**
13:             let $x = (d, f)$
14:             $d_{ij}^f \leftarrow d$
15:         **else**
16:             $d_{ij}^f \leftarrow \infty$

    /* Cruising phase */
17: $\ell \leftarrow r$
18: **while** $\ell < n$ **do**
19:     $\ell_1 \leftarrow \lceil \ell * 3/2 \rceil$
20:     **for all** maximal flow $f$ **do**
21:         **for** $i \leftarrow 1$ to $n$ **do**
22:             Scan $i^{th}$ row of $D^f$ with $j$ and find the smallest set of equal $d_{ij}^f$
23:                 such that $\lceil \ell/2 \rceil \leq d_{ij}^f \leq \ell$ and let the set of corresponding $j$ be $B_i$

24:         **for** $i \leftarrow 1$ to $n$; $j \leftarrow 1$ to $n$ such that $i \neq j$ **do**
25:             $m_{ij} \leftarrow \min_{k \in B_i}\{d_{ik}^f + d_{kj}^f\}$
26:             **if** $m_{ij} \leq \ell_1$ **then**
27:                 $d_{ij}^f \leftarrow m_{ij}$

28:     $\ell \leftarrow \ell_1$
    /* Finalization */
29: **for** $i \leftarrow 1$ to $n$; $j \leftarrow 1$ to $n$ such that $i \neq j$ **do**
30:     **for all** maximal flow $f$ in increasing order **do**
31:         $d \leftarrow d_{ij}^f$
32:         Let the last pair of $s_{ij}$ be $x = (d', f')$
33:         **if** $x = \phi$ or $(f > f'$ and $d < \infty)$ **then**
34:             **if** $d = d'$ **then**
35:                 Replace $x$ with $(d, f)$
36:             **else**
37:                 $s_{ij} \leftarrow s_{ij}||(d, f)$ /* append $(d, f)$ to $s_{ij}$ */

---

*Proof.* We compute the $(max, min)$-products in the acceleration phase, multiplying the capacity matrix $C$ one by one. The $\ell^{th}$ iteration of the acceleration phase, therefore, finds the maximum bottleneck for all paths of lengths up to $\ell$. In other words, after the acceleration phase, for all $s_{ij}$, we have found all relevant $(d, f)$ pairs such that $d \leq \ell$.

After the acceleration phase, based on the current sets of $(d, f)$ pairs in $S$, we initialize a total of $t$ distance matrices as $D^f$, one for each distinct maximal flow value. At this stage, $d_{ij}^f$ is the distance of the shortest path that can push flow $f$, if the path length is $r$ or less. In the cruising phase, we perform repeated squaring on all distance matrices in parallel with the help of the bridging set $B_i$, for each row $i$. At the end of the cruising phase we thus have the shortest distances for all maximal flow values for all pairs of vertices. Retrieving sets of $(d, f)$ from the resulting $D^f$ matrices is simply a reverse process of the initialization for the cruising phase.  □

**Lemma 2.** *Algorithm 3 runs in $O(\sqrt{t}n^{(\omega+9)/4})$ time when $r < n$.*

*Proof.* For the acceleration phase we use the the current best known algorithm given by Duan and Pettie [5] to compute the $(max, min)$-product in each iteration, giving us $O(rn^{(3+\omega)/2})$. The time complexity for the cruising phase is $O(dn^3/r)$. This is because $|B_i|$ is $O(n/r)$ as proven in [2], and no logarithmic factor is required for repeated squaring because the path length $\ell$ increases by a factor of $\frac{3}{2}$ in each iteration resulting in a geometric series i.e. the first term dominates the time complexity. The time complexity for the initialization for the cruising phase is $O(rn^2)$ since at most $r$ $(d, f)$ pairs exist in any $s_{ij}$ after the acceleration phase, and this time complexity is absorbed by the time complexity of the acceleration phase. The time complexity of finalization is $O(tn^2)$, which is absorbed by $O(tn^3/r)$ of the cruising phase since $n/r > 1$. We balance the time complexities of the acceleration phase and the cruising phase by setting $r = \sqrt{t}n^{(3-\omega)/4}$, which gives us the total time complexity of $O(\sqrt{t}n^{(\omega+9)/4})$.  □

Note that the value we choose for $r$ must be less than $n$, otherwise our algorithm is equivalent to simply computing the $(max, min)$-product $n$ times (i.e. staying in the acceleration phase until the problem is solved), resulting in the time complexity of $O(n^{(5+\omega)/2})$. ($r = \sqrt{t}n^{(3-\omega)/4}) \geq n$ when $t \geq n^{(1+\omega)/2}$. This can happen when the graph is relatively dense and most of the edge capacities are distinct. Therefore a more accurate worst case time complexity of Algorithm 3 is actually $O(\min\{n^{(5+\omega)/2}, \sqrt{t}n^{(\omega+9)/4}\})$.

We now compare the time complexity of Algorithm 3 to various straightforward methods of solving the APSP-AF problem. One straightforward method is to repeatedly solve the SSSP-AF problem for all $v \in V$ as the source vertex. This can be done in $O(tmn)$ time, or in $O(mn^2)$ by running Algorithm 2 $n$ times. Clearly Algorithm 3 is faster for dense graphs. Another straightforward method of solving the APSP-AF problem is to solve the APSP problem $t$ times. Using Zwick's algorithm [17], the time complexity of this second straightforward method is $O(tn^{2.53})$. For most (larger) values of $t$, Algorithm 3 is faster.

**Theorem 2.** *The APSP-AF problem can be solved in $\tilde{O}(\sqrt{t}n^{(\omega+9)/4})$ time.*

*Proof.* There can be $O(n)$ $(d, f)$ pairs in each $s_{ij}$. Since the lengths of each path can be $O(n)$, explicitly listing all paths takes $O(n^4)$ time, which is too expensive. As is common and widely accepted in graph paths problems, we get around this by extending the pair $(d, f)$ to the triplet $(d, f, u)$, where $u$ is the *successor* node. In the acceleration phase witnesses can be retrieved with an extra polylog factor [5], and the successor nodes can be computed from the witnesses at each iteration in $O(n^2)$ time [17]. In the cruising phase retrieving $u$ is a trivial exercise since ordinary matrix multiplication is performed. We can generate explicit paths in time linear to the path length by using $d$ for looking up subsequent successor nodes. That is, we can still retrieve each successor node in $O(1)$ time even with $O(n)$ triplets of $(d, f, u)$ in each $s_{ij}$ because we know that the path length decrements by 1 as we step through each successor node.    □

## 6  Practical Applications of the SP-AF Problem

Computer networks can be accurately modeled by unweighted directed graphs with edge capacities, by representing each hop (e.g. router) as a vertex, each network link as an edge, and the bandwidth of each link as edge capacities. However, routing protocols that are commonly used today are based on less accurate models. For example, the Routing Information Protocol (RIP) computes routes based solely on the hop counts, while the Open Shortest Path First (OSPF) protocol, by default, computes routes based solely on the bandwidths.

In today's computer networks each router is autonomous, and therefore each router computes SSSP. RIP is often implemented with Bellman-Ford algorithm [6,3] and OSPF is often implemented with Dijkstra's algorithm [4]. We present SSSP-AF as a better solution that uses both the hop count and the bandwidth at the same time. Advanced routers are able to gather information such as the current flow amount from one IP subnet to another. With SSSP-AF, a router can make a better routing decision for a given flow based on the flow amount by choosing a route that minimizes the latency without causing congestion.

Furthermore, we introduce APSP-AF as a potential routing algorithm for Software Defined Networking (SDN) [18]. SDN is a new paradigm in computer networking where routers are no longer autonomous and the whole network can be controlled at a centralized location. The central controller has in-depth knowledge of the network and as a result SDN can benefit from more sophisticated routing algorithms. By solving APSP-AF for the whole network, the fastest routes can be determined for all flow requirements for all sources and destinations. Algorithm 3 can be easily turned into a practical $O(\sqrt{t}n^3)$ algorithm by using the traditional $O(n^3)$ matrix multiplication method to compute the $(max, min)$-product in the acceleration phase. This is very much relevant in computer networks where distinct bandwidth values are defined (e.g. 100Mbps, 1Gbps).

# 7    Concluding Remarks

We have extended the well known bottleneck paths problems, introducing new problems that clearly have many practical applications. We provided non-trivial algorithms to solve the problems more efficiently than straightforward methods.

This paper only considered directed unweighted graphs. In enterprise computer networks, most links are bi-directional, meaning undirected graphs are adequate for modeling those networks. Also for computer networks involving low latency switches and long cables with repeaters, introducing edge costs may enable more accurate modeling of the networks. Hence solving the SP-AF problem on other types of graphs would not only be a natural extension to this paper, but also allow further practical applications.

Trivial lower bounds of $\Omega(n^2)$ and $\Omega(n^3)$ exist for SSSP-AF and APSP-AF, respectively, on graphs with unit edge costs. Most current researches in the APSP problem focus on breaking the cubic barrier of $O(n^3)$ to get closer to the trivial lower bound of $\Omega(n^2)$. With the APSP-AF problem we have effectively shifted the focus in time complexities from "cubic-to-quadratic" to "quartic-to-cubic". We anticipate many future contributions to take us closer to the lower bounds of this interesting new problem.

# References

1. Aho, A.V., Hopcroft, J.E., Ullman, J.D.: The Design and Analysis of Computer Algorithms. Addison-Wesley (1974)
2. Alon, N., Galil, Z., Margalit, O.: On the Exponent of the All Pairs Shortest Path Problem. In: Proc. 32nd IEEE FOCS, pp. 569–575 (1991)
3. Bellman, R.: On a Routing Problem. Quart. Appl. Math. 16, 87–90 (1958)
4. Dijkstra, E.: A Note on Two Problems in Connexion With Graphs. Numerische Mathematik 1, 269–271 (1959)
5. Duan, R., Pettie, S.: Fast Algorithms for (max,min)-matrix multiplication and bottleneck shortest paths. In: Proc. 19th SODA, pp. 384–391 (2009)
6. Ford, L.: Network Flow Theory. RAND Paper, p. 923 (1956)
7. Le Gall, F.: Faster Algorithms for Rectangular Matrix Multiplication. In: Proc. 53rd FOCS, pp. 514–523 (2012)
8. Galil, Z., Margalit, O.: All Pairs Shortest Paths for Graphs with Small Integer Length Edges. Journal of Computer and System Sciences 54, 243–254 (1997)
9. Robinson, S.: Toward an Optimal Algorithm for Matrix Multiplication. SIAM News 38, 9 (2005)
10. Schönhage, A., Strassen, V.: Schnelle Multiplikation Großer Zahlen. Computing 7, 281–292 (1971)
11. Seidel, R.: On the all-pairs-shortest-path problem. In: Proc. 24th ACM STOC, pp. 213–223 (1990)
12. Takaoka, T.: Sub-cubic Cost Algorithms for the All Pairs Shortest Path Problem. Algorithmica 20, 309–318 (1995)

13. Takaoka, T.: Efficient Algorithms for the 2-Center Problems. In: Taniar, D., Gervasi, O., Murgante, B., Pardede, E., Apduhan, B.O. (eds.) ICCSA 2010, Part II. LNCS, vol. 6017, pp. 519–532. Springer, Heidelberg (2010)
14. Tarjan, R.: Depth-first search and linear graph algorithms. Jour. SIAM 1 2, 146–160 (1972)
15. Vassilevska, V., Williams, R., Yuster, R.: All Pairs Bottleneck Paths and Max-Min Matrix Products in Truly Subcubic Time. Journal of Theory of Computing 5, 173–189 (2009)
16. Williams, V.: Breaking the Coppersmith-Winograd barrier. In: STOC (2012)
17. Zwick, U.: All Pairs Shortest Paths using Bridging Sets and Rectangular Matrix Multiplication. Journal of the ACM 49, 289–317 (2002)
18. Open Networking Foundation: Software-Defined Networking: The New Norm for Networks ONF White Paper (2012)

# I/O Efficient Algorithms for the Minimum Cut Problem on Unweighted Undirected Graphs

Alka Bhushan and G. Sajith

Department of Computer Science and Engineering,
Indian Institute of Technology Guwahati, India-781039
{alka,sajith}@iitg.ac.in

**Abstract.** The problem of finding the minimum cut of an undirected unweighted graph is studied on the external memory model. First, a lower bound of $\Omega((E/V)\text{Sort}(V))$ on the number of I/Os is shown for the problem, where $V$ is the number of vertices and $E$ is the number of edges. Then the following are presented, for $M = \Omega(B^2)$, (1) a minimum cut algorithm that uses $O(c\log E(\text{MSF}(V,E) + \frac{V}{B}\text{Sort}(V)))$ I/Os; here $\text{MSF}(V,E)$ is the number of I/Os needed to compute a minimum spanning tree of the graph, and $c$ is the value of the minimum cut. The algorithm performs better on dense graphs than the algorithm of [7], which requires $O(E + c^2V\log(V/c))$ I/Os, when executed on the external memory model. For a $\delta$-fat graph (for $\delta > 0$, the maximum tree packing of the graph is at least $(1 + \delta)c/2$), our algorithm computes a minimum cut in $O(c\log E(\text{MSF}(V,E) + \text{Sort}(E)))$ I/Os. (2) a randomized algorithm that computes minimum cut with high probability in $O(c\log E \cdot \text{MSF}(V,E) + \text{Sort}(E)\log^2 V + \frac{V}{B}\text{Sort}(V)\log V)$ I/Os. (3) a $(2+\epsilon)$-minimum cut algorithm that requires $O((E/V)\text{MSF}(V,E))$ I/Os and performs better on sparse graphs than our exact minimum cut algorithm.

## 1 Introduction

The minimum cut problem on an undirected unweighted graph seeks to partition the vertices into two sets while minimizing the number of edges from one side of the partition to the other. While efficient in-core and parallel algorithms for the problem are known [4,10,11], this problem has not been explored much from the perspective of massive data sets. However, it is shown in [2] that the minimum cut can be computed in a polylogarithmic number of passes using only a polylogarithmic sized main memory on the stream sort model.

In this paper we consider the minimum cut problem on the external memory model proposed in [1]. To the best of our knowledge, this problem has so far not been investigated on the external memory model. This model has been used to design algorithms intended to work on large data sets that do not fit in the main memory. The external memory model defines the following parameters: $N$ ($= V + E$) is the input size, $M$ is the size of the main memory and $B$ is the size of a disk block. It is assumed that $2B < M < N$. In an I/O operation one block

S.P. Pal and K. Sadakane (Eds.): WALCOM 2014, LNCS 8344, pp. 188–199, 2014.
© Springer International Publishing Switzerland 2014

of data is transferred between the disk and the internal memory. The measure of performance of an algorithm on this model is the number of I/Os it performs. The number of I/Os needed to read (write) $N$ contiguous items from (to) the disk is $\text{Scan}(N) = \Theta(N/B)$. The number of I/Os required to sort $N$ items is $\text{Sort}(N) = \Theta((N/B) \cdot \log_{M/B}(N/B))$ [1]. For all realistic values of $N$, $B$, and $M$, $\text{Scan}(N) < \text{Sort}(N) \ll N$.

For an undirected unweighted graph $G = (V, E)$, a cut $X = (S, V - S)$ is defined as a partition of the vertices of the graph into two nonempty sets $S$ and $V - S$. An edge with one endpoint in $S$ and the other endpoint in $(V - S)$ is called a crossing edge of $X$. The value $c$ of the cut $X$ is the total number of crossing edges of $X$. The minimum cut problem is to find a cut of minimum value. We assume that the input graph is connected, since otherwise the problem is trivial. A cut in $G$ is $\alpha$-minimum, for $\alpha > 0$, if its value is at most $\alpha$ times the minimum cut value of $G$.

A tree packing is a set of spanning trees, each with a weight assigned to it, such that the total weight of the trees containing a given edge is at most one. The value of a tree packing is the total weight of the trees in it. A maximum tree packing is a tree packing of largest value. (When there is no ambiguity, we will use "maximum tree packing" to refer also to the value of a maximum tree packing, and a "minimum cut" to the value of a minimum cut.) A graph $G$ is called a $\delta$-fat graph for $\delta > 0$, if the maximum tree packing of $G$ is at least $\frac{(1+\delta)c}{2}$, where $c$ is the minimum cut [10].

As in [10], we say that a cut $X$ $k$-respects a tree $T$ (equivalently, $T$ $k$-constrains $X$), if $X$ cuts at most $k$ edges of $T$.

Several approaches have been tried in designing in-core algorithms for the minimum cut problem [6,7,9,10,14,17]. Significant progress has been made in designing parallel algorithms as well [8,10,11]. The current best in-core algorithm computes the minimum cut in $O(E + c^2V \log(V/c))$ time [7] and when executed on the external memory model, performs $O(E + c^2V \log(V/c))$ I/Os.

Karger [10] presents a near linear time randomised algorithm that computes a minimum cut with high probability in $O(\min\{E \log^3 V, V^2 \log V\})$ time. This algorithm also computes all $\alpha$-minimum cuts, for $\alpha < 3/2$. These cuts can be stored in a data structure that uses $O(k + V \log V)$ space, where $k$ is the total number of cuts found, in $O(V^2 \log V)$ time. With this data structure, we can verify whether a given cut is $\alpha$-minimum in $O(V)$ time [10]. If we execute this algorithm on the external memory model then the I/O complexity of the cuts computation is $O(\min\{E \log^3 V, V^2 \log V\})$, the construction of the data structure is $O(V^2 \log V)$, and the answering of a query is $O(V)$.

A linear time algorithm for computing a $(2 + \epsilon)$-minimum cut [12] is also known. This executes on the external memory model in $O(V + E)$ I/Os.

The best known RNC algorithm presented in [10] can be simulated in $O(\log^3 V \cdot \text{Sort}(V^2/\log^2 V))$ I/Os in the external memory model using the PRAM simulation presented in [5]. The poly-logarithmic passes stream-sort algorithm presented in [2] implies an external memory algorithm that uses $O(\text{Sort}(E)\text{polylog}(V))$ I/Os under the assumption of $M = \Omega(\text{polylog}(V))$.

In this paper, we present a lower bound of $\Omega((E/V)\text{Sort}(V))$ on the number of I/Os for the minimum cut problem. Next, for $M = \Omega(B^2)$, we present:

- A minimum cut algorithm that runs in $O(c\log E(\text{MSF}(V, E) + \frac{V}{B}\text{Sort}(V)))$ I/Os, and performs better on dense graphs than the algorithm of [7], which requires $O(E + c^2V\log(V/c))$ I/Os, where $\text{MSF}(V, E)$ is the number of I/Os required in computing a minimum spanning tree. The currently best known minimum spanning tree algorithm executes in $O(\text{Sort}(E) \cdot \max\{1, \log\log_{E/V} B\})$ I/Os [3]. For a $\delta$-fat graph, our algorithm mentioned above computes a minimum cut in $O(c\log E(\text{MSF}(V, E) + \text{Sort}(E)))$ I/Os.
- A randomised algorithm that computes minimum cut with high probability in $O(c\log E \cdot \text{MSF}(V, E) + \text{Sort}(E)\log^2 V + \frac{V}{B}\text{Sort}(V)\log V)$ I/Os.
- A $(2 + \epsilon)$-minimum cut algorithm that requires $O((E/V)\text{MSF}(V, E))$ I/Os and performs better on sparse graphs than our exact minimum cut algorithm.

The rest of the paper is organised as follows: sec. 2, defines notations used in this paper, sec. 3 gives a lower bound result for the minimum cut problem, sec. 4 presents an external memory algorithm for the minimum cut problem, sec. 5 improves the I/O complexity of our algorithm by using randomisation, sec. 6 discusses a special class of graphs for which a minimum cut can be computed very efficiently and section 7 presents a $(2 + \epsilon)$-minimum cut algorithm.

## 2    Some Notations

For a cut $X$ of graph $G$, $E(X)$ is the set of crossing edges of $X$. $d(v)$ is the degree of a vertex $v$ in $G$. For a spanning tree $T$ of $G$, $E(T)$ is the set of edges of $T$, $\rho(v)$ is the number of edges whose endpoints' least common ancestor is $v$ and $p(v)$ is the parent of $v$.

Let $v \downarrow$ denote the set of vertices that are descendants of $v$ in the rooted tree, and $v \uparrow$ denote the set of vertices that are ancestors of $v$ in the rooted tree. Note that $v \in v \downarrow$ and $v \in v \uparrow$. Let $C(A, B)$ be the total number of edges with one endpoint in vertex set $A$ and the other in vertex set $B$. An edge with both endpoints in both sets is counted twice. Thus, $C(u, v)$ is 1, if $(u, v)$ is an edge, 0 otherwise. For a vertex set $S$, let $C(S)$ denote $C(S, V - S)$. For a function $f$ on the vertices, $f^{\downarrow}(v) = \sum_{w \in v\downarrow} f(w)$. See [10].

## 3    A Lower Bound for the Minimum Cut Problem

We define a decision problem $\mathbf{P}_1$ as follows: Given as input a set $S$ of $N$ elements, each with an integer key drawn from the range $[1, P]$, say "yes" when $S$ contains either every odd element or at least one even element in the range $[1, P]$, and say "no" otherwise (that is, when $S$ does not contain at least one odd element and any even element in the range $[1, P]$.) Then $(\lceil P/2\rceil - 1)^N$ and $\lceil \frac{P}{2}\rceil(\lceil P/2\rceil - 1)^N$ are, respectively, lower and upper bounds on the number of different "no" instances. We prove that in any decision tree for $\mathbf{P}_1$, the "no" instances and leaves that decide "no" correspond one to one. Therefore, the depth of any decision tree for $\mathbf{P}_1$ is $\Omega(N \log P)$.

Our proof, similar to the ones in [13], considers tertiary decision trees with three outcomes: $<, =, >$, at every decision node. Each node, and in particular each leaf, corresponds to a partial order on $S \cup \{1, \ldots, P\}$. Consider a "no" leaf $l$ and the partial order $PO(l)$ corresponding to $l$. All inputs visiting $l$ must satisfy $PO(l)$. Let $C_1, \ldots C_k$ be the equivalence classes of $PO(l)$. Let $u_i$ and $d_i$ be the maximum and minimum values respectively of the elements of equivalence class $C_i$ over all "no" inputs that visit $l$. Exactly one input instance visits $l$ if and only if $u_i = d_i$ for all $i$. If $u_i \neq d_i$ for some $C_i$ then, pick a "no" input $I$ that visits $l$ and fabricate a "yes" input $I'$ as follows: assign an even integer $e$, $d_i < e < u_i$, to every element in $C_i$, and consistent with this choice and $PO(l)$ change the other elements of $I$ if necessary. Note that this fabrication is always possible. Since $I'$ is consistent with $PO(l)$, it visits $l$; a contradiction. Hence our claim.

Now we consider a restriction $\mathbf{P}'_1$ of $\mathbf{P}_1$. Suppose the input $S$ is divided into $P$ subsets each of size $N/P$ and containing distinct elements from the range $[P+1, 2P]$, where $P < N < P(\lceil P/2 \rceil - 1)$. $\mathbf{P}'_1$ is to decide whether $S$ contains either every odd element or at least one even element in the range $[P+1, 2P]$. A lower bound on the number of "no" instances of $\mathbf{P}'_1$ is $(P' \cdot (P' - 1) \cdot (P' - 2) \cdot \ldots \cdot (P' - N/P))^P = \Omega(P^N)$, where $P' = \lceil P/2 \rceil - 1$. An argument similar to the above shows that in any decision tree for $\mathbf{P}'_1$, the "no" instances and leaves that decide "no" correspond one to one. Therefore, the depth of any decision tree for $\mathbf{P}'_1$ is $\Omega(N \log P)$. Construct an undirected graph $G = (V, E)$ from an input instance $I$ of $\mathbf{P}'_1$ as follows.

1. Let the integers in $[1, 2P]$ constitute the vertices.
2. Make a pass through $I$ to decide if it contains an even element. If it does, then for each $i \in [P, 2P - 1]$, add an edge $\{i, i + 1\}$ to $G$. Otherwise, remove all even integers (vertices) $> P$ from $G$.
3. Make a second pass through $I$. If the $j$th subset of $I$ contains $P + i$, then add an edge $\{j, P + i\}$ to $G$.
4. For $i \in [1, P - 1]$, add an edge $\{i, i + 1\}$ to $G$.

Here $|V| = \Theta(P)$ and $|E| = \Theta(N)$. The construction of the graph requires $O(N/B)$ I/Os. It needs looking at the least significant bits (LSBs) of the keys of the elements; if the LSBs are assumed to be given separately from the rest of the keys, this will not violate the decision tree requirements. Note that the decision tree arguments for $\mathbf{P}1$ and $\mathbf{P}'1$ will remain unchanged, even if we assume that each has a composite key of the form $\langle i, lsb(i) \rangle$. Any comparison between two elements will depend only on the first components of their keys. The value of the minimum cut of $G$ is at least 1 iff $\mathbf{P}'_1$ answers "yes" on $I$.

From the following theorem, given in [13] we prove the lemma given below.

**Theorem 1.** *[13] Let $X$ be the problem solved by a $P$-way indexed I/O tree $T$, with $N$ the number of records in the input. There exists a decision tree $T_c$ solving $X$, such that:*

$$Path_{T_c} \leq N \log B + I/O_T \cdot O(B \log \frac{M - B}{B} + \log P)$$

where, $Path_{T_c}$ denote the number of comparison nodes on the path from the root to a leaf node in $T_c$ and $I/O_T$ denote the maximum number of $I/O$ nodes from the root to a leaf node in $T$.

**Lemma 1.** *The lower bound of the minimum cut problem is $\Omega(\frac{E}{V} Sort(V))$ on I/Os.*

# 4    The Minimum Cut Algorithm

We present an I/O efficient deterministic algorithm for finding minimum cuts on undirected unweighted graphs. Our I/O efficient algorithm is based on the semi-duality between minimum cut and tree packing. The duality was used in designing a deterministic minimum cut in-core algorithm for both directed and undirected unweighted graphs [7] and in designing a faster but randomized in-core algorithm for undirected weighted graphs [10]. Our algorithm uses Karger's ideas [10].

Nash-Williams theorem [15] states that any undirected graph with minimum cut $c$ has at least $c/2$ edge disjoint spanning trees. It follows that in such a packing, for any minimum cut, there is at least one spanning tree that 2-constrains the minimum cut. Once we compute such a packing, the problem reduces to finding a minimum cut that is 2-constrained by some tree in the packing. The assumption on edge disjointness is relaxed by Karger [10] in the following lemma.

**Lemma 2.** *[10] For any graph $G$, for any tree packing $P$ of $G$ of value $\beta c$, and any cut $X$ of $G$ of value $\alpha c$ ($\alpha \geq \beta$), at least $(1/2)(3 - \alpha/\beta)$ fraction (by weight) of trees of $P$ 2-constrains $X$.*

If an approximate algorithm guarantees a $\beta c$ packing $P$, for $\beta > 1/3$, at least $\frac{1}{2}(3 - \frac{1}{\beta})$ fraction (by weight) of the trees in $P$ 2-constrains any given minimum cut $X$. In particular, there is at least one tree in $P$ that 2-constrains any given minimum cut $X$.

## 4.1    The Algorithm

From the above discussion, we can conclude that the minimum cut problem can be divided into two subproblems, (i) compute an approximate maximal tree packing $P$ of value $\beta c$, for $\beta > 1/3$, and (ii) compute a minimum cut of the graph $G$ that is 2-constrained by some tree in $P$.

**Subproblem 1.** We use the greedy tree packing algorithm given in [16,18] as described here. A tree packing $P$ in $G$ is an assignment of weights to the spanning trees of $G$ so that each edge gets a load of $l(u,v) = \sum_{T \in P:(u,v) \in T} w(T) \leq 1$. The value of tree packing $P$ is $W = \sum_{T \in P} w(T)$. In this algorithm, initial weights on all spanning trees, load on all edges, and W are set to zero. While no edge has load 1, following steps are executed:

1. Pick a load minimal spanning tree $T$;
2. $w(T) = w(T) + \epsilon^2/3 \log E$;
3. $W = W + \epsilon^2/3 \log E$;
4. For all edges $(u, v)$ in $T$, $l(u, v) = l(u, v) + w(T)$;

As mentioned in [18], the algorithm obtains the following result.

**Theorem 2.** *[16] The greedy tree packing algorithm given above, when run on a graph $G$, computes a $(1 - \epsilon)$-approximate tree packing of value $W$ for $G$; that is, $(1 - \epsilon)\tau \le W \le \tau$, where $\tau$ is the maximum value of any tree packing of $G$.*

Since each iteration increases the packing value by $\epsilon^2/(3 \log E)$, and the packing value can be at most $c$, the algorithm terminates in $O(c \log E/\epsilon^2)$ iterations. The I/O complexity of each iteration is dominated by the minimal spanning tree computation. Thus, number of I/Os required is $O(c \log E \cdot \mathrm{MSF}(V, E))$, where $\mathrm{MSF}(V, E)$ is the number of I/Os required to compute a minimal spanning tree. The currently best known minimum spanning tree algorithm executes in $O(\mathrm{Sort}(E) \max\{1, \log \log_{E/V} B\}$ I/Os [3]. Since in each iteration, at most one new spanning tree is added in the tree packing therefore upper bound on the total number of trees in the computed tree packing is $O(c \log E)$.

Since the value of the maximum tree packing is at least $c/2$, the size of the computed tree packing is at least $(1 - \epsilon)c/2$. From Lemma 2, it follows that, for $\epsilon < 1/3$, and any minimum cut $X$, the computed tree packing contains at least one tree that 2-constrains $X$.

**Subproblem 2.** Let $T = (V, E')$ be a spanning tree of graph $G = (V, E)$. For every $K \subseteq E'$ there is unique cut $X$ so that $K = E(T) \cap E(X)$. $X$ can be constructed as follows: Let $A = \emptyset$. For some $s \in V$, for each vertex $v$ in $V$, add $v$ to set $A$, iff the path in $T$ from $s$ to $v$ has an even number of edges from $K$; clearly $X = (A, V - A)$ is a cut of $G$.

A spanning tree in the packing produced by Subproblem 1 2-constrains every minimum cut of $G$. We compute the following: (1) for each tree $T$ of the packing, and for each tree edge $(u, v)$ in $T$, a cut $X$ such that $(u, v)$ is the only edge of $T$ crossing $X$, (2) for each tree $T$ of the packing, and for each pair of tree edges $(u_1, v_1)$ and $(u_2, v_2)$, a cut $X$ such that $(u_1, v_1)$ and $(u_2, v_2)$ are the only edges of $T$ crossing $X$. A smallest of all the cuts found is a minimum cut of $G$.

First we describe the computation in (1). Root tree $T$ at some vertex $r$ in $O(\mathrm{Sort}(V))$ I/Os [5]. (See Section 2 for notations.) $C(v \downarrow)$ is the number of edges whose one endpoint is a descendent of $v$, and the other endpoint is a nondescendent of $v$. If $(v, p(v))$ is the only tree edge crossing a cut $X$, then $C(v \downarrow)$ is the value of cut $X$. As given in [10], $C(v \downarrow) = d^{\downarrow}(v) - 2\rho^{\downarrow}(v)$. $C(v \downarrow)$ is to be computed for all vertices $v$ in tree $T$, except for the root $r$.

$d^{\downarrow}(v)$ can be computed by using expression tree evaluation, if the degree of each vertex $v$ is stored with $v$. $\rho^{\downarrow}(v)$ can be computed using least common ancestor queries and expression tree evaluation. Once $d^{\downarrow}(v)$ and $\rho^{\downarrow}(v)$ are known for every vertex $v \in T$, $C(v \downarrow)$ can be computed for every vertex $v$ using expression tree evaluation. If we use the I/O efficient least common ancestor and expression tree evaluation algorithms of [5], the total number of I/Os needed for the computation in (1) is $O(\mathrm{Sort}(V))$.

For the computation in (2), consider two tree edges $(u, p(u))$ and $(v, p(v))$, the edges from two vertices $u$ and $v$ to their respective parents. Let $X$ be the cut characterised by these two edges (being the only tree edges crossing $X$). We say vertices $u$ and $v$ are incomparable, if $u \notin v \downarrow$ and $v \notin u \downarrow$; that is, if they are not on the same root-leaf path. If $u \in v \downarrow$ or $v \in u \downarrow$, then $u$ and $v$ are called comparable and both are in the same root-leaf path. In the following, when we say the cut of $u$ and $v$, we mean the cut defined by edges $(p(u), u)$ and $(p(v), v)$. As given in [10], if $u$ and $v$ are incomparable then the value of cut $X$ is

$$C(u \downarrow \cup v \downarrow) = C(u \downarrow) + C(v \downarrow) - 2C(u \downarrow, v \downarrow)$$

If vertices $u$ and $v$ are comparable then the value of $X$ is

$$C(u \downarrow - v \downarrow) = C(u \downarrow) - C(v \downarrow) + 2(C(u \downarrow, v \downarrow) - 2\rho^{\downarrow}(v))$$

For each tree in the packing, and for each pair of vertices in the tree we need to compute the cuts using the above formulae. We preprocess each tree $T$ as follows. Partition the vertices of $T$ into clusters $V_1, V_2, \ldots, V_N$ (where $N = \Theta(V/B)$), each of size $\Theta(B)$, except for the last one, which can of a smaller size. Our intention is to process the clusters one at a time by reading each $V_i$ into the main memory to compute the cut values for every pair with at least one of the vertices in $V_i$. We assume that $T$ is rooted at some vertex $r$.

*Partitioning of vertices:* For each vertex $v \in V$, a variable $\text{Var}(v)$ is initialised to 1. The following steps are executed for grouping the vertices into clusters.

Compute the depth of each vertex from the root $r$. Sort the vertices $u \in V$ in the decreasing order of the composite key $\langle \text{depth}(u), p(u) \rangle$. Depth of $r$ is 0. Access the vertices in the order computed above. Let $v$ be the current vertex.

- Compute $Y = \text{Var}(v) + \text{Var}(v_1) + \ldots + \text{Var}(v_k)$, where $v_1, v_2, \ldots, v_k$ are the children of $v$. If $Y < B$ then set $\text{Var}(v) = Y$.
- Send the value $\text{Var}(v)$ to the parent of $v$, if $v$ is not the root $r$.
- If $Y \geq B$, divide the children of $v$ into clusters $\mathcal{Q} = Q_1, Q_2, \ldots, Q_l$ such that for each cluster $Q_i$, $\sum_{u \in Q_i} \text{Var}(u) = \Theta(B)$. If $v = r$, it joins one of the clusters $Q_i$.

After executing the above steps for all vertices, consider the vertices $u$, one by one, in the reverse order, that is, in increasing order of the composite key $\langle \text{depth}(u), p(u) \rangle$. Let $v$ be the current vertex. If $v$ is the root, then it labels itself with the label of the cluster to which it belongs. Otherwise, $v$ labels itself with the label received from its parent. If $v$ has not created any clusters, then it sends its label to all its children. Otherwise, let the clusters created by $v$ be $Q_1, Q_2, \ldots, Q_l$; $v$ labels each cluster uniquely and sends to each child $v_i$ the label of the cluster that contains $v_i$. At the end, every vertex has got the label of the cluster that contains it.

In one sort, all vertices belonging to the same cluster $V_i$ can be brought together. Since $T$ is a rooted tree, each vertex $u$ knows its parent $p(u)$. We store $p(u)$ with $u$ in $V_i$. Thus, $V_i$ along with the parent pointers, forms a subforest $T[V_i]$ of $T$, and we obtain the following lemma.

**Lemma 3.** *The vertices of a tree $T$ can be partitioned into clusters $V_1, \ldots V_N$ (where $N = \Theta(V/B)$), of size $\Theta(B)$ each, in $O(\text{Sort}(V))$ I/Os, with the clusters satisfying the following property: for any two roots $u$ and $v$ in $T[V_i]$, $p(u) = p(v)$.*

*Proof.* The partitioning procedure uses the time forward processing method for sending values from one vertex to another and can be computed in $O(\text{Sort}(V))$ I/Os [5]. The depth of the nodes can be found by computing an Euler Tour of $T$ and applying list ranking on it [5] in $O(\text{Sort}(V))$ I/Os. Thus, a total of $O(\text{Sort}(V))$ I/Os are required for the partitioning procedure.

The property of the clusters mentioned in the lemma follows from the way the clusters are formed. Each cluster $V_i$ is authored by one vertex $x$, and therefore each root in $T[V_i]$ is a child of $x$.

Connect every pair $V_i, V_j$ of clusters by an edge, if there exists an edge $e \in E'$ such that one of its endpoint is in $V_i$ and the other endpoint is in $V_j$. The resulting graph $G'$ must be a tree, denoted as cluster tree $T'$. Note that $T'$ can be computed in $O(\text{Sort}(V))$ I/Os. Do a level order traversal of the cluster tree: sort the clusters by depth, and then by key (parent of a cluster) such that (i) deeper clusters come first, and (ii) the children of each cluster are contiguous. We label the clusters in this sorted order: $V_1, V_2, \ldots, V_N$. Within the clusters the vertices are also numbered the same way.

Form an array $S_1$ that lists $V_1, \ldots, V_N$ in that order; after $V_i$ and before $V_{i+1}$ are listed nonempty $E_{ij}$'s, in the increasing order of $j$; $E_{ij} \subseteq E - E'$, is the set of non-$T$ edges of $G$ with one endpoint in $V_i$ and the other in $V_j$. With $V_i$ are stored the tree edges of $T[V_i]$. Another array $S2$ stores the clusters $V_i$ in the increasing order of $i$. The depth of each cluster in the cluster tree can be computed in $O(\text{Sort}(V))$ I/Os [5], and arrays $S_1$ and $S_2$ can be obtained in $O(\text{Sort}(V + E))$ I/Os.

*Computing cut values for all pair of vertices:* Now, we describe how to compute cut values for all pair of vertices. Recall that the value of the cut for two incomparable vertices $u$ and $v$ is

$$C(u \downarrow \cup v \downarrow) = C(u \downarrow) + C(v \downarrow) - 2C(u \downarrow, v \downarrow)$$

and for two comparable vertices $u$ and $v$ is

$$C(u \downarrow - v \downarrow) = C(u \downarrow) - C(v \downarrow) + 2(C(u \downarrow, v \downarrow) - 2\rho^{\downarrow}(v))$$

Except for $C(u \downarrow, v \downarrow)$, all the other values of both expressions have already been computed. $C(u \downarrow, v \downarrow)$ can be computed using the following expression.

$$C(u \downarrow, v \downarrow) = \sum_{\hat{u}} C(\hat{u} \downarrow, v \downarrow) + \sum_{\hat{v}} C(u, \hat{v} \downarrow) + C(u, v)$$

where, $\hat{u}$ and $\hat{v}$ vary over the children of $u$ and $v$ respectively. In Figure 1, we give the procedure for computing $C(u \downarrow, v \downarrow)$ and cut values for all pairs $u$ and

Let binary $L_i$ be 1 iff "$V_i$ is a leaf of the cluster tree"; $\bar{L}_i$ is its negation
Let binary $l_u$ be 1 iff "$u$ is a leaf in its cluster"; $\bar{l}_u$ is its negation
For $1 \leq i \leq N$, For $1 \leq j \leq N$
    For each $(u,v) \in V_i \times V_j$ considered in lexicographic ordering
        if $l_v$ and $\neg L_j$
            Deletemin($Q_1$) to get $\langle j, u, v, Y \rangle$; add $Y$ to $Y_v$;
        if $l_u$ and $\neg L_i$
            Deletemin($Q_2$) to get $\langle i, j, u, v, X \rangle$; add $X$ to $X_u$;
    For each $(u,v) \in V_i \times V_j$ considered in lexicographic ordering
        $A = \sum_{\hat{u}} C(\hat{u} \downarrow, v \downarrow)$
        $B = \sum_{\hat{v}} C(u, \hat{v} \downarrow)$
        $C(u \downarrow, v \downarrow) = A\bar{l}_u + X_u \bar{L}_i l_u + B\bar{l}_v + Y_v \bar{L}_j l_v + C(u, v)$
        if $u$ and $v$ are incomparable vertices
            $C(u \downarrow \cup v \downarrow) = C(u \downarrow) + C(v \downarrow) - 2C(u \downarrow, v \downarrow)$
        if $u$ and $v$ are comparable vertices
            $C(u \downarrow - v \downarrow) = C(u \downarrow) - C(v \downarrow) + 2(C(u \downarrow, v \downarrow) - 2\rho^{\downarrow}(v))$
    Let $r_{i1}, \ldots r_{ik}$ be the roots in $T[V_i]$
    Let $r_{j1}, \ldots r_{jl}$ be the roots in $T[V_j]$
    For each vertex $u \in V_i$
        $Y^u = C(u, r_{j1} \downarrow) + \ldots + C(u, r_{jl} \downarrow)$
        Store $\langle P(V_j), u, p(r_{j1}), Y^u \rangle$ in $Q_1$
    For each vertex $v \in V_j$
        $X^v = C(r_{i1} \downarrow, v \downarrow) + \ldots + C(r_{ik} \downarrow, v \downarrow)$
        Store $\langle P(V_i), j, p(r_{i1}), v, X^v \rangle$ in $Q_2$

**Fig. 1.** Procedure to compute cut values for all pair of vertices

$v$. In the procedure, $p(u)$ is the parent of $u$ in $T$, $P(V_i)$ is the parent of $V_i$ in the cluster tree.

For each $i, j \in N$, $V_i$, $V_j$ and $E_{ij}$ are brought in main memory. Note that the size of $E_{ij}$ can be at most $O(B^2)$. We assume that size of main memory is $\Omega(B^2)$. $C(u \downarrow)$ and $\rho^{\downarrow}(u)$ are stored with vertex $u$. We mark all ancestors of each vertex $u \in V_i$. For any two $u, u' \in V_i$, the sets of ancestors of $u$ and $u'$ that are in cluster $V_j$ for $j \neq i$, are the same. If vertices in $V_i$ have ancestor(s) in $V_j$, then $V_j$ is an ancestor of $V_i$ in the cluster tree $T'$. We can mark all the ancestors in additional $O(V/B)$ I/Os.

Two priority queues $Q_1$ and $Q_2$ are maintained during the execution of the algorithm. $Q_1$ holds value $Y_{ij}^{uv} = C(u, v_1 \downarrow) + \ldots + C(u, v_l \downarrow)$ with key value $\langle j, u, v \rangle$ for each vertex $u \in V_i$ and $v \in V_j$, while cluster $V_j$ is yet to be accessed for $V_i$, and after $V_k$ (with $k < j$, and containing exactly $v_1, \ldots v_l$ among the children of $v$), has been processed for $V_i$, and $C(u, v_1 \downarrow), \ldots C(u, v_l \downarrow)$ have been computed. Note that it is not necessary that all children of $v$ are in one cluster $V_j$. Similarly $Q_2$ holds value $X_{ij}^{uv} = C(u_1 \downarrow, v \downarrow) + \ldots + C(u_l \downarrow, v \downarrow)$ with key value $\langle i, j, u, v \rangle$ for each vertex $u \in V_i$ and $v \in V_j$, while cluster $V_j$ is yet to be accessed for $V_i$, and after $V_j$ has been processed for $V_k$ (with $k < j$, and containing exactly

$u_1, \ldots u_l$ among the children of $u$), and $C(u_1 \downarrow, v \downarrow), \ldots, C(u_l \downarrow, v \downarrow)$ have been computed. Note that it is not necessary that all children of $u$ are in one node $V_k$.

The correctness of the algorithm is easy to prove. Since, for each cluster we perform $O(V/B)$ I/Os and $O(V)$ insertions in each priority queue $Q_1$ and $Q_2$ and the vertices are partitioned into $\Theta(V/B)$ clusters, the total I/O cost is $O(\frac{V}{B}\text{Sort}(V))$. We obtain the following lemma.

**Lemma 4.** *For a tree $T$, a minimum cut can be computed in $O(\text{Sort}(E) + (V/B)\text{Sort}(V))$ I/Os, if at most two edges of it are in $T$.*

We execute the above operations for all trees in packing and hence obtain the following theorem.

**Theorem 3.** *We can compute a minimum cut in $O(c \log E(MSF(V,E) + (V/B)\text{Sort}(V)))$ I/Os for the given undirected unweighted graph $G$, where $c$ is the minimum cut value.*

## 5   The Randomised Algorithm

The I/O complexity of computing the minimum cut can be improved, if spanning trees from the packing are chosen randomly. We assume that the minimum cut is large and $c > \log^2 V$. We use ideas from [10].

The maximum tree packing $\tau$ is at least $c/2$. Consider a minimum cut $X$ and a packing $P$ of size $\tau' = \beta c$. Suppose $X$ cuts exactly one tree edge of $\eta \tau'$ trees in $P$, and cuts exactly 2 tree edges of $\nu \tau'$ trees in $P$. Since $X$ cuts at least three edges of the remaining trees in $P$, $\eta \tau' + 2\nu \tau' + 3(1-\eta-\nu)\tau' \leq c$. Hence, $3 - 2\eta - \nu \leq 1/\beta$, and $\nu \geq 3 - 1/\beta - 2\eta$. First assume that $\eta > \frac{1}{2\log V}$. Uniformly randomly we pick a tree $T$ from our approximate maximal tree packing. The probability is $1/(2\log V)$ that we pick a tree so that exactly one edge of it crosses particular minimum cut $X$. If we choose $2\log^2 V$ trees, then the probability of not selecting a tree that crosses the minimum cut exactly once is

$$\left(1 - \frac{1}{2\log V}\right)^{2\log^2 V} < 2^{-\frac{\log^2 V}{\log V}} < \frac{1}{V}$$

Thus, with probability $(1 - 1/V)$, we compute a minimum cut. Now suppose that $\eta \leq \frac{1}{2\log V}$. Then, we have

$$\nu \geq 3 - \frac{1}{\beta} - \frac{1}{\log V}$$

Randomly pick a tree. The probability is $\nu$ that we will pick a tree whose exactly two tree edges crosses the minimum cut. If we select $\log V$ trees from the packing then the probability of not selecting the right tree is

$$(1 - \nu)^{\log V} \leq \left(\left(\frac{1}{\beta} - 2\right) + \frac{1}{\log V}\right)^{\log V} \leq \frac{1}{V}$$

1. $k = \frac{\lambda_{\min}}{2+\epsilon}$ for $\epsilon > 0$, where $\lambda_{\min}$ is the minimum degree of graph $G$.
2. find sparse $k$-edge connected certificate $H$ (see below for definition).
3. construct graph $G'$ from $G$ by contracting edges not in $H$ and recursively find the approximate minimum cut in the contracted graph $G'$.
4. return the minimum of $\lambda_{\min}$ and cut returned from step 4.

**Fig. 2.** Approximate minimum cut algorithm

If $\beta$ is small enough; for example $\beta = 1/2.2$. Therefore, we compute a minimum cut with probability $1 - 1/V$. This reduces the I/O complexity to $O(c \log E \cdot \text{MSF}(V, E) + \text{Sort}(E) \log^2 V + \frac{V}{B} \text{Sort}(V) \log V)$.

## 6    On a $\delta$-fat Graph

A graph $G$ is called a $\delta$-fat graph for $\delta > 0$, if the maximum tree packing of $G$ is at least $\frac{(1+\delta)c}{2}$ [10]. We can compute an approximate maximal tree packing of size at least $(1 + \delta/2)(c/2)$ from our tree packing algorithm by choosing $\epsilon = \frac{\delta}{2(1+\delta)}$. Since $c$ is the minimum cut, a tree shares on an average $2/(1 + \delta/2)$ edges with a minimum cut which is less than 2, and thus is 1. Hence, for a $\delta$-fat graph, for each tree $T$ we need only to investigate cuts that contain exactly one edge of $T$. Hence, the minimum cut algorithm takes only $O(c \log E(\text{MSF}(V, E) + \text{Sort}(E)))$ I/Os; this is dominated by the complexity of the tree packing algorithm.

## 7    The $(2 + \epsilon)$-minimum Cut Algorithm

In this section, we show that a near minimum cut can be computed more efficiently than an exact minimum cut. The algorithm given in Figure 2 is based on the algorithm of [11,12], and computes a cut of value between $c$ and $(2 + \epsilon)c$, if the minimum cut of the graph is $c$. The proof of correctness can be found in [11]. The depth of recursion for the algorithm is $O(\log E)$ [11] and in each iteration the number of edges are reduced by a constant factor. Except for step 3, each step can be executed in $\text{Sort}(E)$ I/Os. Next, we show that step 3 can be executed in $O(k \cdot \text{MSF}(V, E))$ I/Os.

A $k$-edge-certificate of $G$ is a spanning subgraph $H$ of $G$ such that for any two vertices $u$ and $v$, and for any positive integer $k' \leq k$, there are $k'$ edge disjoint paths between $u$ and $v$ in $H$ if and only if there are $k'$ edge disjoint paths between $u$ and $v$ in $G$. It is called sparse, if $E(H) = O(kV)$. There is one simple algorithm, given in [14], which computes a sparse $k$-edge connectivity certificate of graph $G$ as follows. Compute a spanning forest $F_1$ in $G$; then compute a spanning forest $F_2$ in $G - F_1$; and so on; continue like this to compute a spanning forest $F_i$ in $G - \cup_{1 \leq j < i} F_j$, until $F_k$ is computed. It is easy to see that connectivity of graph $H = \cup_{1 \leq i \leq k} F_i$ is at most $k$ and the number of edges in $H$ is $O(kV)$. Thus, we can compute a sparse $k$-edge connectivity certificate of graph $G$ in $O(k(\text{MSF}(V, E) + \text{Sort}(E)))$ I/Os.

Since $\lambda_{min}$ is $O(E/V)$ and the number of edges is reduced by a constant factor in each iteration, a total of $O(\frac{E}{V}\text{MSF}(V,E))$ I/Os are required to compute a cut of value between $c$ and $(2+\epsilon)c$.

# References

1. Aggarwal, A., Vitter, J.S.: The input/output complexity of sorting and related problems. Commun. ACM 31(9), 1116–1127 (1988)
2. Aggarwal, G., Datar, M., Rajagopalan, S., Ruhl, M.: On the streaming model augmented with a sorting primitive. In: Proc. IEEE Symposium on Foundations of Computer Science, pp. 540–549 (2004)
3. Alka: Efficient algorithms and data structures for massive data sets, Ph.D. thesis, Indian Institute of Technology Guwahati, India (2009), http://arxiv.org/abs/1005.3473
4. Brinkmeier, M.: A simple and fast min-cut algorithm. Theor. Comp. Sys. 41(2), 369–380 (2007)
5. Chiang, Y.J., Goodrich, M.T., Grove, E.F., Tamassia, R., Vengroff, D.E., Vitter, J.S.: External memory graph algorithms. In: Proc. ACM-SIAM Symposium on Discrete Algorithms, pp. 139–149 (1995)
6. Ford, L.R., Fulkerson, D.R.: Maximal flow through a network. Can. J. Math. 8, 399–404 (1956)
7. Gabow, H.N.: A matroid approach to finding edge connectivity and packing arborescences. J. Comput. Syst. Sci. 50(2), 259–273 (1995)
8. Goldschlager, L.M., Shaw, R.A., Staples, J.: The maximum flow problem is LOGSPACE complete for P. Theoret. Comput. Sci. 21(1), 105–111 (1982)
9. Hao, J., Orlin, J.: A faster algorithm for finding the minimum cut in a graph. J. Algorithms 17(3), 424–446 (1994)
10. Karger, D.R.: Minimum cuts in near linear time. J. ACM 47(1), 46–76 (2000)
11. Karger, D.R., Motwani, R.: An NC algorithm for minimum cuts. SIAM J. Comput. 26(1), 255–272 (1997)
12. Matula, D.W.: A linear time $2 + \epsilon$ approximation algorithm for edge connectivity. In: Proc. ACM-SIAM Symposium on Discrete Algorithms, pp. 500–504 (1993)
13. Mungala, K., Ranade, A.: I/O-complexity of graph algorithms. In: Proc. ACM-SIAM Symposium on Discrete Algorithms, pp. 687–694 (1999)
14. Nagamochi, H., Ibaraki, T.: A linear-time algorithm for finding a sparse $k$-connected spanning subgraph of a $k$-connected graph. Algorithmica 7(5-6), 583–596 (1992)
15. Nash-Williams, C.S.J.A.: Edge-disjoint spanning tree of finite graphs. J. London Math. Soc. 36, 445–450 (1961)
16. Plotkin, S.A., Shmoys, D.B., Tardos, É.: Fast approximation algorithms for fractional packing and covering problems. Math. Oper. Res. 20(2), 257–301 (1995)
17. Stoer, M., Wagner, F.: A simple min-cut algorithm. J. ACM 44(4), 585–591 (1997)
18. Thorup, M., Karger, D.R.: Dynamic graph algorithms with applications. In: Halldórsson, M.M. (ed.) SWAT 2000. LNCS, vol. 1851, pp. 1–9. Springer, Heidelberg (2000)

# On Some $\mathcal{NP}$-complete SEFE Problems [*]

Patrizio Angelini[1], Giordano Da Lozzo[1], and Daniel Neuwirth[2]

[1] Dipartimento di Ingegneria, Roma Tre University, Italy
{angelini,dalozzo}@dia.uniroma3.it
[2] Universität Passau, Germany
daniel.neuwirth@uni-passau.de

**Abstract.** We investigate the complexity of some problems related to the *Simultaneous Embedding with Fixed Edges* (SEFE) problem which, given $k$ planar graphs $G_1, \ldots, G_k$ on the same set of vertices, asks whether they can be simultaneously embedded so that the embedding of each graph be planar and common edges be drawn the same. While the computational complexity of SEFE with $k = 2$ is a central open question in Graph Drawing, the problem is $\mathcal{NP}$-complete for $k \geq 3$ [Gassner *et al.*, WG '06], even if the intersection graph is the same for each pair of graphs (*sunflower intersection*) [Schaefer, JGAA (2013)].

We improve on these results by proving that SEFE with $k \geq 3$ and sunflower intersection is $\mathcal{NP}$-complete even when (i) the intersection graph is connected and (ii) two of the three input graphs are biconnected. This result implies that the Partitioned T-Coherent $k$-Page Book-Embedding is $\mathcal{NP}$-complete with $k \geq 3$, which was only known for $k$ unbounded [Hoske, Bachelor Thesis (2012)]. Further, we prove that the problem of maximizing the number of edges that are drawn the same in a SEFE of two graphs is $\mathcal{NP}$-complete (*optimization of SEFE*, Open Problem 9, Chapter 11 of the Handbook of Graph Drawing and Visualization).

## 1 Introduction

Let $G_1, \ldots, G_k$ be $k$ graphs on the same set $V$ of vertices. A *simultaneous embedding with fixed edges* (SEFE) of $G_1, \ldots, G_k$ consists of $k$ planar drawings $\Gamma_1, \ldots, \Gamma_k$ of $G_1, \ldots, G_k$, respectively, such that each vertex $v \in V$ is mapped to the same point in every drawing $\Gamma_i$, and each edge that is common to more than one graph is represented by the same open Jordan curve in the drawings of all such graphs. The *SEFE problem* is the problem of testing whether $k$ input graphs $G_1, \ldots, G_k$ admit a SEFE [11].

The possibility of drawing together a set of graphs gives the opportunity to represent at the same time a set of different binary relationships among the same objects, hence making this topic an important tool in Information Visualization [12]. Motivated by such applications and by their theoretical aspects, simultaneous graph embeddings received wide research attention in the last few years. For an up-to-date survey, see [6].

Recently, a new major milestone to assert the importance of SEFE has been provided by Schaefer [21], who discussed its relationships with some other famous problems in Graph Drawing. In particular, he proved a reduction to SEFE with $k = 2$ from the

---

[*] Part of the research was conducted in the framework of ESF project 10-EuroGIGA-OP-003 GraDR "Graph Drawings and Representations" and of 'EU FP7 STREP Project "Leone: From Global Measurements to Local Management", grant no. 317647'.

S.P. Pal and K. Sadakane (Eds.): WALCOM 2014, LNCS 8344, pp. 200–212, 2014.

*clustered planarity testing* problem [9,10], that can be arguably considered as one of the most important open problems in the field.

The SEFE problem has been proved $\mathcal{NP}$-complete for $k \geq 3$ by Gassner *et al.* [14]. On the other hand, if the embedding of the input graphs is fixed, the SEFE problem becomes polynomial-time solvable for $k = 3$, but remains $\mathcal{NP}$-complete for $k \geq 14$ [3].

In Chapter 11 of the Handbook of Graph Drawing and Visualization [6], the SEFE problem with *sunflower intersection* (SUNFLOWER SEFE) is cited as an open question (Open Problem 7). In this setting, the *intersection graph* $G_\cap$ (that is, the graph composed of the edges that are common to at least two graphs) is such that, if an edge belongs to $G_\cap$, then it belongs to all the input graphs. Haeupler *et al.* [15] conjectured that SUNFLOWER SEFE is polynomial-time solvable. However, Schaefer [21] recently proved that this problem is $\mathcal{NP}$-complete for $k \geq 3$. The reduction is from the $\mathcal{NP}$-complete [16] problem PARTITIONED T-COHERENT K-PAGE BOOK EMBEDDING (PTCKPBE), defined [5] as follows. Given a set $X$ of elements, a tree $T$ whose leaves are the elements of $X$, and a collection of sets $S_i \subseteq X \times X$, for $i = 1, \ldots, k$, is there a $k$-page book-embedding such that the edges in $S_i$ are placed on the $i$-th page and the ordering of the elements of $X$ on the spine is represented by $T$? Note that, the $\mathcal{NP}$-completeness of PTCKPBE holds for $k$ unbounded [16], which implies that the $\mathcal{NP}$-completeness of SUNFLOWER SEFE holds for instances in which the intersection graph is a spanning forest composed of an unbounded number of star graphs.

In this paper, we improve on this result by proving that SUNFLOWER SEFE is $\mathcal{NP}$-complete with $k \geq 3$ even if $G_\cap$ is a spanning tree (actually, a caterpillar), and two of the input graphs are biconnected. Note that, when the intersection graph is connected, SUNFLOWER SEFE is equivalent to PTCKPBE. Hence, our result implies $\mathcal{NP}$-completeness also for PTCKPBE when $k \geq 3$.

For $k = 2$, the complexity of SEFE and of PTCKPBE is still unknown (note that every instance of SEFE with $k = 2$ obviously has sunflower intersection). However, polynomial-time algorithms are known for instances in which: (i) one of $G_1$ and $G_2$ has a fixed embedding [4]; (ii) the intersection graph $G_\cap$ is biconnected [5,15], a star graph [5], or a subcubic graph [21]; (iii) each connected component of $G_\cap$ has a fixed embedding [7]; or (iv) $G_1$ and $G_2$ are biconnected and $G_\cap$ is connected [8].

In this setting, we study the optimization version of SEFE, that we call MAX SEFE, which is cited as an open question (Open Problem 9) in Chapter 11 of the Handbook of Graph Drawing and Visualization [6]. In this problem, one asks for drawings of $G_1$ and $G_2$ such that as many edges of $G_\cap$ as possible are drawn the same. We prove that MAX SEFE is $\mathcal{NP}$-complete, even under some strong constraints. Namely, the problem is $\mathcal{NP}$-complete if $G_1$ and $G_2$ are triconnected, and $G_\cap$ is composed of a triconnected component plus a set of isolated vertices. This implies that the problem is computationally hard both in the fixed and in the variable embedding case. In the latter case, however, we can prove that MAX SEFE is $\mathcal{NP}$-complete even if $G_\cap$ has degree at most 2. Observe that any of these constraints would be sufficient to obtain polynomial-time algorithms for the original SEFE problem.

In Sect. 2 we give some preliminary definitions. In Sect. 3 we deal with the sunflower intersection scenario, while in Sect. 4 we study the MAX SEFE problem. Finally, in

Sect. 5 we discuss some open problems. For space reasons, some proofs are sketched or omitted; complete proofs can be found in the full version of the paper [2].

## 2   Preliminaries

A *drawing* of a graph is a mapping of each vertex to a point of the plane and of each edge to a simple Jordan curve connecting its endpoints. A drawing is *planar* if the curves representing its edges do not cross except, possibly, at common endpoints. A graph is *planar* if it admits a planar drawing. A planar drawing $\Gamma$ determines a subdivision of the plane into connected regions, called *faces*, and a clockwise ordering of the edges incident to each vertex, called *rotation scheme*. The unique unbounded face is the *outer face*. Two drawings are *equivalent* if they have the same rotation schemes. A *planar embedding* is an equivalence class of planar drawings.

The SEFE problem can be studied both in terms of embeddings and in terms of drawings, since edges can be represented by arbitrary curves without geometric restrictions, and since Jünger and Schulz [17] proved that two graphs $G_1$ and $G_2$ with intersection graph $G_\cap$ have a SEFE if and only if there exists a planar embedding $\Gamma_1$ of $G_1$ and a planar embedding $\Gamma_2$ of $G_2$ inducing the same embedding of $G_\cap$. This condition extends to more than two graphs in the sunflower intersection setting.

A graph is *connected* if every pair of vertices is connected by a path. A *$k$-connected* graph $G$ is such that removing any $k - 1$ vertices leaves $G$ connected; 3-connected and 2-connected graphs are also called *triconnected* and *biconnected*, respectively. A *tree* is a graph with no cycle. A *caterpillar* is a tree such that the removal of all the leaves yields a path. A subgraph $H$ of a graph $G$ is *spanning* if for each vertex $v \in G$ there exists an edge of $H$ incident to $v$. The *dual* of a graph $G$ with respect to an embedding $\Gamma$ of $G$ is the graph $G^\star$ having a vertex $v_f$ for each face $f$ of $\Gamma$ and an edge $(v_{f_1}, v_{f_2})$ if and only if faces $f_1$ and $f_2$ of $\Gamma$ have a common edge $e$ in $G$. We say that edge $(v_{f_1}, v_{f_2})$ is the *dual edge* of $e$, and vice versa.

## 3   Sunflower SEFE

In this section we study the Sunflower SEFE problem for $k \geq 3$ graphs, in which the intersection graph $G_\cap$ is such that $G_\cap = G_i \cap G_j$ for each $1 \leq i < j \leq k$. We prove that Sunflower SEFE is $\mathcal{NP}$-complete with $k \geq 3$ even if $G_\cap$ is a spanning caterpillar and two input graphs are biconnected. This implies that Partitioned T-Coherent k-Page Book Embedding [5] is $\mathcal{NP}$-complete with $k \geq 3$.

The proof is based on a polynomial-time reduction from the $\mathcal{NP}$-complete [19] problem Betweenness, that takes as input a finite set $A$ of $n$ objects and a set $C$ of $m$ ordered triples of distinct elements of $A$, and asks whether a linear ordering $\mathcal{O}$ of the elements of $A$ exists such that for each triple $\langle \alpha, \beta, \gamma \rangle$ of $C$, we have either $\mathcal{O} =< \ldots, \alpha, \ldots, \beta, \ldots, \gamma, \ldots >$ or $\mathcal{O} =< \ldots, \gamma, \ldots, \beta, \ldots, \alpha, \ldots >$.

**Theorem 1.** Sunflower SEFE *is $\mathcal{NP}$-complete even if two input graphs are biconnected and the intersection graph is a spanning caterpillar.*

**Proof:** The membership in $\mathcal{NP}$ has been proved in [14] by reducing SEFE to the *Weak Realizability* Problem.

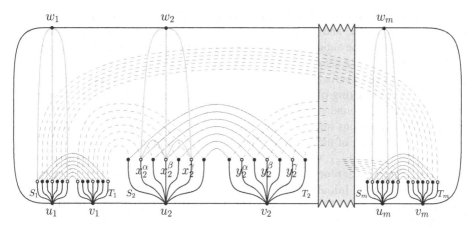

**Fig. 1.** Illustration of the composition of $G_\cap$, $G_1$, $G_2$, and $G_3$, focused on the $i$-th triple $t_i = \langle \alpha, \beta, \gamma \rangle$ of $C$ with $i = 2$

The $\mathcal{NP}$-hardness is proved by means of a polynomial-time reduction from problem BETWEENNESS. Given an instance $\langle A, C \rangle$ of BETWEENNESS, we construct an instance $\langle G_1, G_2, G_3 \rangle$ of SUNFLOWER SEFE that admits a SEFE if and only if $\langle A, C \rangle$ is a positive instance of BETWEENNESS, as follows.

In order to simplify the description, we first describe a reduction in which the produced instance of SUNFLOWER SEFE is such that $G_1$ and $G_2$ are biconnected and the intersection graph $G_\cap$ is a spanning *pseudo-tree*, that is, a connected graph containing only one cycle. In [2] it is described how to modify such instance in order to make $G_\cap$ a spanning caterpillar while maintaining biconnectivity of $G_1$ and $G_2$.

Refer to Fig. 1 for an illustration of the composition of $G_\cap$, $G_1$, $G_2$, and $G_3$.

Graph $G_\cap$ contains a cycle $C = u_1, v_1, u_2, v_2, \ldots, u_m, v_m, w_m, \ldots, w_1$ of $3m$ vertices. Also, for each $i = 1, \ldots, m$, $G_\cap$ contains a star $S_i$ with $n$ leaves centered at $u_i$ and a star $T_i$ with $n$ leaves centered at $v_i$. For each $i = 1, \ldots, m$, the leaves of $S_i$ are labeled $x_i^j$ and the leaves of $T_i$ are labeled $y_i^j$, for $j = 1, \ldots, n$. Graph $G_1$ contains all the edges of $G_\cap$ plus a set of edges $(y_i^j, x_{i+1}^j)$, for $i = 1, \ldots, m$ and $j = 1, \ldots, n$. Here and in the following, $i + 1$ is computed modulo $m$. Graph $G_2$ contains all the edges of $G_\cap$ plus a set of edges $(x_i^j, y_i^j)$, for $i = 1, \ldots, m$ and $j = 1, \ldots, n$. Graph $G_3$ contains all the edges of $G_\cap$ plus a set of edges defined as follows. For each $i = 1, \ldots, m$, consider the $i$-th triple $t_i = \langle \alpha, \beta, \gamma \rangle$ of $C$, and the corresponding vertices $x_i^\alpha$, $x_i^\beta$, and $x_i^\gamma$ of $S_i$; graph $G_3$ contains edges $(w_i, x_i^\alpha)$, $(w_i, x_i^\beta)$, $(w_i, x_i^\gamma)$, $(x_i^\alpha, x_i^\beta)$, and $(x_i^\beta, x_i^\gamma)$.

First note that, by construction, $\langle G_1, G_2, G_3 \rangle$ is an instance of SUNFLOWER SEFE, and graph $G_\cap$ is a spanning pseudo-tree. Also, one can easily verify that $G_1$ and $G_2$ are biconnected. In the following we prove that $\langle G_1, G_2, G_3 \rangle$ is a positive instance if and only if $\langle A, C \rangle$ is a positive instance of BETWEENNESS.

Suppose that $\langle G_1, G_2, G_3 \rangle$ is a positive instance, that is, $G_1$, $G_2$, and $G_3$ admit a SEFE $\langle \Gamma_1, \Gamma_2, \Gamma_3 \rangle$. Observe that, for each $i = 1, \ldots, m$, the subgraph of $G_1$ induced by the vertices of $T_i$ and the vertices of $S_{i+1}$ is composed of a set of $n$ paths of length 3 between $v_i$ and $u_{i+1}$, where the $j$-th path contains internal vertices $y_i^j$ and $x_{i+1}^j$, for $i = 1, \ldots, n$. Hence, in any SEFE of $\langle G_1, G_2, G_3 \rangle$, the ordering of the edges of $T_i$

around $v_i$ is reversed with respect to the ordering of the edges of $S_{i+1}$ around $u_{i+1}$, where the vertices of $T_i$ and $S_{i+1}$ are identified based on index $j$. Also observe that, for each $i = 1, \ldots, m$, the subgraph of $G_2$ induced by the vertices of $S_i$ and the vertices of $T_i$ is composed of a set of $n$ paths of length 3 between $u_i$ and $v_i$, where the $j$-th path contains internal vertices $x_i^j$ and $y_i^j$, for $i = 1, \ldots, n$. Hence, in any SEFE of $G_1$, $G_2$, and $G_3$, the ordering of the edges of $S_i$ around $u_i$ is the reverse of the ordering of the edges of $T_i$ around $v_i$, where the vertices of $S_i$ and $T_i$ are identified based on $j$. The two observations imply that, in any SEFE of $G_1$, $G_2$, and $G_3$, for each $i = 1, \ldots, m$ the ordering of the edges of $S_i$ around $u_i$ is the same as the ordering of the edges of $S_{i+1}$ around $v_{i+1}$, where the vertices of $S_i$ and $S_{i+1}$ are identified based on $j$.

We construct a linear ordering $\mathcal{O}$ of the elements of $A$ from the ordering of the leaves of $S_1$ in $\langle \Gamma_1, \Gamma_2, \Gamma_3 \rangle$. Initialize $\mathcal{O} = \emptyset$; then, starting from the edge of $S_1$ clockwise following $(u_1, w_1)$ around $u_1$, consider all the leaves of $S_1$ in clockwise order. For each considered leaf $x_1^j$, append $j$ as the last element of $\mathcal{O}$. We prove that $\mathcal{O}$ is a solution of $\langle A, C \rangle$. For each $i = 1, \ldots, m$, the subgraph of $G_3$ induced by vertices $w_i$, $u_i$, $x_i^\alpha$, $x_i^\beta$, and $x_i^\gamma$ is such that adding edge $(u_i, w_i)$ would make it triconnected. Hence, it admits two planar embeddings, which differ by a flip. Thus, in any SEFE of $G_1$, $G_2$, and $G_3$, edges $(u_i, x_i^\alpha)$, $(u_i, x_i^\beta)$, and $(u_i, x_i^\gamma)$ appear either in this order or in the reverse order around $u_i$. Since for each triple $t_i = \langle \alpha, \beta, \gamma \rangle$ in $C$ there exists vertices $w_i$, $u_i$, $x_i^\alpha$, $x_i^\beta$, and $x_i^\gamma$ inducing a subgraph of $G_3$ with the above properties, and since the clockwise ordering of the leaves of $S_i$ is the same for every $i$, $\mathcal{O}$ is a solution of $\langle A, C \rangle$.

Suppose that $\langle A, C \rangle$ is a positive instance, that is, there exists an ordering $\mathcal{O}$ of the elements of $A$ in which for each triple $t_i$ of $C$, the three elements of $t_i$ appear in one of their two admissible orderings. We construct an embedding for $G_1$, $G_2$, and $G_3$. For each $i = 1, \ldots, m$, the rotation schemes of $u_i$ and $v_i$ are constructed as follows. Initialize $first = v_{i-1}$ if $i > 1$, otherwise $first = w_1$. Also, initialize $last = u_{i+1}$ if $i < m$, otherwise $last = w_m$. For each element $j$ of $\mathcal{O}$, place $(u_i, x_i^j)$ between $(u_i, first)$ and $(u_i, v_i)$ in the rotation scheme of $u_i$, and set $first = x_i^j$. Also, place $(v_i, x_i^j)$ between $(v_i, last)$ and $(v_i, u_i)$ in the rotation scheme of $v_i$, and set $last = x_i^j$. Since all the vertices of $G_1$ and of $G_2$ different from $u_i$ and $v_i$ ($i = 1, \ldots, m$) have degree 2, the embeddings $\Gamma_1$ and $\Gamma_2$ of $G_1$ and $G_2$, are completely specified. To obtain the embedding $\Gamma_3$ of $G_3$, we have to specify the rotation scheme of $w_i$ and of the three leaves of $S_i$ adjacent to $w_i$, for $i = 1, \ldots, m$. Consider a triple $t_i = \langle \alpha, \beta, \gamma \rangle$ of $C$. Initialize $first = w_{i-1}$, if $i > 1$, and $first = u_1$ otherwise. Also, initialize $last = w_{i+1}$, if $i < m$, and $last = v_m$ otherwise. Recall that $\alpha$, $\beta$, and $\gamma$ appear in $\mathcal{O}$ either in this order or in the reverse one. In the former case, the rotation scheme of $w_i$ is $(w_i, last)$, $(w_i, x_i^\gamma)$, $(w_i, x_i^\beta)$, $(w_i, x_i^\alpha)$, $(w_i, first)$; the rotation scheme of $x_i^\alpha$ is $(x_i^\alpha, w_i)$, $(x_i^\alpha, x_i^\beta)$, $(x_i^\alpha, u_i)$; the rotation scheme of $x_i^\beta$ is $(x_i^\beta, x_i^\alpha)$, $(x_i^\beta, w_i)$, $(x_i^\beta, x_i^\gamma)$, $(x_i^\beta, u_i)$; and the rotation scheme of $x_i^\gamma$ is $(x_i^\gamma, x_i^\beta)$, $(x_i^\gamma, w_i)$, $(x_i^\gamma, u_i)$. In the latter case, the rotation scheme of $w_i$ is $(w_i, last)$, $(w_i, x_i^\alpha)$, $(w_i, x_i^\beta)$, $(w_i, x_i^\gamma)$, $(w_i, first)$; the rotation scheme of $x_i^\alpha$ is $(x_i^\alpha, x_i^\beta)$, $(x_i^\alpha, w_i)$, $(x_i^\alpha, u_i)$; the rotation scheme of $x_i^\beta$ is $(x_i^\beta, x_i^\gamma)$, $(x_i^\beta, w_i)$, $(x_i^\beta, x_i^\alpha)$, $(x_i^\beta, u_i)$; and the rotation scheme of $x_i^\gamma$ is $(x_i^\gamma, w_i)$, $(x_i^\gamma, x_i^\beta)$, $(x_i^\gamma, u_i)$. In order to prove that $\langle \Gamma_1, \Gamma_2, \Gamma_3 \rangle$ is a SEFE, we first observe that the embeddings of $G_\cap$ obtained by restricting $\Gamma_1$, $\Gamma_2$, and $\Gamma_3$ to the edges of $G_\cap$, respectively, coincide by construction. The planarity of $\Gamma_1$ and $\Gamma_2$ descends from the fact that the orderings of the edges incident to $u_i$ and $v_i$, for $i = 1, \ldots, m$, is one the reverse of the other (where vertices are identified based on index $j$). The planarity of $\Gamma_3$ is due to the fact that, by construction, for each $i = 1, \ldots, m$, the subgraph induced by $w_i$, $u_i$, $x_i^{\alpha}$, $x_i^{\beta}$, and $x_i^{\gamma}$ is planar in $\Gamma_3$. This concludes the proof of the theorem.    □

# 4   MAX SEFE

In this section we study the optimization version of the SEFE problem, in which two embeddings of the input graphs $G_1$ and $G_2$ are searched so that as many edges of $G_{\cap}$ as possible are drawn the same. We study the problem in its decision version and call it MAX SEFE. Namely, given a triple $\langle G_1, G_2, k^* \rangle$ composed of two planar graphs $G_1$ and $G_2$, and an integer $k^*$, the MAX SEFE problem asks whether $G_1$ and $G_2$ admit a simultaneous embedding $\langle \Gamma_1, \Gamma_2 \rangle$ in which at most $k^*$ edges of $G_{\cap}$ have a different drawing in $\Gamma_1$ and in $\Gamma_2$. First, in Lemma 1, we state the membership of MAX SEFE to $\mathcal{NP}$, which descends from the fact that SEFE belongs to $\mathcal{NP}$. Then, in Theorem 2 we prove the $\mathcal{NP}$-completeness in the general case. Finally, in Theorem 3, we prove that the problem remains $\mathcal{NP}$-complete even if stronger restrictions are imposed on the intersection graph $G_{\cap}$ of $G_1$ and $G_2$.

**Lemma 1.** MAX SEFE *is in* $\mathcal{NP}$.

In order to prove that MAX SEFE is $\mathcal{NP}$-complete, we show a reduction from a variant of the $\mathcal{NP}$-complete problem PLANAR STEINER TREE (PST) [13], defined as follows: Given an instance $\langle G(V, E), S, k \rangle$ of PST, where $G(V, E)$ is a planar graph whose edges have weights $\omega : E \rightarrow \mathbb{N}$, $S \subset V$ is a set of *terminals*, and $k > 0$ is an integer; does a tree $T^*(V^*, E^*)$ exist such that (1) $V^* \subseteq V$, (2) $E^* \subseteq E$, (3) $S \subseteq V^*$, and (4) $\sum_{e \in E^*} \omega(e) \leq k$? The edge weights in $\omega$ are bounded by a polynomial function $p(n)$ (see [13]). In our variant, that we call UNIFORM TRICONNECTED PST (UTPST), graph $G$ is a triconnected planar graph and all the edge weights are equal to 1. We remark that a variant of PST in which all the edge weights are equal to 1 and in which $G$ is a *subdivision* of a triconnected planar graph (and no subdivision vertex is a terminal) is known to be $\mathcal{NP}$-complete [1]. However, using this variant of the problem would create multiple edges in our reduction. Actually, the presence of multiple edges might be handled by replacing them in the constructed instance with a set of length-2 paths. However, we think that an $\mathcal{NP}$-completeness proof for the PST problem with $G$ triconnected and uniform edge weights may be of independent interest.

**Lemma 2.** UNIFORM TRICONNECTED PST *is* $\mathcal{NP}$-complete.

**Proof sketch:** Since instance of UTPST is also an instance of PST, we have that UTPST belongs to $\mathcal{NP}$. The $\mathcal{NP}$-hardness is proved by means of a polynomial-time reduction from PST. Namely, given an instance $\langle G, S, k \rangle$ of PST, we construct an equivalent instance $\langle G', S', k' \rangle$ of UTPST as follows. First, augment $G$ to a triconnected planar graph $G'$ by adding dummy edges whose weight is the sum of the weight of the edges of $G$, that is bounded by function $p(n)$. Then, in order to obtain uniform

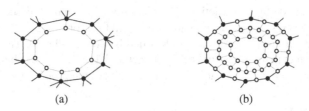

Fig. 2. (a) Gadget added inside a face to make $G^*$ triconnected. (b) Gadget replacing a vertex of degree greater than 3 to make $G_\cap$ subcubic.

weights, replace each edge $e$ in $G'$ with a path $P(e)$ of $\omega(e)$ weight-1 edges. Finally, since the subdivision vertices have degree 2, add inside each face of the unique planar embedding of $G'$ a gadget as the one in Fig. 2(a), all of whose edges have weight 1. □

Then, based on the previous lemma, we prove the main result of this section.

**Theorem 2.** MAX SEFE *is $\mathcal{NP}$-complete.*

**Proof:** The membership in $\mathcal{NP}$ follows from Lemma 1.

The $\mathcal{NP}$-hardness is proved by means of a polynomial-time reduction from problem UTPST. Let $\langle G, S, k \rangle$ be an instance of UTPST. We construct an instance $\langle G_1, G_2, k^* \rangle$ of MAX SEFE as follows (refer to Fig. 3).

Since $G$ is triconnected, it admits a unique planar embedding $\Gamma_G$, up to a flip. We now construct $G_\cap$, $G_1$, and $G_2$. Initialize $G_\cap = G_1 \cap G_2$ as the dual of $G$ with respect to $\Gamma_G$. Since $G$ is triconnected, its dual is triconnected. Consider a terminal vertex $s^* \in S$, the set $E_G(s^*)$ of the edges incident to $s^*$ in $G$, and the face $f_{s^*}$ of $G_\cap$ composed of the edges that are dual to the edges in $E_G(s^*)$. Let $v^*$ be any vertex incident to $f_{s^*}$, and let $v_1^*$ and $v_2^*$ be the neighbors of $v^*$ on $f_{s^*}$. Subdivide edges $(v^*, v_1^*)$ and $(v^*, v_2^*)$ with dummy vertices $u_1^*$ and $u_2^*$, respectively. Add to $G_\cap$ vertex $s^*$ and edges $(s^*, u_1^*)$, $(s^*, u_2^*)$, and $(s^*, v^*)$. Since $v^*$ has at least a neighbor not incident to $f_{s^*}$, vertices $u_1^*$ and $u_2^*$ do not create a separation pair. Hence, $G_\cap$ remains triconnected. See Fig. 3(a).

Graph $G_1$ contains all the vertices and edges of $G_\cap$ plus a set of vertices and edges defined as follows. For each terminal $s \in S$, consider the set $E_G(s)$ of edges incident to $s$ in $G$ and the face $f_s$ of $G_\cap$ composed of the edges dual to the edges in $E_G(s)$. Add to $G_1$ vertex $s$ and an edge $(s, v_i)$ for each vertex $v_i$ incident to $f_s$, without introducing multiple edges. Note that, graph $G_1$ is triconnected. Hence, the rotation scheme of each vertex is the one induced by the unique planar embedding of $G_1$. See Fig. 3(b).

Graph $G_2$ contains all the vertices and edges of $G_\cap$ plus a set of vertices and edges defined as follows. Rename the terminal vertices in $S$ as $x_1, \ldots, x_{|S|}$, in such a way that $s^* = x_1$. For $i = 1, \ldots, |S| - 1$, add edge $(x_i, x_{i+1})$ to $G_2$. The rotation scheme of the vertices of $G_2$ different from $x_1, \ldots, x_{|S|}$ is induced by the embedding of $G_\cap$. The rotation scheme of vertices $x_2, \ldots, x_{|S|}$ is unique, as they have degree less or equal to 2. Finally, the rotation scheme of $s^*$ is obtained by extending the rotation scheme induced by the planar embedding of $G_\cap$, in such a way that edges $(s^*, v^*)$ and $(s^*, x_2)$ are not consecutive. In order to obtain an instance of MAX SEFE in which both graphs are triconnected, we can augment $G_2$ to triconnected by only adding edges among vertices $\{u_1^*, u_2^*\} \cup \{x_1, \ldots, x_{|S|}\}$. See Fig. 3(b). Finally, set $k^* = k$.

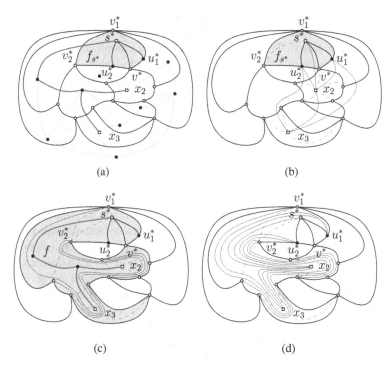

**Fig. 3.** Illustration for the proof of Theorem 2. Black lines are edges of $G_\cap$; grey lines are edges of $G$; dashed red and solid blue lines are edges of $G_1$ and $G_2$, respectively; green edges compose the Steiner tree $T$; white squares and white circles are terminal vertices and non-terminal vertices of $G$, respectively. (a) $G_\cap$, $G$ and $T$; (b) $G_1 \cup G_2$; (c) a drawing of $G_\cap$ where 4 edges have two different drawings; and (d) a solution $\langle \Gamma_1, \Gamma_2 \rangle$ of $\langle G_1, G_2, 4 \rangle$

We show that $\langle G_1, G_2, k^* \rangle$ admits a solution if and only if $\langle G, S, k \rangle$ does.

Suppose that $\langle G, S, k \rangle$ admits a solution $T$. Construct a planar drawing $\Gamma_1$ of $G_1$. The drawing $\Gamma_2$ of $G_2$ is constructed as follows. The edges of $G_\cap$ that are not dual to edges of $T$ are drawn in $\Gamma_2$ with the same curve as in $\Gamma_1$. Observe that, in the current drawing $\Gamma_2$ all the terminal vertices in $S$ lie inside the same face $f$ (see Fig. 3(c)). Hence, all the remaining edges of $G_2$ can be drawn [20] inside $f$ without intersections, as the subgraph of $G_2$ induced by the vertices incident to $f$ and by the vertices of $S$ is planar (see Fig. 3(d)). Since the only edges of $G_\cap$ that have a different drawing in $\Gamma_1$ and $\Gamma_2$ are those that are dual to edges of $T$, $\langle \Gamma_1, \Gamma_2 \rangle$ is a solution for $\langle G_1, G_2, k^* \rangle$.

Suppose that $\langle G_1, G_2, k^* \rangle$ admits a solution $\langle \Gamma_1, \Gamma_2 \rangle$ and assume that $\langle \Gamma_1, \Gamma_2 \rangle$ is optimal (that is, there exists no solution with fewer edges of $G_\cap$ not drawn the same). Consider the graph $T$ composed of the dual edges of the edges of $G_\cap$ that are not drawn the same. We claim that $T$ has at least one edge incident to each terminal in $S$ and that $T$ is connected. The claim implies that $T$ is a solution to the instance $\langle G, S, k \rangle$ of UTPST, since $T$ has at most $k$ edges and since $\langle \Gamma_1, \Gamma_2 \rangle$ is optimal.

Suppose for a contradiction that there exist two connected components $T_1$ and $T_2$ of $T$ (possibly composed of a single vertex). Consider the edges of $G$ incident to vertices of $T_1$ and not belonging to $T_1$, and consider the face $f_1$ composed of their dual edges. Note that, $f_1$ is a cycle of $G_\cap$. By definition of $T$, all the edges incident to $f_1$ have the same drawing in $\Gamma_1$ and in $\Gamma_2$. Finally, there exists at least one vertex of $S$ that lies inside $f_1$ and at least one that lies outside $f_1$. Since all the vertices in $S$ belong to a connected subgraph of $G_2$ not containing any vertex incident to $f_1$, there exist two terminal vertices $s'$ and $s''$ such that $s'$ lies inside $f_1$, $s''$ lies outside $f_1$, and edge $(s', s'')$ belongs to $G_2$. This implies that $(s', s'')$ crosses an edge incident to $f_1$ in $\Gamma_2$, a contradiction. This concludes the proof of the theorem. □

We note from Theorem 2 that MAX SEFE is $\mathcal{NP}$-complete even if $G_1$ and $G_2$ are triconnected, and $G_\cap$ is composed of a triconnected component and a set of isolated vertices (those corresponding to terminal vertices). We remark that, under these conditions, the original SEFE problem is polynomial-time solvable (actually, this is true even if only one of the input graphs has a unique embedding [4]). Further, it is possible to transform the constructed instances so that all the vertices of $G_\cap$ have degree at most 3, by replacing each vertex $v$ of degree $d(v) > 3$ in $G_\cap$ with a gadget as in Fig. 2(b). Such a gadget is composed of a cycle of $2d(v)$ vertices and of an internal grid with degree-3 vertices whose size depends on $d(v)$. Edges incident to $v$ are assigned to non-adjacent vertices of the cycle, in the order defined by the rotation scheme of $v$. Hence, the MAX SEFE problem remains $\mathcal{NP}$-complete even for instances in which $G_\cap$ is subcubic, that is another sufficient condition to make SEFE polynomial-time solvable [21].

In the following we go farther in this direction and prove that MAX SEFE remains $\mathcal{NP}$-complete even if the degree of the vertices in $G_\cap$ is at most 2. The proof is based on a reduction from the $\mathcal{NP}$-complete problem MAX 2-XORSAT [18], which takes as input (i) a set of Boolean variables $B = \{x_1, ..., x_l\}$, (ii) a 2-XorSat formula $F = \bigwedge_{x_i, x_j \in B}(l_i \oplus l_j)$, where $l_i$ is either $x_i$ or $\overline{x_i}$ and $l_j$ is either $x_j$ or $\overline{x_j}$, and (iii) an integer $k > 0$, and asks whether there exists a truth assignment $A$ for the variables in $B$ such that at most $k$ of the clauses in $F$ are not satisfied by $A$.

**Theorem 3.** MAX SEFE *is $\mathcal{NP}$-complete even if the intersection graph $G_\cap$ of the two input graphs $G_1$ and $G_2$ is composed of a set of cycles of length 3.*

**Proof:** The membership in $\mathcal{NP}$ follows from Lemma 1.

The $\mathcal{NP}$-hardness is proved by means of a polynomial-time reduction from problem MAX 2-XORSAT. Let $\langle B, F, k \rangle$ be an instance of MAX 2-XORSAT. We construct an instance $\langle G_1, G_2, k^* \rangle$ of MAX SEFE as follows. Refer to Fig. 4(a).

Graph $G_1$ is composed of a cycle $C$ with $2l$ vertices $v_1, v_2, \ldots, v_l, u_l, u_{l-1}, \ldots, u_1$. Also, for each variable $x_i \in B$, with $i = 1, \ldots, l$, $G_1$ contains a set of vertices and edges defined as follows. First, $G_1$ contains a 4-cycle $V_i = (a_i, b_i, c_i, d_i)$, that we call *variable gadget*, connected to $C$ through edge $(a_i, v_i)$. Further, for each clause $(l_i \oplus l_j) \in F$ (or $(l_j \oplus l_i) \in F$) such that $l_i \in \{x_i, \overline{x_i}\}$, $G_1$ contains (i) a 3-cycle $V_{i,j} = (a_{i,j}, b_{i,j}, c_{i,j})$, that we call *clause-variable gadget*, (ii) an edge $(b_{i,j}, w)$, where either $w = b_i$, if $l_i = x_i$, or $w = d_i$, if $l_i = \overline{x_i}$, and (iii) an edge $(a_{i,j}, c_{i,h})$, where $(l_i \oplus l_h)$ (or $(l_h \oplus l_i)$) is the last considered clause to which $l_i$ participates; if $(l_i \oplus l_j)$ (or $(l_j \oplus l_i)$) is the first considered clause containing $l_i$, then $c_{i,h} = c_i$. When the last clause $(l_i \oplus l_q)$ (or $(l_q \oplus l_i)$)

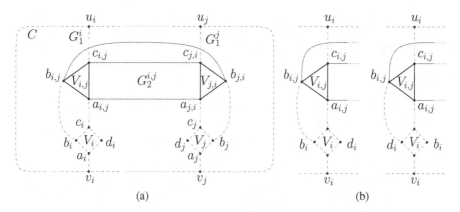

**Fig. 4.** (a) Illustration of the construction of instance $\langle G_1, G_2, k^* \rangle$ of MAX SEFE. (b) Illustration of the two cases in which $l_i$ evaluates to true in $A$.

has been considered, an edge $(c_{i,q}, u_i)$ is added to $G_1$. Note that, the subgraph $G_1^i$ of $G_1$ induced by the vertices of the variable gadget $V_i$ and of all the clause-variable gadgets $V_{i,j}$ to which $l_i$ participates would result in a subdivision of a triconnected planar graph by adding edge $(c_{i,q}, a_i)$, and hence it has a unique planar embedding (up to a flip). Graph $G_2$ is composed as follows. For each clause $(l_i \oplus l_j) \in F$, with $l_i \in \{x_i, \overline{x_i}\}$ and $l_j \in \{x_j, \overline{x_j}\}$, graph $G_2$ contains a triconnected graph $G_2^{i,j}$, that we call *clause gadget*, composed of all the vertices and edges of the clause-variable gadgets $V_{i,j}$ and $V_{j,i}$, plus three edges $(a_{i,j}, a_{j,i})$, $(b_{i,j}, b_{j,i})$, and $(c_{i,j}, c_{j,i})$. Finally, set $k^* = k$.

Note that, with this construction, graph $G_\cap$ is composed of a set of $2|F|$ cycles of length 3, namely the two clause-variable gadgets $V_{i,j}$ and $V_{j,i}$ for each clause $(l_i \oplus l_j)$.

We show that $\langle G_1, G_2, k^* \rangle$ admits a solution if and only if $\langle B, F, k \rangle$ does.

Suppose that $\langle B, F, k \rangle$ admits a solution, that is, an assignment $A$ of truth values for the variables of $B$ not satisfying at most $k$ clauses of $F$. We construct a solution $\langle \Gamma_1, \Gamma_2 \rangle$ of $\langle G_1, G_2, k^* \rangle$. First, we construct $\Gamma_1$. Let the face composed only of the edges of $C$ be the outer face. For each variable $x_i$, with $i = 1, \ldots, l$, if $x_i$ is true in $A$, then the rotation scheme of $a_i$ in $\Gamma_1$ is $(a_i, v_i), (a_i, b_i), (a_i, d_i)$ (as in Fig. 4(a)). Otherwise, $x_i$ is false in $A$, and the rotation scheme of $a_i$ is the reverse (as for $a_j$ in Fig. 4(a)). Since $G_1^i$ has a unique planar embedding, the rotation scheme of all its vertices is univocally determined. Second, we construct $\Gamma_2$. Consider each clause $(l_i \oplus l_j) \in F$, with $l_i \in \{x_i, \overline{x_i}\}$ and $l_j \in \{x_j, \overline{x_j}\}$. If $l_i$ evaluates to true in $A$, then the embedding of $G_2^{i,j}$ is such that the rotation scheme of $a_{i,j}$ in $\Gamma_2$ is $(a_{i,j}, b_{i,j}), (a_{i,j}, c_{i,j}), (a_{i,j}, a_{j,i})$ (as in Fig. 4(a)). Otherwise, $l_i$ is false in $A$ and the rotation scheme of $a_{i,j}$ is the reverse (as for $a_{j,i}$ in Fig. 4(a)). Since $G_2^{i,j}$ is triconnected, this determines the rotation scheme of all its vertices. To obtain $\Gamma_2$, compose the embeddings of all the clause gadgets in such a way that each clause gadget lies on the outer face of all the others.

We prove that $\langle \Gamma_1, \Gamma_2 \rangle$ is a solution of the MAX SEFE instance, namely that at most $k^*$ edges of $G_\cap$ have a different drawing in $\Gamma_1$ and in $\Gamma_2$. Since $G_\cap$ is composed of 3-cycles, this corresponds to saying that at most $k^*$ of such 3-cycles have a different embedding in $\Gamma_1$ and in $\Gamma_2$ (where the embedding of a 3-cycle is defined by the clockwise order of the vertices on its boundary). In fact, a 3-cycle with a different embedding in $\Gamma_1$ and in $\Gamma_2$ can always be realized by drawing only one of its edges with a different curve in the two drawings. By this observation and by the fact that at most $k = k^*$ clauses are not satisfied by $A$, the following claim is sufficient to prove the statement.

**Claim 1.** *For each clause $(l_i \oplus l_j) \in F$, if $(l_i \oplus l_j)$ is satisfied by $A$, then both $V_{i,j}$ and $V_{j,i}$ have the same embedding in $\Gamma_1$ and in $\Gamma_2$, while if $(l_i \oplus l_j)$ is not satisfied by $A$, then exactly one of them has the same embedding in $\Gamma_1$ and in $\Gamma_2$.*

**Proof sketch:** Consider a clause $(l_i \oplus l_j) \in F$, where $l_i \in \{x_i, \overline{x_i}\}$ and $l_j \in \{x_j, \overline{x_j}\}$. First note that $V_{i,j}$ has the same embedding in $\Gamma_1$ and in $\Gamma_2$, independently of whether $(l_i \oplus l_j)$ is satisfied or not, by construction of $\Gamma_2$ and by the fact that the flip of $V_{i,j}$ in $\Gamma_1$ is the same in the two cases in which $l_i$ evaluates to true in $A$, that are depicted in Fig. 4(b). Hence, it remains to prove that, if $(l_i \oplus l_j)$ is satisfied by $A$, then also $V_{j,i}$ has the same embedding in $\Gamma_1$ and in $\Gamma_2$. First, with a case analysis analogous to the one depicted in Fig. 4(b), one can observe that the flip of $V_{i,j}$ and of $V_{j,i}$ in $\Gamma_1$ only depend on the evaluation of $l_i$ and $l_j$, respectively, in $A$. Hence, since one of $l_i$ and $l_j$ evaluates to true in $A$ and the other one to false, the flip of $V_{i,j}$ and of $V_{j,i}$ in $\Gamma_1$ are "opposite" to each other. Further, by the construction of the triconnected clause gadget $G_2^{i,j}$, 3-cycles $V_{i,j}$ and $V_{j,i}$ have "opposite" flips also in $\Gamma_2$. Since $V_{i,j}$ has the same embedding in $\Gamma_1$ and in $\Gamma_2$, the statement of the claim follows.                    □

Suppose that $\langle G_1, G_2, k^* \rangle$ admits a solution $\langle \Gamma_1, \Gamma_2 \rangle$. Assume that $\langle \Gamma_1, \Gamma_2 \rangle$ is optimal, that is, there exists no solution of $\langle G_1, G_2, k^* \rangle$ with fewer edges of $G_\cap$ drawn differently. We construct a truth assignment $A$ that is a solution of $\langle B, F, k \rangle$, as follows. For each variable $x_i$, with $i = 1, \dots, l$, assign true to $x_i$ if the rotation scheme of $a_i$ in $\Gamma_1$ is $(a_i, v_i)$, $(a_i, b_i)$, $(a_i, d_i)$. Otherwise, assign false to $x_i$.

We prove that $A$ is a solution of the MAX 2-XORSAT instance, namely that at most $k$ clauses of $B$ are not satisfied by $A$. Since $\langle \Gamma_1, \Gamma_2 \rangle$ is optimal, for any 3-cycle $V_{i,j}$ of $G_\cap$, at most one edge has a different drawing in $\Gamma_1$ and in $\Gamma_2$. Also, for any clause $(l_i \oplus l_j)$, at most one of $V_{i,j}$ and $V_{j,i}$ has an edge drawn differently in $\Gamma_1$ and in $\Gamma_2$, as otherwise one could flip $G_2^{i,j}$ in $\Gamma_2$ (that is, revert the rotation scheme of all its vertices) and draw all the edges of $V_{i,j}$ and $V_{j,i}$ with the same curves as in $\Gamma_1$. Since $k = k^*$, the following claim is sufficient to prove the statement, and hence the whole theorem.

**Claim 2.** *For each clause gadget $G_2^{i,j}$ such that $V_{i,j}$ and $V_{j,i}$ have the same drawing in $\Gamma_1$ and in $\Gamma_2$, the corresponding clause $(l_i \oplus l_j)$ is satisfied by $A$.*

                                                                                    □

## 5  Conclusions

In this paper we proved that the SUNFLOWER SEFE problem with $k \geq 3$ remains $\mathcal{NP}$-complete even if the intersection graph is a spanning caterpillar and two of the input graphs are biconnected. As a corollary, we have that the PARTITIONED T-COHERENT K-PAGE BOOK EMBEDDING problem is $\mathcal{NP}$-complete with $k \geq 3$.

We also proved the $\mathcal{NP}$-completeness of the optimization version MAX SEFE of SEFE with $k = 2$, in which one wants to draw as many common edges as possible with the same curve in the drawings of the two input graphs, even under strong restrictions on the embedding of the input graphs and on the degree of the intersection graph.

Determining the complexity of the SEFE problem and of the PARTITIONED T-COHERENT K-PAGE BOOK EMBEDDING problem when $k = 2$ remain the main open questions in this context. For $k \geq 3$, an interesting open question is the complexity of SUNFLOWER SEFE when the intersection graph is just a star graph and $k$ is bounded by a constant. Note that, this problem is equivalent to the PTCKPBE when tree $T$ is a star, and hence to the $k$-page book embedding problem with fixed page assignments [16].

**Acknowledgements.** Daniel Neuwirth would like to thank Christopher Auer and Andreas Gaßner for useful discussions about the topics of this paper.

# References

1. Angelini, P., Da Lozzo, G., Di Battista, G., Frati, F., Roselli, V.: Beyond clustered planarity. CoRR abs/1207.3934 (2012)
2. Angelini, P., Da Lozzo, G., Neuwirth, D.: On the complexity of some problems related to SEFE. CoRR abs/1311.3607 (2013)
3. Angelini, P., Di Battista, G., Frati, F.: Simultaneous embedding of embedded planar graphs. Int. J. on Comp. Geom. and Appl. (2013), Special Issue from Asano, T., Nakano, S.-I., Okamoto, Y., Watanabe, O. (eds.): ISAAC 2011. LNCS, vol. 7074. Springer, Heidelberg (2011)
4. Angelini, P., Di Battista, G., Frati, F., Jelínek, V., Kratochvíl, J., Patrignani, M., Rutter, I.: Testing planarity of partially embedded graphs. In: SODA 2010, pp. 202–221 (2010)
5. Angelini, P., Di Battista, G., Frati, F., Patrignani, M., Rutter, I.: Testing the simultaneous embeddability of two graphs whose intersection is a biconnected or a connected graph. J. of Discrete Algorithms 14, 150–172 (2012)
6. Blasiüs, T., Kobourov, S.G., Rutter, I.: Simultaneous embedding of planar graphs. In: Tamassia, R. (ed.) Handbook of Graph Drawing and Visualization. CRC Press (2013)
7. Bläsius, T., Rutter, I.: Disconnectivity and relative positions in simultaneous embeddings. In: Didimo, W., Patrignani, M. (eds.) GD 2012. LNCS, vol. 7704, pp. 31–42. Springer, Heidelberg (2013)
8. Bläsius, T., Rutter, I.: Simultaneous PQ-ordering with applications to constrained embedding problems. In: SODA 2013, pp. 1030–1043 (2013)
9. Di Battista, G., Frati, F.: Efficient c-planarity testing for embedded flat clustered graphs with small faces. J. of Graph Alg. and App. 13(3), 349–378 (2009), Special Issue from Hong, S.-H., Nishizeki, T., Quan, W. (eds.): GD 2007. LNCS, vol. 4875. Springer, Heidelberg (2008)
10. Eades, P., Feng, Q.W., Lin, X., Nagamochi, H.: Straight-line drawing algorithms for hierarchical graphs and clustered graphs. Algorithmica 44(1), 1–32 (2006)
11. Erten, C., Kobourov, S.G.: Simultaneous embedding of planar graphs with few bends. J. of Graph Alg. and App. 9(3), 347–364 (2005)
12. Erten, C., Kobourov, S.G., Le, V., Navabi, A.: Simultaneous graph drawing: Layout algorithms and visualization schemes. J. of Graph Alg. and App. 9(1), 165–182 (2005)
13. Garey, M.R., Johnson, D.S.: The Rectilinear Steiner Tree Problem is NP-Complete. SIAM J. Appl. Math. 32, 826–834 (1977)

14. Gassner, E., Jünger, M., Percan, M., Schaefer, M., Schulz, M.: Simultaneous graph embeddings with fixed edges. In: Fomin, F.V. (ed.) WG 2006. LNCS, vol. 4271, pp. 325–335. Springer, Heidelberg (2006)

15. Haeupler, B., Jampani, K.R., Lubiw, A.: Testing simultaneous planarity when the common graph is 2-connected. In: Cheong, O., Chwa, K.-Y., Park, K. (eds.) ISAAC 2010, Part II. LNCS, vol. 6507, pp. 410–421. Springer, Heidelberg (2010)

16. Hoske, D.: Book embedding with fixed page assignments. Bachelor thesis, Karlsruhe Institute of Technology, Karlsruhe, Germany (2012)

17. Jünger, M., Schulz, M.: Intersection graphs in simultaneous embedding with fixed edges. J. of Graph Alg. and Appl. 13(2), 205–218 (2009)

18. Moore, C., Mertens, S.: The Nature of Computation. Oxford University Press, USA (2011)

19. Opatrny, J.: Total ordering problem. SIAM J. Comput. 8(1), 111–114 (1979)

20. Pach, J., Wenger, R.: Embedding planar graphs at fixed vertex locations. Graphs and Combinatorics 17(4), 717–728 (2001)

21. Schaefer, M.: Toward a theory of planarity: Hanani-tutte and planarity variants. J. of Graph Alg. and Appl. 17(4), 367–440 (2013)

# On Dilworth $k$ Graphs
# and Their Pairwise Compatibility

Tiziana Calamoneri[*] and Rossella Petreschi[*]

Department of Computer Science, "Sapienza" University of Rome, Italy
{calamo,petreschi}@di.uniroma1.it

**Abstract.** The *Dilworth number* of a graph is the size of the largest subset of its nodes in which the close neighborhood of no node contains the neighborhood of another one. In this paper we give a new characterization of Dilworth $k$ graphs, for each value of $k$, exactly defining their structure. Moreover, we put these graphs in relation with *pairwise compatibility graphs (PCGs)*, i.e. graphs on $n$ nodes that can be generated from an edge-weighted tree $T$ that has $n$ leaves, each representing a different node of the graph; two nodes are adjacent in the graph if and only if the weighted distance in the corresponding $T$ is between two given non-negative real numbers, $m$ and $M$. When either $m$ or $M$ are not used to eliminate edges from $G$, the two subclasses *leaf power* and *minimum leaf power* graphs (LPGs and mLPGs, respectively) are defined. Here we prove that graphs that are either LPGs or mLPGs of trees obtained connecting the centers of $k$ stars with a path are Dilworth $k$ graphs. We show that the opposite is true when $k = 1, 2$, but not when $k \geq 3$. Finally, we show that the relations we proved between Dilworth $k$ graphs and chains of $k$ stars hold only for LPGs and mLPGs, but not for PCGs.

**Keywords:** Graphs with Dilworth number $k$, leaf power graphs, minimum leaf power graphs, pairwise compatibility graphs.

## 1   Introduction and Preliminary Definitions

A graph $G = (V, E)$ is said to be a *difference graph* [15] if there is a positive real number $T$, the threshold, and for every node $v$ there is a real weight $|a(v)| < T$, such that $(v, w)$ is an edge if and only if $|a(v) - a(w)| \geq T$. Difference graphs are bipartite, so the nodes are partitioned into two stable sets. An example of a difference graph is shown in Figure 1.a.

A graph $G = (V, E)$ is a *threshold graph* [11] if there is a positive real number $S$, the threshold, and $n$ real weights of the same sign, $|a(v)| < S$, each one associated to a single node $v$ in $V$, such that $(v, w)$ is an edge if and only if $|a(v) + a(w)| \geq S$. The nodes of a threshold graph can be always partitioned into a clique and a stable set and it is trivial to observe that all the edges connecting the clique and the stable set induce a difference graph. An example of a threshold graph is shown in Figure 1.b.

---

[*] Partially supported by "Sapienza" University of Rome projects "Using Graphs to Model Phylogenetic Problems" and "Graphs and their applications to Differential Equations and to Phylogenetics".

S.P. Pal and K. Sadakane (Eds.): WALCOM 2014, LNCS 8344, pp. 213–224, 2014.
© Springer International Publishing Switzerland 2014

A graph $G = (V, E)$ is a *threshold signed graph* [2] if there are two positive real numbers $S, T$, the thresholds, and for every node $v$ there is a real weight $|a(v)| < min(S, T)$ such that $(v, w)$ is an edge if and only if either $|a(v) + a(w)| \geq S$ or $|a(v) - a(w)| \geq T$. If $S = T$ then the threshold signed graph is simply a threshold graph [20]. A threshold signed graph is constituted by two threshold graphs connected by a set of edges inducing a difference graph. An example of a threshold signed graph is shown in Figure 1.c.

In the following, when we want to highlight function $a$ and thresholds $S$ and $T$, we will express a threshold and a threshold signed graph as $G = (V, a, S)$ and $G = (V, a, S, T)$, respectively.

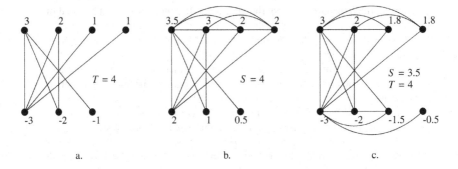

**Fig. 1.** a. A difference graph; b. A threshold graph; c. A threshold signed graph

Given a graph $G = (V, E)$, for each node $v \in V$, we call its *open neighborhood* the set $N(v) = \{u \mid (u, v) \in E\}$ and its *closed neighborhood* the set $N[v] = N(v) \cup \{v\}$. Two nodes $x$ and $y$ are said to be *comparable* if either $N(y) \subseteq N[x]$ or $N(x) \subseteq N[y]$, otherwise they are said to be *incomparable*. A *chain* is a set of pairwise comparable nodes and the *Dilworth number* of $G$ is the largest number of pairwise incomparable nodes of the graph or, in other words, the minimum size of a partition of its nodes into chains [14].

Graphs with Dilworth number 1 and 2 have been deeply studied as they correspond to the classes of threshold and of threshold signed graphs, respectively [20]. Graphs with Dilworth number at most 3 have received some attention as subclasses of a class of graphs defined by a special elimination ordering scheme [16] and graphs with Dilworth number 4 have been shown to be a subset of perfectly orderable graphs [21]. Finally, the authors of [1] proved that a large class of vertex subset and vertex partitioning problems can be solved in polynomial time on Dilworth $k$ graphs. To the best of our knowledge, nothing else is known about Dilworth $k$ graphs, when $k \geq 5$, and, unfortunately, it does not seem possible to define Dilworth $k$ graphs by using thresholds and node-weights for $k \geq 3$.

In this paper, we characterize Dilworth $k$ graphs as those graphs whose node set can be partitioned in order to form $k$ threshold graphs and the set of edges between each pair of threshold graphs induces a special difference graph. Although not difficult to obtain, this result is interesting by itself because no other characterization of Dilworth $k$ graphs is known. This is the content of Section 2.

A *star* $S_i$ is the complete bipartite graph $K_{1,i}$: a tree with one internal node $c$, called *center*, and $i$ leaves. We define a *$k$-star path* $S_{i_1,\ldots,i_k}$ to be a tree that consists of $k$ stars $S_{i_1}, \ldots, S_{i_k}$ whose centers induce a path (i.e. the centers of stars $S_{i_j}$ and $S_{i_{j+1}}$, for each $j = 1, \ldots, k-1$, are connected by an edge). In other words, $S_{i_1,\ldots,i_k}$ is a caterpillar whose the $j$-th node on the spine has $i_j$ leaves.

The *merge* of two star-paths $S_{i_1,\ldots,i_r}$ and $S_{j_1,\ldots,j_s}$ is a star-path $S_{k_1,\ldots,k_{r+s}}$ where $i_t = k_t$ for each $t = 1, \ldots, r$ and $j_t = k_{r+t}$ for each $t = 1, \ldots, s$. In other words, $S_{k_1,\ldots,k_{r+s}}$ is obtained by connecting with an edge the center of the $r$-th star of $S_{i_1,\ldots,i_r}$ with the center of the first star of $S_{j_1,\ldots,j_s}$.

A graph $G = (V, E)$ is a *pairwise compatibility graph (PCG)* [18, 22] if there exists a tree $T$, a positive edge-weight function $w$ on $T$ and two non-negative real numbers $m$ and $M$, $m \leq M$, such that $V$ coincides with the set of leaves of $T$, and the edge $(u, v)$ is in $G$ if and only if $m \leq \mathrm{dist}_{T,w}(u, v) \leq M$, where $\mathrm{dist}_{T,w}(u, v)$ is the sum of the weights of the edges on the unique path from $u$ to $v$ in $T$. In such a case, we say that $G$ is a PCG of $T$ for $m$ and $M$; in symbols, $G = PCG(T, w, m, M)$. When the constraints on the distance between the pairs of leaves deal only with $M$ or $m$, the definition of *leaf power graphs* or *minimum leaf power graphs* arise, respectively. More precisely, a graph $G = (V, E)$ is called a *leaf power graph (LPG)*[3], (respectively *min-leaf power graph (mLPG)*[10]) if there exists a tree $T$, a positive edge-weight function $w$ on $T$, and a non-negative real number $M$ (respectively $m$) such that $V$ coincides with the set of leaves of $T$, and the edge $(u, v)$ is in $G$ if and only if $\mathrm{dist}_{T,w}(u, v) \leq M$ (respectively $\mathrm{dist}_{T,w}(u, v) \geq m$).

The notions of PCG and LPG have been proposed in relation to sampling problems in phylogenetics but, since then, these classes of graphs have received great interest even from a merely graph-theoretic point of view (e.g. see [5, 6, 10, 12, 17, 19, 22–25]).

It is still unknown an algorithm testing whether a graph is a PCG (or a LPG, or an mLPG) or not. So researchers have concentrated on single classes of graphs, trying to identify their relation with respect to the class of PCGs (or LPGs or mLPGs). Here we try to improve the knowledge on PCGs, LPGs, and mLPGs studying their relations with the class of Dilworth $k$ graphs. It is known that Dilworth 1 graphs are obtained as LPGs of stars [10] and Dilworth 2 graphs are obtained as LPGs of 2-star paths [8]. Here, in Section 3 we prove that a graph that is either LPG or mLPG of a $k$-star path has Dilworth number at most $k$. Moreover we show that the opposite is true only when $k = 1, 2$, providing a Dilworth 3 graph that is neither a LPG nor a mLPG.

In Section 4, we highlight that the relations we proved between Dilworth $k$ graphs and $k$-star paths hold only for LPGs and mLPGs, but not for PCGs, since the PCG of a star can have arbitrarily high Dilworth number.

A last section, listing some open problem concludes the paper.

The reader can refer to [4] for all terminology and definitions not explicitly given in this paper.

## 2   A Characterization of Dilworth $k$ Graphs

We start this section deepening the knowledge of the structure of a Dilworth 2 graph. Namely, we prove a new characterization for Dilworth $k$ graphs as a partition into threshold graphs, each pair of which is connected by an edge set that induces a difference graph with a special ordering.

First of all, let us recall some known results from [20] on difference, threshold and threshold signed graphs.

**Lemma 1.** *Let u and w be either any two nodes of a threshold graph or any two nodes in the same set of the partition of a difference graph. It holds that deg(u) ≤ deg(w) if and only if $N(u) \subseteq N[w]$.*

Consider an edge $(u, v)$ of a threshold signed graph $G = (V, a, S, T)$. It is not difficult to see that only one of the two conditions concerning the thresholds can be satisfied. Indeed, when $a(u)$ and $a(v)$ have the same sign, i.e. $a(u) \cdot a(v) > 0$ it can only be satisfied $|a(u) + a(v)| \geq S$; in this case the edge $(u, v)$ is called *S-edge*. On the contrary, when $a(u)$ and $a(v)$ have different sign, i.e. $a(u) \cdot a(v) < 0$, it can only hold that $|a(u) - a(v)| \geq T$ and the edge is called *T-edge*. In the following, we will consider the two chains, $V_1$ and $V_2$ derived from the partition of the nodes of $G$ into the two sets $V_1 = \{x \in V$ s.t. $a(x) < 0\}$ and $V_2 = \{y \in V$ s.t. $a(y) > 0\}$.

Next lemma formalizes some concepts already presented in the previous section:

**Lemma 2.** *Given a Dilworth 2 graph G, its two chains $V_1$ and $V_2$ induce each a threshold graph, $G_1$ and $G_2$, while the edges connecting nodes in different chains induce a difference graph D. So the graph can be expressed as $G = (G_1, G_2, D)$.*

We have already underlined that the definition of threshold signed graphs and of Dilworth 2 graphs are equivalent. In the following, we will use the terms "threshold signed" and "Dilworth 2" depending on what we want to highlight in the graph (either the weight function $a$ and the thresholds, or the structure of two threshold graphs with a difference graph in between).

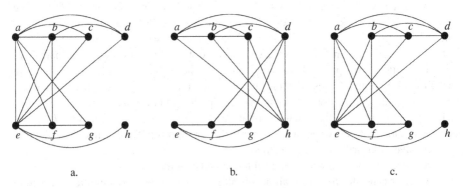

a.                    b.                    c.

**Fig. 2.** Three graphs composed by two threshold graphs unified by a difference graph. Only graph in a. is a threshold signed graph.

The opposite of Lemma 2 is not always true, as graphs in Figure 2 show. Indeed, all the three graphs are composed by two threshold graphs unified by a difference graph, but only the graph in Figure 2.a has Dilworth number 2, while graphs in Figure 2.b and 2.c, have higher Dilworth numbers (this can be easily seen by constructing their chains).

This fact suggests us that in a Dilworth 2 graph the difference graph structure must be enriched with a special order guaranteeing that the higher/lower degree nodes in the threshold graphs are also the higher/lower degree nodes in the difference graph.

Given a graph $G$ constituted by two threshold graphs $G_1$ and $G_2$ connected by a difference graph $D$, $G = (G_1, G_2, D)$, let $S_{G_1} = (v_1^1, \ldots, v_{|V_1|}^1)$, $S_{G_2} = (v_1^2, \ldots, v_{|V_2|}^2)$ and $S_D = (d_1, \ldots, d_n)$ be the ordered sequences of nodes $V_1$ and $V_2$ with respect to their degree within $G_1$, $G_2$ and $D$, respectively. So, $\deg|_{G_i}(v_1^i) \geq \ldots \geq \deg|_{G_i}(v_{|V_i|}^i)$, $i = 1, 2$, and $\deg|_D(v_1) \geq \ldots \geq \deg|_D(v_n)$. The sequence $S_D$ can be split into two subsequences $S_{D_1} = S_D \cap V_1 = (d_1^1, \ldots, d_{|V_1|}^1)$ and $S_{D_2} = S_D \cap V_2 = (d_1^2, \ldots, d_{|V_2|}^2)$. As an example, in Figure 2.a the sequences are: $S_{G_1} = abcd$, $S_{G_2} = efgh$, $S_{D_1} = abcd$ and $S_{D_2} = efgh$; in Figure 2.b the sequences are: $S_{G_1} = abcd$, $S_{G_2} = efgh$, $S_{D_1} = dcba$ and $S_{D_2} = hgfe$; finally, in Figure 2.c the sequences are: $S_{G_1} = dcba$, $S_{G_2} = efgh$, $S_{D_1} = abcd$ and $S_{D_2} = efgh$.

Observe that these orderings are not unique, as nodes with the same degree can be put in any relative order. As a particular case, isolated nodes can be added at the end of a sequence in any order. In the following, two sequences $S_{G_i}$ and $S_D \cap V_i$ are *equal* if $v_k^i = d_k^i$ for all $k = 1, \ldots, |V_i|$.

**Lemma 3.** *Given a graph $G$, the following two claims are equivalent:*

a. *$G$ has Dilworth number at most 2;*
b. *The nodes of $G$ can be partitioned into two sets $V_1$ and $V_2$, such that $V_i$, $i = 1, 2$ induces a threshold graph $G_i$ and the edges between $V_1$ and $V_2$ induce a difference graph $D$; furthermore, there exist three ordered sequences $S_{G_1}$, $S_{G_2}$ and $S_D$ such that $S_{G_i} = S_D \cap V_i$, for $i = 1, 2$.*

*Proof.* First observe that the claim is trivially true for Dilworth 1 graphs because $V_2 = \emptyset$, so in the following we consider the case in which $G$ has Dilworth number exactly 2.

$a. \Rightarrow b.$ $G$ is a Dilworth 2 graph, so Lemma 2 states that the nodes of $G$ can be partitioned into two sets $V_1$ and $V_2$, each one inducing a threshold graph, while the edges in between induce a difference graph. So it remains to prove that it is possible to find the three ordered sequences $S_{G_i}$, $i = 1, 2$ and $S_D$ such that $S_{G_i} = S_{D_i}$, $i = 1, 2$.

By contradiction, assume that it is not possible to find three sequences as required by the theorem. W.l.o.g., let us focus on $G_1$ and on the two sequences $S_{G_1}$ and $S_{D_1} = S_D \cap V_1$ related to it. Among all such possible sequences, consider those such that $v_j^1 = d_j^1$ for every $j < s$, $v_s^1 \neq d_s^1$ and $s$ is as large as possible (although, by hypothesis, $s \leq |V_1|$). Let $u = v_s^1$ and $w = d_s^1$.

There must exist an index $r > s$ such that $w = d_s^1 = v_r^1$. It follows that $\deg|_{G_1}(u) \geq \deg|_{G_1}(w)$. From the other side, since a Dilworth 2 graph is a threshold signed graph [20], all nodes belonging to the same partition are characterized by having weight values with the same signs and, inside it, two nodes $x$ and $y$ are connected by an edge if $|a(x) + a(y)| \geq S$. From these facts it follows that $a(u) \geq a(w)$.

Analogously, there must exist an index $t > s$ such that $u = v_s^1 = d_t^1$. Furthermore, an edge of $D$ connects two nodes $x$ and $y$ if they have weight values with different signs and if $|a(x) - a(y)| \geq T$. From these inequalities, it follows that $a(w) \geq a(u)$.

The two inequalities $a(u) \geq a(w)$ and $a(w) \geq a(u)$ imply $a(u) = a(w)$ and consequently $\deg|_D(u) = \deg|_D(w)$ and hence $\deg|_D(d_s^1) = \ldots = \deg|_D(d_t^1)$. So, in the sequence $S_{D_1}$, it is possible to swap $d_s^1 = w$ with $d_t^1 = u$ and now $v_s^1 = d_s^1 = u$; but this contradicts that $s$ is as large as possible.

$b. \Rightarrow a.$ Consider any two nodes $u$ and $w$ in the same set $V_1$. By hypothesis, each of them occupies the same position into the two orderings $S_{G_1}$ and $S_{D_1}$: let it be $u = v_r^1 = d_r^1$ and $w = v_s^1 = d_s^1$. It is not restrictive to assume $r < s$, that is $\deg|_{G_1}(w) \le \deg|_{G_1}(u)$ and $\deg|_D(w) \le \deg|_D(u)$. Since $G_1$ is a threshold graph, and $D$ is a difference graph, from Lemma 1, it follows that $N|_{G_1}(w) \subseteq N|_{G_1}[u]$ and $N|_D(w) \subseteq N|_D[u]$. In view of the general choice of $u$ and $w$ inside $V_1$ and observing that $N(w) = N|_{G_1}(w) \cup N|_D(w)$ and $N[u] = N|_{G_1}[u] \cup N|_D[u]$, it follows that $V_1$ constitutes a unique chain. □

Now we are ready to prove a characterization for Dilworth $k$ graphs.

**Theorem 1.** *Given a graph $G = (V, E)$, the following two claims are equivalent:*

a. *$G$ has Dilworth number at most $k$;*
b. *The nodes of $G$ can be partitioned into $k$ sets $V_1, \ldots, V_k$, such that $V_i$, $i = 1, \ldots, k$ induces a threshold graph and the edges between $V_r$ and $V_s$, for any $r, s = 1, \ldots, k$, $r < s$, induce a difference graph $D_{r,s}$; furthermore, there exist $\frac{k^2+k}{2}$ ordered sequences $S_{G_i}$, $i = 1, \ldots, k$ and $S_{D_{r,s}}$, for all $r, s = 1, \ldots k$, $r < s$, such that $S_{G_i} = S_{D_{r,s}} \cap V_i$, for all $r, s = 1, \ldots k$, $r \ne s$ and $i = r, s$.*

*Proof.* $a. \Rightarrow b.$ If $G$ is a graph with Dilworth number at most $k$, it is possible to determine its $k$ chains in polynomial time [13]; let $V_1, \ldots, V_k$ be such chains. Consider the graphs induced by these chains. Trivially they have Dilworth number 1 and are threshold graphs. Each graph induced by any pair of two chains $V_r$ and $V_s$, $r, s = 1, \ldots k, r < s$ is a Dilworth 2 graph and, by Lemma 2, the edges connecting nodes of $V_r$ with nodes of $V_s$ induce a difference graph.

In order to prove the claim about the ordered sequences, observe that Lemma 3 can be applied to each pair of chains $V_r$ and $V_s$. Since $r$ and $s$ vary in all possible ways, we have that $S_{G_i} = S_{D_{r,s}} \cap V_i$, for all $r, s = 1, \ldots k, r < s$ and $i = r, s$. Finally, the ordered sequences we considered are one for each threshold graph $G_i$, $i = 1, \ldots, k$ and one for each difference graph $D_{r,s}$, $r, s = 1, \ldots, k, r < s$, that is $k + \frac{k(k-1)}{2} = \frac{k^2+k}{2}$.

$b. \Rightarrow a.$ Consider the nodes of $G$ as partitioned into $k$ sets, $V_1, \ldots, V_k$. In view of the hypothesis, the graph induced by $V_i \cup V_j$, for each pair $i, j = 1, \ldots, k, i < j$ satisfies condition b. of Lemma 3 and hence is a Dilworth 2 graph.

We claim that, even when we put together all the Dilworth 2 graphs into the whole graph $G$, each set $V_i$, $i = 1, \ldots k$, forms a unique chain, so deducing that $G$ is a graph with Dilworth number at most $k$. In order to prove this claim, consider two nodes $u$ and $w$ of the same set $V_i$, $u = v_x^i$ and $w = v_y^i$. W.l.o.g. let us assume $x < y$. By hypothesis, $u$ precedes $w$ also in all the sequences $S_{D_{i,j}}$, for each $j = 1, \ldots, k, i \ne j$. [1] For any other set $V_j$, $V_i \cup V_j$ induces a Dilworth 2 graph, so it also holds that $N|_{V_i \cup V_j}(v_y^i) \subseteq N|_{V_i \cup V_j}[v_x^i]$. Since the neighborhood of a node $v$ of set $V_i$ can be expressed as $N(v) = \cup N|_{V_i \cup V_j}(v)$, and the same holds for the closed neighborhood, we consequently have that:

---

[1] For the sake of simplicity, here and in the following we are omitting to differentiate the two cases $i < j$ and $i > j$, leading to the nomenclature $D_{i,j}$ and $D_{j,i}$, respectively. Indeed, we use the condition $i < j$ only to avoid to consider the same set twice, once as $D_{i,j}$ and once as $D_{j,i}$.

$$N(v_y^i) \bigcup_{j=1,\ldots,k} N|_{V_i \cup V_j}(v_y^i) \subseteq \bigcup_{j=1,\ldots,k} N|_{V_i \cup V_j}[v_x^i] = N[v_x^i]$$

so showing that set $V_i$ constitutes a unique chain of $G$ and proving that $G$ is a graph with Dilworth number at most $k$. □

**Corollary 1.** *The class of Dilworth $k$ graphs is self-complementary.*

*Proof.* When passing from a Dilworth $k$ graph to its complement, the sequences $S_{G_i}$, $i = 1, \ldots, k$ and $D_{r,s}$, $r, s = 1, \ldots, k$, $r < s$, as characterized in item b. of Theorem 1, reverse their order. The proof of the claim is then easily derived from this property and from the self-complementarity of threshold and difference graphs. □

## 3 LPGs and mLPGs of $k$-star Paths Are Dilworth $k$ Graphs

In this section we highlight the relations between Dilworth $k$ graphs, LPGs and mLPGs of $k$-star chains.

**Theorem 2.** *Given a $k$-star path $S_{i_1,\ldots,i_k}$, $i_1 + \ldots + i_k = n$, an edge-weight function $w$ on $S_{i_1,\ldots,i_k}$ and a value $M$, the graph $G = LPG(S_{i_1,\ldots,i_k}, w, M)$ is a graph with Dilworth number at most $k$ and the set of leaves of each star induces at most one chain.*

*Proof.* We will prove the claim by induction on $k$.

When $k = 1$, the $k$-star path degenerates in a single star $S_n$. Consider two leaves of $S_n$, $u$ and $v$, such that $w(u) \leq w(v)$. For any other leaf $x$, if $\text{dist}_{S_n,w}(v, x) = (w(v)+w(x)) \leq M$ then $\text{dist}_{S_n,w}(u, x) \leq M$. It follows that $N(v) \subseteq N[u]$. In general, all nodes belong to the same chain, i.e. $G$ has Dilworth number 1.

By inductive hypothesis, assume now that $LPG(S_{i_1,\ldots,i_{k-1}}, w, M)$ is a graph with Dilworth number at most $k - 1$, where the set of leaves of each star induces at most one chain, and consider the $k$-star path $S_{i_1,\ldots,i_k}$ obtained by merging a star $S_{i_k}$ and the previous $(k - 1)$-star path. We have to prove that the addition of $S_{i_k}$ and of the edge $(c_{k-1}, c_k)$ does not modify in any way the existing relations between any pair of nodes in $S_{i_1,\ldots,i_{k-1}}$ and add at most one new chain.

In other words, for each $u$ and $v$ in $S_{i_1,\ldots,i_k}$, considering w.l.o.g. $w(u) \leq w(v)$, we have to prove the following assertions:

- if $u$ and $v$ are both in $S_{i_k}$ then $N(v) \subseteq N[u]$;
- if $u$ and $v$ are both in $S_{i_1,\ldots,i_{k-1}}$ and $N|_{V_1 \cup \ldots V_{k-1}}(v) \subseteq N|_{V_1 \cup \ldots V_{k-1}}[u]$ then $N|_{V_1 \cup \ldots V_k}(v) \subseteq N|_{V_1 \cup \ldots V_k}[u]$;
- if $u$ in $S_{i_1,\ldots,i_{k-1}}$ and $v$ in $S_{i_k}$ (or, vice-versa, $v$ in $S_{i_1,\ldots,i_{k-1}}$ and $u$ in $S_{i_k}$) then either $N(v) \subseteq N[u]$ or the neighborhoods of $u$ and $v$ are incomparable.

Let $u$ and $v$ be two leaves in $S_{i_k}$. For any other leaf $x$ in $S_{i_k}$, if $\text{dist}_{S_{i_k},w}(v, x) = (w(v) + w(x)) \leq M$ then $\text{dist}_{S_{i_k},w}(u, x) \leq M$ and $N(v) \subseteq N[u]$. In general, all nodes in $S_{i_k}$ belong to the same chain. For any other node $x \in V_{\bar{i}}$ (where $\bar{i} \neq i_k$), if $x \in N(v)$ then $\text{dist}_{S_{i_1,\ldots,i_k},w}(x, v) = w(x) + w(\{c_{\bar{i}}, c_{\bar{i}+1}\}) + \ldots + w(\{c_{k-1}, c_k\}) + w(v) \leq M$ and so

$dist_{S_{i_1,\ldots,i_k},w}(x,u) \le M$, that is $x \in N[u]$, so confirming that $u$ and $v$ belong to the same chain and that at most one new chain is introduced by the nodes in $S_{i_k}$.

Let $u$ and $v$ be two leaves in $S_{i_1,\ldots,i_{k-1}}$. If all nodes in $S_{i_k}$ are not adjacent to any node in $S_{i_1,\ldots,i_{k-1}}$, then trivially the second assertion is true. Let $v$ belong to $S_{\bar{\imath}}$, $1 \le \bar{\imath} \le k - 1$. Now, let us consider $x \in V_k$, s.t. $x \in N(v)$. This means that $dist_{S_{i_1,\ldots,i_k},w}(x,v) = w(v) + w(\{c_{\bar{\imath}}, c_{\bar{\imath}+1}\})+\ldots+w(\{c_{k-1}, c_k\})+w(x) \le M$, and hence it holds that $dist_{S_{i_1,\ldots,i_k},w}(x,u) \le M$.

The third assertion is trivially true. We want only to point out that the incomparability implies that the Dilworth number of $LPG(S_{i_1,\ldots,i_k}, w, M)$ is strictly greater than the Dilworth number of $LPG(S_{i_1,\ldots,i_{k-1}}, w, M)$. □

The following theorem can be proved with considerations that are very similar to those exploited in the previous proof. Nevertheless, we prefer to present a different approach.

**Theorem 3.** *Given a k-star path $S_{i_1,\ldots,i_k}$, $i_1 + \ldots + i_k = n$, an edge-weight function $w$ on $S_{i_1,\ldots,i_k}$ and a value $m$, the graph $G = mLPG(S_{i_1,\ldots,i_k}, w, m)$ is a graph with Dilworth number at most k.*

*Proof.* W.l.o.g., let $w$ assume integer values (if not, it is known [7] that it is possible to find a new edge-weight function $w'$ and a new value $m'$ such that $G = mLPG(S_{i_1,\ldots,i_k}, w', m')$ and $w'$ has integer values). Consider the graph $\bar{G} = LPG(S_{i_1,\ldots,i_k}, w, m - 1)$. It is easy to see that $\bar{G}$ is in fact the complement of $G$.

From Theorem 2, $\bar{G}$ is a graph with Dilworth number at most $k$ because it is a LPG of a $k$-star path. In view of Corollary 1, the class of Dilworth $k$ graphs is self-complementary, and so $G$ is a graph with Dilworth number at most $k$. □

The opposite of the previous theorems holds when $k = 1, 2$, in view of the following constructions:

- Let $G$ be a threshold graph on $n$ nodes and let $B_i$, $i = 1, \ldots, r$ be the set of its nodes having degree $i$. $G = (V, a, S)$ is a leaf power graph of a star $S_n$, the weight of each edge $\{c, v_i\}$, $w(\{c, v_i\})$, is equal to $j$ if $v_i \in B_j$, $1 \le j \le r$ and $M$ coincides with $r + 1$, i.e. $G = LPG(S_n, w, M)$. $G$ is even a minimum leaf power graph of the same star $S_n$, the wight of each edge $\{c, v_i\}$, $w'(\{c, v_i\})$ is equal to $r + 1 - j$ if $v_i \in B_j$, $1 \le j \le r$ and $m = r + 1$, i.e. $G = mLPG(S_n, w', m)$. [10].
- A threshold signed graph $G = (V, a, S, T)$, $V = V_1 \cup V_2$, is a leaf power graph of a 2-star path $S_{|V_1|,|V_2|}$ whose nodes are $c, v_{i_1}, \ldots, v_{i_{|V_1|}}$ and $d, u_{i_1}, \ldots, u_{i_{|V_2|}}$. The weight of each edge $\{c, v_i\}$, $w(\{c, v_i\})$, is equal to $-a(v_i)$ and the weight of each edge $\{d, u_i\}$, $w(\{d, u_i\})$, is equal to $a(u_i)$. Finally, the weight of the edge $\{c, d\}$ is equal to $S - T$ and $M$ coincides with $S$, i.e. $G = LPG(S_{|V_1|,|V_2|}, w, S)$. It is even possible to determine an edge-weight function $w'$ and a value $m$ such that $G = mLPG(S_{|V_1|,|V_2|}, w', m)$ [8].

Hence we have:

**Corollary 2.** *Given an n node graph $G$ and a k-star path, with $k = 1, 2$, it holds:*

- *$G$ is a Dilworth 1 graph if and only if $G = LPG(S_n, w, M)$ ($G = mLPG(S_n, w', m)$) for some edge-weight function $w$ ($w'$) on $S_n$ and some value of $M$ ($m$).*

– $G$ is a Dilworth 2 graph if and only if $G = LPG(S_{i_1,i_2}, w, M)$ ($G = mLPG(S_{i_1,i_2}, w', m)$) for some edge-weight function $w$ ($w'$) on $S_{i_1,i_2}$ and some value of $M$ ($m$).

In the following we will show that the opposite of Theorem 2 is not true for $k \geq 3$, because there exists a Dilworth 3 graph that is not LPG. Before proving the existence of such graph, let us recall two lemmas from [12].

**Lemma 4.** *Let G, be the cycle of four nodes a,b,c,d. $G = PCG(T, w, m, M)$ for some tree T, edge-weight w and non-negative real numbers m and M. Then $dist_{T,w}(a, c)$ and $dist_{T,w}(b, d)$ cannot be both greater than M.*

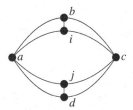

**Fig. 3.** The graph $H$

**Lemma 5.** *Let H be the graph depicted in Figure 3, where the nodes are a,b,c,d,i,j. $H = PCG(T, w, m, M)$ for some tree T, edge-weight w and non-negative real numbers m and M. Then at least one of $dist_{T,w}(a, c)$, $dist_{T,w}(b, d)$, $dist_{T,w}(i, d)$, $dist_{T,w}(j, b)$, $dist_{T,w}(i, j)$ must be greater than M.*

**Lemma 6.** *Graph H has Dilworth number 3 and is neither a LPG nor a mLPG.*

*Proof.* It is easy to see that $H$ is a Dilworth 3 graph, indeed its three chains are $\{a, c\}$, $\{b, i\}$ and $\{d, j\}$ and it is not possible to merge them into only two chains.

Assume now, by contradiction that $H$ is either a LPG or a mLPG, that is either $m$ or $M$ are set in order not to exclude any edge. Lemma 4 implies that at least one of the non-existing chords $\{a, c\}$ and $\{b, d\}$ must necessarily be excluded from $H$ by using $m$; from the other side, Lemma 5 ensures that at least one of the non-edges of $H$ must be excluded by using $M$. It follows that both $m$ and $M$ are necessary and hence $H$ is neither a LPG nor a mLPG. □

## 4  PCGs of a Star Can Have Arbitrarily Large Dilworth Number

In the previous section we have proved that LPGs and mLPGs of $k$-star paths are Dilworth $k$ graphs. We wonder whether PCGs of $k$-star paths are Dilworth $k$ graphs, too. The answer is negative, indeed even the PCG of a single star can have arbitrarily high Dilworth number, as proved in the following theorem.

**Theorem 4.** *There exists an edge-weight function $w$ of an $n$ leaf star, and two non-negative numbers $m$ and $M$ such that the $n$ node graph $G = PCG(S_n, w, m, M)$ has Dilworth number at least $n/3$.*

*Proof.* Consider an $n$ leaf star where, for the sake of simplicity, $n$ is a multiple of 3, and partition its leaves into three equally sized sets: $K = \{k_1, \ldots, k_{n/3}\}$, $S_1 = \{s_1, \ldots, s_{n/3}\}$ and $S_2 = \{t_1, \ldots, t_{n/3}\}$. Define the edge-weight function $w$ as follows: $w(\{c, s_i\}) = i$, $w(\{c, k_i\}) = n/3 + i$, and $w(\{c, t_i\}) = 2n/3 + i$, for each $i = 1, \ldots n/3$. Let $m = 2n/3 + 1$ and $M = 4n/3 + 1$.

In agreement with [9], where PCGs of stars are studied, $K$ induces a clique while $S_1$ and $S_2$ induce a stable set; these sets are pairwise connected by difference graphs.

In order to prove our claim, we focus on the two difference graphs between $K$ and $S_1$ and between $K$ and $S_2$. For any two nodes in $K$, $k_i$ and $k_j$, with $i < j$, it holds:

$$N|_{K \cup S_1}(k_i) \subseteq N|_{K \cup S_1}[k_j]$$
$$N|_{K \cup S_2}(k_j) \subseteq N|_{K \cup S_2}[k_i].$$

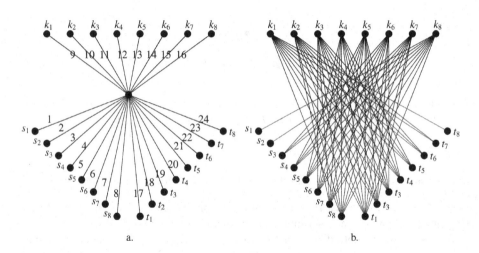

**Fig. 4.** a. An edge-weighted star $S_{24}$ where $w$ is defined according to the proof of Theorem 4; b. The resulting PCG when $m = 17$ and $M = 33$, having Dilworth number at least 8. For the sake of clarity, the edges between nodes in $K$ and between $S_1$ and $S_2$ have been omitted.

It is easy to see that both the set inclusions are in fact strict, i.e there exist two nodes $s_p \in S_1$ and $t_q \in S_2$ such that $s_p \in N|_{K \cup S_1}(k_j) \setminus N|_{K \cup S_1}(k_i)$ while $t_q \in N|_{K \cup S_2}(k_i) \setminus N|_{K \cup S_2}(k_j)$. For example, choose $p = n/3 - i$ and $q = n - j + 2$. It holds that $N(k_i) = N|_{K \cup S_1}(k_i) \cup N|_{K \cup S_2}(k_i)$ and $N(k_j) = N|_{K \cup S_1}(k_j) \cup N|_{K \cup S_2}(k_j)$, hence $s_p \in N(k_j)$ but $s_p \notin N(k_i)$, while $t_q \in N(k_i)$ but $t_q \notin N(k_j)$; it follows that $k_i$ and $k_j$ do not belong to the same chain.

In general, each node of $K$ is in a different chain, so that the Dilworth number of $G$ is at least $|K| = n/3$.

In Figure 4.a a graph with $n = 24$ nodes and Dilworth number at least $n/3 = 8$ is depicted. Figure 4.b represents the star that witnesses that $G$ is PCG.    □

## 5  Open Problems

From the results of this paper, many questions naturally arise. We list here some of them, sure that this is not an exhaustive list.

In this paper, we preliminary provide a characterization of Dilworth $k$ graphs, a result that is interesting by itself because no other characterization is known in the literature for this class of graphs. It deserves to be investigated whether this characterization is useful to provide a fast recognition algorithm for Dilworth $k$ graphs.

Then, we exploit this characterization to prove that graphs that are either LPGs or mLPGs of $k$-star paths are Dilworth $k$ graphs, but the opposite is not true, unless $k = 1$ or $k = 2$, indeed we provide a simple counterexample (a graph with Dilworth number 3 that is neither a LPG nor a mLPG). It is natural to wonder if graphs having Dilworth number equal to 3 are Pairwise Compatibility Graphs ore not. In fact, it would be of interest to understand which is the smallest Dilworth number of a graph that is not PCG. It is neither clear whether there exists a Dilworth 3 graph that is a LPG but not a LPG of a 3-star path.

Finally, we show that the relation we highlighted between Dilworth $k$ graphs and $k$-star paths hold only for LPGs and mLPGS, but not for PCGs; indeed the PCG of a star can have arbitrarily high Dilworth number. We wonder if the Dilworth number of an arbitrary LPG is unbounded or not.

## References

1. Belmonte, R., Vatshelle, M.: Graph Classes with Structured Neighborhoods and Algorithmic Applications. Theoretical Computer Science (2013), doi:10.1016/j.tcs.2013.01.011
2. Benzaken, C., Hammer, P.L., de Werra, D.: Threshold characterization of graphs with Dilworth number 2. Journal of Graph Theory 9, 245–267 (1985)
3. Brandstädt, A.: On Leaf Powers. Technical report, University of Rostock (2010)
4. Brandstädt, A., Le, V.B., Spinrad, J.: Graph classes: a survey. SIAM Monographs on Discrete Mathematics and Applications (1999)
5. Brandstädt, A., Hundt, C.: Ptolemaic Graphs and Interval Graphs Are Leaf Powers. In: Laber, E.S., Bornstein, C., Nogueira, L.T., Faria, L. (eds.) LATIN 2008. LNCS, vol. 4957, pp. 479–491. Springer, Heidelberg (2008)
6. Calamoneri, T., Frascaria, D., Sinaimeri, B.: All graphs with at most seven vertices are Pairwise Compatibility Graphs. The Computer Journal 56(7), 882–886 (2013)
7. Calamoneri, T., Montefusco, E., Petreschi, R., Sinaimeri, B.: Exploring Pairwise Compatibility Graphs. Theoretical Computer Science 468, 23–36 (2013)
8. Calamoneri, T., Petreschi, R.: Graphs with Dilworth Number Two are Pairwise Compatibility Graphs. Lagos 2013, Electronic Notes in Discrete Mathematics 44, 31–38 (2013)
9. Calamoneri, T., Petreschi, R., Sinaimeri, B.: On the Pairwise Compatibility Property of some Superclasses of Threshold Graphs. In: Special Issue of WALCOM 2012 on Discrete Mathematics, Algorithms and Applications (DMAA), vol. 5(2) (2013) (to appear)

10. Calamoneri, T., Petreschi, R., Sinaimeri, B.: On relaxing the constraints in pairwise compatibility graphs. In: Rahman, M.S., Nakano, S.-I. (eds.) WALCOM 2012. LNCS, vol. 7157, pp. 124–135. Springer, Heidelberg (2012)
11. Chvatal, V., Hammer, P.L.: Aggregation of inequalities in integer programming. Annals of Discrete Math. 1, 145–162 (1977)
12. Durocher, S., Mondal, D., Rahman, M.S.: On Graphs that are not PCGs. In: Ghosh, S.K., Tokuyama, T. (eds.) WALCOM 2013. LNCS, vol. 7748, pp. 310–321. Springer, Heidelberg (2013)
13. Felsner, S., Raghavan, V., Spinrad, J.: Recognition Algorithms for Orders of Small Width and Graphs of Small Dilworth Number. Order 20(4), 351–364 (2003)
14. Foldes, S., Hammer, P.L.: The Dilworth number of a graph. Annals of Discrete Math. 2, 211–219 (1978)
15. Hammer, P., Peled, U.N., Sun, X.: Difference graphs. Discrete Applied Math. 28(1), 35–44 (1990)
16. Hoang, C.T., Mahadev, N.V.R.: A note on perfect orders. Discrete Math. 74, 77–84 (1989)
17. Kearney, P.E., Corneil, D.G.: Tree powers. J. Algorithms 29(1) (1998)
18. Kearney, P.E., Munro, J.I., Phillips, D.: Efficient generation of uniform samples from phylogenetic trees. In: Benson, G., Page, R.D.M. (eds.) WABI 2003. LNCS (LNBI), vol. 2812, pp. 177–189. Springer, Heidelberg (2003)
19. Lin, G., Kearney, P.E., Jiang, T.: Phylogenetic $k$-Root and Steiner $k$-Root. In: Lee, D.T., Teng, S.-H. (eds.) ISAAC 2000. LNCS, vol. 1969, pp. 539–551. Springer, Heidelberg (2000)
20. Mahadev, N.V.R., Peled, U.N.: Threshold Graphs and Related Topics. Ann. Discrete Math., vol. 56. North-Holland, Amsterdam (1995)
21. Payan, C.: Perfectness and Dilworth number. Discrete Math. 44, 229–230 (1983)
22. Phillips, D.: Uniform sampling from phylogenetics trees. Masters Thesis, University of Waterloo (2002)
23. Yanhaona, M.N., Hossain, K.S.M.T., Rahman, M.S.: Pairwise compatibility graphs. J. Appl. Math. Comput. 30, 479–503 (2009)
24. Yanhaona, M.N., Hossain, K.S.M.T., Rahman, M.S.: Ladder graphs are pairwise compatibility graphs. In: AAAC 2011 (2011)
25. Yanhaona, M.N., Bayzid, M.S., Rahman, M.S.: Discovering Pairwise compatibility graphs. Discrete Mathematics, Algorithms and Applications 2(4), 607–623 (2010)

# Efficient Algorithms for Sorting $k$-Sets in Bins

Atsuki Nagao[1], Kazuhisa Seto[2], and Junichi Teruyama[3,4]

[1] Kyoto University
a-nagao@kuis.kyoto-u.ac.jp
[2] Seikei University
seto@st.seikei.ac.jp
[3] National Institute of Informatics
[4] JST, ERATO, Kawarabayashi Large Graph Project
c/o Global Research Center for Big Data Mathematics
teruyama@nii.ac.jp

**Abstract.** We give efficient algorithms for Sorting $k$-Sets in Bins. The Sorting $k$-Sets in Bins problem can be described as follows: We are given numbered $n$ bins with $k$ balls in each bin. Balls in the $i$-th bin are numbered $n - i + 1$. We can only swap balls between adjacent bins. How many swaps are needed to move all balls to the same numbered bins. For this problem, we design an efficient greedy algorithm with $\frac{k+1}{4}n^2 + O(kn)$ swaps. As $k$ and $n$ increase, this approaches the lower bound of $\lceil \binom{kn}{2}/(2k-1) \rceil$. In addition, we design a more efficient recursive algorithm using $\frac{15}{16}n^2 + O(n)$ swaps for the $k = 3$ case.

**Keywords:** Greedy, Mathematical puzzle, Recursion, Sorting, Swap.

## 1 Introduction

The Sorting problem is a classical fundamental problem in theoretical computer science. Various kinds of sorting problems have been studied [5–7]. One of the most basic sorting problems is the Swap-Sort problem : Given an list with $n$ integer numbers in non-increasing order, if we are only allowed to swap two adjacent rows, how many swaps do we need to sort them in non-decreasing order? It is well-known fact that $\binom{n}{2}$ swaps are necessary and sufficient. In this paper, we consider a natural extension of this problem. Peter Winkler [14] introduced "Sorting Pairs in Bins" and Ito, Teruyama and Yoshida [12] extended it to more general problem "Sorting $k$-Sets in Bins".

**Sorting $k$-Sets in Bins:** We are given $n$ numbered bins each with $k$ numbered balls, such that bin $i$ is adjacent to bins $i - 1$ and $i + 1$, bin $n$ is not adjacent to bin 1, and the balls in bin $i$ are each numbered $n + 1 - i$. We may swap any two balls between adjacent bins. How many swaps are necessary to get every ball into the bin carrying its number?

For $k = 2$, Winkler [14] showed a lower bound $\lceil \binom{2n}{2}/3 \rceil = \lceil \frac{n(2n-1)}{3} \rceil$ and asked whether this lower bound is optimal or not. It is easy to see $n^2$ swaps is

S.P. Pal and K. Sadakane (Eds.): WALCOM 2014, LNCS 8344, pp. 225–236, 2014.

sufficient by using bubble sort twice. West [15] proposed an algorithm with $\frac{4}{5}n^2$ swaps. Ito, Teruyama and Yoshida [12] gave an affirmative answer to Winkler's question by designing an almost optimal algorithm with $\frac{2}{3}n^2 - O(n)$ swaps, and Püttman [13] independently showed a similar result. Ito, Teruyama and Yoshida also stated (without proof) a lower bound for any $k$ and proposed the question of whether there exists an algorithm satisfying the lower bound when $k \geq 3$. When $k = 3$, by using bubble sort three times, we can easily get an algorithm with at most $\frac{3}{2}n^2$ swaps. Moreover, by combining bubble sort with the algorithm in [12], the number of swaps decreases to $\frac{7}{6}n^2$. However, the lower bound is $\frac{9}{10}n^2$. There still remains a large gap between the upper bound and the lower bound even when $k = 3$. Therefore, a natural problem is to design a more efficient algorithm for $k = 3$. In this paper, we give two efficient algorithms, one greedy, one recursive, for this problem when $k = 3$ and show that our greedy algorithm is applicable to the problem for any $k$ and its performance approaches to the lower bound as $k$ and $n$ increase.

## 1.1   Our Contribution

We show two algorithms for Sorting 3-Sets in Bins, one of which can be applied to Sorting $k$-Sets in Bins. We call the algorithms $Greedy(n, k)$ and $Recursive(n)$. For the Sorting 3-Sets in Bins problem, the number of swaps is $n^2 + O(n)$ for $Greedy(n, 3)$ and $\frac{15}{16}n^2 + O(n)(= 0.9375n^2 + O(n))$ for $Recursive(n)$. These values are close to the $\frac{9}{10}n^2(= 0.9n^2)$ lower bound shown in [12]. For Sorting $k$-Sets in Bins problem, $Greedy(n, k)$ achieves $\frac{k+1}{4}n^2 + O(kn)$ swaps. This result asymptotically approaches to the lower bound $(1 - \frac{k-1}{2k^2+k-1})\frac{k+1}{4}n^2 + O(n)$ as $k$ and $n$ increase. Formally, we prove the following two theorems.

**Theorem 1.** *There exists a greedy algorithm $Greedy(n, k)$ which solves Sorting $k$-Sets in Bins with $\frac{k+1}{4}n^2 + O(kn)$ swaps.*

**Theorem 2.** *There exists a recursive algorithm $Recursive(n)$ which solves Sorting 3-Sets in Bins with $\frac{15}{16}n^2 + O(n)$ swaps.*

## 1.2   Related Work

There are very few results to our problem. However, many other kinds of sorting problems have been well studied. For example, Partial Quicksort is the problem that given a list with integer, how many comparison do we need to sort the first $l$-th smallest elements in the list? It was introduced by [11] and shown a tight upper bound in [2]. Another example is the sorting problem for partially ordered sets, in which some pairs of elements are incomparable. Faigle and Turán introduced this problem [8] and very recently [4] made substantial advances. For other sorting problems, e.g. [1, 3, 5–7, 9, 10].

## 1.3   Paper Organization

In section 2, we give a proof of the lower bound of Sorting $k$-Sets in Bins and some notation needed for the description of our algorithms. In section 3, we present the algorithm $Greedy(n, k)$ and provide an analysis of its performance. In section 4, we present a more efficient algorithm $Recursive(n)$ and analyze its performance.

## 2 Preliminaries

### 2.1 Lower Bound

In this section, we give a precise proof of the lower bound for any $k$ [12]. Our proof is based on the point system used by Winkler [14]. An algorithm starts with 0 points. With each swap, it gets some number of points corresponding to the change in the states of the bins. We can easily calculate the total points $t$ needed to complete the sorting and the maximum points $m$ obtained by any one swap in the algorithm. Thus, we get the lower bound $t/m$. We prove the following theorem:

**Theorem 3.** *To solve Sorting $k$-Sets in Bins, we need at least* $\left\lceil \binom{kn}{2} / (2k-1) \right\rceil$ *swaps if $n$ is even, and* $\left\lceil \left( \binom{kn}{2} - \binom{k-1}{2} \right) / (2k-1) \right\rceil$ *swaps if $n$ is odd.*

*Proof.* We first construct our point system. Given two balls numbered $x$ and $y$, $x \neq y$, (without loss of generality $x > y$), we will refer to these balls as "ball $x$", "ball $y$" respectively.

**Passing (1 point) :**  Ball $x$ passes ball $y$ from left to right. In other words, we swap balls $x$ and $y$.

**Catching up (1/2 point) :**  Before the swap, ball $x$ and $y$ are in different bins. After the swap, they are in the same bin. (See Fig. 1.)

**Moving on (1/2 point) :**  Let $u > v$. Before the swap, balls $x$ and $y$ are in the same bin. After swap, ball $x$ is in the $u$-th bin and ball $y$ is in the $v$-th bin. (See Fig. 2.)

**Fig. 1.** Catching up                **Fig. 2.** Moving on

It is easy to see that for any $x$ and $y$, a set of "catching up" and "moving on" operations is the same as a "passing" operation.

Next, we consider the case where $x = y$. Because we refer to two different balls, we will continue to refer to "ball $x$" and "ball $y$".

**Separate (1/2 point) :** Before the swap, balls $x$ and $y$ are in the same bin. After the swap, they are in different bins.

**Recombine (1/2 point) :** Before the swap, balls $x$ and $y$ are in different bins. After the swap, they are in the same bin.

Now, we calculate the total points needed to complete sorting. We first assume the initial state is in reverse order. If balls $x$ and $y$ have different numbers and $x > y$, clearly ball $x$ must pass ball $y$. Thus, for any pair of different numbered balls, either a passing operation or a set of moving on and catching up operations must occur at least once. If balls $x$ and $y$ have the same number, then balls $x$ and $y$ must separate at some swap and recombine at some other swap. In each case, a 1 point charge is incurred. Thus, to complete sorting, we need at least $\binom{kn}{2}$ points. Note that if $n$ is odd, $k - 1$ balls labeled $\lceil n/2 \rceil$ may not move, so the point total is $\binom{k-1}{2}$ less than the case when $n$ is even. Therefore, we charge at least $\binom{kn}{2}$ points if $n$ is even, and $\binom{kn}{2} - \binom{k-1}{2}$ points if $n$ is odd.

Let us consider how many points can be obtained in one swap. Suppose that ball $x$ is in the $v$-th bin and ball $y$ is in the $u$-th bin, where $u > v$. The maximum amount of points that we can get by swapping $x$ and $y$ is 1 (by passing). Therefore, it suffices to consider the case of $x > y$. In this case, we focus on the other balls in the $v$-th bin and $u$-th bin. Let ball $z_i$, $i \in \{1, \ldots k - 1\}$, be a non-$y$ ball in the $u$-th bin. If each ball $z_i$ satisfies $z_i \le x$, for each $z_i$ we can get $\frac{1}{2}$ point by catching up or recombine. In addition, we consider the relation between ball $y$ and ball $z$, for each $z_i$ satisfying $y \le z$ we obtain $\frac{1}{2}$ point by applying moving on or separate operations between ball $y$ and ball $z_i$. The algorithm gets at most $2 \cdot \frac{1}{2}(k - 1)$ points by these arguments. By a similar argument on the balls in the $v$-th bin, it is possible to get at most $2 \cdot \frac{1}{2}(k - 1)$ points. Thus, the maximum total point we can get is $1 + 4 \cdot \frac{1}{2}(k - 1) = 2k - 1$. Let $T(n)$ be the number of swaps necessary to complete sorting. $T(n) \ge \lceil \binom{kn}{2}/(2k - 1) \rceil$ if $n$ is even, $T(n) \ge \lceil (\binom{kn}{2} - \binom{k-1}{2})/(2k - 1) \rceil$ if $n$ is odd.     □

## 2.2   Notations and Definitions Used in Our Algorithms

In this section, we provide some notations and definitions used to explain our algorithms. To be distinguishable and comparable, we assign all same-numbered balls an index from 1 to $k$. Let $x_i$ be the ball labeled $x$ and indexed $i$. We define the total order of balls as $x_i < y_j$ if $x < y$ or if $x = y$ and $i < j$. We represent the state of balls and bins as in Fig 3. The numbers in the bottom row are the labels of bins and the other numbers correspond to balls.

The *initial state* of the $n$ bins is the state where the $i$-th bin has $k$ balls labeled $n + 1 - i$. (See Fig. 3.) The *target state* of the $n$ bins is the state where the $i$-th bin has $k$ balls labeled $i$. (See Fig. 4.) The *solution for $n$ bins* begins with an initial state and ends with a target state for $n$ bins.

If we *move* a ball $x_i$ to the $u$-th bin, we swap $x_i$ for the lowest-labeled ball in the bin to $x_i$'s right until $x_i$ arrives at the $u$-th bin. We know the following fact directly from the properties of "move".

| $n_1$ | $n-1_1$ | | $2_1$ | $1_1$ |
|---|---|---|---|---|
| $n_2$ | $n-1_2$ | $\cdots$ | $2_2$ | $1_2$ |
| $n_3$ | $n-1_3$ | | $2_3$ | $1_3$ |
| 1 | 2 | $\cdots$ | $n-1$ | $n$ |

| $1_1$ | $2_1$ | | $n-1_1$ | $n_1$ |
|---|---|---|---|---|
| $1_2$ | $2_2$ | $\cdots$ | $n-1_2$ | $n_2$ |
| $1_3$ | $2_3$ | | $n-1_3$ | $n_3$ |
| 1 | 2 | $\cdots$ | $n-1$ | $n$ |

**Fig. 3.** Initial state of $n$ bin as $k=3$         **Fig. 4.** Target state of $n$ bin as $k=3$

**Fact 1.** *Suppose that the ball $x_i$ is in the $u$-th bin and the ball $y_j$ is in the $v$-th bin with $v < u$. If the ball $x_i$ shifts to the $(u-1)$-th bin by moving the ball $y_j$ to the $w$-th bin (where $w \geq u$), then the ball $x_i$ must have the minimum label in the $u$-th bin.*

## 3  Greedy Algorithm

### 3.1  Algorithm

Fig. 5 introduces a greedy algorithm $Greedy(n,k)$ which solves Sorting $k$-Sets in Bins with $n$ bins for any $n,k$.

---

$Greedy(n,k)$, $n, k$: **integer**
1: **for** $x = n$ to 2
2:     **for** $i = k$ to 1
3:         Move the ball $x_i$ to the $x$-th bin
4:     **end for**
5: **end for**

---

**Fig. 5.** Greedy Algorithm

**Lemma 1.** *The algorithm $Greedy(n,k)$ solves Sorting $k$-Sets in Bins with $n$ bins.*

*Proof.* Let us consider a ball $n_i (i \in [k])$. At first, we see that the $n$-th bin contains all balls labeled $n_i$ at the end of the loop on line 1 in this algorithm when $x = n$. All balls labeled $n_i$ go to the $n$-th bin on lines 2–4. If a ball $n_j$ ($j \in [k]$) leaves from the the $n$-th bin, it must be the minimum number in the $n$-th bin by Fact 1. This situation occurs only when the $n$-th bin has $k$ balls labeled $n_i$ and $j$ is 1. This means the loop on line 1 with $x = n$ is over. Thus, the $n$-th bin must contain all balls $n_i$ and the other balls are in the other bins at the end of the loop on line 1 with $x = n$. Next, let us consider the loop on line 1 with $x = n-1$. By substituting $n$ with $n-1$ and using the fact that this algorithm does not swap the balls in the $(n-1)$-th bin with the $n$-th bin, we get a similar result. By repeating this argument, we conclude that $Greedy(n,k)$ correctly solves Sorting $k$-Sets in Bins with $n$ bins. □

## 3.2    Number of Swaps

In this section, we estimate the number of swaps used by $Greedy(n, k)$ to solve Sorting $k$-Sets in Bins with $n$ bins.

**Lemma 2.** *The number of swaps performed by the algorithm $Greedy(n, k)$ is at most $\frac{k+1}{4}n^2 + O(kn)$.*

*Proof.* To prove Lemma 2, we prove the following lemma.

**Lemma 3.** *For any $x$ and any $i$, where $1 \leq x \leq n$ and $2 \leq i \leq k$, if a ball $x_i$ shifts left from its initial $(n - x + 1)$-th bin, then $x_i$ has the minimum label of all balls in any bin from $(n - x + 1)$ to $n$.*

*Proof.* Let us consider the case where a ball $x_i$ shifts left from its initial $(n-x+1)$-th bin. From Fact 1, the ball $x_i$ is the minimum-labeled ball in the $(n - x + 1)$-th bin. Because $i \geq 2$, the $(n - x + 1)$-th bin must have a ball $y_j$, where $y > x$ and $j \in [k]$. Initially, that ball $y_j$ started in the $(n - y + 1)$-th bin, which is further left than the $(n - x + 1)$-th bin. In each iteration of the loop on line 1, no balls shift right except when they go to their target bin. So, if the above situation happen, $y$ must be $(n - x + 1)$ and loop iterations from $n$ to $y + 1$ on line 1 are over. By the proof of Lemma 1, for all $u > n - x + 1$, the $u$-th bin contains all balls $u_l$ ($l \in [k]$). Thus, the proof is completed.    □

First, we count the total number of swaps needed for a ball labeled in $[\lfloor n/2 \rfloor + 1, n]$ to go to its target bin.

From Lemma 3, for $\lfloor n/2 \rfloor + 1 \leq x \leq n$ and $2 \leq i \leq k$, a ball $x_i$ does not shift left until it moves to the $x$-th bin. This means that these $x_i$ balls are in the $(n - x + 1)$-th bin. The number of swaps needed for a ball $x_i$ to move to the $x$-th bin is $2x - n - 1$. In addition, the number of swaps needed for a ball $x_1$ moves to $x$-th bin is trivially at most $x - 1$.

Thus, the total number of swaps needed for all balls $x_i$ ($x \in \{\lfloor n/2 \rfloor + 1, \ldots, n\}, i \in [k]$) to move to their target bins is at most

$$\sum_{x=\lfloor n/2 \rfloor + 1}^{n} \{x - 1 + (k - 1) \times (2x - n - 1)\} = \frac{2k + 1}{8}n^2 + O(kn). \quad (1)$$

Next, we count the total number of swaps required to move the remaining balls to their target bins. To analyse this, we need some lemmas. From Fact 1 and Lemma 3, we have Lemma 4 as follows.

**Lemma 4.** *For $x$, $1 \leq x \leq \lfloor n/2 \rfloor$ and $i \in \{2, \ldots, k\}$, suppose that a ball $x_i$ has shifted left at least once and is currently in the $l$-th bin, where $l < n - x + 1$. Then, all balls in the $u$-th bin have greater labels than $x_i$ for all $u > l$.*

**Lemma 5.** *For $x$, $1 \leq x \leq \lfloor n/2 \rfloor$ and $i \in \{2, \ldots, k\}$, a ball $x_i$ does not shift left from the $x$-th bin.*

*Proof.* Assume that a ball $x_i$ shifts left from the $x$-th bin. From Fact 1, this ball $x_i$ must have the minimum label in the $x$-th bin. Moreover, from Lemma 4, all balls in the $u$-th bin are greater than $x_i$ for all $u > x$. This means that the number of balls which are greater than $x_i$ is at least $(n-x+1)k-1$. However, the number of balls which have greater labels than $x_i$ is $(n-x)k + k - i = (n-x+1)k - i$. It leads to a contradiction, because $i \geq 2$.                                                              □

Now we have all the tools required to count the total swaps to move the remaining balls to their target bins. From the proofs of Lemmas 1 and 5, for $y \leq \lfloor n/2 \rfloor$, $k-1$ balls $y_i$ ($i \in \{2, \ldots, k\}$) are in the $y$-th bin after the iteration of loop on line 1 with $x = y$. Thus we only need to move the ball $y_1$ to complete the loop iteration on line 1 with $x = y$. That is, the number of swaps we need to move ball $y_1$ is trivially at most $y - 1$. Therefore, in total, the number of swaps required for the loop on line 1 with $y = \lfloor n/2 \rfloor$ to 2 is at most

$$\sum_{y=2}^{\lfloor n/2 \rfloor} (y - 1) = \frac{1}{8}n^2 + O(n). \tag{2}$$

From equations 1 and 2, in total, the number of swaps $Greedy(n, k)$ needs is at most

$$\frac{2k+1}{8}n^2 + O(kn) + \frac{1}{8}n^2 + O(n) = \frac{k+1}{4}n^2 + O(kn). \tag{3}$$

Thus the Lemma 2 is proved.                                                                              □

*Proof of Theorem 1.* Directly follows from Lemmas 1, 2.                                      □

## 4  Recursive Algorithm

### 4.1  Outline of Our Algorithm

In this section, we present a recursive algorithm which is more efficient for the case $k = 3$. Before describing the details of our recursive algorithm, we outline it using the case where $n$ is a multiple of 6 as an example.

First, we move the ball $n_3$ from the 1st bin to the $n$-th bin by using $n - 1$ swaps. Next, we move the ball $n_2$ from the 1st bin to the $n$-th bin by using $n - 1$ swaps, and the state becomes as in Fig. 6. Then, we move the ball $n - 1_3$ from the 2nd bin to the $(n-1)$-th bin using $n - 3$ swaps, and move the ball $n - 1_2$ from the 2nd bin to the $(n-1)$-th bin using $n - 3$ swaps. Now, the state is as in Fig. 7. For $x = n$ to $n/2 + 1$, we similarly move the ball labeled $x_3$ to the $x$-th bin, and then move the ball labeled $x_2$ to the $x$-th bin. Now, the resulting state is as in Fig. 8.

Here, for the first $n/3$ bins (i.e. from the 1st bin to the $(n/3)$-th bin), we relabel the ball $x_1$ to $\lceil x/3 \rceil$ where $x \in [n]$. Using this state as the initial state of the $n/3$ bins problem, we can recurse. After the recursive call, returning the labels of the balls in the first $n/3$ bins to their original labels yields a state where the 1st bin contains balls $1_1$, $2_1$ and $3_1$, the 2nd bin contains balls $4_1$, $5_1$ and $6_1$, and so on. (See Fig. 9.)

Finally, for $x = n$ to 2, we move ball $x_1$ to the $x$-th bin as follows. First, when we move the ball $n_1$ from the $(n/3)$-th bin to the $n$-th bin, each ball

| $n-2_1$ | $n-3_1$ | $n-4_1$ | | $1_1$ | $1_2$ | $1_3$ |
|---|---|---|---|---|---|---|
| $n-1_1$ | $n-1_2$ | $n-2_2$ | $\cdots$ | $3_2$ | $2_2$ | $n_2$ |
| $n_1$ | $n-1_3$ | $n-2_3$ | | $3_3$ | $2_3$ | $n_3$ |
| 1 | 2 | 3 | $\cdots$ $n-2$ | $n-1$ | $n$ |

| $n-2_1$ | $n-5_1$ | $n-6_1$ | | $2_2$ | $2_3$ | $1_3$ |
|---|---|---|---|---|---|---|
| $n-1_1$ | $n-4_1$ | $n-2_2$ | $\cdots$ | $3_2$ | $n-1_2$ | $n_2$ |
| $n_1$ | $n-3_1$ | $n-2_3$ | | $3_3$ | $n-1_3$ | $n_3$ |
| 1 | 2 | 3 | $\cdots$ $n-2$ | $n-1$ | $n$ |

**Fig. 6.** After moving balls $n_3$ and $n_2$ to the $n$-th bin

**Fig. 7.** After moving balls $n-1_3$ and $n-1_2$ to the $(n-1)$-th bin

| $n-2_1$ | | $1_1$ | $1_2$ | $4_2$ | | $n/2-2_2$ | $n/2_3$ | | $2_3$ | $1_3$ |
|---|---|---|---|---|---|---|---|---|---|---|
| $n-1_1$ | $\cdots$ | $2_1$ | $2_2$ | $5_2$ | $\cdots$ | $n/2-1_2$ | $n/2+1_2$ | $\cdots$ | $n-1_2$ | $n_2$ |
| $n_1$ | | $3_1$ | $3_3$ | $6_2$ | | $n/2_2$ | $n/2+1_3$ | | $n-1_3$ | $n_3$ |
| 1 | $\cdots$ | $n/3$ | $n/3+1$ | $n/3+2$ | $\cdots$ | $n/2$ | $n/2+1$ | $\cdots$ | $n-1$ | $n$ |

**Fig. 8.** After moving all balls $x_3$ and $x_2$ to $x$-th bin for $x = n$ to $n/2+1$

$1_2, 4_2, 7_2, \ldots n/2 - 2_2$ and $n/2_3, n/2 - 1_3, \ldots, 1_3$ is shifted left. (See Fig. 10.) Next, when moving the ball $n - 1_1$ from the $(n/3)$-th bin to the $(n-1)$-th bin, each ball $2_2, 5_2, 8_2, \ldots, n/2 - 1_2$ and $n/2 - 1_3, \ldots, 1_3$ is shifted left. (See Fig. 11.) Similarly, we move balls $n - 2_1, n - 3_1, \ldots, 2_1$ to their target bins respectively in this order, the sorting completing.

For $n$ which are not multiples of 6, the same strategy can be applied. The details are shown in the next subsection.

## 4.2 Algorithm

We present a recursive algorithm $Recursive(n)$. (See Fig 12.) This algorithm solves for $n$ bins with $k = 3$.

Now, we prove the correctness of our algorithm.

**Lemma 6.** *The algorithm $Recursive(n)$ solves for $n$ bins.*

*Proof.* We divide the algorithm into three steps as follows: Step 1 is lines 1–4, Step 2 is lines 5–11 and Step 3 is lines 12–14, respectively. We will show which bins have which balls after each step.

**[Step 1]** We will consider three cases, where each case is with respect to the index of the balls (e.g. $x_1, x_2, x_3$).

| $1_1$ | $4_1$ | | $n-2_1$ | $1_2$ | | $n/2-2_2$ | $n/2_3$ | | $2_3$ | $1_3$ |
|---|---|---|---|---|---|---|---|---|---|---|
| $2_1$ | $5_1$ | $\cdots$ | $n-1_1$ | $2_2$ | $\cdots$ | $n/2-1_2$ | $n/2+1_2$ | $\cdots$ | $n-1_2$ | $n_2$ |
| $3_1$ | $6_1$ | | $n_1$ | $3_3$ | | $n/2_2$ | $n/2+1_3$ | | $n-1_3$ | $n_3$ |
| 1 | 2 | $\cdots$ | $n/3$ | $n/3+1$ | $\cdots$ | $n/2$ | $n/2+1$ | $\cdots$ | $n-1$ | $n$ |

**Fig. 9.** The state after sorting for a recursive structure

| $1_1$ | | $1_2$ | $2_2$ | | $n/2-1_2$ | $n/2-1_3$ | | $1_3$ | $n_1$ |
|---|---|---|---|---|---|---|---|---|---|
| $2_1$ | $\cdots$ | $n-2_1$ | $3_2$ | $\cdots$ | $n/2_2$ | $n/2+1_2$ | $\cdots$ | $n-1_2$ | $n_2$ |
| $3_1$ | | $n-1_1$ | $4_3$ | | $n/2_3$ | $n/2+1_3$ | | $n-1_3$ | $n_3$ |
| $1$ | $\cdots$ | $n/3$ | $n/3+1$ | $\cdots$ | $n/2$ | $n/2+1$ | $\cdots$ | $n-1$ | $n$ |

**Fig. 10.** After moving the ball $n_1$ to the $n$-th bin

| $1_1$ | | $1_2$ | $3_2$ | | $n/2-1_3$ | $n/2-2_3$ | | $n-1_1$ | $n_1$ |
|---|---|---|---|---|---|---|---|---|---|
| $2_1$ | $\cdots$ | $2_2$ | $4_2$ | $\cdots$ | $n/2_2$ | $n/2+1_2$ | $\cdots$ | $n-1_2$ | $n_2$ |
| $3_1$ | | $n-2_1$ | $5_2$ | | $n/2_3$ | $n/2+1_3$ | | $n-1_3$ | $n_3$ |
| $1$ | $\cdots$ | $n/3$ | $n/3+1$ | $\cdots$ | $n/2$ | $n/2+1$ | $\cdots$ | $n-1$ | $n$ |

**Fig. 11.** After moving the ball $n-1_1$ to the $(n-1)$-th bin

**A.** Let $x_1$ be an arbitrary ball with index 1 and $x \in [n]$. Let $m$ be the number of times the ball $x_1$ moves left. Note that the ball $x_1$ is initially in the $(n-x+1)$-th bin. We consider what happens when we move balls $n-y+1_3$ or $n-y+1_2$ from the $y$-th bin to the $(n-y+1)$-th bin for each $y$. If the ball $x_1$ is located in the $u$-th bin, where $u > y$, then $x_1$ shifts left once during this movement. After iterating $2 \cdot \lfloor n/2 \rfloor$ times, the ball $x_1$ goes to the $(n-x+1-m)$-th bin, where $m$ is the largest integer which satisfies $\lceil m/2 \rceil < n-x+2-m$. That is,

$$m = \max\{0 \le i \le n-1 \mid i + \lceil i/2 \rceil < n-x+2\}.$$

We set $r$ and $q$ such that $n-x+2 = 3r+q, q \in \{0,1,2\}$. Then, $m$ is $2r-1$ if $q = 0$ and $2r$ if $q = 1,2$. After Step 1, the ball $x_1$ is in the bin numbered

$$n-x+1-m = 3r+q-1-m = \begin{cases} r & (q = 0, 1) \\ r+1 & (q = 2) \end{cases}.$$

As $x = n-3r-q+2$, the $r$-th bin contains the three balls $n-3r+1_1$, $n-3r+2_1$, and $n-3r+3_1$ after Step 1, where $x \in [n]$, $1 \le r \le \lceil n/3 \rceil$.

**B.** Let $x_2$ be an arbitrary ball with the index 2 and $x \in \{1, \ldots, \lceil n/2 \rceil\}$. By analysis similar to the proof of Lemma 3, one can see the following fact.

**Fact 2.** *The ball $x_2$ does not shift before the ball $n-x+1_2$ moves to the $(n-x+1)$-th bin in Step 1.*

We consider what happens to $x_2$ when we move balls $n-y+1_2$ or $n-y+1_3$ from the $y$-th bin to the $(n-y+1)$-th bin, where $n-y+1 \le n-x+1$. If the ball $x_2$ is in the $u$-th bin, where $u > y$, $x_2$ shifts left during this movement. By a similar argument to case A, the ball $x_2$ goes to the $(n+x-m)$-th bin, where $m$ is the maximum integer which satisfies $\lceil m/2 \rceil < (n-x+1)-(m-1-2x+1)$; that is,

$$m = \max\{0 \le i \le n-1 \mid i + \lceil i/2 \rceil < n+x+1\}.$$

*Recursive(n)*, n: **integer**
1: **for** $x = n$ to $\lceil n/2 \rceil + 1$
2:    Move the ball $x_3$ to the $x$-th bin
3:    Move the ball $x_2$ to the $x$-th bin
4: **end for**
5: **for** $x = 1$ to $n$
6:    Relabel the ball $x_1$ (in the 1st bin to $\lceil n/3 \rceil$-th bin) to $\lceil x/3 \rceil$
7: **end for**
8: **If** the $\lceil n/3 \rceil$-th bin has any ball $y_2$ or $y_3$ ($y \in [n]$)
9: **then** Relabel this ball with $\lceil n/3 \rceil$
10: Solve for the first $\lceil n/3 \rceil$ bins by applying *Recursive*($\lceil n/3 \rceil$)
(By lines 5–9, all balls in the first $\lceil n/3 \rceil$ bins are numbered at most $\lceil n/3 \rceil$. Therefore the first $\lceil n/3 \rceil$ bins are reduced to an instance of Sorting 3-Sets in Bins with $\lceil n/3 \rceil$ bins.)
11: relabel all balls in the first $\lceil n/3 \rceil$ bins with their original numbers
12: **for** $x = n$ to 2
13:    Move the ball $x_1$ to the $x$-th bin
14: **end for**

**Fig. 12.** Recursive Algorithm

We set $n + x + 1 = 3r + q, q \in \{0, 1, 2\}$. By calculation, $m$ is $2r - 1$ if $q = 0$ and $2r$ if $q = 1, 2$. Thus, after Step 1, the ball $x_2$ is in bin

$$n + x - m = 3r + q - 1 - m = \begin{cases} r & (q = 0, 1) \\ r + 1 & (q = 2) \end{cases}.$$

As $x = 3r + q - n - 1$, the $r$-th bin contains the three balls $3r - n - 2_2$, $3r - n - 1_2$, and $3r - n_2$ after Step 1, where $\lceil (n+1)/3 \rceil \leq r \leq \lceil n/2 \rceil$.

**C.** Any ball $x_3$ ($x = 1, \ldots, \lceil n/2 \rceil$) does not shift.

[**Step 2**] Applying the recursive algorithm, balls $x_1$ ($x = 1, \ldots, n$) are sorted in the first $\lceil n/3 \rceil$ bins. Each ball $x_1$ is in the $\lceil x/3 \rceil$-th bin.

[**Step 3**] Let us observe the state after moving the ball $n_1$ to the $n$-th bin. Note that the ball $n_1$ is in the $\lceil n/3 \rceil$-th bin after Step 2.

Balls of the form $x_1$ ($x = 1, \ldots, n - 1$) do not shift, since their bin numbers are less than or equal to $\lceil n/3 \rceil$.

Balls of the form $x_2$($x = 1, \ldots, \lceil n/2 \rceil$) are each in some bin between $\lceil (n+1)/3 \rceil$ and $\lceil n/2 \rceil$ after Step 2. From the analysis of Step 1 for $x_2$, the $r$-th bin has three balls $3r - n - 2_2$, $3r - n - 1_2$, and $3r - n_2$. As the ball $3r - n - 2_2$ shifts left once while the ball $n_1$ moves to the right, the $r$-th bin will have the three balls $3r - n - 1_2$, $3r - n_2$, and $3r - n + 1_2$ after completing this movement.

Before moving ball $n_1$, each ball of the form $x_3$ ($x = 1, \ldots, \lfloor n/2 \rfloor$) is in the $(n - x)$-th bin, which is not its target bin. These balls each shift left once as the ball $n_1$ moves to the $n$-th bin. That is, every ball $x_3$ ($x = 1, \ldots, \lfloor n/2 \rfloor$) shifts to the $(n - x - 1)$-th bin.

From the above arguments, the state of the first $n - 1$ bins after moving the ball $n_1$ to the $n$-th bin is the same as the state after Step 2 when we apply $Recursive(n-1)$ to Sorting 3-Sets in Bins with $n-1$ bins. After the second-last execution of the Step 3 loop (when $x = 3$), the state of the first two bins is the same as the state after Step 2 in the solution of Sorting 3-Sets in Bins with two bins. In other words, the state is as follows:

$$
\begin{array}{|c|c|}
\hline
1_1 & 1_3 \\
1_2 & 2_2 \\
2_1 & 2_3 \\
\hline
1 & 2 \\
\hline
\end{array}
$$

We can complete the sorting by moving $2_1$ to the second bin.    □

### 4.3   Number of Swaps

We count the number of swaps for $Recursive(n)$.

**Lemma 7.** *The number of swaps performed by $Recursive(n)$ is less than $\frac{15}{16}n^2 + 12n$.*

*Proof.* We will use induction. Let $S(n)$ be the number of swaps performed by $Recursive(n)$. When $n = 2$, $S(2) = 3 < \frac{15}{16}2^2 + 24$.

Suppose that $S(l) < \frac{15}{16}l^2 + 12l$ holds for any $l < n$. We count the number of swaps performed by $Recursive(n)$, where $n \geq 3$.

In Step 1, we need $2i - n - 1$ swaps to move each ball $i_3$ and $i_2$ to the $i$-th bin, so the total number of swaps in Step 1 is at most

$$
\sum_{i=n}^{\lceil n/2 \rceil + 1} (2i - n - 1) \times 2 = 2 \cdot \lceil n/2 \rceil \cdot (n - \lceil n/2 \rceil) \leq \frac{1}{2}n^2.
$$

Step 2 needs $S\left(\lceil n/3 \rceil\right) < \frac{15}{16}\left(\lceil n/3 \rceil\right)^2 + 12\lceil n/3 \rceil$ swaps.

It remains to count the number of swaps in Step 3. We consider the state of each bin. The $r$-th bin ($r \in \{1, \ldots, \lceil n/3 \rceil\}$) has three balls each labeled $3r - 3 + j_1$ ($j \in \{1, 2, 3\}$.) The total number of swaps needed for these three balls to move to the target bin is at most

$$
\sum_{j=1}^{3}(3r - 3 + j - r) = 6r - 3.
$$

Therefore, Step 3 needs at most

$$
\sum_{r=1}^{\lceil n/3 \rceil} 6r - 3 = 3(\lceil n/3 \rceil)^2
$$

swaps. Thus,

$$
\begin{aligned}
S(n) &< \frac{1}{2}n^2 + \frac{15}{16}(\lceil n/3 \rceil)^2 + 12\lceil n/3 \rceil + 3(\lceil n/3 \rceil)^2 \\
&< \frac{15}{16}n^2 + 12n, \qquad\qquad (\because \lceil n/3 \rceil < n/3 + 1 \text{ and } n \geq 3)
\end{aligned}
$$

as required.    □

*Proof of Theorem 2.* Directly follows from Lemmas 6 and 7.                    □

**Acknowledgement.** This work is supported in part by the ELC project (Grant-in-Aid for Scientific Research on Innovative Areas MEXT Japan, KAKENHI No. 24106003). This research is supported by JST, ERATO, Kawarabayashi Large Graph Project. And the authors wishes to thank Lyle Waldman for him help in editing the article.

# References

1. Bóna, M., Flynn, R.: Sorting a Permutation with Block Moves. arXiv:0806.2787v1
2. Martínez, C., Rösler, U.: Partial quicksort and quickpartitionsort. In: DMTCS Proceedings 2001, pp. 505–512 (2010)
3. Cranston, D., Sudborough, I.H., West, D.B.: Short Proofs for Cut-and-Paste Sorting of Permutations. Discrete Mathematics 307, 2866–2870 (2007)
4. Daskalakis, C., Karp, R.M., Mossel, E., Eiesenfeld, S.J., Verbin, E.: Sorting and selection in posets. SIAM Journal on Computing 40(3), 597–622 (2011)
5. Dweighter, H.: Elementary Problems. American Mathematical Monthly 82, 1010 (1975)
6. Elizalde, S., Winkler, P.: Sorting by Placement and Shift. In: Proc. ACM/SIAM Symp. on Discrete Algorithms (SODA), pp. 68–75 (2009)
7. Eriksson, H., Eriksson, K., Karlander, J., Svensson, L., Wástlund, J.: Sorting a bridge hand. Discrete Math. 241, 289–300 (2001)
8. Faigle, U., Turán, G.: Sorting and Recognition Problems for Ordered Sets. SIAM Journal on Computing 17(1), 100–113 (1988)
9. Gates, W.H., Papadimitriou, C.H.: Bounds for sorting by prefix reversal. Discrete Math. 27, 47–57 (1979)
10. Heydari, M.H., Sudborough, I.H.: On the diameter of pancake network. J. Algorithms 25, 67–94 (1997)
11. Hoare, C.A.R.: PARTITION (Algorithm 63);QUICKSORT (Algorithm 64);FIND (Algorithm 65). Communication of the Association for Computing Machinery 4, 321–322 (1961)
12. Ito, H., Teruyama, J., Yoshida, Y.: An almost optimal algorithm for winkler's sorting pairs in bins. Progress in Informatics (9), 3–7 (2012)
13. Püttmann, A.: Krawattenproblem, http://www.springer.com/cda/content/document/cda_downloaddocument/SAV_Krawattenraetsel_Loesung_Puettmann
14. Winkler, P.: Mathematical Puzzles: A Connoisseur's Collection. A K Peters 143, 149–151 (2004)
15. West, D.B.: (2008), http://www.math.uiuc.edu/~west/regs/sortpair.html

# Results on Independent Sets in Categorical Products of Graphs, the Ultimate Categorical Independence Ratio and the Ultimate Categorical Independent Domination Ratio*

Wing-Kai Hon[1], Ton Kloks[1], Ching-Hao Liu[1], Hsiang-Hsuan Liu[1],
Sheung-Hung Poon[1], and Yue-Li Wang[2]

[1] Department of Computer Science
National Tsing Hua University, Taiwan
{wkhon,hhliu,spoon}@cs.nthu.edu.tw, chinghao.liu@gmail.com
[2] Department of Information Management
National Taiwan University of Science and Technology
ylwang@cs.ntust.edu.tw

**Abstract.** We first present polynomial algorithms to compute maximum independent sets in the categorical products of two cographs or two splitgraphs, respectively. Then we prove that computing the independent set of the categorical product of a planar graph of maximal degree three and $K_4$ is NP-complete. The ultimate categorical independence ratio of a graph $G$ is defined as $\lim_{k \to \infty} \alpha(G^k)/n^k$. The ultimate categorical independence ratio can be computed in polynomial time for cographs, permutation graphs, interval graphs, graphs of bounded treewidth and splitgraphs. Also, we present an $O^*(3^{n/3})$ exact, exponential algorithm for the ultimate categorical independence ratio of general graphs. We further present a PTAS for the ultimate categorical independence ratio of planar graphs. Lastly, we show that the ultimate categorical independent domination ratio for complete multipartite graphs is zero, except when the graph is complete bipartite with color classes of equal size (in which case it is $1/2$).

## 1 Introduction

Let $G$ and $H$ be two graphs. The categorical product also travels under the guise of tensor product, or direct product, or Kronecker product, and even more names have been given to it. It is defined as follows. It is a graph, denoted by $G \times H$. Its vertices are the ordered pairs $(g, h)$ where $g \in V(G)$ and $h \in V(H)$. Two of its vertices, say $(g_1, h_1)$ and $(g_2, h_2)$ are adjacent if $\{g_1, g_2\} \in E(G)$ and $\{h_1, h_2\} \in E(H)$. One of the reasons for its popularity is Hedetniemi's conjecture, which is now more than 40 years old [10].

* Supported in part by grants 100-2628-E-007-020-MY3 and 101-2218-E-007-001 of the National Science Council (NSC), Taiwan, R.O.C.

S.P. Pal and K. Sadakane (Eds.): WALCOM 2014, LNCS 8344, pp. 237–248, 2014.

*Conjecture 1 ([10]).* For any two graphs $G$ and $H$

> Hedetniemi's conjecture:    $\chi(G \times H) = \min \{ \chi(G), \chi(H) \},$

where $\chi(I)$ denotes the chromatic number of a graph $I$. It is easy to see that the right-hand side is an upperbound. Namely, if $f$ is a vertex coloring of $G$ then one can color $G \times H$ by defining a coloring $f'$ as follows $f'((g,h)) = f(g)$, for all $g \in V(G)$ and $h \in V(H)$. Recently, Zhu showed that the fractional version of Hedetniemi's conjecture is true [25].

When restricted to perfect graphs, say $G$ and $H$, Hedetniemi's conjecture is true. Namely, let $K$ be a clique of cardinality at most $|K| \leq \min\{\omega(G), \omega(H)\}$, where $\omega(A)$ denotes the clique number of a graph $A$. It is easy to check that $G \times H$ has a clique of cardinality $|K|$. One obtains an 'elegant' proof via homomorphisms as follows. By assumption, there exist homomorphisms $K \to G$ and $K \to H$. This implies that there is also a homomorphism $K \to G \times H$ (see, eg, [9,11]). (Actually, if $W$, $P$ and $Q$ are any graphs, then there exist homomorphisms $W \to P$ and $W \to Q$ if and only if there exists a homomorphism $W \to P \times Q$.) In other words [9, Observation 5.1], $\omega(G \times H) \geq \min\{\omega(G), \omega(H)\}$. Since $G$ and $H$ are perfect, $\omega(G) = \chi(G)$ and $\omega(H) = \chi(H)$. This proves the claim, since $\chi(G \times H) \geq \omega(G \times H) \geq \min\{\omega(G), \omega(H)\} = \min\{\chi(G), \chi(H)\} \geq \chi(G \times H)$.

Since much less is known about the independence number $\alpha(G \times H)$ of the categorical products of two graphs $G$ and $H$, we are motivated to study this problem. It is easy to see that

$$\alpha(G \times H) \geq \max \{ \alpha(G) \cdot |V(H)|, \ \alpha(H) \cdot |V(G)| \}. \tag{1}$$

But this lowerbound can be arbitrarily bad, even for threshold graphs [13]. For any graph $G$ and any natural number $k$ there exists a threshold graph $H$ such that $\alpha(G \times H) \geq k + L(G, H)$, where $L(G, H)$ is the lowerbound shown in (1). Zhang recently proved that, when $G$ and $H$ are vertex transitive then equality holds in (1) [24]. We consider the computation of the independence number of the categorical product $G \times H$ for cographs, splitgraphs, or other graph classes, respectively. The formal definitions of cographs and splitgraphs are as follows.

**Definition 1.** *A graph is a cograph if it contains no induced $P_4$, which is a path with four vertices.*

**Definition 2.** *A graph $G$ is a splitgraph if there is a partition $\{S, C\}$ of its vertices such that $G[C]$ is a clique and $G[S]$ is an independent set.*

We then proceed to consider a more general product, the *k-fold categorical product* $G^k = G \times \cdots \times G$ of $k$ copies of $G$ for $k \to \infty$. Notice that, when $G$ is vertex transitive then $G^k$ is also vertex transitive and so, by the "no-homomorphism" lemma of Albertson and Collins [1], $\alpha(G^k) = \alpha(G) \cdot n^{k-1}$.

Since the independence number $\alpha(G^k)$ may not converge when $k \to \infty$, the target is, instead, to compute the ratio of the independence number $\alpha(G^k)$ versus the number of vertices of $G^k$. By (1) for any two graphs $G$ and $H$ we have

$r(G \times H) \geq \max\{r(G), r(H)\}$. It follows that $r(G^k)$ is non-decreasing. Also, it is bounded from above by 1 and so the limit when $k \to \infty$ exists. This limit was introduced in [4] as the *ultimate categorical independence ratio*. See also [2,8,12,16]. For simplicity we call it the *tensor capacity* of a graph. Such a concept is analogous to the Shannon's capacity [4], which arose from a real-world problem of transmitting words over a noisy communication channel and has a number applications. Alon and Lubetzky, and also Tóth claim that computing the tensor capacity is NP-complete but, unfortunately neither provides a proof [2,16,19]. In the following, we give the formal definitions of the independence ratio and the tensor capacity.

**Definition 3.** *The independence ratio of a graph $G$ is $r(G) = \frac{\alpha(G)}{|V(G)|}$.*

**Definition 4.** *The tensor capacity of a graph $G$ is $\Theta^T(G) = \lim_{k \to \infty} r(G^k)$.*

To provide readers some feeling about the value of $\Theta^T(G)$, we mention a related result by Tóth [20] as follows. Let $G$ be a complete multipartite graph where $n$ is the number of vertices and $\ell$ is the size of the largest partite class. If $\ell \leq \frac{n}{2}$, then $\Theta^T(G) = \frac{\ell}{n}$; otherwise, $\Theta^T(G) = 1$.

We consider the computation of the tensor capacity of the $k$-fold categorical products of some subclasses of perfect graphs, such as cographs and splitgraphs.

Moreover, by incorporating the domination constraint into the concept of ultimate categorical independence ratio, we further investigate the *ultimate categorical independent domination ratio* problem. Such a ratio $\Delta(G)$ is defined to be the *independent domination ratio* $r_i(G^k)$ of the $k$-fold categorical product $G^k$ for $k \to \infty$. In the following, we give the formal definitions of the *independent domination number*, the *independent domination ratio* and the *ultimate categorical independent domination ratio*.

**Definition 5.** *Let $G$ be a graph. The independent domination number $i(G)$ is the smallest cardinality of an independent dominating set in $G$. That is, $i(G)$ is the cardinality of a smallest maximal independent set in $G$.*

**Definition 6.** *The independent domination ratio of a graph $G$ is $r_i(G) = \frac{i(G)}{|V(G)|}$.*

**Definition 7.** *The ultimate categorical independent domination ratio of a graph $G$ is $\Delta(G) = \lim_{k \to \infty} r_i(G^k)$.*

Then in our paper, we study the problem of computing $\Delta(G)$ for complete multipartite graphs, which is a subclass of cographs.

The rest of this paper is organized as follows. In Section 2, we show that $\alpha(G \times H)$ for two cographs or two splitgraphs $G$ and $H$ can be computed in polynomial time, respectively, and that it is NP-complete to compute the maximum independent set of $G \times K_4$, where $G$ is a planar graph of maximal degree 3. In Section 3, we show that the tensor capacity for cographs, splitgraphs and three other graph classes can be computed in polynomial time, respectively, and that the tensor capacity for graphs with $n$ vertices can be computed in $O^*(3^{n/3})$ time. Also, we present a PTAS for the tensor capacity of planar graphs. Lastly, in Section 4, we determine $\Delta(G)$ for complete multipartite graphs.

# 2    Algorithms for Independence in Categorical Products

In the section, we first show that an equation for the computation of the categorical product of two complete multipartite graphs. Then we show that $\alpha(G \times H)$ for cographs $G$ and $H$ can be computed in $O(n^2)$ time. Lastly, we show that $\alpha(G \times H)$ for splitgraphs $G$ and $H$ can be computed in polynomial time.

Cographs are perfect, see, eg, [15, Section 3.3]. When $G$ and $H$ are cographs then $G \times H$ is not necessarily perfect. For example, when $G$ is the paw, ie, $G \simeq K_1 \otimes (K_2 \oplus K_1)$ then $G \times K_3$ contains an induced $C_5$ [18]. Ravindra and Parthasarathy characterize the pairs $G$ and $H$ for which $G \times H$ is perfect [18, Theorem 3.2]. (See the proof of Lemma 1 below.)

## 2.1    Complete Multipartite Graphs

It is well-known that $G \times H$ is connected if and only if both $G$ and $H$ are connected and at least one of them is not bipartite [22]. When $G$ and $H$ are connected and bipartite, then $G \times H$ consists of two components. In that case, two vertices $(g_1, h_1)$ and $(g_2, h_2)$ belong to the same component if the distances $d_G(g_1, g_2)$ and $d_H(h_1, h_2)$ have the same parity.

**Definition 8.** *The rook's graph $R(m, n)$ is the linegraph of the complete bipartite graph $K_{m,n}$.*

The rook's graph $R(m, n)$ has as its vertices the vertices of the grid, $(i, j)$, with $1 \le i \le m$ and $1 \le j \le n$. Two vertices are adjacent if they are in the same row or column of the grid. The rook's graph is perfect, since all linegraphs of bipartite graphs are perfect (see, eg, [15]). By Lovász' perfect graph theorem, also the complement of rook's graph is perfect.

**Proposition 1.** *Let $m, n \in \mathbb{N}$. Then $K_m \times K_n \simeq \bar{R}$, where $\bar{R}$ is the complement of the rook's graph $R = R(m, n)$.*

**Lemma 1.** *Let $G$ and $H$ be complete multipartite. Then $G \times H$ is perfect.*

*Proof.* Ravindra and Parthasarathy prove in [18] that $G \times H$ is perfect if and only if either

(a) $G$ or $H$ is bipartite, or
(b) Neither $G$ nor $H$ contains an induced odd cycle of length at least 5 nor an induced paw.

Since $G$ and $H$ are perfect, they do not contain an odd hole. Furthermore, the complement of $G$ and $H$ is a union of cliques, and so the complements are $P_3$-free. The complement of a paw is $K_1 \oplus P_3$ and so it has an induced $P_3$. This proves the claim.    □

Let $G$ and $H$ be complete multipartite. Let $G$ be the join of $m$ independent sets, say with $p_1, \ldots, p_m$ vertices, and let $H$ be the join of $n$ independent sets, say with $q_1, \ldots, q_n$ vertices. We shortly describe how $G \times H$ is obtained from the complement of a generalized rook's graph $R(m, n)$.

Each vertex $(i, j)$ in $R(m, n)$ is replaced by an independent set $I(i, j)$ of cardinality $p_i \cdot q_j$. Denote the vertices of this independent set as $(i_s, j_t)$ where $1 \le s \le p_i$ and $1 \le t \le q_j$. The graph $G \times H$ is obtained from the partial complement of this 'generalized rook's graph.' One can refer [5] for the definition of partial complement.

Let $\kappa(G)$ denote the clique cover number of a graph $G$.

**Theorem 1.** *Let $G$ and $H$ be complete multipartite graphs. Then*

$$\alpha(G \times H) = \kappa(G \times H) = \max \{ \alpha(G) \cdot |V(H)|, \ \alpha(H) \cdot |V(G)| \}. \quad (2)$$

*Proof.* Two vertices $(g_1, h_1)$ and $(g_2, h_2)$ are adjacent if $g_1$ and $g_2$ are not in a common independent set in $G$ and $h_1$ and $h_2$ are not in a common independent set in $H$.

Let $\Omega$ be a maximum independent set of $G$. Then $\{(g, h) \mid g \in \Omega$ and $h \in V(H)\}$ is an independent set in $G \times H$. We show that all maximal independent sets are of this form or of the symmetric form with $G$ and $H$ interchanged.

It is easy to see that if a vertex $(g, h)$ of $G \times H$ is in a maximal independent set, then any vertex $(g', h')$ where $g$ and $g'$ are in the same partite of $G$ or $h$ and $h'$ are in the same partite of $H$ must be also in the same set. Thus we may assume that $G$ and $H$ are cliques and consider a maximum independent set of the complement of the rook's graph as follows.

Any independent set must have all its vertices in one row or in one column. This shows that every maximal independent set in $G \times H$ is a generalized row or column in the rook's graph. Since the graphs are perfect, the number of cliques in a clique cover of $G \times H$ equals $\alpha(G \times H)$. This completes the proof.  □

## 2.2   Cographs

Cographs are characterized by the property that every induced subgraph $H$ satisfies one of the following three conditions, that is, $H$ has only one vertex, or $H$ is disconnected, or $\bar{H}$ is disconnected. It follows that cographs can be represented by a cotree [6].

A cotree is a pair $(T, f)$ where $T$ is a rooted tree and $f$ is a 1-1 map from the vertices of $G$ to the leaves of $T$. Each internal node of $T$, including the root, is labeled as $\otimes$ or $\oplus$. When the label is $\oplus$ then the subgraph $H$, induced by the vertices in the leaves, is disconnected. Each child of the node represents one component. When the node is labeled as $\otimes$ then the complement of the induced subgraph $H$ is disconnected. In that case, each component of the complement is represented by one child of the node. When $G$ is a cograph then a cotree for $G$ can be obtained in linear time [6]. We have the following theorem for cographs by using the structure of cotrees, whose proof is omitted due to lack of space.

**Theorem 2.** *There exists an $O(n^2)$ algorithm which computes $\alpha(G \times H)$ when $G$ and $H$ are cographs.*

## 2.3  Splitgraphs

**Theorem 3.** *Let $G$ and $H$ be splitgraphs. There exists a polynomial-time algorithm to compute the independence number of $G \times H$.*

*Proof.* Let $\{S_1, C_1\}$ and $\{S_2, C_2\}$ be the partition of $V(G)$ and $V(H)$, respectively, into independent sets and cliques. Let $c_i = |C_i|$ and $s_i = |S_i|$ for $i \in \{1,2\}$. See Figure 1(c) for an example of $G \times H$. The vertices of $C_1 \times C_2$ form, of course, a rook's graph. Note that $C_1 \times C_2$ is the complement of a rook's graph.

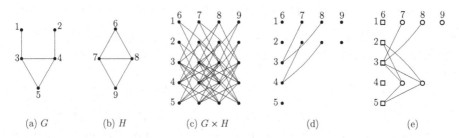

(a) $G$          (b) $H$          (c) $G \times H$          (d)          (e)

**Fig. 1.** (a) A split graph $G$ with $S_1 = \{1,2\}$ and $C_1 = \{3,4,5\}$. (b) A split graph $H$ with $S_2 = \{6\}$ and $C_2 = \{7,8,9\}$. (c) $G \times H$. (d) The subgraph of $G \times H$ induced by the vertices in $V(S_1) \times V(C_2) \cup V(C_1) \times V(S_2) \cup V(S_1) \times V(S_2)$. (e) A bipartite graph with the two color classes, where circles and boxes represent the vertices in the color classes defined in (3) and (4), respectively.

We consider three cases. First consider the maximum independent sets without any vertex of $V(C_1) \times V(C_2)$. Notice that the subgraph of $G \times H$ induced by the vertices of $V(S_1) \times V(C_2) \cup V(C_1) \times V(S_2) \cup V(S_1) \times V(S_2)$ is bipartite. See Figure 1(d). A maximum independent set in a bipartite graph can be computed in polynomial time.

Consider maximum independent sets that contain exactly one vertex $(c_1, c_2)$ of $V(C_1) \times V(C_2)$. The maximum independent set of this type can be computed as follows. Consider the bipartite graph of the previous case and remove the neighbors of $(c_1, c_2)$ from this graph, where a *neighbor* of a vertex $x$ is a vertex adjacent to $x$. The remaining graph is bipartite. Maximizing over all pairs $(c_1, c_2)$ gives the maximum independent set of this type.

Consider maximum independent sets that contain at least two vertices of the rook's graph $V(C_1) \times V(C_2)$. Then the two vertices must be in one row or in one column of the grid, since otherwise they are adjacent. For example, let the vertices of the independent set be contained in row $c_1 \in V(C_1)$. Then the vertices of $V(S_1) \times V(C_2)$ of the independent set are contained in $W = \{(s_1, c_2) \mid s_1 \notin N_G(c_1)$ and $c_2 \in V(C_2)\}$. Consider the bipartite graph $I$ with one color class defined as the following set of vertices

$$W \cup \{\, (c_1, c_2) \mid c_2 \in V(C_2) \,\}, \tag{3}$$

and the other color class defined as

$$\{ (v, s_2) \mid v \in V(C_1) \cup V(S_1) \text{ and } s_2 \in V(S_2) \}. \tag{4}$$

See Figure 1(e) for an example. Notice that since $S_1$ and $S_2$ are independent sets of $G \times H$, the subgraphs of $G \times H$ induced by the two color classes are independent, respectively, and thus the graph $I$ is bipartite. Then the maximum independent set of this type can be computed in polynomial time by maximizing over the rows $c_1 \in V(C_1)$ and columns $c_2 \in V(C_2)$. This proves the theorem.  □

### 2.4  Hardness for Planar Graphs

In this section, we show that it is NP-complete to compute the maximum independent set of $G \times K_4$, where $G$ is a planar graph of maximal degree 3 in the following theorem, whose proof is omitted due to lack of space.

**Theorem 4.** *Let $G$ be a planar graph of maximal degree 3. It is NP-complete to compute the maximum independent set of $G \times K_4$.*

## 3  Algorithms for Computing Tensor Capacity

In this section, we consider the powers of a graph under the categorical product. We first describe a polynomial time algorithm to compute the tensor capacity for cographs and splitgraphs, respectively. Then, the technique for splitgraphs is extended to handle the general graphs, giving an exact algorithm.

We start with some related works for tensor capacity as follows. First we mention a result for a fundamental graph product, say Cartesian product [3]. Hahn, Hell and Poljak prove that for the Cartesian product, $\frac{1}{\chi(G)} \leq \lim_{k \to \infty} r(\Box^k G) \leq \frac{1}{\chi_f(G)}$, where $\chi_f(G)$ is the fractional chromatic number of $G$ [8]. This shows that it is computable in polynomial time for graphs that satisfy $\omega(G) = \chi(G)$.

The *neighborhood* $N(x)$ of a vertex $x$ is the set of vertices $y$ such that $x$ and $y$ are adjacent in a graph. Then the *closed neighborhood* of a vertex $x$ is defined as $N[x] = N(x) \cup \{x\}$. The *neighborhood* of a set $X$ of vertices is the union of $N(x)$ for $x \in X$. Similarly, the *closed neighborhood* of a set $X$ of vertices is the union of $N[x]$ for $x \in X$.

Brown et al. [4, Theorem 3.3] obtain the following lowerbound for the tensor capacity.

$$\Theta^T(G) \geq a(G) \quad \text{where} \quad a(G) = \max_{I \text{ is an independent set}} \frac{|I|}{|I| + |N(I)|}. \tag{5}$$

Notice that $\frac{|I|}{|I|+|N(I)|}$ is the same as $\frac{|I|}{|N[I]|}$ in [4, Theorem 3.3].

It is related to the binding number $b(G)$ of the graph $G = (V, E)$, where $b(G) = \min_{A \subseteq V} \{ \frac{|N(A)|}{|A|} \mid A \neq \emptyset, N(A) \neq V \}$. In fact, the binding number is less than 1 if and only if $a(G) > \frac{1}{2}$. In that case, the binding number is realized by an independent set and it is equal to $b(G) = \frac{1-a(G)}{a(G)}$ [14,19]. The binding number is computable in polynomial time [7,14,23]. See also Corollary 1 below.

The following proposition was proved in [4].

**Proposition 2 ([4]).** If $r(G) > \frac{1}{2}$ then $\Theta^T(G) = 1$.

Therefore, a better lowerbound for $\Theta^T(G)$ is provided by

$$\Theta^T(G) \geq a^*(G) = \begin{cases} a(G) & \text{if } a(G) \leq \frac{1}{2} \\ 1 & \text{if } a(G) > \frac{1}{2}. \end{cases} \tag{6}$$

**Definition 9.** Let $G = (V, E)$ be a graph. A fractional matching is a function $f : E \to \mathbb{R}^+$, which assigns a non-negative real number to each edge, such that for every vertex $x$, $\sum_{e \ni x} f(e) \leq 1$. A fractional matching $f$ is perfect if it achieves the maximum $f(E) = \sum_{e \in E} f(e) = \frac{|V|}{2}$.

Alon and Lubetzky proved the following theorem in [2] (see also [14]).

**Theorem 5 ([2]).** For every graph $G$

$$\Theta^T(G) = 1 \Leftrightarrow a^*(G) = 1 \Leftrightarrow G \text{ has no fractional perfect matching.} \tag{7}$$

**Corollary 1.** There exists a polynomial-time algorithm to decide whether $\Theta^T(G) = 1$ or $\Theta^T(G) \leq \frac{1}{2}$.

The following theorem was raised as a question by Alon and Lubetzky in [2,16]. The theorem was proved by Ágnes Tóth [19].

**Theorem 6 ([19]).** For every graph $G$, $\Theta^T(G) = a^*(G)$. Equivalently, every graph $G$ satisfies

$$a^*(G^2) = a^*(G). \tag{8}$$

Tóth proves that

$$\text{if } a(G) \leq \frac{1}{2} \text{ or } a(H) \leq \frac{1}{2} \text{ then } a(G \times H) \leq \max\{a(G), a(H)\}. \tag{9}$$

Actually, Tóth shows that, if $I$ is an independent set in $G \times H$ then

$$|N_{G \times H}(I)| \geq |I| \cdot \min\{b(G), b(H)\}.$$

From this, Theorem 6 easily follows. As a corollary (see [2,16,19]) one obtains that, for any two graphs $G$ and $H$

$$r(G \times H) \leq \max\{a^*(G), a^*(H)\}.$$

Tóth also proves the following theorem in [19]. This was conjectured by Brown et al. [4].

**Theorem 7** ([19]). *For any two graphs $G$ and $H$,*

$$\Theta^T(G \oplus H) = \max \{ \Theta^T(G), \Theta^T(H) \}. \tag{10}$$

The analogue of this statement, with $a^*$ instead of $\Theta^T$, is straightforward. The first part of the following theorem was proved by Alon and Lubetzky in [2].

**Theorem 8.** *For any two graphs $G$ and $H$,*

$$\Theta^T(G \oplus H) = \Theta^T(G \times H) = \max \{ \Theta^T(G), \Theta^T(H) \}. \tag{11}$$

### 3.1   Cographs

For cographs we obtain the following theorem.

**Theorem 9.** *There exists an efficient algorithm to compute the tensor capacity for cographs.*

*Proof.* By Theorem 6 it is sufficient to compute $a(G)$, as defined in (5).

Consider a cotree for $G$. For each node the algorithm computes a table. The table contains numbers $\ell(k)$, for $k \in \mathbb{N}$, where

$$\ell(k) = min \{ |N(I)| \mid I \text{ is an independent set with } |I| = k \}.$$

Notice that $a(G)$ can be obtained from the table at the root node via $a(G) = \max_k \frac{k}{k+\ell(k)}$.

Assume $G$ is the union of two cographs $G_1 \oplus G_2$. An independent set $I$ is the union of two independent sets $I_1$ in $G_1$ and $I_2$ in $G_2$. Let the table entries for $G_1$ and $G_2$ be denoted by the functions $\ell_1$ and $\ell_2$. Then

$$\ell(k) = min \{ \ell_1(k_1) + \ell_2(k_2) \mid k_1 + k_2 = k \}.$$

Assume that $G$ is the join of two cographs, say $G = G_1 \otimes G_2$. An independent set in $G$ can have vertices in at most one of $G_1$ and $G_2$. Therefore,

$$\ell(k) = min \{ \ell_1(k) + |V(G_2)|, \ell_2(k) + |V(G_1)| \}.$$

This proves the theorem.     □

### 3.2   Splitgraphs

Let $G$ be a splitgraph with a partition $\{S, C\}$ of its $n$ vertices such that $G[C]$ is a clique and $G[S]$ is an independent set. For any independent set $I$ of $G$, $I$ can contain at most one vertex from $C$. Define, for $i \in \{0,1\}$,

$$a_i(G) = \max \left\{ \frac{|I|}{|I| + |N(I)|} \mid I \text{ is an independent set with } |C \cap I| = i \right\}$$

Then $a(G) = \max \{ a_0(G), a_1(G) \}$.

To compute $a_0(G)$, we shall make use of the following, simple observation.

If $S$ can be partitioned into two sets $S_1$ and $S_2$, such that their neighbor sets $N(S_1)$ and $N(S_2)$ are disjoint, then there exists an optimal $I^*$ for $a_0(G)$, such that $I^* \subseteq S_1$ or $I^* \subseteq S_2$.

To see this, suppose that it is not the case. Then, by assumption we can partition $I^*$ into non-empty sets $I_1 = I^* \cap S_1$ and $I_2 = I^* \cap S_2$, and we have $|I^*| = |I_1| + |I_2|$ and $|N(I^*)| = |N(I_1)| + |N(I_2)|$. Then

$$a_0(G) = \frac{|I^*|}{|I^*| + |N(I^*)|} \leq \max \left\{ \frac{|I_1|}{|I_1| + |N(I_1)|}, \frac{|I_2|}{|I_2| + |N(I_2)|} \right\} \leq a_0(G).$$

This proves the claim.

Based on this observation, we modify a technique, maybe first described by Cunningham [7], that transforms the problem into a max-flow (min-cut) problem. We construct a flow network $F$ with vertices corresponding to each vertex of $S$ and $C$, a source vertex $s$ and a sink vertex $t$. We make the source $s$ adjacent to each vertex in $S$, with capacity 1, and the sink $t$ adjacent to each vertex in $C$, with capacity 1 as well. In addition, if $u \in S$ and $v \in C$ are adjacent in the original graph $G$, the corresponding vertices are adjacent in $F$, with capacity set to $\infty$. Note that we omit the edges between vertices in $C$.

Consider a minimum $s$-$t$ cut in $F$. Let $S_1$ be the subset of $S$ whose vertices are in the same partition as $s$, and $S_2 = S - S_1$. The weight of such a cut must be finite, as the maximum $s$-$t$ flow is bounded by $\min \{ |S|, |C| \}$. Thus, we have that $N(S_1)$ and $N(S_2)$ are disjoint. Moreover, the total weight of the edges in the cut-set is $|S| - |S_1| + |N(S_1)|$, which implies that

$$S_1 = \arg\min_{S'} \{ |N(S')| - |S'| \mid S' \subseteq S \}.$$

So after running the flow algorithm to obtain $S_1$, there will be three cases:

**Case 1:** the optimal $I^*$ for $a_0(G)$ is exactly $S_1$;
**Case 2:** the optimal $I^*$ for $a_0(G)$ is a proper subset of $S_1$;
**Case 3:** the optimal $I^*$ for $a_0(G)$ is a subset of $S_2$;

Note that Case 2 is impossible, since for any such proper subset $S_1'$, we have

$$|N(S_1')| - |S_1'| \geq |N(S_1)| - |S_1| \qquad \text{(by min-cut)}$$

which implies

$$\frac{|N(S_1')| - |S_1'|}{|S_1'|} > \frac{|N(S_1')| - |S_1'|}{|S_1|} \geq \frac{|N(S_1)| - |S_1|}{|S_1|} \quad \Rightarrow \quad \frac{|N(S_1')|}{|S_1'|} > \frac{|N(S_1)|}{|S_1|}.$$

Consequently, $S_1'$ cannot be an optimal set that achieves $a_0(G)$.

Thus, we have either Case 1 or Case 3. To handle Case 3, we simply remove $S_1$ and $N(S_1)$ from the graph, and solve it recursively. In total, finding $a_0(G)$ requires $O(|S|)$ runs of the max-flow algorithm, and can be solved in polynomial time.

Finally, to compute $a_1(G)$, notice that, if an independent set $I$ contains some vertex $v \in C$ then $N(I)$ contains all vertices of $C$. When $|I|/(|I| + |N(I)|)$ is maximal, $I$ will contain all the vertices in $S$ that are nonadjacent to $v$. Hence $a_1(G) = \frac{n-d}{n}$, where $d$ denotes the minimum degree of a vertex in $C$. It follows that $a_1(G)$ can be obtained in linear time. This proves the following theorem.

**Theorem 10.** *There exists a polynomial-time algorithm to compute the tensor capacity for splitgraphs.*

### 3.3 An Exact Exponential Algorithm for General Graphs

We modify the approach in the previous subsection to obtain an exact algorithm for the tensor capacity of a general graph $H$. Let $n$ be the number of vertices in $H$. Let $I$ be a maximal independent set of $H$. We let $I$ play the role of $S$ and $N(I)$ play the role of $C$ in the analysis above. Then, by the flow algorithm, we obtain a subset $I_1 \subseteq I$ with $I_1 = \arg\max_{I'} \left\{ \frac{|I'|}{|I'|+|N(I')|} \mid I' \subseteq I \right\}$.

The algorithm generates all the maximal independent sets $I$ in $H$ and finds the optimal subset $I_1$ in each of them. This yields the value $a(H)$. By Moon and Moser's classic result [17], $H$ contains at most $3^{n/3}$ maximal independent sets. Furthermore, by, eg, the algorithm of Tsukiyama et al. [21], they can be generated in polynomial time per maximal independent set. This proves the following theorem.

**Theorem 11.** *There exists an $O^*(3^{n/3})$ algorithm to compute the tensor capacity for a graph with $n$ vertices.*

### 3.4 Other Classes of Graphs

Furthermore, as the additional results, we show that the tensor capacity of permutation graphs, interval graphs and graphs of bounded treewidth can be computed in polynomial time, whose proofs are omitted due to lack of space.

**Theorem 12.** *There exists $O(n^3)$ algorithms to compute the tensor capacity for permutation graphs and interval graphs, respectively. Moreover, the tensor capacity for the class of graphs that have treewidth at most $k$ where $k \in \mathbb{N}$ can be computed in polynomial-time.*

**Theorem 13.** *There exists a PTAS to approximate the ultimate categorical independence ratio in planar graphs.*

## 4 The Ultimate Categorical Independent Domination Ratio

In this section, we determine $\Delta(G)$ for complete multipartite graphs $G$ in the following theorem, whose proof is omitted due to lack of space.

**Theorem 14.** *Let $G$ be a complete multipartite graph with $t$ color classes of size $n_1 \leq \cdots \leq n_t$. Then $\Delta(G) = 0$ unless $t = 2$ and $n_1 = n_2$, in which case $\Delta(G) = \frac{1}{2}$.*

# References

1. Albertson, M., Collins, K.: Homomorphisms of 3-chromatic graphs. Discrete Mathematics 54, 127–132 (1985)
2. Alon, N., Lubetzky, E.: Independent sets in tensor graph powers. Journal of Graph Theory 54, 73–87 (2007)
3. Aurenhammer, F., Hagauer, J., Imrich, W.: Cartesian graph factorization at logarithmic cost per edge. Computational Complexity 2, 331–349 (1992)
4. Brown, J., Nowakowski, R., Rall, D.: The ultimate categorical independence ratio of a graph. SIAM Journal on Discrete Mathematics 9, 290–300 (1996)
5. Chartrand, G., Kapoor, S.F., Lick, D.R., Schuster, S.: The partial complement of graphs. Periodica Mathematica Hungarica 16, 83–95 (1985)
6. Corneil, D., Perl, Y., Stuwart, L.: A linear recognition algorithm for cographs. SIAM Journal on Computing 14, 926–934 (1985)
7. Cunningham, W.: Computing the binding number of a graph. Discrete Applied Mathematics 27, 283–285 (1990)
8. Hahn, G., Hell, P., Poljak, S.: On the ultimate independence ratio of a graph. European Journal of Combinatorics 16, 253–261 (1995)
9. Hahn, G., Tardif, C.: Graph homomorphisms: structure and symmetry. In: Graph Symmetry – Algebraic Methods and Applications. NATO ASI Series C: Mathematical and Physical Sciences, vol. 497, pp. 107–166. Kluwer (1997)
10. Hedetniemi, S.: Homomorphisms of graphs and automata. Technical report 03105-44-T, University of Michigan (1966)
11. Hell, P., Nešetřil, J.: Graphs and homomorphisms. Oxford Univ. Press (2004)
12. Hell, P., Yu, X., Zhou, H.: Independence ratios of graph powers. Discrete Mathematics 27, 213–220 (1994)
13. Jha, P., Klavžar, S.: Independence in direct-product graphs. Ars Combinatoria 50, 53–63 (1998)
14. Kloks, T., Lee, C., Liu, J.: Stickiness, edge-thickness, and clique-thickness in graphs. Journal of Information Science and Engineering 20, 207–217 (2004)
15. Kloks, T., Wang, Y.: Advances in graph algorithms (Manuscript 2013)
16. Lubetzky, E.: Graph powers and related extremal problems. PhD Thesis, Tel Aviv University, Israel (2007)
17. Moon, J., Moser, L.: On cliques in graphs. Israel Journal of Mathematics 3, 23–28 (1965)
18. Ravindra, G., Parthasarathy, K.: Perfect product graphs. Discrete Mathematics 20, 177–186 (1977)
19. Tóth, Á.: Answer to a question of Alon and Lubetzky about the ultimate categorical independence ratio. Manuscript on arXiv:1112.6172v1 (2011)
20. Tóth, Á.: The ultimate categorical independence ratio of complet multipartite graphs. SIAM Journal on Discrete Mathematics 23, 1900–1904 (2009)
21. Tsukiyama, S., Ide, M., Ariyoshi, H., Shirakawa, I.: A new algorithm for generating all the maximal independent sets. SIAM Journal on Computing 6, 505–517 (1977)
22. Weichsel, P.: The Kronecker product of graphs. Proceedings of the American mathematical Society 13, 47–52 (1962)
23. Woodall, D.: The binding number of a graph and its Anderson number. Journal of Combinatorial Theory, Series B 15, 225–255 (1973)
24. Zhang, H.: Independent sets in direct products of vertex-transitive graphs. Journal of Combinatorial Theory, Series B 102, 832–838 (2012)
25. Zhu, X.: The fractional version of Hedetniemi's conjecture is true. European Journal of Combinatorics 32, 1168–1175 (2011)

# Editing the Simplest Graphs

Peter Damaschke and Olof Mogren

Department of Computer Science and Engineering
Chalmers University, 41296 Göteborg, Sweden
{ptr,mogren}@chalmers.se

**Abstract.** We study the complexity of editing a graph into a target graph with any fixed critical-clique graph. The problem came up in practice, in mining a huge word similarity graph for well structured word clusters. It also adds to the rich field of graph modification problems. We show in a generic way that several variants of this problem are in SUBEPT. As a special case, we give a tight time bound for edge deletion to obtain a single clique and isolated vertices, and we round up this study with NP-completeness results for a number of target graphs.

## 1 Introduction

Graphs in this paper are undirected and have no loops or multiple edges. In an edge modification problem, an input graph must be modified by edge insertions or deletions or both, into a target graph with some prescribed property. Edge editing means both insertions and deletions. Edge insertion is also known as fill-in. The computational problem is to use a minimum number $k$ of edits. There is a rich literature on the complexity for a number of target graph properties, and on their various applications. Here we cannot survey them, we only refer to a few representative papers on hardness results [1,10]. Ironically, results are missing on edge modification problems for some structurally very simple target graphs. Informally, "simple" means that the graph becomes small after identification of its twin vertices (see Section 2). For any fixed graph $H$, our target graphs are all graphs obtained from $H$ by replacing vertices with bags of true twins.

Our motivation is the concise description of graphs with very few cliques (that may overlap) and some extra or missing edges. They appear, e.g., as subgraphs in co-occurence graphs of words, and constitute meaningful word clusters there. Within a data mining project we examined a similarity matrix of some 26,000 words, where similarity is defined by co-occurence in English Wikipedia. By thresholding we obtain similarity graphs, and we consider subgraphs that have small diameter and only few cut edges to the rest of the graph. Words occurring in the same contexts form nearly cliques. These are often not disjoint, as words appear in several contexts. Synonyms may not always co-occur (as different authors prefer different expressions), but they co-occur with other words. Relations like this give rise to various cluster structures. As opposed to partitioning entire graphs into overlapping clusters (as in [6]), we want to single out simple subgraphs of the aforementioned type. Experience in our project shows that some

S.P. Pal and K. Sadakane (Eds.): WALCOM 2014, LNCS 8344, pp. 249–260, 2014.

existing standard clustering methods generate poor word clusters which are either too small or dragged out and not internally dense. This suggested the idea to define the clusters directly by the desired properties, and then to determine them by edge editing of candidate subgraphs. Next, instead of describing the clusters naively as edge lists we list their vertices along with the few edited edges (to achieve cliques). Altogether this yields very natural word clusters, and by varying the threshold we also obtain different granularities. Applications of word clusters include sentence similarity measures for text summarization, search query result diversification, and word sense disambiguation. Thus, we believe that the problems are of importance, but they are also interesting as pure graph-algorithmic problems.

For any fixed $H$, our edge modification problems are (easily) in FPT. As our main result we get in Section 3 that they even belong to SUBEPT. (Not very many natural SUBEPT problems are known so far, as discussed in [7].) Every such problem has a $2^{\sqrt{k}\log k}$ time bound. The special case of $p$-CLUSTER EDITING, where $H$ is the graph with $p$ vertices and no edges, was recently treated in [7], using techniques like enumeration of small cuts. Our result is more general, and the quite different algorithm looks conceptually simpler, at the price of a somewhat worse time for the special case. Therefore it remains interesting to tighten the time bounds for other specific graphs $H$ as well. Consequently, we then turn to the absolutely simplest graphs $H$: In Section 4 we study the (NP-complete) edge deletion problem towards a single clique plus isolated vertices. We give a refined FPT time bound where the target clique size $c$ appears explicitly. Intuitively, $2k/c^2$ is an "edit density". Using an evident relationship to vertex covers we achieve, for small edit densities, essentially $O^*(1.2738^{k/c})$ time. For large enough $k/c$ we invoke a naive algorithm instead, and the time can also be bounded by $O(1.6355^{\sqrt{k\ln k}})$. The base 1.2738 is due to the best known VERTEX COVER algorithm from [3]. Moreover, the bound is tight: We show that the base of $k/c$ cannot beat the base in the best FPT algorithm for VERTEX COVER. Section 5 gives a similar FPT time bound for edge editing towards a single clique plus isolated vertices. Here, NP-completeness is an intriguing open question. However, in Section 6 we make some progress in proving NP-completeness systematically, for many graphs $H$. The results indicate that almost all our modification problems, with rather few exceptions, might be NP-complete. But recall that, on the positive side, they are in SUBEPT.

## 2    Preliminaries

The number of vertices and edges of a graph $G = (V, E)$ is denoted $n$ and $m$, respectively. For a graph $G$, the complement graph $\bar{G}$ is obtained by replacing all edges with non-edges, and vice versa. We also use standard notation for some specific graphs: $K_n$, $C_n$, $P_n$ is the complete graph (clique), the chordless cycle, the chordless path, respectively, on $n$ vertices, and $K_{n_1,n_2,\ldots,n_p}$ is the complete multipartite graph with $p$ partite sets of $n_i$ vertices. The disjoint union $G + H$ of graphs $G$ and $H$ consists of a copy of $G$ and a copy of $H$ on disjoint vertex sets. $G - v$ denotes the graph $G$ after removal of vertex $v$ and all incident edges. Notation $G - X$ is similarly defined for any subset $X \subseteq V$. The subgraph of $G$ induced by $X \subseteq V$ is denoted $G[X]$.

By definition, a graph class $\mathcal{G}$ is hereditary if, for every graph $G \in \mathcal{G}$, all induced subgraphs of $G$ are also members of $\mathcal{G}$. Any hereditary graph class $\mathcal{G}$ can be characterized by its forbidden induced subgraphs: $F$ is a forbidden induced subgraph if $F \notin \mathcal{G}$, but $F - v \in \mathcal{G}$ for every vertex $v$.

The open neighborhood of a vertex $v$ is the set $N(v)$ of all vertices adjacent to $v$, and the closed neighborhood is $N[v] := N(v) \cup \{v\}$. For a subset $X$ of vertices, $N[X]$ is the union of all $N(v)$, $v \in X$. Vertices $u$ and $v$ are called true twins if $uv$ is an edge and $N[u] = N[v]$. Vertices $u$ and $v$ are called false twins if $uv$ is a non-edge and $N(u) = N(v)$. The true twin relation is an equivalence relation whose equivalence classes are known as the critical cliques of the graph. (The false twin relation is an equivalence relation as well.) In the critical-clique graph $H$ of a graph $G$, every critical clique of $G$ is represented by one vertex of $H$, and two vertices of $H$ are adjacent if and only if some edge exists (hence all possible edges exist) between the corresponding critical cliques of $G$. For brevity we refer to the critical cliques as bags, and we say that $G$ is a "graph $H$ of bags". (Similarly one could also consider target graphs with small modular decompositions.)

For every graph $H$ we define three edge modification problems called $H$-BAG INSERTION, $H$-BAG DELETION, $H$-BAG EDITING, as follows: Given an input graph $G$ and a parameter $k$, change $G$ by at most $k$ edge insertions, deletions, or edits, respectively, such that the resulting graph has $H$ or an induced subgraph of $H$ as its critical-clique graph. We allow induced subgraphs of $H$ in order to allow bags to be empty. Similarly we define the problems $H[0]$-BAG DELETION and $H[0]$-BAG EDITING. The difference is that the target graph may additionally contain isolated vertices, that is, false twins with no edges. Thus, not all vertices are forced into the bags. Problem $H[0]$-BAG INSERTION easily reduces to $H$-BAG INSERTION. (As only insertions are permitted, the isolated vertices in an optimal solution are exactly the isolated vertices of $G$.) We also consider problem variants where the bags have prescribed sizes. We sometimes refer to all the mentioned problems collectively as bag modification problems. We say that editing an edge $uv$ affects its end vertices $u$ and $v$. A vertex is called unaffected if it is not affected by any edit. Without loss of generality we can always assume that $H$ has no true twins, because they could be merged, which leads to the same problems with a smaller graph in the role of $H$. For a fixed graph $H$ understood from context, let $\mathcal{H}$ be the class all graphs whose critical-clique graph is $H$ or an induced subgraph thereof. Let $\mathcal{H}[0]$ be the class of graphs consisting of all graphs of $\mathcal{H}$ with, possibly, additional isolated vertices. All these classes are hereditary.

We assume that the reader is familiar with the notion of fixed-parameter tractability (FPT) and basic facts, otherwise we refer to [5,12]. A problem with input size $n$ and an input parameter $k$ is in FPT if some algorithm can solve it in $f(k) \cdot p(n)$ for some computable function $f$ and some polynomial $p$. We use the $O^*(f(k))$ notation that suppresses $p(n)$. The subexponential parameterized tractable problems where $f(k) = 2^{o(k)}$ form the subclass SUBEPT. In our time analysis we will encounter branching vectors of a special form. The proof of the branching number, by standard algebra, is omitted due to lack of space.

**Lemma 1.** *The branching vector* $(1, r, \ldots, r)$ *with* $q$ *entries* $r$ *has a branching number bounded by* $1 + \frac{\log_2 r}{r}$, *if* $r$ *is large enough compared to the fixed* $q$.

## 3    Fixed-Parameter Tractability

Some bag modification problems (in different terminology) are known to be NP-complete, among them cases with very simple graphs $H$. Specifically, for $H = K_1$, problem $H[0]$-BAG DELETION can be stated as follows. Given a graph $G$, delete at most $k$ edges so as to obtain a clique $C$ and a set $I$ of isolated vertices. Equivalently, delete a set $I$ of vertices incident to at most $k$ edges, and delete all these incident edges, so as to retain a clique. The problem is NP-complete due to an obvious reduction from MAXIMUM CLIQUE. Next, for any fixed $p$, the $p$-CLUSTER EDITING problem asks to turn a graph, by editing at most $k$ edges, into a disjoint union of at most $p$ cliques. $p$-CLUSTER INSERTION and $p$-CLUSTER DELETION are similarly defined. In other words, these are the bag modification problems where $H = \bar{K}_p$. It is known that $p$-CLUSTER INSERTION is polynomial for every $p$, and so is $p$-CLUSTER DELETION for $p = 2$, but it is NP-complete for every $p \geq 3$, whereas $p$-CLUSTER EDITING is NP-complete for every $p \geq 2$ [13].

The hardness results provoke the question on fixed-parameter tractability. By a well-quasi ordering argument based on Dickson's lemma [4] one can show that $\mathcal{H}$ and $\mathcal{H}[0]$ have only finitely many induced subgraphs, and then the general result from [2] implies that the bag modification problems are in FPT. Although the argument is neat, we omit the details, because we will prove a stronger statement: membership in SUBEPT. The following observation is known for CLUSTER EDITING ($H = \bar{K}_p$) due to [8]; here we show it for general $H$.

**Proposition 1.** *Any bag modification problem has an optimal solution where any two true twins of the input graph belong to the same bag (or both are isolated) in the target graph.*

*Proof.* First we consider $H$-BAG EDITING. For a vertex $v$, an input graph, and a solution, we define the edit degree of $v$ to be the number of edits that affect $v$. For any class $T$ of true twins, let $v \in T$ be some vertex with minimum edit degree. Consider any $u \in T \setminus \{v\}$. If $u$ is put in a different bag than $v$, we undo all edits that affect $u$, and instead edit each edge $uw$, $w \neq v$, if and only if $vw$ is edited. We also move $u$ to the bag of $v$ and undo the deletion of edge $uv$ (if it happened). Clearly, this yields a valid solution and does not increase the number of edits between $u$ and vertices $w \neq v$. Since we do not incur an additional edit of $uv$ either, the new solution is no worse. We proceed in this way for all $u \in T \setminus \{v\}$, and also for all $T$. This proves the assertion for $H$-BAG EDITING.

For $H[0]$-BAG EDITING we treat the set of isolated vertices as yet another bag. The same arguments apply. What is not covered in the previous reasoning is the case when $v$ is isolated and $u$ is in a bag. But then $u$ and $v$ are not adjacent, neither before nor after the move, hence the number of edits does not increase. For the INSERTION and DELETION problems, again the same arguments go through in all cases. Just replace "edit" with "insert" or "delete". The only change is that, in INSERTION, the edge $uv$ cannot have been deleted.     □

We make another simple observation. In the following let $p$ always denote the number of vertices of our fixed $H$.

**Lemma 2.** *In any bag modification problem, the input graph has at most $2k + p$ critical cliques (isolated vertices not counted), or the instance has no solution.*

*Proof.* The unaffected vertices induce a subgraph that belongs to $\mathcal{H}$ or $\mathcal{H}[0]$, respectively, hence it has at most $p$ bags. Any affected vertex is adjacent to either all or none of the vertices of any of these bags (since the latter ones are unaffected). In the worst case, $k$ edits affect $2k$ vertices, and each of them becomes a critical clique of its own. Together this yields the bound.     □

Lemma 2 implies again that all bag modification problems for fixed $H$ are in FPT: Assign every critical clique in the input graph to some bag of the target graph (or make its vertices isolated, in the $H[0]$ case). These are at most $p + 1$ options. For isolated vertices it suffices to decide how many of them we put in each bag, which are $O(n^p)$ options. Hence the time for this naive branching is $O^*((p + 1)^{2k+p})$. Instead of this poor bound we now show:

**Theorem 1.** *Any bag modification problem with a fixed graph $H$ can be solved in $2^{\sqrt{k}\log k}$ time, hence it belongs to SUBEPT.*

*Proof.* First we focus on $H$-BAG EDITING. Let $a$, $0 < a < 1$, be some fixed number to be specified later. To avoid bulky notation, we omit rounding brackets and work with terms like $k^a$ as if they were integers.

One difficulty is that the sizes of the $p$ bags are not known in advance. A *preprocessing phase* takes care of that. Initially all bags are *open*. In every bag we create $k^a$ "places" that we successively treat as follows. At every place we branch: Either we *close* the bag and leave it, or we decide on a critical clique of the input graph and put any of its vertices in the bag. (Clearly, the latter choice is arbitrary. By Proposition 1 we can even immediately fill further places with the entire critical clique, but our analysis will not take advantage of that.) Due to Lemma 2 these are at most $2k + p + 1$ branches, hence the total number of branches is $(2k + p + 1)^{pk^a} = O(k)^{pk^a} = 2^{k^a \log k}$. Note that $p$ is fixed, and constant factors are captured by the base of log. Every open bag has now $k^a$ vertices. We will not add any further vertices to closed bags. Vertices that are not yet added to bags are called *undecided*. We also do all necessary edits of edges between different bags, to stick to the given graph $H$, and reduce $k$ accordingly.

In the *main phase* we do branchings that further reduce the parameter $k$ by edits. The branching rules are applied exhaustively in the given order. In the following we first consider the special case that all bags are open. Later we show how to handle closed bags, too.

If there exists an undecided vertex $u$ and a bag $B$ such that $u$ is adjacent to some but not all vertices of $B$, then we branch and either insert all missing edges between $u$ and $B$, or delete all edges between $u$ and $B$. (But for now, $u$ is not yet added to any bag.) The branching vector is some $(i, k^a - i)$ with two positive entries, or a better vector if already more than $k^a$ vertices got into $B$.

Now every undecided vertex $u$ is either completely adjacent or completely non-adjacent to each bag $B$. We say that $u$ fits in $B$, if $u$ is adjacent to exactly those bags that belong to $N[B]$. Remember that $H$ has no true twins. It follows that every vertex $u$ fits in at most one bag.

If there exists an undecided vertex $u$ that fits in no bag, we branch and decide on a bag for $u$, put $u$ in this bag, and do the necessary edits. Since $u$ does not fit anywhere, we need at least $k^a$ edits, thus the branching vector, of length $p$, is $(k^a, \ldots, k^a)$ or better.

After that, every undecided vertex $u$ fits in exactly one bag $B(u)$. Suppose that two undecided vertices $u$ and $v$ have the wrong adjacency relation. That is, either $uv$ is an edge but $B(u)$ and $B(v)$ are not adjacent, or $uv$ is not an edge but $B(u)$ and $B(v)$ are adjacent or $B(u) = B(v)$. We branch as follows. Either we edit $uv$ or not. If we don't, $u$ and $v$ cannot be both added to their designated bags. Then we also decide on $u$ or $v$ and put that vertex in one of the other $p-1$ bags, which again costs at least $k^a$ edits. Thus, the worst-case branching vector is $(1, k^a, \ldots, k^a)$ with $2p - 2$ entries $k^a$. Finally, all undecided vertices have their correct adjacency relations, hence the graph belongs to $\mathcal{H}$.

The difficulty with closed bags is that they do not guarantee at least $k^a$ edits. Let $U$ be the set of vertices of $H$ corresponding to the open bags. Note that $H[U]$ may have true twins. In that case we merge every critical clique of $H[U]$ into one superbag. Trivially, each superbag is larger than $k^a$. On $H[U]$ and the superbags we perform exactly the same branching rules as above. Since we have fewer branches, the branching vectors do not get worse. A new twist is needed only when we actually add a vertex $u$ to a superbag $S$. In every such event we also decide on the bag within $S$ that will host $u$. This latter choice does not change the adjacency relations within the union of open bags and undecided vertices any more. Therefore we can take these decisions independently for all $u$, and always choose some bag in $S$ that causes the minimum number of edits of edges between $u$ and the closed bags.

The worst branching vector we encounter is $(1, k^a, \ldots, k^a)$ with $2p - 2$ entries $k^a$. From Lemma 1 we obtain the bound $(1 + \frac{a \log_2 k}{k^a})^k = 2^{k^{1-a} \log k}$ for some suitable logarithm base. We must multiply this with the bound $2^{k^a \log k}$ from the first phase. Choosing $a = 1/2$ yields the product $2^{\sqrt{k} \log k}$.

For $H$-BAG DELETION and $H$-BAG INSERTION we proceed similarly. Since only one type of edits is permitted, some of the branches are disabled, which cannot make the branching vectors worse. In $H[0]$-BAG DELETION and $H[0]$-BAG EDITING we can treat the set of isolated vertices like a bag; some necessary adjustments are straightforward.                                                □

# 4    Clique Deletion

If $H$ is the one-vertex graph, then the $H[0]$ edge modification problems aim at a single clique plus isolated vertices. Instead of "$H[0]$-BAG ..." we speak in this case of CLIQUE INSERTION, CLIQUE DELETION, and CLIQUE EDITING, which is more suggestive. CLIQUE INSERTION is a trivial problem. In this section we study CLIQUE DELETION: given a graph $G$, delete at most $k$ edges so as to obtain a clique $C$ and a set $I$ of isolated vertices. An equivalent formulation is to delete a set $I$ of vertices incident to at most $k$ edges, and delete all these incident edges as well, so as to retain a clique. This vertex-deletion interpretation is sometimes more convenient. The problem is NP-complete due to an obvious reduction from CLIQUE or VERTEX COVER, and in SUBEPT by Theorem 1.

Besides the generic time bound with unspecified constants, we are now aiming at an FPT algorithm with a tight time bound, as a function of $k$ and $c = |C|$. We remark that the smallest possible $c$ can be calculated from the number $m$ of edges in the input graph. Clearly, we must have $m - k \leq \frac{1}{2}c(c - 1)$, thus $c \geq \frac{1}{2} + \sqrt{\frac{1}{4} + 2(m - k)}$. We may even guess the exact clique size $c$ above this threshold and try all possible sizes, which adds at most a factor $n - c$ to the time bound.

**Lemma 3.** *A partitioning of the vertex set of a graph $G$ into sets $C$ and $I$ is a valid solution to* CLIQUE DELETION *if and only if $I$ is a vertex cover of $\bar{G}$. Moreover, a minimum vertex cover $I$ of $\bar{G}$ also yields a minimum number of edge deletions in $G$.*

*Proof.* The first assertion is evident. For the second assertion, note that CLIQUE DELETION requests a vertex cover $I$ of $\bar{G}$ being incident to the minimum number of edges of $G$. Since $C$ is a clique, and every edge of $G$ is either in $C$ or incident to $I$, we get the following chain of equivalent optimization problems: minimize the number of edges incident to $I$, maximize the number of edges in $C$, maximize $|C|$, minimize $|I|$. □

Before we turn to an upper complexity bound, we first give an implicit lower bound. Let us join our input graph $G$ with a clique $K$, and define $c^* := |K|$. Joining means that all possible edges between $K$ and $G$ are created. Observe that an optimal solution for the joined graph consists of an optimal solution for $G$, with $K$ added to $C$. Thus, if $k$ edges are deleted in $G$, then $k + (n - c)c^*$ edges are deleted in the joined graph, the size of the solution clique is $c^* + c$. Furthermore, the size of the vertex cover in $\bar{G}$ is $n - c$. If we choose $c^*$ "large" compared to $n$, but still polynomial in $n$, then the number of deleted edges and the clique size are essentially $(n - c)c^*$ and $c^*$, respectively. Their ratio is the vertex cover size $n - c$. Back to the original notations $k$ and $c$ for these numbers, it follows that any FPT algorithm for CLIQUE DELETION, that runs in time bounded by some function $f(k/c)$, could be used to solve also VERTEX COVER on $\bar{G}$ within time $f(n - c)$. Therefore, the best we can hope for is a CLIQUE DELETION algorithm with a time bound $O^*(b^{k/c})$, with some constant base $b > 1$ that cannot be better than in the state-of-the-art VERTEX COVER algorithm. This bound is also tight in a sense, as we will see below.

The exponent $k/c$ can be rewritten as $c(k/c^2)$, where the second factor has a natural interpretation: Since the number of edges in $C$ is roughly $c^2/2$, we can view $2k/c^2$ as an "edit density", the ratio of deleted edges and remaining edges in the target graph. It will be convenient to define $c' := c - 1$, and to define the edit density slightly differently as $d := 2k/c'^2$. In applications we are mainly interested in instances that are already nearly cliques, thus we keep a special focus on the case $d < 1$ in the following.

Our algorithm for CLIQUE DELETION preprocesses the input graph with a single reduction rule: Remove each vertex $v$ of degree smaller than $c'$, along with all incident edges. After exhaustive application, there remains a graph with minimum degree $c'$. From now on we can suppose without loss of generality that $G$ has already minimum degree $c'$. This also bounds $i := |I|$ as follows.

**Lemma 4.** *With the above denotations we have $i \leq 2k/c'$, and in the case $d < 1$ this can be improved to $i \leq \frac{2}{1+\sqrt{1-d}} \cdot k/c'$.*

*Proof.* Let $h$ be the number of edges in $I$. Since at most $k$ edge deletions are permitted, we have $ic' - h \leq k$. Since $h \leq k$ (or we must delete too many edges already in $I$), it follows $i \leq 2k/c' = dc'$.

For $d < 1$, this further implies $i \leq c'$. Using $h < i^2/2$, the previous inequality $ic' - h \leq k$ yields $ic' - i^2/2 \leq k$, thus $i^2 - 2c'i + 2k \geq 0$ with the solution $i \leq c' - \sqrt{c'^2 - 2k}$. (We have excluded the case $i > c'$.) By simple algebra this can be rewritten as $i \leq \frac{2}{1+\sqrt{1-d}} \cdot k/c'$.                              □

Note that the factor in front of $k/c'$ grows from 1 to 2 when $d$ grows from 0 to 1. To make this factor more comprehensible, we may also simplify it to a slightly worse upper bound: Since $\sqrt{1-d} > 1-d$, we have $i \leq \frac{2}{2-d} \cdot k/c'$. We also remark that CLIQUE DELETION is trivial if $k < c'$, because, after the reduction phase, either there remains a clique, or the instance has no solution.

**Theorem 2.** CLIQUE DELETION *can be solved in $O^*(1.2738^{\frac{2}{1+\sqrt{1-d}} \cdot k/c'})$ time.*

*Proof.* After applying our reduction rule, due to Lemma 3 it suffices to compute a vertex cover of minimum size in $\bar{G}$. As for the time bound, the base comes from [3] and the exponent comes from Lemma 4.                              □

For large edit densities we may also express the time bound as a function of $k$ only, as in the previous section, but with a specific base. The algorithm of Theorem 2 with the simpler bound from Lemma 4 has the running time $O^*(1.2738^{2k/c})$. (We replace $c'$ with $c$, which does not make a difference asymptotically.) If $c$ is small, we can instead use a brute-force approach and check all subsets of $c$ vertices for being cliques. This runs in $O^*(2k^c)$ time, since at most $2k + c$ non-isolated vertices exist, and $k$ is large compared to $c$ in the considered case. The two expressions decrease and increase, resepctively, as $c$ grows. Hence their minimum is maximized if, approximately, $c = 0.492\sqrt{k/\ln k}$. Plugging in this $c$ yields $1.6355^{\sqrt{k \ln k}}$ time. The naive $O^*(2k^c)$ bound can certainly be improved by excluding most $c$-vertex subsets as candidates for the final clique.

# 5    Clique Editing

Recall that CLIQUE EDITING is the problem of editing at most $k$ edges so as to obtain a clique $C$, say of size $c$, and a set $I$ of $n - c$ isolated vertices.

**Theorem 3.** CLIQUE EDITING *with prescribed size c of the target clique is W[1]-complete in parameter n − c, hence also NP-complete.*

*Proof.* We argue with the optimization versions and show that minimizing the number of edited edges is equivalent to finding a set $I$ of $n - c$ vertices being incident to the minimum number of edges: Simply note that the edges incident to $I$ are exactly those to be deleted, and minimizing deletions means maximizing the number of remaining edges. Since $c$ is fixed, this also minimizes the number of edge insertions needed to make $C$ a clique. Due to [9], finding at least $s$ vertices that cover at most $t$ edges, known as MINIMUM PARTIAL VERTEX COVER, is W[1]-complete in parameter $s$, thus our assertion follows with $s := n - c$.    □

Note that Theorem 3 does not immediately imply NP-completeness of CLIQUE EDITING with free size $c$, since the prescribed clique sizes $c$ in the reduction graphs may be different from $c$ in optimal solutions to CLIQUE EDITING on these graphs, and our problem might still be polynomial for the "right" $c$, albeit this is hard to imagine. We conjecture that CLIQUE EDITING is NP-complete. Another equivalent formulation of CLIQUE EDITING is: Given a graph $G$, find a subset $C$ of vertices that induces a subgraph that maximizes the number of edges minus the number of non-edges. Denoting the number of edges by $m(G)$, the objective can be written as $m(G[C]) - m(\bar{G}[C])$. This becomes also interesting in a weighted version. For a given real number $w > 0$, maximize $m(G[C]) - w \cdot m(\bar{G}[C])$. This problem is trivial for $w = 0$ (the whole vertex set is an optimal $C$), and NP-complete if $w$ is part of the input (since a maximum clique is an optimal $C$ if $w$ is large enough). What happens in between? For any constant $w > 0$? In particular, for $w = 1$? We must leave this question open.

Next we propose an FPT algorithm CLIQUE EDITING when $k$ is the parameter. It works if $c$ is part of the input (cf. Theorem 3), and hence also for free $c$, by trying all values. Membership in SUBEPT follows from Theorem 1, but as earlier we are also interested in the dependency of the time bound on $c$. The following algorithm that uses similar ideas as the earlier ones is omitted due to lack of space.

**Theorem 4.** CLIQUE EDITING *can be solved in $2^{\log c \cdot k/c}$ time.*

# 6    Some Hardness Results

All bag modification problems are trivially in NP. In this section we prove the NP-completeness of bag modification problems for many target graphs $H$. We give a general construction that "lifts" NP-completeness from some $H$ to larger graphs $H'$. To be specific, suppose that $H$-BAG EDITING is already known to be NP-complete. We will reduce it in polynomial time to $H'$-BAG EDITING, for certain graphs $H'$ specified later on.

Let the graph $G$ and parameter $k$ be an instance of $H$-BAG EDITING. Let $H'$ be a graph that contains $H$ as an induced subgraph. We choose a particular subset $S$ of vertices of $H'$ such that $H'[S]$ is isomorphic to $H$. (Note that the same graph may have several occurrences as induced subgraph, hence we must fix some $S$.) Let $S_0$ and $S_1$ be some set of vertices of $H' - S$ being adjacent to no vertices of $S$, and to all vertices of $S$, respectively. We construct a graph $G'$ as follows, in polynomial time. We replace every vertex of $S_0 \cup S_1$ with a bag of size $c > 2k$. Two bags are joined by all possible edges (by no edges) if the corresponding vertices of $H'$ are adjacent (not adjacent). Then we add $G$ and insert all possible edges between $S_1$ and the vertices of $G$.

If $G$ with parameter $k$ is a yes-instance of $H$-BAG EDITING, then we can mimic the same, at most $k$, edits also in the subgraph $G$ of $G'$, which implies that $G'$ with parameter $k$ is a yes-instance of $H'$-BAG EDITING. Our aim in the following is to show the converse, under some conditions on $H$ and $H'$. The equivalence will then establish the desired reduction. Specifically, suppose that the following technically looking condition is fulfilled. Here, an embedding of a graph into another graph means that edges are mapped to edges, and non-edges are mapped to non-edges.

(*) Let $J$ be any induced subgraph of $H'$ isomorphic to $H'[S_0 \cup S_1]$. Accordingly, we embed $J$ into any graph of $\mathcal{H}'$ and divide the vertex set of $J$ in two sets $U_0$ and $U_1$, of those vertices coming from $S_0$ and $S_1$, respectively. For any such embedding, let $T$ be the set of vertices $t$ such that $N[t]$ contains all vertices of $U_1$ and no vertex of $U_0$. Then the subgraph induced by $T$ is always in $\mathcal{H}$.

Note that there may exist many possible embeddings of $J$, and our condition must hold for each of them. Also, $T$ may contain some vertices of $U_1$.

Now suppose that at most $k$ edits of edges in $G'$ have produced a graph in $\mathcal{H}'$. Since $k$ edits affect at most $2k$ vertices, but $c > 2k$, clearly every bag in the edited graph corresponding to a vertex of $S_0$ or $S_1$ still has at least one unaffected vertex. We select one from each bag and obtain a set $U$ of unaffected vertices. The subgraph induced by $U$ is isomorphic to $H'[S_0 \cup S_1]$. Let $U_0$ and $U_1$ be the subset of vertices of $U$ corresponding to vertices of $S_0$ and $S_1$, respectively. Then we have $U = U_0 \cup U_1$, and all vertices of $G$ are still adjacent (non-adjacent) to all vertices of $U_1$ ($U_0$). Thus (*) implies that, after editing, the vertices of $G$ form a graph in $\mathcal{H}$. Since at most $k$ edits have been done in the whole graph, we get that $G$ with parameter $k$ is a yes-instance of $H$-BAG EDITING.

Condition (*) looks more complicated than it is, when it comes to specific graphs $H$. In the following we give some examples. We refer to vertices in $S_0$ and $S_1$ as 0-vertices and 1-vertices, respectively, and we call any graph in $\mathcal{H}$ a graph $H$ of bags.

**Theorem 5.** $H'$-BAG EDITING *is NP-complete for, at least, the following graphs* $H'$: *complete multipartite graphs with some partite set of at least* 3 *vertices; the* *complete multipartite graph with partite sets of exactly 2 vertices; $K_3$-free graphs with maximum degree at least 3.*

*Proof.* $H$-BAG EDITING for $H = \bar{K}_p$ is $p$-CLUSTER EDITING, which is known to be NP-complete for every $p \geq 2$ [13]. We reduce the case $H = \bar{K}_p$ for a suitable $p \geq 2$ to the case $H'$.

In a complete multipartite graph $H'$, let $b \geq 3$ be the size of some largest partite set. We choose $p = b - 1 \geq 2$. We let $S_1$ be empty, and let $S_0$ consist of a single vertex in a partite set of size $b$. The vertices of $H'$ being non-adjacent to this 0-vertex induce a $\bar{K}_{b-1} = \bar{K}_p$. No matter where else we embed our 0-vertex in a graph $H'$ of bags, the set $T$ as defined in (*) forms a $\bar{K}_{b-1}$ of bags (note that bags are allowed to be empty), hence (*) is satisfied.

Consider $H' = K_{2,2} = C_4$. We choose $p = 2$. We let $S_1$ consist of two non-adjacent vertices, while $S_0$ is empty. Clearly, their common neighbors induce $H = \bar{K}_2$. The only possible embedding of our two non-adjacent 1-vertices in a $C_4$ of bags is to put them in two non-adjacent bags, such that the set $T$ forms a graph $H = \bar{K}_2$ of bags, hence (*) is satisfied. For $H' = K_{2,...,2}$ we proceed by induction on the number of partite sets. Let $H = K_{2,...,2}$ with two vertices less. Then the same choice of $S_1$ and $S_0$ and the same arguments establish the induction step.

Let $H'$ be $K_3$-free, $v$ a vertex of maximum degree $d \geq 3$, and $u$ some neighbor of $v$. We choose $p = d - 1$, $S_1 = \{v\}$ and $S_0 = \{u\}$. The vertices adjacent to $v$ and non-adjacent to $u$ induce $\bar{K}_{d-1} = \bar{K}_p$. For any embedding of an adjacent pair of a 1-vertex and a 0-vertex into an $H'$ of bags, the set $T$ forms a graph $H = \bar{K}_p$ of bags, since in $H'$ every vertex has at most $d - 1$ neighbors, and they are pairwise non-adjacent. □

The same construction also lifts NP-completeness results from $H[0]$ to $H'[0]$, whenever we can choose $S_1 = \emptyset$ and a suitable $S_0$. Our construction also works for $H'$-BAG DELETION and $H'$-BAG INSERTION, however, note that we need an NP-complete case to start with. For edge deletions we can use $\bar{K}_p$ with $p \geq 3$. Remember that $\bar{K}_p$-BAG INSERTION is polynomial [13] for every $p$. However, we can start from $P_3$ instead:

**Theorem 6.** $H'$-BAG INSERTION *is NP-complete for, at least, the graphs* $H' = P_3$, *and* $H' = P_p$ *and* $H' = C_p$ *for each* $p \geq 6$.

*Proof.* $P_3$-BAG INSERTION in $G$ means to delete in $\bar{G}$ a minimum number of edges so as to reach a complete bipartite graph (biclique) and isolated vertices. This is equivalent to finding a biclique with maximum number of edges. The latter problem is NP-complete (even in bipartite graphs and hence in general graphs) due to [11]. We reduce $P_3$-BAG INSERTION to $P_n$-BAG INSERTION for each $n \geq 6$ by setting $S_1 = \emptyset$ and $S_0$ isomorphic to $P_{n-4}$. Similarly, we reduce $P_3$-BAG INSERTION to $C_n$-BAG INSERTION for each $n \geq 6$ by setting $S_1 = \emptyset$ and $S_0$ isomorphic to $P_{n-5}$. It is easy to verify condition (*) in the equivalence proofs of the reductions. □

These Theorems are only illustrations of a few cases. The conditions on $H'$ can be weakened, and even more cases $H'$ proved to be NP-complete, however we want to avoid a tedious list of applications of one particular technique. On the negative side, the current construction fails for other graphs. The "smallest" open cases are $K_1[0]$-Bag Editing and $P_3$-Bag Editing. We also remark that $P_3$-Bag Deletion is polynomial: consider the complement graph and proceed similarly as in [13].

**Acknowledgments.** This work has been done within the projects "Generalized and fast search strategies for parameterized problems", grant 2010-4661 from the Swedish Research Council (Vetenskapsrådet), and "Data-driven secure business intelligence", grant IIS11-0089 from the Swedish Foundation for Strategic Research (SSF). The authors would also like to thank Gabriele Capannini for sharing the data and discussing the applications, and Henning Fernau for intensive discussions of some open problems during a short visit at Chalmers.

# References

1. Burzyn, P., Bonomo, F., Durán, G.: NP-completeness Results for Edge Modification Problems. Discr. Appl. Math. 154, 1824–1844 (2006)
2. Cai, L.: Fixed-Parameter Tractability of Graph Modification Problems for Hereditary Properties. Inf. Proc. Lett. 58, 171–176 (1996)
3. Chen, J., Kanj, I.A., Xia, G.: Improved Upper Bounds for Vertex Cover. Theor. Comp. Sci. 411, 3736–3756 (2010)
4. Dickson, L.E.: Finiteness of the Odd Perfect and Primitive Abundant Numbers with $n$ Distinct Prime Factors. Am. J. Math. 35, 413–422 (1913)
5. Downey, R.G., Fellows, M.R.: Parameterized Complexity. Springer, New York (1999)
6. Fellows, M.R., Guo, J., Komusiewicz, C., Niedermeier, R., Uhlmann, J.: Graph-Based Data Clustering with Overlaps. Discr. Optim. 8, 2–17 (2011)
7. Fomin, F.V., Kratsch, S., Pilipczuk, M., Pilipczuk, M., Villanger, Y.: Tight Bounds for Parameterized Complexity of Cluster Editing. In: Portier, N., Wilke, T. (eds.) STACS 2013, Dagstuhl. LIPIcs, vol. 20, pp. 32–43 (2013)
8. Guo, J.: A More Effective Linear Kernelization for Cluster Editing. In: Chen, B., Paterson, M., Zhang, G. (eds.) ESCAPE 2007. LNCS, vol. 4614, pp. 36–47. Springer, Heidelberg (2007)
9. Guo, J., Niedermeier, R., Wernicke, S.: Parameterized Complexity of Vertex Cover Variants. Theory Comput. Syst. 41, 501–520 (2007)
10. Natanzon, A., Shamir, R., Sharan, R.: Complexity Classification of some Edge Modification Problems. Discr. Appl. Math. 113, 109–128 (2001)
11. Peeters, R.: The Maximum Edge Biclique Problem is NP-complete. Discr. Appl. Math. 131, 651–654 (2003)
12. Niedermeier, R.: Invitation to Fixed-Parameter Algorithms. Oxford Lecture Series in Math. and its Appl. Oxford Univ. Press (2006)
13. Shamir, R., Sharan, R., Tsur, D.: Cluster Graph Modification Problems. Discr. Appl. Math. 144, 173–182 (2004)

# Alignment with Non-overlapping Inversions on Two Strings[*]

Da-Jung Cho, Yo-Sub Han, and Hwee Kim

Department of Computer Science, Yonsei University
50, Yonsei-Ro, Seodaemun-Gu, Seoul 120-749, Republic of Korea
{dajung,emmous,kimhwee}@cs.yonsei.ac.kr

**Abstract.** The inversion is one of the important operations in bio sequence analysis and the sequence alignment problem is well-studied for efficient bio sequence comparisons. Based on inversion operations, we introduce the alignment with non-overlapping inversion problem: Given two strings $x$ and $y$, does there exist an alignment with non-overlapping inversions for $x$ and $y$. We, in particular, consider the alignment problem when non-overlapping inversions are allowed for both $x$ and $y$. We design an efficient algorithm that determines the existence of non-overlapping inversions and present another efficient algorithm that retrieves such an alignment, if exists.

## 1 Introduction

In modern biology, it is important to determine exact orders of DNA sequences, retrieve relevant information of DNA sequences and align these sequences [1, 7, 12, 13]. For a DNA sequence, a *chromosomal translocation* is to relocate a piece of the DNA sequence from one place to another and, thus, rearrange the sequence [9]. The chromosomal translocation is a crucial operation in DNAs since it alters a DNA sequence and often causes genetic diseases [10]. A *chromosomal inversion* occurs when a single chromosome undergoes breakage and rearrangement within itself [11]. Based on the important biological events such as translocation and inversion, there is a well-defined string matching problem: given two strings and translocation or inversion, the string matching problem is finding all matched strings allowing translocations or inversions. Moreover, people proposed an alignment with translocation or inversion problem, which is closely related to find similarity between two given strings; that is to obtain minimal occurrences of translocation or inversion that transform one to the other. Many researchers investigated efficient algorithms for this problem [1–4, 6, 8, 13, 14].

The inversions, which are one of the important biological operations, are not automatically detected by the traditional alignment algorithms [14]. Schöniger

---

[*] This research was supported by the Basic Science Research Program through NRF funded by MEST (2012R1A1A2044562).

S.P. Pal and K. Sadakane (Eds.): WALCOM 2014, LNCS 8344, pp. 261–272, 2014.

and Waterman [13] introduced the alignment problem with non-overlapping inversions, defined a simplification hypothesis that all regions will not be allowed to overlap, and showed an $O(n^6)$ algorithm that computes local alignments with inversions between two strings of length $n$ and $m$ based on the dynamic programming, where $n \geq m$. Vellozo et al. [14] presented an $O(n^2 m)$ algorithm, which improved the previous algorithm by Schöniger and Waterman. They built a matrix for one string and partially inverted string using table filling method with regard to the extended edit graph. Recently, Cantone et al. [1] introduced an $O(nm)$ algorithm using $O(m^2)$ space for the string matching problem, which is to find all locations of a pattern of length $m$ with respect to a text of length $n$ based on non-overlapping inversions.

Many diseases are often caused by genetic mutations, which can be inherited through generations and can result in new sequences from a normal gene [5]. In other words, we may have two different sequences from a normal gene by different mutations. This motivates us to examine the problem of deciding whether or not two gene sequences are mutated from the same gene sequence. In particular, we consider an inversion mutation. See Fig. 1 for an example.

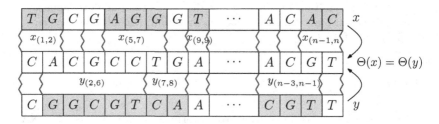

**Fig. 1.** An example of non-overlapping inversions on both strings $x$ and $y$, where $\Theta_x = (1,2)(3,3)'(4,4)'(5,7)(8,8)'(9,9) \cdots (n-1,n)$ and $\Theta_y = (1,1)'(2,6)(7,8)(9,9)' \cdots (n-3, n-1)(n,n)'$. Note that $(i,i)'$ denotes the alignment at position $i$ without complementing $x[i]$.

Note that this problem is different from the previous problem [13, 14], where a non-overlapping inversion occurs only in one string and transforms the string to the other string; namely $\Theta(x) = y$ for a set $\Theta$ of non-overlapping inversions. On the other hand, we consider more general case where inversions can occur in both $x$ and $y$ simultaneously. The problem is also equivalent to the string alignment problem allowing the inversions occurring at most two times at the same positions.

## 2    Preliminaries

Let $A[a_1][a_2] \cdots [a_n]$ be an $n$-dimensional array, where the size of each dimension is $a_i$ for $1 \leq i \leq n$. Let $A[i_1][i_2] \cdots [i_n]$ be the element of $A$ with indices $(i_1, i_2, \ldots, i_n)$. Given a finite set $\Sigma$ of character and a string $s$ over $\Sigma$, we use

$|s|$ to denote the length of $s$ and $s[i]$ to denote the symbol of $s$ at position $i$. We use $s_{(i,j)}$ to denote a substring $s[i]s[i+1]\cdots s[j]$, where $1 \le i \le j \le |s|$. We consider biological operation *inversion* $\theta$ and denote by $\theta(s)$ the reverse and complement of a string $s$. For example, $\theta(A) = T$ and $\theta(AGG) = \theta(G)\theta(G)\theta(A) = CCT$. We define an inversion operation $\theta_{(i,j)}$ for a given range $i, j$ as follows:

$$\theta_{(i,j)}(s) = \theta(s_{(i,j)}).$$

For simplicity, we use $(i, j)$ instead of $\theta_{(i,j)}$ if the notation is clear in the context. We say that $\frac{i+j}{2}$ is the *center of the inversion* for $(i, j)$. We define a *set* $\Theta$ *of non-overlapping inversion* to be

$$\Theta = \{(i,j) \mid 1 \le i \le j \text{ and for } \forall (i',j') \ne (i,j) \in \Theta, j < i' \text{ or } j' < i\}.$$

Then, for a set $\Theta$ of non-overlapping inversions and a string $s$, we have $\Theta(s) = s'$, where

$$s'[i] = \begin{cases} \theta(s[j + k - i]) & \text{if } (j,k) \in \Theta \text{ and } j \le i \le k \\ s[i] & \text{otherwise.} \end{cases}$$

For example, given $\Theta = \{(1,1),(2,3)\}$ and $s = AGCC$, we have $\Theta(s) = \theta(A)\theta(GC)C = TGCC$. From now on, we use a set of inversions instead of a set of non-overlapping inversions since we only consider sets of non-overlapping inversions.

**Definition 1.** *We define a new alignment problem with non-overlapping inversions on two strings as follows: Given two strings $x$ and $y$ of the same length, can we determine whether or not there exist two sets $\Theta_x$ and $\Theta_y$ of inversions such that $\Theta_x(x) = \Theta_y(y)$?*

## 3   The Algorithm

We use $x = AGCT$ and $y = CGAA$ as our example strings for explaining the algorithm. Remark that $\theta(AG)C\theta(T) = CTCA = C\theta(GA)A$ and, thus, we have two sets $\Theta_x = \{(1,2),(4,4)\}$ and $\Theta_y = \{(2,3)\}$.

We start from building a table in which each cell contains a pair of a range and a character. We define an array $T_x[n][n + 1]$ for $x$ as follows:

$$T_x[i][j] = \begin{cases} ((j,i), \theta(x[j])) & \text{if } j < i, \\ ((i,i)', x[i]) & \text{if } j = i, \\ ((i,j-1), \theta(x[j-1])) & \text{if } j > i. \end{cases}$$

We call all elements in $T_x$ *inversion fragments* of $x$. For an inversion fragment $\mathbb{F} = ((p,q), \sigma)$ or $((p,p)', \sigma)$, we say that $\mathbb{F}$ *yields* the character $\sigma$ and $\frac{p+q}{2}$ is the *center* of the inversion fragment. For a sequence of inversion fragments $\mathbb{F}_1, \ldots, \mathbb{F}_n$, where $\mathbb{F}_i$ yields $\sigma_i$, we say that the sequence yields a string $\sigma_1 \cdots \sigma_n$.

Inversion fragments become useful to compute a substring created by any inversion because of the following property of the inversion operation:

$$\theta_{(i-1,j+1)}(x) = \theta(x[j+1])\theta_{(i,j)}(x)\theta(x[i-1]).$$

From a string $x$ and its table $T_x$, we make the following observation:

**Observation 1.** *For a string $x$ and its $T_x$,*

*(1) $\theta_{(i,j)}(x) = \theta(x[j])\theta(x[j-1])\cdots\theta(x[i+1])\theta(x[i])$,*
*(2) $((i,j),\theta(x[j]))$,$((i+1,j-1),\theta(x[j-1]))$,...,$((i+1,j-1),\theta(x[i+1]))$,$((i,j),\theta(x[i]))$ are all inversion fragments in the $i$th, $i+1$th, ..., $j-1$th, $j$th columns of $T_x$ and have the same center.*

It is easy to verify from the construction that we can construct $T_x$ in $O(n^2)$ time and the size of $T_x$ is $O(n^2)$, where $|x| = n$. See Fig. 2 for an example. We also construct $T_y$ for $y$.

| $T_x$ | 1 | 2 | 3 | 4 |
|---|---|---|---|---|
| 1 | $((1,1)',A)$ | $((1,2),T)$ | $((1,3),T)$ | $((1,4),T)$ |
| 2 | $((1,1),T)$ | $((2,2)',G)$ | $((2,3),C)$ | $((2,4),C)$ |
| 3 | $((1,2),C)$ | $((2,2),C)$ | $((3,3)',C)$ | $((3,4),G)$ |
| 4 | $((1,3),G)$ | $((2,3),G)$ | $((3,3),G)$ | $((4,4)',T)$ |
| 5 | $((1,4),A)$ | $((2,4),A)$ | $((3,4),A)$ | $((4,4),A)$ |

**Fig. 2.** An example table $T_x$ for $x = AGCT$. In this example, $T_x[3][3] = ((3,3)',C)$ means that we put $C$ instead of $\theta(C) = G$ since the range is $(3,3)'$. On the other hand, we have $T_x[3][4] = ((3,3),G)$ because the range is $(3,3)$. Shaded cells denote inversion fragments that represent the inversion $\theta_{(1,3)}$.

Given a pair $(((p_1,p_2),\sigma_1),((q_1,q_2),\sigma_2))$ of two inversion fragments, we say that the pair is an *agreed pair* if $q_1 = p_2 + 1$ or $p_1 + p_2 = q_1 + q_2$. Otherwise, we call it a *disagreed pair*. Then, for two agreed pairs $(((p_1,p_2),\sigma_1),((q_1,q_2),\sigma_2))$ and $(((q_1,q_2),\sigma_2),((r_1,r_2),\sigma_3))$, we say that two pairs are *connected* by $((q_1,q_2),\sigma_2)$. We define an *agreed sequence* $\mathbb{S}$ to be a sequence of inversion fragments $((a_i,b_i),\sigma_i)$ for $1 \le i \le n$, where $a_1 = 1$, $b_n = n$ and $(((a_i,b_i),\sigma_i),((a_{i+1},b_{i+1}),\sigma_{i+1}))$ is an agreed pair for $1 \le i \le n-1$.

Given an agreed sequence $\mathbb{S}$, we define a set $\mathcal{F}_\Theta(\mathbb{S})$ of inversions from $\mathbb{S}$ as follows:

$$\mathcal{F}_\Theta(\mathbb{S}) = \{(p,q) \mid \exists \mathbb{S}[p] \text{ such that } \mathbb{S}[p] = ((p,q),\sigma)\}.$$

Given a set $\Theta$ of inversions, we can return an agreed sequence $\mathbb{S}$ from $\Theta$ as follows: (namely, $\mathbb{S} = \mathcal{F}_\mathbb{S}(\Theta)$.)

$$\mathbb{S}[i] = \begin{cases} ((i,j+k-i),\theta(x[j+k-i])) & \text{if } \exists(j,k) \in \Theta \text{ s.t. } j \le i \le k \text{ and } i < \frac{j+k}{2}, \\ ((j+k-i,i),\theta(x[j+k-i])) & \text{if } \exists(j,k) \in \Theta \text{ s.t. } j \le i \le k \text{ and } i \ge \frac{j+k}{2}, \\ ((i,i)',x[i]) & \text{otherwise.} \end{cases}$$

**Observation 2.** *Given a string $x$ and its set $\Theta_x$ of non-overlapping inversions, $\mathcal{F}_{\mathbb{S}}(\Theta_x)$ yields $\Theta_x(x)$.*

Note that an agreed pair of two inversion fragments with the same center represents the same inversion in $\Theta$. We define an agreed sequence $\mathbb{S}_x$ ($\mathbb{S}_y$, respectively) to be *legal* if there exist $\Theta_x$ and $\Theta_y$ such that $\Theta_x(x) = \Theta_y(y)$ and $\mathcal{F}_{\Theta}(\mathbb{S}_x) = \Theta_x$ ($\mathcal{F}_{\Theta}(\mathbb{S}_y) = \Theta_y$, respectively). Then, our problem is to determine whether or not there exist legal sequences $\mathbb{S}_x$ and $\mathbb{S}_y$ for two strings $x$ and $y$.

The main idea of our algorithm is to keep tracking of all possible agreed pairs for adjacent indices and check if there exist two connected pairs $\mathbb{P}_x$ and $\mathbb{P}_y$ for $x$ and $y$, that generate a common substring of $x$ and $y$; namely, we check if there exist a common substring $\sigma_1\sigma_2\sigma_3$ and $\mathbb{P}_x = (\mathbb{F}_1, \mathbb{F}_2), (\mathbb{F}_2, \mathbb{F}_3)$ for $x$ and $\mathbb{P}_y$ for $y$ such that $\mathbb{F}_j$ yields $\sigma_j$. For instance, for $x = AGCT$ and $y = CGAA$, there exists a common substring $CTC$ from index 1 to 3 such that

$$\mathbb{P}_x = (((1,2), C), ((1,2), T)), (((1,2), T), ((3,3)', C))$$

and

$$\mathbb{P}_y = (((1,1)', C), ((2,3), T)), (((2,3), T), ((2,3), C)).$$

Next, we define the following four sets for each index $i$:

**Definition 2.** *For a string $x$, its $T_x$ and an index $i$, we define four sets as follows:*

(1) $AH_x^i = \{((p,i), \sigma) = T_x[i][p] \mid 1 \le p \le i \le n-1\} \cup \{((i,i)', x[i])\}$, which is a set of all inversion fragments that end at $i$.

(2) $AT_x^i = \{((i+1,q), \sigma) = T_x[i+1][q+1] \mid i < q \le n\} \cup \{((i+1,i+1)', x[i+1])\}$, which is a set of all inversion fragments that start from $i+1$.

(3) $BH_x^i = \{((p,q), \sigma) = T_x[i][j] \mid p > 1, 1 \le j \le n\}$, which is a set of all inversion fragments that start before or from $i$.

(4) $BT_x^i = \{((p,q), \sigma) = T_x[i+1][j] \mid q < n, 1 \le j \le n\}$, which is a set of all inversion fragments that end after or at $i+1$.

From these four sets, we establish the following observations:

**Observation 3.** *Given a string $x$ and its four sets $AH_x^i, AT_x^i, BH_x^i$ and $BT_x^i$, the following statements hold: For two inversion fragments $\mathbb{F}_1, \mathbb{F}_2$, if $(\mathbb{F}_1, \mathbb{F}_2)$ is an agreed pair, then*

(1) $\mathbb{F}_1 \in AH_x^i$ and $\mathbb{F}_2 \in AT_x^i$, or

(2) $\mathbb{F}_1 \in BH_x^i$ and $\mathbb{F}_2 \in BT_x^i$.

Based on Observation 3, it is possible to create all agreed pairs for an index $i$ by comparing $AH_x^i$ and $AT_x^i$, and $BH_x^i$ and $BT_x^i$. If we can connect pairs through all indices, then we are able to generate all agreed sequences, which are essentially all sets of non-overlapping inversions.

We need additional tables $C_x^i[|\Sigma|][|\Sigma|][|\Sigma|]$ and $C_y^i[|\Sigma|][|\Sigma|][|\Sigma|]$ to record a sequence of three characters generated by connected pairs. For an index $i$,

$$C_x^i[\sigma_1][\sigma_2][\sigma_3] = \begin{cases} true & \text{if } \exists \mathbb{S}_x \text{ such that } \mathbb{S}_x[i-1]\mathbb{S}_x[i]\mathbb{S}_x[i+1] \text{ yields } \sigma_1\sigma_2\sigma_3, \\ false & \text{otherwise.} \end{cases}$$

We also define $S_x^i[2][n+1]$ and $S_y^i[2][n+1]$ as follows. For an index $i$ and $1 \leq j \leq 2, 1 \leq k \leq n+1$, we say that $S_x^i[j][k] \ni (t, \sigma_1\sigma_2\sigma_3)$ if there exists $\mathbb{S}_x$ such that

- $T_x[i+j-1][k] = \mathbb{S}_x[i+j-1] = ((p_1, p_2), \sigma_3)$,
- $t \leq i + j - 1$,
- $\mathbb{S}_x[t] = ((t, p_1 + p_2 - t), \omega)$ for some $\omega$.

In other words, an element $(t, \sigma_1\sigma_2\sigma_3)$ in the first column (second column, respectively) of $S_x^i$ for an index $i$ represents that there exists an agreed sequence $\mathbb{S}_x$, where inversion $(t, s)$ (or the identity function at index $t$) creates a suffix of the string yielded by the sequence.

We design an algorithm that computes $S_x^i$ and $S_y^i$ for each index and checks whether or not $S_x^{n-1}$ and $S_y^{n-1}$ are empty. If $T_x[1][j]$ yields $\sigma$, then we set $S_x^1[1][j] = \{(1, AA\sigma)\}$ as initial data. We also set $S_y^1$ similarly. Note that each cell in $S_x^i$ and $S_y^i$ has $O(n)$ elements. Now we execute the following steps from index 1 to $n-1$ for $x$ and $y$. We only illustrate the case for $x$. (The case for $y$ is similar.)

**STEP-1:** We check all inversion fragments in the $i$th column of $T_x$. For $T_x[i][j] = ((j,i), \sigma_2)$ (or $((i,i)', \sigma_2)$), if $(j, \sigma_0\sigma_1\sigma_2) \in S_x^i[1][j]$, then we add $((j,i), \sigma_1\sigma_2)$ (or $((i,i)', \sigma_1\sigma_2)$) to a set $AH_x^i$. Namely, $AH_x^i$ contains every inversion fragment that can be the $i$th element in a legal sequence and ends at $i$. We need to check $i+1$ inversion fragments for this step, and for each inversion fragment, we need to examine $O(n)$ elements. Therefore, we need $O(n^2)$ time for the step.

**STEP-2:** We check all inversion fragments in the $i+1$th column of $T_x$. For $T_x[i+1][i+1] = ((i+1, i+1),' \sigma_3)$, we add $((i+1, i+1)', \sigma_3)$ to a set $AT_x^i$. Moreover, for each $T_x[i+1][j] = ((i+1, j-1), \sigma_3)$, we add $((i+1, j-1), \sigma_3)$ to a set $AT_x^i$. In other words, $AT_x^i$ contains every inversion fragment that can be the $i+1$th element in a legal sequence and starts from $i+1$. Note that $AT_x^i$ has $n-i$ inversion fragments. Therefore, the total process takes $O(n^2)$ time and requires $O(n)$ space for storing all inversion fragments.

Once we have two set $AH_x^i$ and $AT_x^i$, we can calculate all agreed pairs generated from $AH_x^i$ and $AT_x^i$.

**STEP-3:** Based on Observation 3(1), for every $((p_1, i), \sigma_1\sigma_2)$ (or $((i,i)', \sigma_1\sigma_2)$) in $AH_x^i$ and $((i+1, p_2), \sigma_3)$ (or $((i+1, i+1)', \sigma_3)$) in $AT_x^i$, we set $C_x^i[\sigma_1][\sigma_2][\sigma_3] = true$. We also add $(i+1, \sigma_1\sigma_2\sigma_3)$ to $S_x^i[2][p_2]$. Since $|AH_x^i| \leq i+1$ and $|AT_x^i| = n - i$, we repeat this step at most $(i+1)(n-i) = O(n^2)$ times. Thus, we can update $C_x^i$ and $S_x^i$ in $O(n^2)$ time.

| $T_x$ | 1 | 2 |
|---|---|---|
| 1 | $((1,1)', A)$ | $((1,2), T)$ |
| 2 | $((1,1), T)$ | $((2,2)', G)$ |
| 3 | $((1,2), C)$ | $((2,2), C)$ |
| 4 | $((1,3), G)$ | $((2,3), G)$ |
| 5 | $((1,4), A)$ | $((2,4), A)$ |

| $S_x^1$ | 1 | 2 |
|---|---|---|
| 1 | $(1, AAA)$ | |
| 2 | $(1, AAT)$ | $(2, AAG), (2, ATG)$ |
| 3 | $(1, AAC)$ | $(2, AAC), (2, ATC)$ |
| 4 | $(1, AAG)$ | $(2, AAG), (2, ATG)$ |
| 5 | $(1, AAA)$ | $(2, AAA), (2, ATA)$ |

**Fig. 3.** An example for $S_x^1$ after **STEP-3**. Shaded cells in the first column of $T_x$ are $AH_x^1$ and shaded cells in the second column of $T_x$ are $AT_x^1$.

We have calculated all agreed pairs generated from $AH_x^i$ and $AT_x^i$ in **STEP-3**, and the other case of generating agreed pairs for an index $i$ is to use the inversion fragments with the same center from $BH_x^i$ and $BT_x^i$. Note that inversion fragments with the same center means the same inversion by Observation 1.

**STEP-4:** Based on Observation 3(2), for two elements $((p_1, p_2), \sigma_2) = T_x[i][j]$ and $((q_1, q_2), \sigma_3) = T_x[i+1][k]$, where $p_1 + p_2 = q_1 + q_2$ and $i \neq j$ and $i+1 \neq k$, if $(t, \sigma_0\sigma_1\sigma_2) \in S_x^i[1][j]$ and $t \leq p_1$, then we set $C_x^i[\sigma_1][\sigma_2][\sigma_3] = true$ and add $(t, \sigma_1\sigma_2\sigma_3)$ to $S_x^i[2][k]$. For an inversion fragment in the $i$th column of $T_x$, we can find the inversion fragment in the $i+1$th column with the same center in the constant time. Since there are $O(n)$ inversion fragments in $i$th column of $T_x$ and for each inversion fragment we need to examine $O(n)$ elements in $S_x^i$, the whole process takes $O(n^2)$ time.

| $T_x$ | 1 | 2 |
|---|---|---|
| 1 | $((1,1)', A)$ | $((1,2), T)$ |
| 2 | $((1,1), T)$ | $((2,2)', G)$ |
| 3 | $((1,2), C)$ | $((2,2), C)$ |
| 4 | $((1,3), G)$ | $((2,3), G)$ |
| 5 | $((1,4), A)$ | $((2,4), A)$ |

| $S_x^1$ | 1 | 2 |
|---|---|---|
| 1 | $(1, AAA)$ | $(1, ACT)$ |
| 2 | $(1, AAT)$ | $(2, AAG), (2, ATG)$ |
| 3 | $(1, AAC)$ | $(2, AAC), (2, ATC), (1, AGC)$ |
| 4 | $(1, AAG)$ | $(2, AAG), (2, ATG), (1, AAG)$ |
| 5 | $(1, AAA)$ | $(2, AAA), (2, ATA)$ |

**Fig. 4.** An example for $S_x^1$ after **STEP-4**. Shaded cells in $T_x$ generates elements for $S_x^1$.

**STEP-5:** For all three letter strings $\sigma_1\sigma_2\sigma_3$ over $\Sigma$,

$$C_x^i[\sigma_1][\sigma_2][\sigma_3] = \begin{cases} true & \text{if } C_x^i[\sigma_1][\sigma_2][\sigma_3] = true \text{ and } C_y^i[\sigma_1][\sigma_2][\sigma_3] = true, \\ false & \text{otherwise.} \end{cases}$$

Once we recompute $C_x^i$, for each $(p, \sigma_1\sigma_2\sigma_3)$ in $S_x$, we remove $(p, \sigma_1\sigma_2\sigma_3)$ from $S_x^i$ if $C_x^i[\sigma_1][\sigma_2][\sigma_3] = false$. The process ensures that $S_x^i$ and $S_y^i$ produce the same sequence of characters by connected pairs. Since the size of $S_x^i$ is $O(n^2)$ and the size of $C_x^i$ is constant, this step takes $O(n^2)$ time.

---

**Algorithm 1**

---

**Input**: Strings $x$ and $y$

**Output**: Boolean (whether or not there exist $\Theta_x$ and $\Theta_y$ s.t. $\Theta_x(x) = \Theta_y(y)$.)

/* time complexity: $O(n^3)$, space complexity: $O(n^2)$ */

1 make $T_x$ and $T_y$.

2 initialize $S_x^1$ and $S_y^1$.

3 **for** $i \leftarrow 1$ **to** $n - 1$ **do**

4    **for** *strings x and y* **do**

5      **for** $j \leftarrow 1$ **to** $i + 1$ **do**          // STEP-1

6        $\sigma_2$ is yielded from $T_x[i][j]$

7        **if** $(j, \sigma_0\sigma_1\sigma_2) \in S_x^i[1][j]$ **then** $((j, i), \sigma_1\sigma_2) \in AH_x^i$

8      **for** $j \leftarrow i + 1$ **to** $n + 1$ **do**          // STEP-2

9        $T_x[i][j] \in AT_x^i$

10      **for each** $((p_1, i), \sigma_1\sigma_2) \in AH_x^i$ and $((i + 1, p_2), \sigma_3) \in AT_x^i$ **do** // STEP-3

11        $C_x^i[\sigma_1][\sigma_2][\sigma_3] = true$

12        $(i + 1, \sigma_1\sigma_2\sigma_3) \in S_x^i[2][p_2]$

13      **for** $j \leftarrow 1$ **to** $n + 1$ **except** $min(i + 1, 1), i$ **do**          // STEP-4

14        **if** $j = i + 1 \vee j = i + 2$ **then**

15          $k \leftarrow j - 2$

16        **else**

17          $k \leftarrow j - 1$

18        $((p_1, p_2), \sigma_2) \leftarrow T_x[i][j], ((q_1, q_2), \sigma_3) \leftarrow T_x[i + 1][k]$

19        **if** $(t, \sigma_0\sigma_1\sigma_2) \in S_x^i[1][j] \wedge t \leq p_1$ **then**

20          $C_x^i[\sigma_1][\sigma_2][\sigma_3] = true$

21          $(r, \sigma_1\sigma_2\sigma_3) \in S_x^i[2][k]$

22    $C_x^i, C_y^i \leftarrow C_x^i \wedge C_y^i$          // STEP-5

23    **for** *strings x and y* **do**

24      **for each** $(p, \sigma_1\sigma_2\sigma_3) \in S_x^i$ **do**

25        **if** $C_x^i[\sigma_1][\sigma_2][\sigma_3] = false$ **then** remove $(p, \sigma_1\sigma_2\sigma_3)$ from $S_x^i$

26    copy the second columns of $S_x^i$ and $S_y^i$ to the first column of $S_x^{i+1}$ and $S_y^{i+1}$.

27 **if** *the second columns of $S_x^{n-1}$ and $S_y^{n-1}$ are not empty* **then**

28    **return true**

29 **else**

30    **return false**

---

We are now ready to present the whole procedure of our algorithm. See **Algorithm 1** that is a pseudo description of the proposed algorithm.

Once we finish calculating $S_x^i$ and $S_y^i$ using **STEPS-1,2,3,4** and **5** from index 1 to $n - 1$, we check whether or not the second columns in $S_x^{n-1}$ and $S_y^{n-1}$ are empty. If they are not empty, then there exist agreed sequences for $x$ and $y$ that generate the same string, which are legal sequences. On the other hand, if they are empty, then there are no legal sequences for $x$ and $y$.

| $T_x$ | 1 | 2 |
|---|---|---|
| 1 | $((1,1)', A)$ | $((1,2), T)$ |
| 2 | $((1,1), T)$ | $((2,2)', G)$ |
| 3 | $((1,2), C)$ | $((2,2), C)$ |
| 4 | $((1,3), G)$ | $((2,3), G)$ |
| 5 | $((1,4), A)$ | $((2,4), A)$ |

| $S_x^1$ | 1 | 2 |
|---|---|---|
| 1 | $(1, AAA)$ | $(1, ACT)$ |
| 2 | $(1, AAT)$ | |
| 3 | $(1, AAC)$ | $(2, ATC), (1, AGC)$ |
| 4 | $(1, AAG)$ | |
| 5 | $(1, AAA)$ | |

| $T_y$ | 1 | 2 |
|---|---|---|
| 1 | $((1,1)', C)$ | $((1,2), G)$ |
| 2 | $((1,1), G)$ | $((2,2)', G)$ |
| 3 | $((1,2), C)$ | $((2,2), C)$ |
| 4 | $((1,3), T)$ | $((2,3), T)$ |
| 5 | $((1,4), T)$ | $((2,4), T)$ |

| $S_y^1$ | 1 | 2 |
|---|---|---|
| 1 | $(1, AAC)$ | |
| 2 | $(1, AAG)$ | |
| 3 | $(1, AAC)$ | $(2, AGC), (1, ATC)$ |
| 4 | $(1, AAT)$ | $(2, ACT)$ |
| 5 | $(1, AAT)$ | $(2, ACT)$ |

**Fig. 5.** An example for $S_x^1$ and $S_y^1$ after **STEP-5**. Note that the second column of $S_x^1$ and $S_y^1$ generate same substrings, $ACT$, $ATC$ and $AGC$.

| $T_x$ | 3 | 4 |
|---|---|---|
| 1 | $((1,3), T)$ | $((1,4), T)$ |
| 2 | $((2,3), C)$ | $((2,4), C)$ |
| 3 | $((3,3)', C)$ | $((3,4), G)$ |
| 4 | $((3,3), G)$ | $((4,4)', T)$ |
| 5 | $((3,4), A)$ | $((4,4), A)$ |

| $S_x^3$ | 1 | 2 |
|---|---|---|
| 1 | $(1, GCT)$ | |
| 2 | | |
| 3 | $(3, CTC)$ | |
| 4 | $(3, TCG)$ | $(4, CTT), (4, TCT), (4, CGT)$ |
| 5 | | $(4, CTA), (4, TCA), (4, CGA)$ |

**Fig. 6.** An example for $S_x^3$ after **STEP-5**. Since the second column of $S_x^3$ (and $S_y^3$) is not empty, the algorithm returns true.

**Theorem 4.** *The proposed algorithm runs in $O(n^3)$ time using $O(n^2)$ space, where $n = |x| = |y|$.*

Lemmas 1 and 2 guarantee the correctness of our algorithm.

**Lemma 1.** *If $(t, \sigma_1\sigma_2\sigma_3) \in S_x^i[2][j]$ after completing **STEP-5**, then there exists $(t', \sigma_0\sigma_1\sigma_2) \in S_x^{i-1}[2][k]$ after completing **STEP-5**.*

**Lemma 2.** *If there exists a string $s = \sigma_0\sigma_1\cdots\sigma_n$ such that $S_x^i[2][j_i] = (t_i, \sigma_{i-1}\sigma_i\sigma_{i+1})$, then there exists a sequence $\mathbb{S}$ of inversion fragments whose $i$th element $\mathbb{S}[i]$ is*

$$\mathbb{S}[i] = T_x[i][k], \text{ where } \begin{cases} k \in \{1,2\} \text{ and } T_x[1][k] \text{ yields } \sigma_1 & \text{if } i = 1 \text{ and } t_1 = 2, \\ T_x[i][k] = ((1, p+q-1), \sigma_1) & \text{if } i = 1 \text{ and } t_1 = 1, \\ k = j_i & \text{otherwise.} \end{cases}$$

*Then, $\mathbb{S}$ is an agreed sequence.*

**Theorem 5.** *We can solve the alignment with non-overlapping inversion problem in $O(n^3)$ time using $O(n^2)$ space, where $n$ is the size of input strings.*

---

**Algorithm 2**

---

**Input:** Strings $x$ and $s$ of length $n$
**Output:** Agreed sequence $\mathbb{S}_x$ for $x$ such that $\mathbb{S}_x$ yields $s$

1  $t \leftarrow 1$                    // $t$ is the length of the comparing substrings
2  **for** $i \leftarrow 1$ **to** $n$ **do**
         // comparing substrings created from identity function
3      **if** $t = 1 \wedge \theta(x)[n+1-i] = \theta(s[i])$ **then**
4          $\lfloor$ add $((i,i)', s[i])$ to $\mathbb{S}_x$

         // comparing substrings created from inversion
5      **else if** $\theta(x)_{(n+1-i,n+t-i)} = s_{(i-t+1,i)}$ **then**
6          **for** $j \leftarrow 1$ **to** $\lfloor t/2 \rfloor$ **do**
7              $\lfloor$ add $((i-t+j, i+1-j), s[i-t+j])$ to $\mathbb{S}_x$
8          **for** $j \leftarrow \lceil t/2 \rceil$ **to** $t$ **do**
9              $\lfloor$ add $((i+1-j, i-t+j), s[i-t+j])$ to $\mathbb{S}_x$
10         $\lfloor$ $t \leftarrow 1$

11     **else**
12         $\lfloor$ $t \leftarrow t+1$

13 **return** $\mathbb{S}_x$

---

Next, we consider the problem of retrieving an alignment when we know that there exist two non-overlapping inversions for $x$ and $y$. Note that **Algorithm 1** determines the existence of $\Theta_x$ and $\Theta_y$.

**Definition 3.** *We define the alignment finding problem with non-overlapping inversions on two strings as follows: Given two strings $x$ and $y$ of the same length, find two sets $\Theta_x$ and $\Theta_y$ of inversions such that $\Theta_x(x) = \Theta_y(y)$.*

We tackle the problem in Definition 3 by retrieving the common string $s$ such that $s = \Theta_x(x) = \Theta_y(y)$. After completing **STEP-5** for each $i$, we store every $\sigma_1 \sigma_2 \sigma_3$ to a set $F^i$, where $C_x^i[\sigma_1][\sigma_2][\sigma_3] = true$.

**Observation 6.** *For such sets $F^i$'s, we have the following two observations:*

1. *The space requirement for $F^i$'s is $O(n)$,*
2. *For any string $s$, where $s_{(i-2,i)} \in F^i$, for $3 \leq i \leq n$, there exist $\Theta_x$ and $\Theta_y$ such that $s = \Theta_x(x) = \Theta_y(y)$.*

Due to Observation 6, the problem becomes to find $\Theta_x$ for $x$ and $s$ such that $\Theta_x(x) = s$ (and $\Theta_y$ for $y$). **Algorithm 2** retrieves $\mathbb{S}_x$ equivalent to $\Theta_x$ from $x$ and $s$. The idea of the algorithm is that every substring generated by an inversion on $x$ is a substring of $\theta(x)$, and substrings do not overlap with each other on $\theta(x)$ if inversions do not overlap. Fig. 7 illustrates this idea.

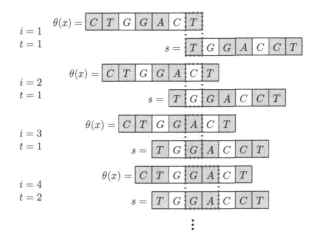

**Fig. 7.** An example of comparing $\theta(x)$ and $s$, where $x = AGTCCAG$ and $s = TGGACCT$. Dotted boxes are compared substrings in each $i$. In $\theta(x)$, substrings matched by inversions are same as substrings in $s$, and substrings matched by identity functions are complements of substrings in $s$. When $i = 3$, since $A \neq G$ and $C$, $t$ is increased and the algorithm compares $GA$ and $GA$ when $i = 4$.

**Theorem 7.** *Once we solve the alignment with non-overlapping inversion prob-lem, we can solve the alignment finding problem in $O(n^2)$ using additional $O(n)$ space.*

## 4   Conclusions

The inversion is an important operation for bio sequences such as DNA or RNA and is closely related to mutations. We have, in particular, considered non-overlapping inversions on both sequences, which is important to find the original common sequence from two mutated sequences. We have proposed a new prob-lem, alignment with non-overlapping inversions on two strings, and presented a polynomial algorithm for the problem. Given two strings $x$ and $y$, based on the inversion properties, our algorithm decides whether or not there exist two sets $\Theta_x$ and $\Theta_y$ of inversions for $x$ and $y$ such that $\Theta_x(x) = \Theta_y(y)$ in $O(n^3)$ time using $O(n^2)$ space, where $n = |x| = |y|$. Once we know the existence of $\Theta_x$ and $\Theta_y$, we can retrieve $\Theta_x$ and $\Theta_y$ in $O(n^2)$ time using additional $O(n)$ space. One future work is to improve the current running time $O(n^3)$. As far as we are aware, this algorithm is the first try to find an alignment with non-overlapping inversions on both strings. The proposed problem is about the sequence align-ment and can be extended to approximate pattern matching or edit distance problem.

# References

1. Cantone, D., Cristofaro, S., Faro, S.: Efficient string-matching allowing for non-overlapping inversions. Theoretical Computer Science 483, 85–95 (2013)
2. Cantone, D., Faro, S., Giaquinta, E.: Approximate string matching allowing for inversions and translocations. In: Proceedings of the Prague Stringology Conference 2010, pp. 37–51 (2010)
3. Chen, Z.-Z., Gao, Y., Lin, G., Niewiadomski, R., Wang, Y., Wu, J.: A space-efficient algorithm for sequence alignment with inversions and reversals. Theoretical Computer Science 325(3), 361–372 (2004)
4. Grabowski, S., Faro, S., Giaquinta, E.: String matching with inversions and translocations in linear average time (most of the time). Information Processing Letters 111(11), 516–520 (2011)
5. Ignatova, Z., Zimmermann, K., Martinez-Perez, I.: DNA Computing Models. Advances in Information Security (2008)
6. Kececioglu, J.D., Sankoff, D.: Exact and approximation algorithms for the inversion distance between two chromosomes. In: Apostolico, A., Crochemore, M., Galil, Z., Manber, U. (eds.) CPM 1993. LNCS, vol. 684, pp. 87–105. Springer, Heidelberg (1993)
7. Li, S.C., Ng, Y.K.: On protein structure alignment under distance constraint. In: Dong, Y., Du, D.-Z., Ibarra, O. (eds.) ISAAC 2009. LNCS, vol. 5878, pp. 65–76. Springer, Heidelberg (2009)
8. Lipsky, O., Porat, B., Porat, E., Shalom, B.R., Tzur, A.: Approximate string matching with swap and mismatch. In: Tokuyama, T. (ed.) ISAAC 2007. LNCS, vol. 4835, pp. 869–880. Springer, Heidelberg (2007)
9. Ogilvie, C.M., Scriven, P.N.: Meiotic outcomes in reciprocal translocation carriers ascertained in 3-day human embryos. European Journal of Human Genetics 10(12), 801–806 (2009)
10. Oliver-Bonet, M., Navarro, J., Carrera, M., Egozcue, J., Benet, J.: Aneuploid and unbalanced sperm in two translocation carriers: evaluation of the genetic risk. Molecular Human Reproduction 8(10), 958–963 (2002)
11. Painter, T.S.: A New Method for the Study of Chromosome Rearrangements and the Plotting of Chromosome Maps. Science 78, 585–586 (1933)
12. Sakai, Y.: A new algorithm for the characteristic string problem under loose similarity criteria. In: Asano, T., Nakano, S.-i., Okamoto, Y., Watanabe, O. (eds.) ISAAC 2011. LNCS, vol. 7074, pp. 663–672. Springer, Heidelberg (2011)
13. Schöniger, M., Waterman, M.S.: A local algorithm for DNA sequence alignment with inversions. Bulletin of Mathematical Biology 54(4), 521–536 (1992)
14. Vellozo, A.F., Alves, C.E.R., do Lago, A.P.: Alignment with non-overlapping inversions in $O(n^3)$-time. In: Bücher, P., Moret, B.M.E. (eds.) WABI 2006. LNCS (LNBI), vol. 4175, pp. 186–196. Springer, Heidelberg (2006)

# Collapsing Exact Arithmetic Hierarchies

Nikhil Balaji and Samir Datta

Chennai Mathematical Institute
{nikhil,sdatta}@cmi.ac.in

**Abstract.** We provide a uniform framework for proving the collapse of the hierarchy $NC^1(\mathcal{C})$ for an exact arithmetic class $\mathcal{C}$ of polynomial degree. These hierarchies collapse all the way down to the third level of the $AC^0$-hierarchy, $AC_3^0(\mathcal{C})$. Our main collapsing exhibits are the classes

$$\mathcal{C} \in \{C_=NC^1, C_=L, C_=SAC^1, C_=P\}.$$

$NC^1(C_=L)$ and $NC^1(C_=P)$ are already known to collapse [1,19,20].

We reiterate that our contribution is a framework that works for *all* these hierarchies. Our proof generalizes a proof from [9] where it is used to prove the collapse of the $AC^0(C_=NC^1)$ hierarchy. It is essentially based on a polynomial degree characterization of each of the base classes.

## 1 Introduction

Collapsing hierarchies has been an important activity for structural complexity theorists through the years [13,22,15,24,19,18,5,12]. We provide a uniform framework for proving the collapse of the $NC^1$ hierarchy over an exact arithmetic class. Using our method, such a hierarchy collapses all the way down to the $AC_3^0$ closure of the class.

Our main collapsing exhibits are the $NC^1$ hierarchies over the classes $C_=NC^1$, $C_=L$, $C_=SAC^1$, $C_=P$. Two of these hierarchies, viz. $NC^1(C_=L), NC^1(C_=P)$, are already known to collapse ([1,20,19]) while a weaker collapse is known for a third one viz. that of $AC^0(C_=NC^1)$. We reiterate that our contribution is a simple proof that works for all these hierarchies. Our proof is a generalization of a proof from [9] who used it to prove the collapse of the $C_=NC^1$ hierarchy and is essentially based on a polynomial degree characterization of each of the corresponding arithmetic classes.

The most well known amongst the exact arithmetic circuit hierarchy collapses is the collapse of the $NC^1$ hierarchy over $C_=L$. This was proved by using the linear algebraic properties of $C_=L$ an elegant and non-trivial argument by Allender, Beals, and Ogihara [1]. We find it remarkable that our proof does not use any linear algebra (apart from the characterization of GapL functions as being exactly the functions computed by weakly skew circuits [25], which involves a certain amount of linear algebra) to prove the collapse to $AC_3^0(C_=L)$.

The collapse of the $NC^1$ hierarchy over $C_=P$ to a constant level of the $AC^0$ hierarchy follows easily from the results by Ogihara from [20,19]. Our proof is quite orthogonal to the proofs there.

We would like to point out a notational quirk that we carry over from [9] by consistently calling the finite level of $AC^0$ circuit to which we collapse the various classes as

S.P. Pal and K. Sadakane (Eds.): WALCOM 2014, LNCS 8344, pp. 273–285, 2014.
© Springer International Publishing Switzerland 2014

$AC_3^0$ where we actually mean two layers of Boolean gates i.e $\wedge, \vee$. The reason for the subscript 3 being that we include negation gates while counting the depth contrary to popular usage.

## 1.1 Historical Perspective

Counting classes have been studied from the early days of complexity theory. Gill and Simon [11,23] were the first to introduce the class PP(Probabilistic polynomial time) which consists of all languages decidable by a nondeterministic Turing machine in polynomial time in which at least half of the computation paths lead to acceptance. This class gained importance due to Toda's theorem (which states that a polynomial time machine with access to a PP oracle is at least as powerful as the entire polynomial hierarchy) and Beigel, Reingold and Spielman [4] who proved that it is closed under intersection.

$C_=P$ is the humbler cousin of PP though it happens to be one of our main protagonists. It was introduced by Simon [23] and consists of languages where the acceptance and rejection probability is the same.

Fenner, Fortnow and Kurtz[10] introduced the notion of Gap to aid the study of structural complexity of these counting classes. For a class $\mathcal{C}$, captured by nondeterministic Turing machines, denote by Gap$\mathcal{C}$ the class of functions expressible as the difference between the number of accepting and rejecting computations of some nondeterministic Turing machine. Then $C_=P$ is just those languages which have a zero gap (with an NP-machine).

The class $C_=L$ and the $NC^1, AC^0$ hierarchies over it have also received attention in literature [1,21,14]. In [1] it was first proved that the $NC^1$ hierarchy over $C_=L$ collapses all the way down to the first level of the hierarchy, $L^{C_=L}$. Further collapse to $C_=L$ is not known because this class is not known to be closed under complement.

The class $C_=NC^1$ was probably mentioned explicitly for the first time in [7] where it is shown that (a uniform version of) it is contained in Logspace. Various hierarchies over this class including the Boolean, $AC^0$, and the arithmetic hierarchy were studied in [9].

To the extent of our knowledge, the only previous known occurence of the class $C_=SAC^1$ is in the context of deciding the properties of monomials computed by arithmetic circuits[17]. The classes $SAC^1$ and its arithmetic analogs $\#SAC^1$, GapSAC$^1$ have been actively investigated in literature [28,2,16].

## 1.2 Our Results

In this paper, we extend and generalize the framework used in [9] to prove that the $NC^1(C_=\mathcal{K})$ for a $NC^1$-well-behaved class[1] $\mathcal{K}$ collapses to $AC_3^0(C_=\mathcal{K})$. As a result, we obtain several results as corollaries:

1. $AC^0(C_=NC^1) = NC^1(C_=NC^1) = AC_3^0(C_=NC^1)$, improving on [9].
2. $AC^0(C_=L) = NC^1(C_=L) = AC_3^0(C_=L)$, which gives an alternative proof of [1]

---

[1] See Definitions 4, 8.

3. $AC^0(C_=SAC^1) = NC^1(C_=SAC^1) = AC_3^0(C_=SAC^1)$.
4. $AC^0(C_=P) = NC^1(C_=P) = AC_3^0(C_=P)$, which gives an alternative proof of [19]

We prove that such a collapse can be made DLOGTIME-uniform. Note that we need to prove this strict uniformity of our collapse in order to exploit Venkateswaran's characterization of NP as exactly those languages that are decidable by DLOGTIME-uniform semi-unbounded circuit families of exponential size.

### 1.3   Proof Idea

The unifying feature of the aforementioned classes($C_=NC^1$, $C_=L$, $C_=P$, $C_=SAC^1$) is that roughly speaking, they can be viewed as languages accepted by the $C_=$-version of an arithmetic circuit family of *polynomial degree*. Thus if we consider arithmetic formulas, arithmetic weakly skew circuits, arithmetic circuits of polynomial size and arithmetic circuits of exponential size respectively, then imposing the polynomial degree bound gives us exactly the classes $GapNC^1$, $GapL$, $GapSAC^1$ and $GapP$ [7,25,26,27]. Applying the $C_=$ operator to these arithmetic classes yields our candidate exact arithmetic classes. We also tacitly use that the class of circuits are closed under composition with formulas.

The other important point to notice is that the [9] proof which just shows the collapse of the $AC^0(C_=NC^1)$ hierarchy to the third level can be extended to a collapse of $NC^1(C_=NC^1)$. This follows by observing that the level by level collapse outlined in [9] does not blow up the degree to more than a polynomial value even though we need to perform it logarithmically many times. This is because of the Cook-Wilson relativization, which can be interpreted as saying that if the oracle circuits have polynomial degree, so is the final circuit. This idea can then be generalized to any circuit family of polynomial degree.

A simple but crucial observation makes it possible to extend the results from bounded fan-in circuits in [9] to circuits where this restriction does not hold e.g. in the cases of $C_=SAC^1$ and $C_=P$. This observation concerning Vandermonde Determinants is described in Proposition 1 and used in the proof of Lemma 1.

### 1.4   Organization of the Paper

We present preliminaries in Section 2. In Section 3, we state our main result and list some immediate corollaries of our result. We discuss possible future directions to our work in Section 5.

## 2   Preliminaries

We mention the standard complexity classes that we will use. For definitions and important results regarding these classes, we refer the reader to a standard text like [29].

An *arithmetic circuit* is a directed acyclic graph(DAG) with nodes labelled by $\{\times, +\} \cup X \cup \{-1\}$, where $X$ is the set of input variables. Note that $-1$ is the only constant necessary, and we can use it to generate $1 = (-1) \times (-1), 0 = (-1) + 1$, and use $-1, 0, 1$ to generate any integer.

Given an arithmetic circuit $C$, the formal degree of the circuit is defined inductively as follows: Every input variable and the constant $-1$ have degree[2] 1. If $C_1$ and $C_2$ are two subcircuits of degree $d_1$ and $d_2$ fed in to a $\times$ gate (respectively $+$ gate), then the degree of the $\times$ gate ($+$ gate) is $d_1 + d_2$ (respectively $\max(d_1, d_2)$).

An $(m \times m)$ Vandermonde matrix $V$ is one where the $(i, j)$-th entry of $V$, $V_{ij} = i^j$.

**Fact 1.** *The absolute value of determinant of $V$ is equal to $\prod_{i<j; i,j\leq m}(j - i) = \prod_{k=1}^{m} k!$*

It is important to note that the determinant of an $(m \times m)$-Vandermonde matrix(with entries that are indeterminates) can be computed by an arithmetic circuit of degree $m$. But since we start out with just the constant $-1$, we will have to construct these numbers $1, \ldots, m$ and use them to find the the entries of the matrix, which can be done via repeated squaring and addition with a circuit of size at most $O(m^2)$, depth $O(\log m)$ and degree $O(m^2 \log m)$, for every entry of the matrix. Also, the value of this determinant is at most $\prod_{k=1}^{m} k! \leq (m!)^m \leq m^{m^2}$ which can be computed by an arithmetic circuit of size at most $O(\binom{m}{2}m^2) = O(m^4)$, depth $O(\log m)$ and degree $O(m^4 \log m)$.

We start with the usual definitions of $\mathsf{AC}^0$ and $\mathsf{NC}^1$ hierarchies over Boolean complexity classes.

**Definition 1.** *Let $C$ be a Boolean complexity class. $\mathsf{AC}^0(C)$ is the class of languages recognized by $\mathsf{AC}^0$ circuits with additional oracle gates(of unbounded fan-in) for $C$.*

Defining the $\mathsf{NC}^1$ hierarchy naively as above will yield a circuit where there could be $O(\log n)$ many oracle gates on any path, which will stand in contrast to the definition of $\mathsf{NC}^1$ as that of circuits with bounded fan-in gates. The definition due to Cook and Wilson[8,30] gives a reasonable model, which avoids these problems:

**Definition 2.** *(Cook-Wilson) Let $C$ be a Boolean complexity class. $\mathsf{NC}^1(C)$ is the class of languages recognized by $\mathsf{NC}^1$ circuits with additional oracle gates for $C$, where an oracle gate of fan-in $k$ is charged $\log k$ towards the depth of the circuit.*

We note that the *small blob chains* property defined in [9] is essentially modelled by the Cook-Wilson relativization for $\mathsf{NC}^1$ as given in Definition 2. An easy inclusion follows from the definitions above: For any boolean complexity class $C$, we have $\mathsf{AC}^0(C) \subseteq \mathsf{NC}^1(C)$. Next, we abstract the classes of base circuits over which we will consider various hierarchies:

**Definition 3.** *Let $K$ be a class of arithmetic circuits. Then $\mathsf{C}_=K$ is the class of languages recognized by the circuits from the class $K$ with an additional gate at the top that compares the output to zero to produce a Boolean output.*

We will abuse notation to identify $\mathsf{C}_=K$ with the class of circuits recognizing the class of represented languages.

---

[2] Note the non-standard convention to account for the degree of a constant. We do this so as to account for the size and degree contribution of the Vandermonde matrices, which we will use extensively in our results.

**Definition 4.** *Let $\mathcal{K}_1, \mathcal{K}_2$ be classes of circuits. Then $\mathcal{K}_1 \circ \mathcal{K}_2$ is the class consisting of circuits with a circuit from $\mathcal{K}_1$ at the top, all of whose inputs are circuits from $\mathcal{K}_2$. We say that the class $\mathcal{K}_2$ is closed under composition with $\mathcal{K}_1$ if $\mathcal{K}_2 \circ \mathcal{K}_1 \subseteq \mathcal{K}_2$.*

**Definition 5.** *(Cook-Wilson for Arithmetic Classes) Let $\mathcal{K}$ be an arithmetic complexity class of polynomial formal degree. A special case when $\mathcal{K}$ is composed with itself $O(\log n)$ many times is denoted by,*

$$\mathcal{K}^{(l)} = \overbrace{\mathcal{K} \circ \mathcal{K} \ldots \circ \mathcal{K}}^{O(\log n) \; times}$$

*where along any path in the $\mathcal{K}^{(l)}$ circuit, the product of the degrees of $\mathcal{K}$ circuits is bounded by a polynomial in the number of input variables to the $\mathcal{K}^{(l)}$ circuit.*

Throughout the paper, we use a very specific type of $AC^0$ circuit, namely, an $AC_3^0$ circuit which is essentially a boolean circuit of depth 3 consisting of an $\vee$ gate at the root, followed by a layer of $\wedge$ gates, which are fed by $C_=\mathcal{K}$ or $coC_=\mathcal{K}$ oracle gates. We will refer to such circuits as $AC_3^0(C_=\mathcal{K})$ circuits.

**Definition 6 (Toda[25]).** *A gate in a circuit is said to be weakly skew if for any multiplication gate $\alpha$ with children $\beta$ and $\gamma$, one of the two sub-circuits $C_\beta$ or $C_\gamma$ is only connected to the rest of the circuit by the wire going to $\alpha$. A weakly skew circuit is one where all the multiplication gates are weakly skew.*

A simple consequence of this definition is that every formula is a weakly skew circuit.

# 3  Exact Arithmetic Hierarchies

**Definition 7.** *Some examples of natural arithmetic circuit classes are as follows:*

- $\mathcal{K}_{form}$ *are formulas of polynomial size*
- $\mathcal{K}_{wskew}$ *are weakly skew circuits of polynomial size*
- $\mathcal{K}_{poly}$ *are circuits of polynomial degree and polynomial size*
- $\mathcal{K}_{exp}$ *are circuits of polynomial degree and exponential $(= 2^{n^{O(1)}})$ size*

Throughout this paper, we will be interested in uniform versions of the classes above. Unless mentioned otherwise, all the arithmetic circuit families we mention are DLOGTIME-uniform.

**Proposition 1.** *We prove the following for the aforementioned circuit classes:*

1. $C_=\mathcal{K}_{form} = C_=NC^1$
2. $C_=\mathcal{K}_{wskew} = C_=L$
3. $C_=\mathcal{K}_{poly} = C_=SAC^1$
4. $C_=\mathcal{K}_{exp} = C_=P$

*Proof.* Some of the above follow from well-known results:

1. Follows from the simulation of formulas by arithmetic straight-line programs due to Ben-Or and Cleve [6].
2. Follows from Toda's characterization of determinant using weakly skew circuits[25].
3. Follows from Venkateswaran's uniform circuit characterization of $SAC^1$ as consisting of languages recognized by circuit families having polynomial degree and polynomial size[27,26].
4. Follows from Venkateswaran's uniform circuit characterization of NP as consisting of languages recognized by circuit families having polynomial degree and exponential size.[27]

$\square$

Notice that skew circuits which yield the usual definition of GapL are not closed under composition with formulas (because formulas are not necessarily skew circuits). Thus we have to use the alternative(and equivalent) definition of GapL in terms of weakly skew circuits and in this case all formulas are indeed weakly skew. One of the key requirements in our collapse is the ability to compose two different circuit families.

**Definition 8.** *Given an arithmetic class $\mathcal{K}$, consider the following:*

*(i)* $\mathcal{K}_{form} \subseteq \mathcal{K}$
*(ii)* $\mathcal{K} \subseteq \mathcal{K}_{exp}.$
*(iii)* $\mathcal{K} \circ \mathcal{K} \subseteq \mathcal{K}$
*(iv)* $\mathcal{K}^{(l)} \subseteq \mathcal{K}.$

*We call $\mathcal{K}$, $AC^0$-well-behaved if it satisfies $(i), (ii), (iii)$ and $NC^1$-well-behaved if it satisfies $(i), (ii), (iv)$.*

It is easy to see that every $NC^1$-well-behaved class of circuits $\mathcal{K}$ is also $AC^0$-well-behaved. Theorem 1 below already entails a collapse of $AC^0(C_=\mathcal{K})$ by the observation that $NC^1$-well-behaved implies $AC^0$-well-behaved. We show that if one is interested in the collapse of $AC^0(C_=\mathcal{K})$ then it is sufficient for $\mathcal{K}$ to be $AC^0$-well-behaved, which is a weaker notion.

Our main goal in this paper is to study the $AC^0$ and $NC^1$ reducibilities to $C_=\mathcal{K}$. $AC^0(C_=\mathcal{K})$(respectively $NC^1(C_=\mathcal{K})$) is the class consisting of languages which can be decided by constant depth(respectively $O(\log n)$-depth), polynomial size circuits consisting of $\wedge, \vee, \neg$ and $C_=\mathcal{K}$ gates. Our main theorem is the following:

**Theorem 1.** *For every $NC^1$-well-behaved class $\mathcal{K}$ the exact hierarchy $NC^1(C_=\mathcal{K})$ collapses to $AC^0_3(C_=\mathcal{K})$.*

*Proof.* First we recall some terminology from [9]. Let $C$ be a circuit from $NC^1(C_=\mathcal{K})$. Then a blob consists of the subcircuit rooted at some equality gate $g$ in the circuit where the equality gates in the sub-circuit are replaced by new formal variables. In other words it consists of a single oracle gate. A blob chain consists of a root to leaf path in the $NC^1$ circuit and includes all the equality gates (aka oracle gates) along the path.

We need a generalization of a lemma from [9] which shows that a circuit of two blob layers i.e. a blob with its input blobs can be replaced by a shallow Boolean circuit ($AC_3^0$) with a single layer of blobs below it. In this "flattening" process the degree and the size of the circuit increase by an extent we explicate in the lemma.

In [9] this lemma was proved in the context of $C_=NC^1$ but can be generalized to any $C_=\mathcal{K}$ since the only property of $GapNC^1$ used in its proof was that it was a class of arithmetic circuits closed under composition with logarithmic depth polynomial size formulas; and this is true for all well-behaved $\mathcal{K}$ by definition.

**Lemma 1.** *(Rephrased from [9]) Let $C_0(x), C_1(y), \ldots, C_m(y)$ be $m+1$ circuits from $C_=\mathcal{K}$. Let $f : y \mapsto C_0(C(y))$ be the function computed by feeding in the circuits $C_i$ for $i > 0$ as inputs of $C_0$. Then, $f$ is also computed by the following $AC_3^0(C_=\mathcal{K})$ circuit:*

$$\bigvee_{i=1}^{m} z_i(y) = \bigvee_{i=1}^{m} [S_{i,0}(C'(y)) \wedge S_{i,1}(C'(y)) \wedge B_i(C'(y))]$$

*Notice that, we have used abbreviations, $x = (x_1, \ldots, x_m), y = (y_1, \ldots, y_n)$, and similarly $C, C'$ to simplify notation. More importantly, $C_i'$ are functions in $\mathcal{K}$, $S_{i,0}$ is a $C_=\mathcal{K}$ circuit, $S_{i,1}$ is a $C_{\neq}\mathcal{K}$ circuit, and $B_i$ from $C_{\neq}\mathcal{K}$. Further, the size and the degree of the entire circuit is bounded by $O(ms_0) + O(m^8 s)$ and $O(m^9 d + d_0 d)$ respectively. Here $s_0$ is the size of $C_0$, $s$ is the sum of sizes of the $C_i$'s (for $i > 0$), $d_0$ is the degree of $C_0$ and $d$ is the sum of degrees of $C_i$'s for $i > 0$.*

Emulating [9] we can use Lemma 1 (We present a complete proof of the lemma in [3]) to collapse two levels of blobs into one with an $AC_3^0$ circuit at the top. The $AC_3^0$ circuit can be converted to an element of $\mathcal{K}_{form}$ (the proof of the lemma shows that the $AC_3^0$ circuit is an *unambiguous* formula and straightaway arithmetization will preserve the value). Assuming we start with a $AC_k^0(C_=\mathcal{K})$ circuit, then the closure of $\mathcal{K}$ under composition with $\mathcal{K}_{form}$ allows us to convert the circuit to the form $AC_{k-1}^0(C_=\mathcal{K})$ i.e. with depth one lesser than earlier. Since we need to repeat this operation $k$ times the size remains bounded by $n^{O(k)}$ times the size of the original circuit and the degree by the degree of the original circuit raised to a power which is $O(k)$. Notice that this seems to work only for $AC^0$ relativizations because the naïve upper bounds on the size and degree become quasipolynomial when we repeat the collapse logarithmically many times as necessitated by $NC^1$ relativizations. We show in Lemma 3 that this not the case.

Now we show how to do a careful analysis of the collapse in case of $NC^1$ relativizations which allows us to conclude that the size and degree do not become too large. The key is to use basic counting along with Cook-Wilson condition on $NC^1$-relativization which allows us to conclude the following:

**Proposition 2.** *Consider an $NC^1(C_=\mathcal{K})$ circuit $C$ with size, height and number of inputs $s(C), h(C)$ and $n(C)$ respectively. Then,*

- *the number of distinct maximal blob chains is upper bounded by $s(C)$.*
- *the product of the fan-ins of the oracle gates along a blob-chain is upper bounded by $(n(C))^{c_1}$ for some constant $c_1$ depending on the circuit family;(a consequence of the Cook-Wilson property)*

Next we bound the total size of the flattened circuit. Lemma 2 shows that, for the $\mathcal{K}^{(l)}$ circuits obtained as a result of the collapse, indeed $\mathcal{K}^{(l)} = \mathcal{K}$.

**Lemma 2.** *The size of the flattened circuit is upper bounded by: $n^c s^2$, where $s$ is the sum of sizes of the circuits corresponding to oracle gates and $c$ is a constant depending on the family of circuits (and $n$ is the number of inputs to the original circuit).*

*Proof.* We use the recurrence:

$$S(g) = 3f_g^9 \sum_{h:\text{parent}(h)=g} S(h) + s_g f_g \tag{1}$$

$$\leq 3f_g^9 \sum_{h:\text{parent}(h)=g} S(h) + s_g^2 \text{ (Since } f_g \leq s_g) \tag{2}$$

Here $S(g)$ is the total size of the sub-circuit below gate $g$ after flattening, $s_g$ is the original size of the $\mathcal{K}$-circuit for just the gate $g$ and $f_g$ is the fan-in of $g$. Let us further denote by $h_g$ the height of $g$. Then the recurrence has the solution:

$$S(g) = 3^{h_g} \sum_{g':g' \text{ is a descendant of } g} \left( \left( \prod_{h \in B(g,g')} f_h^9 \right) s_{g'}^2 \right)$$

where $B(g, g')$ denotes the blob-chain between $g$ and $g'$.

Let $g_0$ be the root of the circuit. From Proposition 2, the products of the fan-ins in the sum above is upper bounded by $n^{c_1}$, where $n = n(C)$ is the number of inputs of the circuit. Also, $h_{g_0}$ is bounded by $c_2' \log n$ for some constant $c_2'$, since the reduction is an $NC^1$ reduction. Hence, $S(g_0) \leq n^c \sum_g s_g^2$. $\square$

Finally we bound the degree of the flattened circuit, which is crucial to proving our collapse:

**Lemma 3.** *The degree of the flattened circuit is bounded by $n^{c'}$ where $c'$ is a constant depending on the original circuit family.*

*Proof.* The degree of a well-behaved circuit on $t$ inputs is at most $t^{c_1'}$ for some constant $c_1'$ depending on the circuit family (since a circuit family is a subclass of $\mathcal{K}_{exp}$ circuits which have a polynomial degree. Now, we have the recurrence:

$$D(g) = f_g^9 d_g \sum_{h:\text{parent}(h)=g} D(h)$$

where we define $D(g)$ to be the degree of the flattened circuit at $g$ and $d_g$ its original degree as a circuit in $\mathcal{K}$. This has the solution:

$$D(g) = \sum_{B:B \text{ is a maximal blob-chain rooted at } g} \left( \prod_{h:h \in B} (f_h^9 f_h^{c_1'}) \right)$$

Thus, $D(g_0) \leq s(C) n^{(c_1'+1)c_1} \leq n^{c'}$, using the bounds from Proposition 2 for the first inequality and observing that the fan-in of any oracle gate and the size of the $NC^1$ circuit is bounded by $n$. $\square$

Hence we have proved that for each equality gate in the final circuit, the degree remains polynomial via composition with formulas which means the circuit rooted at the equality gate is in $C_=\mathcal{K}$, and the overall circuit in $AC_3^0(C_=\mathcal{K})$. This completes the proof of the theorem. $\qquad\square$

We now document some consequences of Theorem 1:

**Corollary 1.** $NC^1(C_=NC^1) = AC_3^0(C_=NC^1)$

*Proof.* Formulas are $AC^0$-well-behaved since they are clearly closed under composition with formulas and have polynomial degree. The fact that they are $NC^1$-well-behaved follows from Definition 5, namely the arithmetic Cook-Wilson property. $\qquad\square$

**Corollary 2.** $NC^1(C_=L) = AC_3^0(C_=L)$

*Proof.* Weakly skew circuits are closed under composition with $\mathcal{K}_{form}$ since $\mathcal{K}_{form}$ circuits are weakly skew(by [25]). Also, the polynomials computable by weakly skew circuits of polynomial size are also computable by determinants of polynomial sized matrices by [25]. Hence the former have polynomial degrees. Thus weakly skew circuits are $AC^0$-well-behaved. To prove that weakly skew circuits are $NC^1$-well-behaved, it is sufficient to note that, weakly skew circuits composed with themselves $O(\log n)$-many times, remain weakly skew due to the arithmetic Cook-Wilson in Definition 5. $\qquad\square$

**Corollary 3.** $NC^1(C_=SAC^1) = AC_3^0(C_=SAC^1)$

*Proof.* Formulas of polynomial size have polynomial degree, thus polynomial degree circuits of polynomial size are closed under composition with them completing the proof that they are $AC^0$-well-behaved. Similarly, polynomial degree circuits composed with themselves $O(\log n)$-many times still have polynomial formal degree, and hence are $NC^1$-well-behaved. $\qquad\square$

**Corollary 4.** $NC^1(C_=P) = AC_3^0(C_=P)$

*Proof.* This also follows exactly from the same reasoning as the one for $C_=SAC^1$ circuits. $\qquad\square$

# 4    Uniformity

We have proved that $NC^1(C_=\mathcal{K})$ collapses to $AC_3^0(C_=\mathcal{K})$. In this section we show that such an $AC_3^0(C_=\mathcal{K})$ family is DLOGTIME-uniform, when the original $NC^1(C_=\mathcal{K})$ circuit family is DLOGTIME-uniform.

Following [29], we call our $AC_3^0$ circuit uniform, if the following language can be decided in $O(\log n)$ time by a deterministic turing machine (random access to its tape) : $L_{DC} = \{\langle y, g, p, b\rangle\}$, where

- $|y| = n$
- $g$ is the number of a gate $v$ in $C_n$

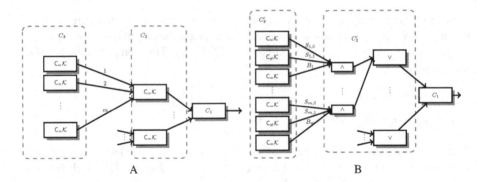

**Fig. 1.** (A) A circuit of the form $C_1 \circ C_2 \circ C_3 = \mathsf{C}_=\mathcal{K} \circ \mathsf{C}_=\mathcal{K} \circ \mathsf{C}_=\mathcal{K}$ (B) The collapsed circuit obtained from the circuit $C_1 \circ C_2 \circ C_3$ above of the form $C_1 \circ C_2' \circ C_3' = \mathsf{C}_=\mathcal{K} \circ \vee \circ \wedge \circ C_=\mathcal{K}$

- $p \in \{0,1\}^*$ such that if $p = \epsilon$, then $b$ encodes the type of the gate from the basis of the circuit family and if $p$ is the binary representation of $k$, then $b$ is the number of the $k$-th predecessor gate to $v$ wrt to a fixed ordering on the gates in the circuit.

We encode every gate $g$ in the given circuit $\mathsf{NC}^1(\mathsf{C}_=\mathcal{K})$ $C$ by a concatenation of two labels $(g_e, g_i)$. $g_e$ is an external encoding which assigns labels lexicographically to every $\mathsf{C}_=\mathcal{K}$ gate in $C$. $g_i$ is an internal encoding which assigns labels to the $+, \times, =$ gates which constitute the $\mathsf{C}_=\mathcal{K}$ gates. More precisely, given $\langle y, g, p, b \rangle$, we have $y = 1^n$, $g = (g_e, g_i)$. $g_e$ is a $O(\log n)$ length string (since the underlying circuit is a $\mathsf{NC}^1$ circuit the total number of bits needed to uniquely address a gate in the circuit is $O(\log n) + \max_\pi \sum_{g \in \pi} \log \operatorname{fanin}(g) = O(\log n)$, where the maximum is taken over all paths $\pi$ in the circuit) by the definition of oracle circuits in the Cook-Wilson model). $g_i$ is of length $O(\log n)$(since each of the $\mathsf{C}_=\mathcal{K}$ gates is poly-sized)[3].

We will analyze a typical scenario in our collapse - Given a circuit of the form $C = C_1 \circ C_2 \circ C_3$ (see Figure 1A), where $C_1$, $C_2$ and $C_3$ are $\mathsf{C}_=\mathcal{K}$ circuits, we collapse $C_2$ and $C_3$ to obtain a circuit $C' = C_1 \circ C_2' \circ C_3'$ (see Figure 1B). Here $C_2'$ is a circuit of the form $\vee \circ \wedge$ and $C_3'$ is a layer of $\mathsf{C}_=\mathcal{K}$ circuits. We prove that the collapse of $C$ to $C'$ can be made DLOGTIME-uniform and then show that under the Cook-Wilson property, such collapses can be combined to make the collapse of $\mathsf{NC}^1(\mathsf{C}_=\mathcal{K})$, DLOGTIME-uniform.

*Claim.* For every $\mathsf{NC}^1$-well-behaved class $\mathcal{K}$, DLOGTIME-uniform $\mathsf{C}_=\mathcal{K} \circ \mathsf{C}_=\mathcal{K} \circ \mathsf{C}_=\mathcal{K} = $ DLOGTIME-uniform $\mathsf{AC}_3^0(\mathsf{C}_=\mathcal{K})$.

*Proof.* Assuming the DLOGTIME-uniformity and the aforementioned labelling scheme of $C$, we will now see how to decide the connection language of $C'$ when we have access to the connection language of $C$. First we explain how we label $C'$ from $C$: The labels of $C_1$ carry over from $C$ to $C'$. For the $\vee \circ \wedge$ circuit $C_2'$ we give new set of external and internal labels similar to $C$. This takes $O(\log n)$ many bits. There

---

[3] Note that from [27] $\mathsf{C}_=\mathsf{P}$ is exactly characterized by DLOGTIME-uniform semi-unbounded circuits of $2^{n^{O(1)}}$ size and logarithmic depth and in this case the labels of the gate themselves will be strings of length $n^{O(1)}$

are two kinds of circuits in $C'_3$: circuits computing the symmetric functions, which are built via interpolation by symmetric polynomials, using the Vandermonde matrix and circuits of the form $C_= \mathcal{K} \circ \mathcal{K}$ (see Lemma 1). From Fact 1, it is clear that the entries of this matrix can be computed by circuits of polynomial size, which are labelled using $O(\log n)$ bits. The circuit of the form $C_= \mathcal{K} \circ \mathcal{K}$ in $C'_3$ is essentially $C_2$ to which the inputs are symmetric functions of $C_3$. Even here we assign a concatenation of external and internal encoding of the circuits from $C_2$ and $C_3$ respectively, which are $O(\log n)$ bits long.

In each of $C_1, C'_2, C'_3$, we identify three kinds of gates which we call $\alpha_0, \alpha_{con}, \alpha_{mid}$. $\alpha_0$ gates are either the output gate or the input gates of $C'$. $\alpha_{con}$ are gates which connect $C_i$ to $C_{i+1}$ for $i \in \{1, 2\}$. For example, the root gate of $C_1$, the $\vee$ gate in $C'_2$ and the $=$ gate in $C'_3$ are $\alpha_{con}$ type gates - their parents and children are not of type $\alpha_{con}$. $\alpha_{mid}$ gates are those whose parents and children are both $\alpha_{mid}$ or $\alpha_{con}$. The internal gates of $C_1$ and $C_3$ are gates of type $\alpha_{mid}$. Note that all the gates in $C'$ belong to one of these three categories.

So now we are ready to describe the label of a gate in $C'$. It is of the form $\langle y, g, p, b \rangle$ as above where $g$ is now the concatenation of the external labels of the path to the nearest $=$ ancestor to the gate and the internal label of the gate in $C$. For example, the label of a gate in $C'_3$ would consist of the concatenation of external labels of $C_1, C'_2$, and the internal label of the gate in $C'_3$. If $p$ is $\epsilon$, then $b$ is the type of the gate, which could be $\wedge, \vee, C_= \mathcal{K}$. Else the binary number encoded by the string $p$ points to the position of the gate according to our fixed ordering. Note that at the end of one collapse, the label lengths increase by atmost $O(\log n)$ bits(namely the bits required to internally label the $\vee \circ \wedge$ system arising out of the collapse. With the above labelling convention, now there are two tasks to be accomplished:

1. Given a label of a gate $g$ in $C'$, we have to verify if it is indeed a valid label.
2. Given labels of gates $g, h$ is $g$ a parent/child of $h$?

The validity of a label is easily checked: One has to check if the concatenation of labels leads to a valid external and internal encoding by querying the DLOGTIME machine for the original circuit $C$ and check if it is a $\alpha_0, \alpha_{mid}$ or $\alpha_{con}$ gate. For example, given two gates, deciding if one is the parent/child of the other can be done by checking if the label of one of them is a prefix of the other, and if yes, verifying if the gate types of the parent and child is valid according to the relation between $\alpha$s specified above.    □

Now we prove that such a labelling convention leads to easily checking the validity of a label and connections in the collapse from Theorem 1.

*Claim.* For every $NC^1$-well-behaved class $\mathcal{K}$, DLOGTIME-uniform $NC^1(C_= \mathcal{K}) =$ DLOGTIME-uniform $AC^0_3(C_= \mathcal{K})$.

*Proof.* To label the final $AC^0(C_= \mathcal{K})$ circuit, we concatenate the labels of the intermediate collapses. This might seem like it requires labels of length $O(\log^2 n)$ since the circuit we started out with had labels of length $O(\log n)$ and the $NC^1(C_= \mathcal{K})$ circuit requires $O(\log n)$ collapses to reduce it to an $AC^0(C_= \mathcal{K})$ circuit. But just as in the proof of Theorem 1, the Cook-Wilson property ensures that the length of the labels is

$O(\log n)$. This is because, even though we do the collapse as many as $O(\log n)$ times, the increase in label lengths is only due to the $\vee \circ \wedge$ block. Recall that the number of $\wedge$ gates created in 1 is exactly equal to the fan-in of the oracle gate above, and the Cook-Wilson relativization model bounds the product of fan-ins along any path to a polynomial in the number of inputs. This also bounds the number of new gates created, and hence they can be labelled by $O(\log n)$-many bits.                    □

## 5    Conclusion and Open Problems

We provide sufficient conditions on arithmetic circuits under which the $NC^1$ hierarchy over the corresponding exact arithmetic class collapses. Natural extensions can include proving similar results for more powerful reducibilities like Boolean Formula reductions or even Logarithmic Boolean Formula reductions (which do not satisfy the "small-blob-chains" property or equivalently do not follow Cook-Wilson relativization). An appropriately defined notion of $SAC^1$-reductions is another intriguing possibility.

It may also be an interesting idea to give similar uniform proofs for the $NC^1$ hierarchies over the better known classes $PNC^1, PL, PSAC^1, PP$ where we already know collapses for the second[18] and the last [5] hierarchies and also of the $AC^0(PNC^1)$-hierarchy [9].

**Acknowledgements.** We thank Eric Allender, Vikraman Arvind, Sourav Chakraborty, Meena Mahajan, Prajakta Nimbhorkar and B.V.Raghavendra Rao for helpful discussions. The second author would like to thank the organisers of the Dagstuhl seminar on "Algebraic and Combinatorial Methods in Computational Complexity" where many of the previously mentioned discussions took place. We thank the anonymous referees for helpful comments.

## References

1. Allender, E., Beals, R., Ogihara, M.: The complexity of matrix rank and feasible systems of linear equations. Computational Complexity 8(2), 99–126 (1999)
2. Allender, E., Jiao, J., Mahajan, M., Vinay, V.: Non-commutative arithmetic circuits: depth reduction and size lower bounds. Theoretical Computer Science 209(1), 47–86 (1998)
3. Balaji, N., Datta, S.: Collapsing exact arithmetic hierarchies. Electronic Colloquium on Computational Complexity (ECCC) 20, 131 (2013)
4. Beigel, R., Reingold, N., Spielman, D.A.: PP is closed under intersection. Journal of Computer and System Sciences 50(2), 191–202 (1995)
5. Beigel, R., Fu, B.: Circuits over pp and pl. J. Comput. Syst. Sci. 60(2), 422–441 (2000)
6. Ben-Or, M., Cleve, R.: Computing algebraic formulas using a constant number of registers. SIAM Journal on Computing 21, 54–58 (1992)
7. Caussinus, H., McKenzie, P., Thérien, D., Vollmer, H.: Nondeterministic $NC^1$ computation. Journal of Computer and System Sciences 57, 200–212 (1998), Preliminary Version in Proceedings of the 11th IEEE Conference on Computational Complexity, 12–21 (1996)
8. Cook, S.: A taxonomy of problems with fast parallel algorithms. Information and Control 64, 2–22 (1985)

9. Datta, S., Mahajan, M., Rao, B.V.R., Thomas, M., Vollmer, H.: Counting classes and the fine structure between $NC^1$ and L. In: Hliněný, P., Kučera, A. (eds.) MFCS 2010. LNCS, vol. 6281, pp. 306–317. Springer, Heidelberg (2010)

10. Fenner, S.A., Fortnow, L.J., Kurtz, S.A.: Gap-definable counting classes. Journal of Computer and System Sciences 48(1), 116–148 (1994)

11. Gill, J.: Computational complexity of probabilistic turing machines. SIAM Journal on Computing 6(4), 675–695 (1977)

12. Gottlob, G.: Collapsing oracle-tape hierarchies. In: IEEE Conference on Computational Complexity, pp. 33–42 (1996)

13. Hemachandra, L.: The strong exponential hierarchy collapses. In: Structure in Complexity Theory Conference. IEEE Computer Society (1987)

14. Hoang, T.M., Thierauf, T.: The complexity of the characteristic and the minimal polynomial. Theor. Comput. Sci. 295, 205–222 (2003)

15. Immerman, N.: Nondeterministic space is closed under complementation. SIAM Journal on Computing 17(5), 935–938 (1988)

16. Limaye, N., Mahajan, M., Rao, B.V.R.: Arithmetizing classes around $NC^1$ and L. In: Thomas, W., Weil, P. (eds.) STACS 2007. LNCS, vol. 4393, pp. 477–488. Springer, Heidelberg (2007)

17. Mengel, S.: Conjunctive Queries, Arithmetic Circuits and Counting Complexity. Ph.D. thesis, Universität Paderborn (2012)

18. Ogihara, M.: The PL hierarchy collapses. SIAM J. Comput. 27(5), 1430–1437 (1998)

19. Ogihara, M.: Equivalence of $NC^k$ and $AC^{k-1}$ closures of $NP$ and Other Classes. Inf. Comput. 120(1), 55–58 (1995)

20. Ogiwara, M.: Generalized theorems on relationships among reducibility notions to certain complexity classes. Mathematical Systems Theory 27(3), 189–200 (1994)

21. Santha, M., Tan, S.: Verifying the determinant in parallel. Computational Complexity 7(2), 128–151 (1998)

22. Schöning, U., Wagner, K.W.: Collapsing oracle hierarchies, census functions and logarithmically many queries. In: Cori, R., Wirsing, M. (eds.) STACS 1988. LNCS, vol. 294, pp. 91–97. Springer, Heidelberg (1988)

23. Simon, J.: On some central problems in computational complexity (1975)

24. Szelepcsényi, R.: The method of forced enumeration for nondeterministic automata. Acta Informatica 26(3), 279–284 (1988)

25. Toda, S.: Classes of arithmetic circuits capturing the complexity of computing the determinant. IEICE Transactions on Information and Systems E75-D, 116–124 (1992)

26. Venkateswaran, H.: Properties that characterize LogCFL. Journal of Computer and System Sciences 42, 380–404 (1991)

27. Venkateswaran, H.: Circuit definitions of nondeterministic complexity classes. SIAM J. on Computing 21, 655–670 (1992)

28. Vinay, V.: Counting auxiliary pushdown automata and semi-unbounded arithmetic circuits. In: Proceedings of 6th Structure in Complexity Theory Conference, pp. 270–284 (1991)

29. Vollmer, H.: Introduction to Circuit Complexity: A Uniform Approach. Springer-Verlag New York Inc. (1999)

30. Wilson, C.B.: Relativized circuit complexity. J. Comput. Syst. Sci. 31(2), 169–181 (1985)

# Complexity of Disjoint $\Pi$-Vertex Deletion for Disconnected Forbidden Subgraphs

Jiong Guo* and Yash Raj Shrestha**

Universität des Saarlandes,
Campus E 1.7, 66123 Saarbrücken, Germany
{jguo,yashraj}@mmci.uni-saarland.de

**Abstract.** We investigate the computational complexity of DISJOINT $\Pi$-VERTEX DELETION. Here, given an input graph $G = (V, E)$ and a vertex set $S \subseteq V$, called a solution set, whose removal results in a graph satisfying a non-trivial, hereditary property $\Pi$, we are asked to find a solution set $S'$ with $|S'| < |S|$ and $S' \cap S = \emptyset$. This problem is partially motivated by the "compression task" occurring in the iterative compression technique. The complexity of this problem has already been studied, with the restriction that $\Pi$ is satisfied by a graph $G$ iff $\Pi$ is satisfied by each connected component of $G$ [7]. In this work, we remove this restriction and show that, except for few cases which are polynomial-time solvable, almost all other cases of DISJOINT $\Pi$-VERTEX DELETION are $\mathcal{NP}$-hard.

## 1 Introduction

A *graph property* $\Pi$ can be considered as a set of graphs. We say that a graph $G$ *satisfies* $\Pi$ if $G \in \Pi$. The classical $\Pi$-Vertex Deletion problem is defined as follows:

$\Pi$-VERTEX DELETION ($\Pi$-VD)
**Input:** An undirected graph $G = (V, E)$ and a non-negative integer $k$.
**Question:** Is there a set $S$ of at most $k$ vertices whose removal results in a graph $G'$ with $G' \in \Pi$?

Many prominent problems are special cases of $\Pi$-VD. For example, VERTEX COVER is the case of $\Pi$ being "edgeless". Lewis and Yannakakis [16] showed that $\Pi$-VD is $\mathcal{NP}$-complete for any non-trivial, hereditary property $\Pi$ that can be verified in polynomial time. A graph property $\Pi$ is *hereditary*, if it is closed under vertex deletion, and *non-trivial* if it is satisfied by infinitely many graphs and it is not satisfied by infinitely many graphs.

In the last 20 years, this $\mathcal{NP}$-completeness result motivated various research directions on $\Pi$-VD, for instance, approximation algorithms [1], its complexity on special input graphs [10], and the edge deletion counterpart [17]. One of the

---

* Supported by the DFG Excellence Cluster MMCI.
** Supported by the DFG research project DARE GU 1023/1.

S.P. Pal and K. Sadakane (Eds.): WALCOM 2014, LNCS 8344, pp. 286–297, 2014.
© Springer International Publishing Switzerland 2014

recently most remarkable approaches to cope with the $\mathcal{NP}$-completeness of $\Pi$-VD is the parameterized algorithms [18,6]. $\Pi$-VD carries with its definition a natural parameter, the solution size $k$.

In 2004, Reed et al. [19] introduced the iterative compression technique [14,18] which turned out to be particularly useful for achieving parameterized algorithms for $\Pi$-VD, for instance, UNDIRECTED/DIRECTED FEEDBACK VERTEX SET where $\Pi$ is "acyclic" [4,13,2,3] and CLUSTER VERTEX DELETION where $\Pi$ be "disjoint union of cliques" [15]. This technique builds on two separate routines, namely, the iterative routine and the compression routine. In the former, we build the instance step by step from an empty instance, while in the latter we are given an instance and a solution, and we endeavor to search for a better solution for the given instance. The compression routines of iterative compression algorithms for $\Pi$-VD basically deal with a disjoint version of $\Pi$-VD which can be defined as follows:

**DISJOINT $\Pi$-VERTEX DELETION (D-$\Pi$-VD)** [7]
**Input:** An undirected graph $G = (V, E)$ and a vertex set $X \subseteq V$ such that $G[V \setminus X] \in \Pi$.
**Question:** Is there a vertex subset $X' \subseteq V$ with $|X'| < |X|$ and $X \cap X' = \emptyset$ such that $G[V \setminus X']$ satisfies $\Pi$?

Fellows et al. [7] initialized the study of D-$\Pi$-VD and gave a complexity dichotomy of this problem for the case that the non-trivial, hereditary property $\Pi$ is determined by components. A graph property $\Pi$ is determined by components, if a graph $G$ satisfy $\Pi$ iff each of $G$'s connected components satisfies $\Pi$. Since every hereditary graph property $\Pi$ can be characterized by a set $\mathcal{H}$ of forbidden induced subgraphs, the dichotomy achieved in [7] holds for D-$\Pi$-VD, where $\Pi$ corresponds to a forbidden subgraph set $\mathcal{H}$ that contains only connected graphs. Fellows et al. [7] proved that as long as the forbidden set $\mathcal{H}$ does not contain a star with at most two leaves, the corresponding D-$\Pi$-VD problem is $\mathcal{NP}$-hard; otherwise, it is polynomial-time solvable.

In this paper, we generalize the results of Fellows et al. to the case that the forbidden set $\mathcal{H}$ allows to contain disconnected graphs. Note that many important graph properties can be characterized by forbidden sets $\mathcal{H}$ containing disconnected graphs, for instance, chain graphs with $\mathcal{H}$ containing two independent edges and threshold graphs with $\mathcal{H}$ containing 4-vertex path, 4-vertex cycle and two independent edges. Recently, parameterized algorithms based on the iterative compression technique for $\Pi$-VD have been derived for $\Pi$ corresponding to disconnected forbidden subgraph characterizations [12,11].

*Our Results.* Let $\mathcal{H}$ be the set of forbidden subgraphs corresponding to a graph property $\Pi$. Note that, if $\mathcal{H}$ or the set $\widetilde{\mathcal{H}}$ of the complement graphs of the graphs in $\mathcal{H}$ contains only connected graphs, the dichotomy in [7] applies. Thus, we only consider the case when both $\mathcal{H}$ and $\widetilde{\mathcal{H}}$ contain disconnected graphs. Our results can be summarized in the following: If $\mathcal{H}$ contain no star with at most two leaves and no disconnected graph whose connected components are all stars with at most two leaves, then D-$\Pi$-VD is $\mathcal{NP}$-complete. Then, for the case that $\mathcal{H}$ contains a star of two leaves, we prove polynomial-time solvability of D-$\Pi$-VD. For $\mathcal{H}$ containing disconnected forbidden subgraphs with stars of at most two leaves as connected components, we achieve $\mathcal{NP}$-hardness as well as polynomial-time solvability results. Further, few cases are left open.

*Preliminaries.* For a graph $G = (V, E)$, let $E(G)$ and $V(G)$ denote the set of edges and vertices of $G$, respectively. Unless specifically mentioned, we follow the graph theoretic notations and definitions from [5]. If we delete a vertex $v$ or a subgraph $S$ from graph $G$, we denote the resulting graph as $G - \{v\}$ and $G - S$, respectively. A *complement* or *inverse* of a graph $G$ is a graph $\widetilde{G}$ such that $\widetilde{G}$ is on the same vertices as $G$ and two vertices of $\widetilde{G}$ are adjacent if and only if they are not adjacent in $G$. For a set $\mathcal{H}$ of graphs, let $\widetilde{\mathcal{H}}$ be the set containing the complements of all graphs in $\mathcal{H}$.

A $P_4$ is a path on four vertices. A *star* $S_l$ is a star with $l$ leaves, while a $S_{\leq l}$ has at most $l$ leaves. We call a graph a *non-star* if it does not satisfy the condition for stars. For a disconnected graph $G$, if each of its connected components is a star, we call $G$ an *all-star*. A $(\geq i)$-all star is an all-star containing a star $S_l$ with $l \geq i$. An $(i)$-all star is an all-star containing a star $S_l$ with $l = i$. A $(\leq i)$-all star is an all-star which consists only of stars $S_{\leq i}$ as its connected components. Clearly, a $(< 1)$-all star is an edgeless graph. A $(P_4, 3)$-all star is a disconnected graph which consists only of $P_4$'s and stars as its connected components, among which the largest star being $S_3$. A $(P_4, \leq 2)$-all star is a disconnected graph which consists only of $P_4$'s and stars $S_l$ with $l \leq 2$ as its connected components.

Given a set $\mathcal{H}$ of graphs, let $\mathcal{H}_c$ and $\mathcal{H}_d$ be the sets which contain all connected and disconnected graphs of $\mathcal{H}$, respectively. Each graph $h_{di} \in \mathcal{H}_d$ can be viewed as a set of connected components, denoted as $h_{di(1)}, h_{di(2)}, \cdots, h_{di(t)}$. We say that $h_{di(j)}$ has a *number of occurrence* $x$ in $h_{di}$ if there exist $x$ many connected components in $h_{di}$ which are isomorphic to $h_{di(j)}$. Due to lack of space, some definitions and proofs are deferred to the full version of the paper.

## 2   No Stars with at Most Two Leaves

We prove here that, if $\mathcal{H}$ contain neither a star with at most two leaves nor an $(\leq 2)$-all-star, then D-$\Pi$-VD is $\mathcal{NP}$-hard. To this end, we distinguish two cases based on the number of leaves: First, we adapt the reduction in [7] to deal with the case that $\mathcal{H}$ contains no $S_{\leq 3}$ and no $(\leq 3)$-all stars. Then, a new reduction is given for the three leaf star case.

Lewis and Yannakakis [16] devised a framework, to prove that $\Pi$-VD for a non-trivial hereditary property $\Pi$ is $\mathcal{NP}$-complete. Later, Fellows et al. [7] modified this framework for the $\mathcal{NP}$-hardness proofs of D-$\Pi$-VD with connected forbidden subgraphs when $\mathcal{H}$ contains no star with at most three leaves. They reduced from VERTEX COVER on triangle-free graphs, where they picked the subgraph $H$ in $\mathcal{H}$, that has the lexicographically smallest $\alpha$-sequence[1], to build

---

[1] For the definition of $\alpha$-sequences, see [16].

the vertex and edge gadgets [7]. We can further extend this adaptation to deal with disconnected forbidden subgraphs for the above case. However, the selection of the subgraph $H$ for the vertex and edge gadgets is more tricky. Here, we have to consider the connected components of each disconnected subgraphs. If a connected component $C$ of a disconnected subgraph $H$ has lexicographically smallest $\alpha$-sequence among all connected subgraphs in $\mathcal{H}_c$ and all connected components of disconnected subgraphs in $\mathcal{H}_d$, then we use $C$ to build the vertex and edge gadgets. Further constructions are also needed for other components of $H$. In particular, we prove the following theorem and lemma.

**Lemma 1.** *If $\mathcal{H}_c$ does not contain $S_{\leq 3}$ or $(\leq 3)$-all star, then* D-$\Pi$-VD *is $\mathcal{NP}$-hard.*

In the following, we consider the case that is, $\mathcal{H}$ contains $S_3$ or 3-all stars. Here, we distinguish two cases and derive completely different reductions compared to [7]. In particular, we prove the following theorem:

**Lemma 2.** *If there exists a star $S_3$ in $\mathcal{H}_c$ or an 3-all-star in $\mathcal{H}_d$ and there exists no $S_{\leq 2}$ or $(\leq 2)$-all star in $\mathcal{H}$, then* D-$\Pi$-VD *is $\mathcal{NP}$-hard.*

*Proof.* Here, we distinguish two cases: 1) $\mathcal{H}$ contains neither $P_4$ nor $(P_4, \leq 2)$-all star, and 2) $\mathcal{H}$ contains $P_4$ or $(P_4, \leq 2)$-all star. We show in the following only 1). The proof for another case is in the full version of the paper.

Assume that there exists a $S_3$ in $\mathcal{H}_c$ or a 3-all star in $\mathcal{H}_d$. Moreover, $P_4$ and $(P_4, \leq 2)$-all star are not present in $\mathcal{H}$. If there exists a $S_3$ in $\mathcal{H}_c$, we set $H$ as $S_3$; otherwise, among the 3-all stars and $(P_4, 3)$-all stars in $\mathcal{H}_d$, we choose the one with the minimum number of occurrence of $S_3$ as $H$. Let this minimum number of occurrence be $x$. We can observe that all other graphs in $\mathcal{H}_d$ have either at least $x$ occurrence of $S_3$ or it contains at least one connected component with a higher $\alpha$-sequence than the one of $S_3$.

The reduction is from the $\mathcal{NP}$-complete 3SAT-2L problem [9], which is defined as follows:

**3SAT-2l**
**Input:** A 3-CNF boolean formula $F$ where each literal appears at most twice in the clauses.
**Output:** A satisfying assignment for $F$.

We assume without loss of generality that each variable appears in each clause at most once. Let $F = c_1 \wedge \cdots \wedge c_q$ be a 3SAT-2L formula over a variable set $Y = \{y_1, \cdots, y_p\}$. We denote the $k$-th literal in clause $c_j$ by $l_j^k$, for $1 \leq k \leq 3$. Starting with an empty graph $G$ and $X := \emptyset$, construct an instance $(G, X)$ for D-$\Pi$-VD as follows. An example of the construction is given in Fig 1. For each variable $y_i$, introduce a star $Y_i$ with three leaves (variable gadget), add one leaf and the center vertex of $Y_i$ to $X$ and label the remaining leaves of $Y_i$ with "$+$" and "$-$", respectively. For each clause $c_j$, add a star $C_j$ with three leaves (clause gadget) and add its center vertex to $X$. Add a degree-1 neighbor to each vertex of $C_j$ and this degree-1 vertex is added to $X$. Each of the three leaves of $C_j$

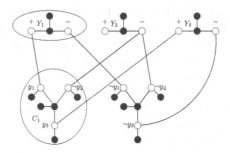

**Fig. 1.** Example for the reduction in the proof of Theorem 2 from the 3SAT-2L instance with formula $(y_1 \wedge \neg y_2 \wedge y_3) \vee (\neg y_1 \wedge \neg y_2 \wedge \neg y_3)$ to DISJOINT $\Pi$-VERTEX DELETION with the given solution X (black vertices). For illustration the the clause gadget $C_1$ and variable gadget $Y_1$ are labelled.

corresponds to a literal in $c_j$, and each leaf is connected to a variable gadget as follows. Suppose that $l_j^k$ is a literal of variable $y_i$, and let $a_k$ be the leaf of $C_j$ corresponding to $l_j^k$. Add an edge (connection gadget) between $a_k$ and the "+"-leaf of the corresponding $Y_i$, if $l_j^k = y_i$; otherwise, to the "−"-leaf of $Y_i$. Finally, if $H$ is a disconnected graph, we create a *satellite gadget* $S(G)$ isomorphic to $H - S_3$, that is, the subgraph of $H$ with one $S_3$ removed. Moreover, we add all vertices of this satellite gadget to $X$.

Obviously, $G[V \setminus X]$ only contains disjoint stars with at most two leaves. We note here that, since each literal can occur in at most two clauses, these two leaf stars can only be centered at "+" or "−" labelled vertices of $Y_i$. Hence, $G[V \setminus X] \in \Pi$. We now show that formula $F$ has a satisfying truth assignment if and only if there exists a size-$(p + 2q)$ set $X'$ with $X' \cap X = \emptyset$, that obstructs all forbidden induced subgraphs in $G$. Clearly, $|X| > p + 2q$.

($\Rightarrow$) Assume that a satisfying truth assignment for $F$ is given. Based on this truth assignment, we construct the disjoint solution $X'$, beginning with $X' := \emptyset$, as follows. For each variable $y_i$, $1 \leq i \leq p$, if $y_i = $ TRUE, then add the vertex labelled "+" in $Y_i$ to $X'$; otherwise, add the vertex labelled "−" to $X'$. This ensures that from each variable gadget, exactly one vertex will be in $X'$. Next, for each clause $c_j$ we have at least one literal set TRUE, say $l_j^i$. Then, we add the two leaves of $C_j$ which do not correspond to $l_j^i$ to $X'$. This procedure obstructs the stars with three or more leaves at the clause gadgets by totally $2q$ vertices. Then, $|X'| = p + 2q$. The connected components of $G[V \setminus (X' \cup V(S(G)))]$ are either isolated vertices, isolated edges, or $P_4$'s. Hence, $G[V \setminus X'] \in \Pi$.

($\Leftarrow$) Let $X'$ with $X' \cap X = \emptyset$ be a size-$(p + 2q)$ vertex set that obstructs all forbidden induced subgraphs in $G$. Due to the satellite gadget $S(G)$, $X'$ must obstruct all $S_3$'s in $G[V \setminus V(S(G))]$. Since there exists a $S_3$ in each variable gadget, at least one vertex from each variable gadget must be in $X'$, which requires in total $p$ vertices. This means that we can construct an assignment for $F$: From a variable gadget $Y_i$, if the vertex labelled "+" is in $X'$, we assign TRUE to $y_i$; otherwise, we assign false to $y_i$. The other $2q$ vertices of $X'$ must be used for obstructing the stars with three or more leaves at clause gadgets, which implies that for each clause gadget $c_j$, exactly one leaf remains in $G[V \setminus X']$. The constructed assignment has to satisfy the formula $F$, since for each connection gadget, there exist some $S_3$'s, each consisting of this connection gadget and a vertex in $X$ and thus, from each connection gadget at least one vertex is in $X'$.

$\square$

Combining Lemmas 1 and 2, we arrive at the following theorem:

**Theorem 1.** *D-$\Pi$-VD is $\mathcal{NP}$-complete, unless $\mathcal{H}$ contains a star $S_{\leq 2}$ or a ($\leq 2$)-all star.*

## 3 Stars with Two Leaves

We examine now the cases that $\mathcal{H}$ contains $S_{\leq 2}$ or ($\leq 2$)-all star. Here, we achieve $\mathcal{NP}$-completeness as well as polynomial-time solvability results. First we show that the case with $S_{\leq 2} \in \mathcal{H}$ is solvable in polynomial time.

### 3.1 Forbidden Stars with Two Leaves

Fellows et al. proved that in the case of connected forbidden subgraphs, if $\mathcal{H}$ contains $S_{\leq 2}$, then D-$\Pi$-VD is polynomial-time solvable. We extend this result to the disconnected case.

**Theorem 2.** *D-$\Pi$-VD can be solved in polynomial time when $\mathcal{H}$ contains a star $S_{\leq 2}$.*

This theorem applies to the graph properties $\Pi$ whose sets of minimal forbidden induced subgraphs $\mathcal{H}$ contain $K_1$ (a single vertex), $P_2$ (a single edge) or $P_3$ (a path on three vertices). D-$\Pi$-VD with $K_1$ being forbidden is not non-trivial, as $\Pi$ is an empty set, and hence solvable in polynomial time. If $\mathcal{H}$ contains $P_2$, then other minimal forbidden subgraphs in $\mathcal{H}$ can only be sets of independent vertices; otherwise, these forbidden subgraphs are not minimal. If $\mathcal{H}$ consists of only $P_2$, then the compression routine in the iterative compression algorithm for VERTEX COVER given in [18] directly gives a polynomial-time algorithm for the corresponding D-$\Pi$-VD problem. If $\mathcal{H}$ contains in addition to $P_2$, an independent set of size $s$, then $\Pi$ is not non-trivial. Now, we give a polynomial-time algorithm for D-$\Pi$-VD when $\mathcal{H}$ contains $P_3$.

Note that a graph is called a *cluster graph* if it contains no induced $P_3$. A cluster graph is a disjoint set of cliques. Since $\mathcal{H}$ is a set of minimal forbidden induced subgraphs and $P_3 \in \mathcal{H}$, all other forbidden subgraphs in $\mathcal{H}$ must also be cluster graphs. Let $H_1, H_2, \ldots, H_l$ be the minimal cluster graphs present in $\mathcal{H}$. Now, for each forbidden cluster graph $H_i$, let $H_i^1, H_i^2, \ldots, H_i^c$ be its connected components arranged in non-ascending order of their sizes. Let $(G = (V, E), X)$ be the input instance of D-$\Pi$-VD with $|V| = n$. Since $X$ is a solution set, $G[X]$ is a collection of cliques and it induces no forbidden cluster graphs from $\mathcal{H}$.

We describe an algorithm that finds a minimum-size vertex set $X'$ such that $X \cap X' = \emptyset$ and $G[V \setminus X'] \in \Pi$, or returns "no-instance". The algorithm is similar to the compression routine of the iterative compression algorithm for CLUSTER VERTEX DELETION [15], but additionally takes into account the forbidden cluster graphs in $\mathcal{H}$. In first step, the instance is simplified by two simple data reduction rules, whose correctness is easy to see [15]:

1. Delete all vertices in $R := V \setminus X$ that are adjacent to more than one clique in $G[X]$.
2. Delete all vertices in $R$ that are adjacent to some, but not all vertices of a clique in $G[X]$.

After these data reduction rules have been exhaustively applied, the instance has the following property. In each clique of $G[R]$, we can divide the vertices into equivalence classes according to their neighborhoods in $X$, where each class then contains vertices either adjacent to all vertices of a particular clique in $G[X]$, or adjacent to no vertex in $X$. This classification is useful because of the following:

**Lemma 3.** *[14] If there exists a solution for D-$\Pi$-VD, then in the cluster graph resulting by this solution, for each clique in $G[R]$ the vertices of at most one equivalence class are present.*

Due to Lemma 3, the remaining task for solving D-$\Pi$-VD is to assign each clique in $G[R]$ to one of its equivalence classes in such a way that the forbidden cluster graphs in $\mathcal{H}$ are also obstructed. Hence, in our algorithm we will enumerate all the possiblities which will not induce any forbidden cluster graph and choose the one with the minimum number of vertex deletions. However, we cannot do this independently for each clique in $G[R]$. The reason is that we cannot choose two classes from different cliques in $G[R]$ that are adjacent to the same clique in $G[X]$, since this would create an induced $P_3$. This assignment problem can be modelled as a *weighted bipartite matching* problem in an auxiliary graph $J = (V', E')$ where each edge corresponds to a possible choice of a clique. Let $|E'| = m$. Moreover, for each edge $e \in E'$, we set two weights $w_1(e)$ and $w_2(e)$ which will be instrumental while enumerating the possibilities. We delete a set $X_1'$ of vertices while eumerating the possibilities and create another set $X_2'$ by the weighted bipartite matching procedure which will be explained later. Among all the resulting graphs which satisfy $\Pi$, we choose the one with the minimum cardinality of $X' = X_1' \cup X_2'$. The graph $J$ is constructed as follows. See Fig 2 for an illustration.

1. For every clique in $G[R]$ which has at least one neighbour in $G[X]$, add a vertex (white vertex) in $J$.
2. For every clique $C_X$ in $G[X]$ which has at least one neighbour in $G[R]$, add a vertex $v$ (black vertex in $X$) in $J$. Moreover, add a new degree-1 vertex $u$ (white vertex in $X$) and an edge $\{u, v\}$. Set the weights $w_1$ and $w_2$ of this edge $\{u, v\}$ to be the size of $C_X$. This edge corresponds to choosing $C_X$ and removing all vertices adjacent to $C_X$ from $G[R]$.

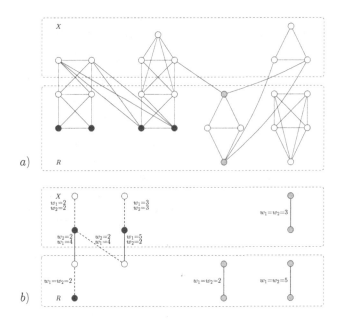

**Fig. 2.** a): Data reduction in the algorithm for D-$\Pi$-VD when $P_3$ is present in $\mathcal{H}$. The grey vertices are deleted by the data reduction rules. The black vertices correspond to the minimal solution $X_2'$ determined by the solution of weighted bipartite matching problem. b) The graph represents the corresponding weighted bipartite matching instance with the edge weights represented by integers next to the edges. The dashed edges represent the cliques that are added into $X_2'$ while the black edges represent the remaining cluster graph.

3. For a clique $C_X$ in $G[X]$ and a clique $C_R$ in $G[R]$, add an edge $e$ between the vertex for $C_X$ and the vertex for $C_R$ if there is an equivalence class in $C_R$ adjacent to $C_X$. This edge corresponds to choosing this class for $C_R$ and is assigned two different weights, $w_1(e)$ is assigned the total number of vertices in the corresponding class of $C_R$ and in $C_X$ and $w_2(e)$ is assigned the total number of vertices in $C_R$.

4. Add a vertex for the class in a clique $C_R$ that is not adjacent to any clique in $G[X]$ (black vertices outside $X$), and connect it to the vertex representing $C_R$. Again, this edge corresponds to choosing this class for $C_R$ and both its weights $w_1$ and $w_2$ are assigned the total number of vertices in this class.

5. For each clique $C_R$ in $G[R]$ which is non-adjacent to any vertex in $G[X]$, add an edge $e$ between two new grey vertices and its weights $w_1(e)$ and $w_2(e)$ are set to the total number of vertices in $C_R$.

6. For each clique $C_X$ in $G[X]$ which is non-adjacent to any vertex in $G[R]$, add an edge $e$ between two new grey vertices and its weights $w_1(e)$ and $w_2(e)$ are equal to the total number of vertices in $C_X$. We denote the set of these edges by $T$.

Since, we only added edges between black and white vertices and isolated edges between two grey vertices, $J$ is bipartite. The task is now to find a *maximum-weight bipartite matching* on $J$ with edge weights $w_2$, that is a set of edges of maximum weight where no two edges have an endpoint in common. However, before that we must take care that the collection of cliques resulting from the matchings does not induce any forbidden cluster graph in $\mathcal{H}$. To this end, we enumerate all matchings in graph $G[V \setminus X']$ which do not induce any forbidden cluster graphs from $\mathcal{H}$ with the help of the edge weights $w_1$.

We use *cluster configurations* and the corresponding *cluster configuration diagrams* for enumerating all such matchings. An example is illustrated in Fig 3 for the case that $\mathcal{H}$ consists of $P_3$ and a cluster graph $h_1$ with four cliques $h_1^1$, $h_1^2$, $h_1^3$ and $h_1^4$ of size $s_1 \geq s_2 \geq s_3 \geq s_4$, respectively. The cluster configuration diagram is a table in which columns represent the *size intervals* of the allowable cliques. For example, in Fig 3, the column between $h_1^1$ and $h_1^2$ represents the size interval of $[s_2, s_1)$. Each row depicts a feasible configuration. Here, "↑" represents a clique in the corresponding interval and "⇒"'s position represents the upper bound on the size of the remaining cliques.

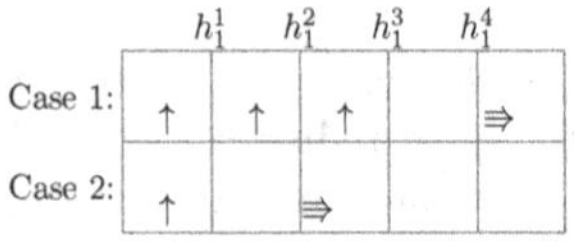

**Fig. 3.** Cluster configuration diagram when the corresponding forbidden subgraphs in $\mathcal{H}$ consists of a $P_3$ and a cluster graph $h_1$ with four cliques as connected components

**Lemma 4.** *The number of different feasible clique configurations of $G[V \setminus X']$ in the case that $\mathcal{H}$ consists of $P_3$ and cluster graphs, is polynomial in the size of $G$.*

*Proof.* Let the number of forbidden cluster graphs in $\mathcal{H}$ be $d$. Let $b$ be the number of disjoint cliques in the forbidden cluster graph in $\mathcal{H}$, which has the maximum number of disjoint cliques among all cluster graphs in $\mathcal{H}$. Then, in the *cluster configuration diagram* there will be at most $bd + 1$ intervals, since the number of distinct-sized cliques, which are connected components of cluster graphs in $\mathcal{H}$, is at most $bd$. Furthermore, we observe that for any fixed position of "⇒", say $s$, there can be at most $b - 1$ cliques represented by entries with "↑" to the left of $s$. Since each forbidden cluster graph in $\mathcal{H}$ has at most $b$ components, any cluster configuration $Y$ with more than $b - 1$ cliques to the left of $s$ will be handled by a different cluster configuration $Z$, such that $Z$ is formend from $Y$ by removing the rightmost "↑"and "⇒" is moved to the beginning of the interval containing the removed "↑". Hence, for each fixed position $s$ for "⇒", from all possible cliques with size at least $s$, we need to pick at most $b - 1$ entries to add the "↑"'s to the left of $s$. The number of the different ways of arranging $b - 1$ ↑'s in $bd + 1$ intervals is bounded by $(b - 1)^{(bd+1)}$. Now, for each such arrangement, we can pick each ↑ from the edges in the reduced graph $J$. We note here that every edge with weight $w_1$ that does not fit into a particular interval can be trimmed down by adding some vertices of the corresponding clique to $X_1'$. This ends up with $m^{(b-1)^{(bd+1)}}$ possibilities. Let $c = (b-1)^{(bd+1)}$. Now, since for each fixed $\Pi$, $\mathcal{H}$ is fixed, hence, $b$ and $d$ are constants, i.e., $c$ is a constant. Since there are at most $(bd+1)$ positions to fix "⇒", the total number of configurations is a polynomial function of $m$, i.e., $\mathcal{O}(m^c)$.    □

From all cluster configurations, we choose only those configurations which are feasible for a set of forbidden subgraphs $\mathcal{H}$, which can be done in polynomial time. Any feasible configuration places the cliques in $G[X]$ with sizes greater than the position of "⇒" at the positions set by "↑". Moreover, in any feasible configuration, if any edge from $J$ is selected to fill the position for "↑", none of its adjacent edges can be picked to fill the positions of other "↑" in the same configuration. Moreover, it also places all cliques represented by edges in $T$ in their corresponding intervals. Next, for each feasible configuration we do the following:

1. We remove the vertices at the end-points of the edges corresponding to the chosen cliques denoted by "↑".
2. In the remaining graph $J$, we maintain an upper-bound corresponding to the position of "⇒" on the weights $w_1$ of the remaining edges. Let the position of the "⇒" be $s$. For each edge $c$ corresponding to clique $C$ with $w_1(c) \geq s$, add $|V(C)| - s$ arbitrary vertices from $V(C) \cap R$ to $X_1'$ and reduce the weights $w_1(c)$ and $w_2(c)$ by $w_1(c) - s$.
3. Now, we run the algorithm for maximum weighted bipartite matching on the remaining graph with weights $w_2$. Since the weight of each edge in $J$ is bounded by $n$, the size of the given instance, we get a running time of $\mathcal{O}(m^4\sqrt{n}\log n)$ [8].

The set $X_2'$ can be directly constructed from a maximum matching returned in Step 3; it contains all vertices in the equivalence classes in $G[R]$ that correspond to the edges not chosen by the matching. Hence, our disjoint solution is $X' = X_1' \cup X_2'$. Clearly, both $X_1'$ and $X_2'$ can be computed in polynomial time. Combined with the polynomial number of configurations we have an overall polynomial-time algorithm.

### 3.2    All-Stars Containing Stars of at Most Two Leaves

For this case, we cannot give a complete dichotomy of the complexity of D-$\Pi$-VD. However, we can present some $\mathcal{NP}$-hard and polynomial cases. For example, we have the following result for 2-all star. See full version of the paper for other $\mathcal{NP}$-hard cases with ($\leq 2$)-all stars.

**Lemma 5.** *If there exists 2-all star in $\mathcal{H}_d$, each of which contains at least two occurrence of star $S_2$ as their connected components, and $\mathcal{H}$ is free from the following graphs, then D-$\Pi$-VD is $\mathcal{NP}$-hard:*

1. A star $S_{\leq 2}$.
2. A $(\leq 2)$-all star with only one occurrence of $S_2$ as its connected component.
3. A $(< 2)$-all star.
4. A subgraph which is an induced subgraph of a graph $G' = (V', E')$ with the following properties:
    i. There exists a set $M \subset V'$ which induces a clique.
    ii. For each vertex $v \in M$, there exists an edge $\{x, y\}$, such that either $\{x, v\} \in E'$ or $\{y, v\} \in E'$ and $N(x) \cup N(y) = \{x, y, v\}$. Let $N \subset V'$ be the set which contains all such vertices $\{x, y\}$ for each $v \in V'$.
    iii. $J := V' \setminus (M \cup N)$ induces an independent set in $G'$ such that for every $v \in J$, $N(v) \subseteq M$.
5. A disconnected forbidden graph which consists of graphs in (1-4) as its connected components.

Finally, we present a polynomial-time solvable case of D-$\Pi$-VD with $\mathcal{H}$ containing $(\leq 1)$-all star; this case generalizes the result for DISJOINT SPLIT VERTEX DELETION, implicitly shown in [11]. We call a graph pseudo-split if the forbidden subgraph set $\mathcal{H}$ is equal to $\{2K_2, C_4\}$. Here $C_4$ is a cycle on four vertices.

**Lemma 6.** D-$\Pi$-VD when $\Pi$ being pseudo-split graphs can be solved in polynomial time.

## 4    Open Problems

First, the computational complexity of D-$\Pi$-VD for the case, when $(\leq 1)$-all stars are present in $\mathcal{H}$, is partially resolved, for instance, for $\Pi$ being threshold graphs. Here, the set $\mathcal{H}$ consists of $2K_2$, $C_4$ and $P_4$. It would also be interesting to study the computational complexity of D-$\Pi$-VD in directed graphs and the variant of D-$\Pi$-VD where the solution comprises of edges to be deleted instead of vertices.

## References

1. Bafna, V., Berman, P., Fujito, T.: A 2-approximation algorithm for the undirected feedback vertex set problem. SIAM J. Discrete Math. 12(3), 289–297 (1999)
2. Chen, J., Fomin, F.V., Liu, Y., Lu, S., Villanger, Y.: Improved algorithms for the feedback vertex set problems. In: Dehne, F., Sack, J.-R., Zeh, N. (eds.) WADS 2007. LNCS, vol. 4619, pp. 422–433. Springer, Heidelberg (2007)
3. Chen, J., Liu, Y., Lu, S., O'Sullivan, B., Razgon, I.: A fixed-parameter algorithm for the directed feedback vertex set problem. In: STOC, pp. 177–186 (2008)
4. Dehne, F.K.H.A., Fellows, M.R., Langston, M.A., Rosamond, F.A., Stevens, K.: An $o(2^{o(k)}n^3)$ fpt algorithm for the undirected feedback vertex set problem. Theory Comput. Syst. 41(3), 479–492 (2007)
5. Diestel, R.: Graph theory, 2nd edn. Springer, New York (2000)

6. Downey, R.G., Fellows, M.R.: Parameterized Complexity, p. 530. Springer, New York (1999)
7. Fellows, M.R., Guo, J., Moser, H., Niedermeier, R.: A complexity dichotomy for finding disjoint solutions of vertex deletion problems. TOCT 2(2), 5 (2011)
8. Gabow, H.N., Tarjan, R.E.: Faster scaling algorithms for network problems. SIAM J. Comput. 18, 1013–1036 (1989)
9. Garey, M.R., Johnson, D.S.: Computers and Intractability: A Guide to the Theory of NP-Completeness. Freeman, San Francisco (1979)
10. Gaspers, S., Mnich, M.: Feedback vertex sets in tournaments. Journal of Graph Theory 72(1), 72–89 (2013)
11. Ghosh, E., Kolay, S., Kumar, M., Misra, P., Panolan, F., Rai, A., Ramanujan, M.S.: Faster parameterized algorithms for deletion to split graphs. In: Fomin, F.V., Kaski, P. (eds.) SWAT 2012. LNCS, vol. 7357, pp. 107–118. Springer, Heidelberg (2012)
12. Guo, J.: Problem kernels for NP-complete edge deletion problems: Split and related graphs. In: Tokuyama, T. (ed.) ISAAC 2007. LNCS, vol. 4835, pp. 915–926. Springer, Heidelberg (2007)
13. Guo, J., Gramm, J., Hüffner, F., Niedermeier, R., Wernicke, S.: Compression-based fixed-parameter algorithms for feedback vertex set and edge bipartization. J. Comput. Syst. Sci. 72(8), 1386–1396 (2006)
14. Guo, J., Moser, H., Niedermeier, R.: Iterative compression for exactly solving NP-hard minimization problems. In: Lerner, J., Wagner, D., Zweig, K.A. (eds.) Algorithmics. LNCS, vol. 5515, pp. 65–80. Springer, Heidelberg (2009)
15. Hüffner, F., Komusiewicz, C., Moser, H., Niedermeier, R.: Fixed-parameter algorithms for cluster vertex deletion. In: Laber, E.S., Bornstein, C., Nogueira, L.T., Faria, L. (eds.) LATIN 2008. LNCS, vol. 4957, pp. 711–722. Springer, Heidelberg (2008)
16. Lewis, J.M., Yannakakis, M.: The node-deletion problem for hereditary properties is np-complete. J. Comput. Syst. Sci. 20(2), 219–230 (1980)
17. Natanzon, A., Shamir, R., Sharan, R.: Complexity classification of some edge modification problems. Discrete Applied Mathematics 113(1), 109–128 (2001)
18. Niedermeier, R.: Invitation to Fixed Parameter Algorithms (Oxford Lecture Series in Mathematics and Its Applications). Oxford University Press, USA (2006)
19. Reed, B.A., Smith, K., Vetta, A.: Finding odd cycle transversals. Oper. Res. Lett. 32(4), 299–301 (2004)

# Quasi-Upward Planar Drawings of Mixed Graphs with Few Bends: Heuristics and Exact Methods

Carla Binucci and Walter Didimo

Dipartimento di Ingegneria Elettronica e dell'Informazione
Università degli Studi di Perugia
{carla.binucci,walter.didimo}@unipg.it

**Abstract.** A *mixed graph* has both directed and undirected edges. We study how to compute a crossing-free drawing of a planar embedded mixed graph, such that it is upward "as much as possible". Roughly speaking, in an *upward drawing* of a mixed graph all edges are monotone in the vertical direction and directed edges flow monotonically from bottom to top according to their orientation. We study *quasi-upward drawings* of mixed graphs, that is, upward drawings where edges can break the vertical monotonicity in a finite number of edge points, called *bends*. We describe both efficient heuristics and exact methods for computing quasi-upward planar drawings of planar embedded mixed graphs with few bends, and we extensively compare them experimentally: the results show the effectiveness of our algorithms in many cases.

## 1 Introduction

An *upward drawing* of a directed graph (digraph) $G = (V, E)$ is a geometric representation of $G$ such that each vertex $v \in V$ is drawn as a distinct point $p_v$ of the plane and each edge $(u, v) \in E$ is drawn as a simple curve with monotonously increasing $y$-coordinates from $p_u$ to $p_v$. The problem of computing upward drawings of digraphs has a long tradition in Graph Drawing. In particular, lots of papers study crossing-free upward drawings of planar digraphs, called *upward planar drawings* (see Fig. 1(a)).

Deciding whether a planar digraph admits an upward planar drawing is NP-complete in the general case [28]. This problem is polynomial-time solvable for few subfamilies of planar digraphs [6,21,30,33] or when the planar embedding of the digraph is fixed [5]; a (di)graph with a fixed planar embedding is called an *embedded planar (di)graph*. Exponential-time upward planarity testing algorithms have been also described [4,12,15,21,29]; among them, the one in [15] works for general planar digraphs and it turns out to be efficient in practice for digraphs with few hundreds of vertices. Extensions of upward planarity to different surfaces are studied in [1,2]. See also [7,16,19] for additional references on upward planar drawings.

For planar digraphs that do not admit upward planar drawings, different approaches have been proposed to compute a drawing that is upward planar "as much as possible" according to some criteria. One approach is based on allowing edge crossings while guaranteeing that each edge is drawn upward; in this case the goal is minimizing the number of edge crossings [13,14,24]. Another approach is to remove from the digraph

S.P. Pal and K. Sadakane (Eds.): WALCOM 2014, LNCS 8344, pp. 298–309, 2014.

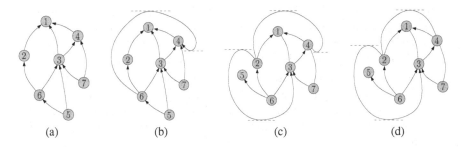

**Fig. 1.** (a) An upward planar drawing of a digraph. (b) A quasi-upward planar drawing of a digraph with 2 bends on edge $(6, 4)$; the bends are the tangent points between the dashed horizontal segments and the edge. (c) A strong quasi-upward planar drawing of a mixed graph with 4 bends; (d) A weak quasi-upward planar drawing of the same mixed graph with 3 bends.

a minimum number of edges so that the resulting digraph can be drawn upward planar [9]. A third approach is to allow edge bends while guaranteeing planarity and while guaranteeing that each edge $(u, v)$ leaves $u$ from above and enters $v$ from below; such a drawing is called a *quasi-upward planar drawing* and the goal is minimizing the number of bends [4]. Namely, a *bend* of a quasi-upward planar drawing corresponds to a point in which an edge inverts its vertical direction; if the edge is represented as a smoothed curve, each bend corresponds to a point with horizontal tangent for the edge (see Fig. 1(b)). Thus, a quasi-upward planar drawing with zero bends is an upward planar drawing.

The concept of upward drawing has been recently extended to *mixed graphs* [8], i.e., graphs with both directed and undirected edges. Mixed graphs arise in many application domains, and have received considerable attention in the scientific literature (see, e.g., [3,11,25,32]). An *upward drawing* of a mixed graph is such that each directed edge $(u, v)$ is drawn upward from $u$ to $v$ while an undirected edge $\{u, v\}$ is just drawn with monotonically increasing $y$-coordinates, moving either from $u$ to $v$ or from $v$ to $u$. Deciding whether a planar embedded mixed graph $G$ has an embedding preserving upward planar drawing is equivalent to finding an orientation of the undirected edges of $G$ such that the resulting embedded digraph has an upward planar drawing. Although the complexity of this problem is still unknown, a testing and drawing algorithm is given in [8]; it is based on integer linear programming (ILP) and it is fast in practice even for graphs with several hundreds of vertices. Also, in [26] it is proven that the problem is polynomially-time solvable for restricted classes of planar embedded mixed graphs.

In this paper we study *quasi-upward planar drawings* of mixed graphs, an extension of the paradigm proposed in [4] for digraphs, where edges can bend. As for digraphs, in a quasi-upward planar drawing of a mixed graph each directed edge $(u, v)$ leaves $u$ from above and enters $v$ from below. Instead, for the geometric representation of the undirected edges $e = \{u, v\}$ two alternative properties can be required: (1) $e$ enters one of its end-vertices from below and leaves the other end-vertex from above, or (2) $e$ is incident to $u$ (resp. to $v$) either from below or from above. Notice that property (1) is more restrictive than (2). A drawing that satisfies (1) is called a *strong* quasi-upward planar drawing; a drawing that satisfies (2) is called a *weak* quasi-upward planar drawing. See

Fig. 1(c) and Fig. 1(d) for an illustration. Observe that property (2) just requires that each undirected edge is not incident to an end-vertex horizontally. Strong quasi-upward planar drawings have been studied in [10], where it is proven that deciding whether a planar embedded mixed graph admits such a drawing is NP-complete.

**Contribution.** Here, we focus on weak quasi-upward planar drawings, which are much less restrictive than strong drawings and we address the new following optimization problem: *Given a planar embedded mixed graph G, compute an embedding preserving weak quasi-upward planar drawing of G with the minimum number of bends.* Our contribution is as follows:

($i$) We provide efficient heuristics for computing weak quasi-upward planar drawings of planar embedded mixed graphs with few edge bends. Examples of drawings computed with our algorithms are in Fig. 4.

($ii$) We describe an ILP model for solving the above optimization problem in an exact way. With few additional constraints, this model can be also used for finding a strong quasi-upward planar drawing with the minimum number of bends, if such a drawing exists.

($iii$) We discuss the results of an extensive experimental analysis that compares the performance of our different algorithms on a large set of randomly generated instances.

The paper is structured as follows. Section 2 recalls basic notions and results about upward and quasi-upward planar drawings of digraphs. Section 3 presents a characterization of the planar digraphs that admit weak quasi-upward planar drawings. Heuristics and exact methods for computing weak quasi-upward planar drawings with minimum number of bends are in Section 4 and Section 5, respectively. Section 6 describes the experimental analysis. Conclusions and open problems are in Section 7.

## 2   Preliminaries

We assume familiarity with basic concepts of graph drawing and planarity [19]. Let $G$ be an embedded planar digraph (i.e., a digraph with a given planar embedding). A *source vertex* (resp. a *sink vertex*) of $G$ is a vertex with only outgoing edges (resp. incoming edges). A source vertex or a sink vertex of $G$ is also called a *switch vertex* of $G$. A vertex $v$ of $G$ is *bimodal* if the circular list of its incident edges can be split into two linear lists, one consisting of all its incoming edges and the other consisting of all its outgoing edges. Digraph $G$ and its embedding are *bimodal* if every vertex of $G$ is bimodal.

The concepts of upward and quasi-upward planar drawings of a planar digraph have been already given in the introduction. The following theorem holds.

**Theorem 1.** [4] *An embedded planar digraph G admits an embedding preserving quasi-upward planar drawing if and only if G is bimodal.*

Bertolazzi et al. [4] also describe a flow network based polynomial-time algorithm that computes a quasi-upward planar drawing of a planar bimodal embedded digraph $G$ with minimum number of bends.

## 3   Quasi-Upward Planar Drawings of Mixed Graphs

Let $G = (V, E)$ be a mixed graph: $E_d$ and $E_u$ denote the subset of directed and undirected edges of $G$, respectively ($E = E_d \cup E_u$). If an edge $e \in E_d$ is oriented from a vertex $u$ to a vertex $v$, we write $e = (u, v)$; to denote an undirected edge $e \in E_u$ with end-vertices $u$ and $v$, we write $e = \{u, v\}$ (or equivalently $e = \{v, u\}$).

The concepts of strong and weak quasi-upward planar drawings of planar mixed graphs have been already given in the introduction. Clearly, each strong quasi-upward planar drawing is also a weak quasi-upward planar drawing, but not vice-versa. Each edge of a strong quasi-upward planar drawing has always an even number of bends (possibly zero bends). In a weak quasi-upward planar drawing this property is still true for directed edges; instead, the number of bends of an undirected edge $e = \{u, v\}$ is odd if $e$ is incident both to $u$ and to $v$ from above or both to $u$ and to $v$ from below, while it is even otherwise. The following result holds.

**Theorem 2.** *A planar embedded mixed drawing $G = (V, E_d \cup E_u)$ admits a weak quasi-upward planar drawing if and only if $G_d = (V, E_d)$ is bimodal.*

*Proof.* If $G = (V, E_d \cup E_u)$ has a weak quasi-upward planar drawing $\Gamma$, then the subdrawing of $G_d$ in $\Gamma$ is quasi-upward planar, and hence $G_d$ is bimodal by Theorem 1. Conversely, if $G_d$ is bimodal, then by Theorem 1, $G_d$ has an embedding preserving quasi-upward planar drawing $\Gamma_d$. A weak quasi-upward planar drawing $\Gamma$ of $G$ can be constructed from $\Gamma_d$ by adding the edges of $E_u$ one by one, while preserving the planar embedding of $G$. Namely, each time a new edge $e = \{u, v\}$ of $G$ is added to the current planar drawing, $u$ and $v$ belong to the same face (since we maintain the planar embedding); hence, it is possible to draw $e$ as a simple curve such that the planar embedding of $G$ is still preserved and $e$ is neither incident to $u$ nor to $v$ horizontally.   □

In the remainder of the paper we address the new following optimization problem:

**Problem** MINBENDWEAKQUASIUPWARD: *Given a planar embedded mixed graph $G = (V, E_d \cup E_u)$ such that $G_d$ is bimodal, compute a weak quasi-upward planar drawing of $G$ with the minimum number of bends.*

Problem MINBENDWEAKQUASIUPWARD is at least as difficult as deciding whether $G$ admits an upward planar drawing, i.e., a weak quasi-upward planar drawing with no bend. This decision problem is still of unknown complexity: a fast ILP approach is described in [8] and polynomial-time testing algorithms are known only for restrictive classes of plane graphs [26]. For the MINBENDWEAKQUASIUPWARD problem, we first describe efficient heuristics (Section 4) and then devise an ILP exact approach, which extends the model given in [8] (Section 5).

## 4   Heuristics

Let $G = (V, E_d \cup E_u)$ be a planar embedded mixed graph such that $G_d$ is bimodal. Theorem 2 implies that a weak quasi-upward planar drawing of $G$ exists. We give a first heuristic, called HEURQUASIUPWARDMIXED, that computes a weak quasi-upward planar drawing with few edge bends. It exploits Theorem 1 and works in two steps (see Fig. 2 for an illustration):

**STEP 1:** It computes a bimodal digraph $G'$ obtained from $G$ by possibly subdividing some edges of $E_u$ and suitably orienting all the undirected edges ($G'$ maintains the same embedding as $G$ for the common vertices). Namely, the edges of $E_u$ are considered one by one; each time an edge $e = \{u, v\} \in E_u$ is considered, the algorithm applies the following operations, in this order: $(i)$ it orients $e$ either from $u$ to $v$ or from $v$ to $u$ if one of these two orientations does not violate the bimodality of $u$ and $v$ for their incident directed edges; $(ii)$ else, it subdivides $e$ with a dummy vertex $w$, and then orients $\{u, w\}$ and $\{w, v\}$ such that they are both outgoing $w$ or both incoming $w$, depending on which of the two choices guarantees bimodality (again looking at the directed edges only).

**STEP 2:** It computes a quasi-upward planar drawing $\Gamma'$ of digraph $G'$, with the minimum number of bends within the planar embedding of $G'$. This is done by applying the algorithm given in [4]. The final drawing $\Gamma$ is computed from $\Gamma'$ by: $(i)$ removing the orientation of the edges incident to dummy vertices and of the edges of $E_u$ that were not subdivided in the previous step; $(ii)$ replacing each dummy vertex with a bend point.

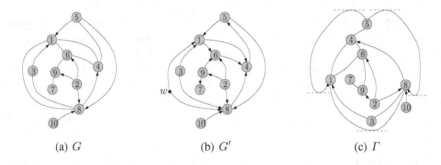

(a) $G$          (b) $G'$          (c) $\Gamma$

**Fig. 2.** Illustration of heuristic HEURQUASIUPWARDMIXED: (a) input: a planar embedded mixed graph $G$ such that $G_d$ is bimodal. (b) output of STEP 1: a bimodal directed graph $G'$ with a dummy vertex $w$ that subdivides edge $\{1, 8\}$. (c) output of STEP 2: an embedding preserving weak quasi-upward planar drawing $\Gamma$ of $G$ with five bends.

**Theorem 3.** *Let $G = (V, E_d \cup E_u)$ be a planar embedded mixed graph such that $G_d$ is bimodal, and let $F$ be the number of faces of $G$. Algorithm* HEURQUASIUPWARD-MIXED *computes an embedding preserving weak quasi-upward planar drawing of $G$ with at most $2(|V| + |E_u|)|F| + |E_u| = O(|V|^2)$ bends in $O(|V|^2)$ time.*

*Proof.* Denoting by $V'$ the vertex set of the bimodal digraph $G'$ computed in STEP 1, we have that $|V'| \leq |V| + |E_u|$. Also, the number of faces of $G'$ equals $|F|$. The algorithm applied in STEP 2, which computes a quasi-upward planar drawing $\Gamma'$ of $G'$ with minimum number of bends, generates at most $2|V'||F|$ bends in the drawing [4]. The number of bends of $\Gamma$ equals that of $\Gamma'$ plus $|E_u|$ (each dummy vertex subdividing an edge of $E_u$ is a bend point in the final drawing). Hence, the number of bends of $\Gamma$ is at most $2(|V| + |E_u|)|F| + |E_u| = O(|V|^2)$, because $O(|E_u|) = O(|F|) = O(|V|)$.

About the time complexity, STEP 1 takes $O(\sum_{\{u,v\}\in E_u}(deg(u) + deg(v))) = O(|V|^2)$, because for each edge $\{u,v\} \in E_u$ we need to scan the edges incident to $u$ and to $v$ to check bimodality for the four possible orientations. The drawing algorithm applied in STEP 2 is based on a minimum-cost flow network formulation, which can be solved in time $O(|V'|^{3/2})$ with the algorithm in [17], and then on a compaction technique that is linear in the number of vertices and bends. Since $O(|V'|) = O(|V|)$ and the number of bends is $O(|V|^2)$, the whole algorithm takes $O(|V|^2)$ time.     □

We observe that HEURQUASIUPWARDMIXED may create several directed cycles in $G'$, and each of these cycles gives rise to at least two bends in the final drawing (see, e.g., cycles $\{5, 1, 4\}$, $\{5, 4, 8\}$ in Fig. 2(b) and the final drawing in Fig. 2(c)). See also [4]. In order to mitigate this problem, we also propose a variant of HEURQUA-SIUPWARDMIXED, which we call HEURQUASIUPWARDMIXEDREFINED. It tries to orient the edges of $E_u$ such that they flow in a common direction as much as possible. To this aim, it executes the following additional operations:

**OPERATION 1**: It is a pre-processing step that assigns to each vertex $v$ of $G$ a distinct integer number $l(v) \in \{1, \ldots, |V|\}$, while trying to minimize the number of edges $(u, v) \in E_d$ such that $l(u) > l(v)$; these edges are called *downward edges*. This minimization problem is equivalent to finding a minimum feedback arc set in $G_d$. This problem is NP-complete [27] for general digraphs, but it is known to be polynomial-time solvable for planar digraph [31]. However, the result in [31] leads to a polynomial-time algorithm with high time complexity and difficult to implement (see, e.g., [34]). To compute a minimal solution, we use an effective linear-time greedy heuristic that guarantees at most $|E_d|/2 - |V|/6$ downward edges [23].

**OPERATION 2**: It modifies STEP 1 in two ways: ($i$) Each edge $e = \{u, v\} \in E_u$ gets a priority $i$ that is equal to the number of the orientations of $e$ (with possible subdivision) that do not violate bimodality (this number varies from 1 to 4): at each iteration, the edge with minimum priority is considered, and once it has been processed the priorities of its adjacent edges are updated; hence, the algorithm considers first the edges with the smallest number of admissible orientations. ($ii$) When an edge $e = \{u, v\} \in E_u$ is considered, assuming $l(u) < l(v)$, the algorithm orients $e$ from $u$ to $v$ if this choice preserves bimodality, otherwise it checks the other possibilities.

Once all the edges have been oriented, HEURQUASIUPWARDMIXEDREFINED applies STEP 2 as for HEURQUASIUPWARDMIXED (see also Fig. 3(a) and 3(b)).

The time complexity of HEURQUASIUPWARDMIXEDREFINED is still $O(|V|^2)$; indeed, if we handle the edge priorities with a binary heap (which allows to insert an element in logarithmic time and to extract the minimum key in constant time), the cost of the algorithm is still dominated by checking bimodality when an edge is considered.

We remark that, as it will be shown in Section 6, we experimentally observed that the actual number of bends generated by each of our heuristics is always much smaller than the theoretical worst case bound. Also, we will see that HEURQUASIUPWARD-MIXEDREFINED usually computes drawings with significantly less bends than those computed by HEURQUASIUPWARDMIXED, while the running time of the two heuristics is very similar.

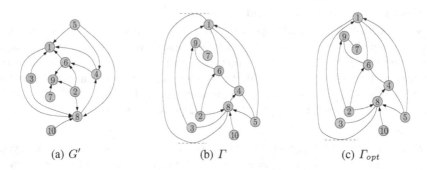

(a) $G'$          (b) $\Gamma$          (c) $\Gamma_{opt}$

**Fig. 3.** (a) Output of STEP 1 of HEURQUASIUPWARDMIXEDREFINED for the graph $G$ in Fig 2(a). (b) Output of STEP 2 of HEURQUASIUPWARDMIXEDREFINED: a weak-quasi upward planar drawing $\Gamma$ of $G$ with two bends. (c) An embedding preserving weak quasi-upward planar drawing $\Gamma_{opt}$ of $G$ with minimum number of bends; it has only one bend.

## 5 Integer Linear Programming Model

In order to deal with weak quasi-upward planar drawings of embedded mixed graphs, we extend the ILP model in [8] by adding suitable variables and constraints. For space reasons, we only sketch the idea behind our extension. Let $\Gamma$ be a weak quasi-upward planar drawing of a planar embedded mixed graph $G$. Denote by $\Gamma'$ the drawing obtained from $\Gamma$ by replacing each bend point with a dummy vertex, and let $G'$ be the planar embedded mixed graph represented by $\Gamma'$. Clearly, $\Gamma'$ is an upward planar drawing of $G'$. Using this observation, and denoting by $k$ an upper bound on the number of bends of a bend-minimum weak quasi-upward planar drawing of $G$, our idea is the following:

$(i)$ Each edge of $G$ is subdivided with a sequence $\{d_1, d_2, \ldots, d_k\}$ of $k$ dummy vertices. Namely, an undirected edge $\{u, v\}$ is split into a path $\{u, d_1\}, \{d_1, d_2\}, \ldots, \{d_k, v\}$, while a directed edge $(u, v)$ is split into a path $(u, d_1), \{d_1, d_2\}, \ldots, \{d_{k-1}, d_k\}, (d_k, v)$. Call $G'$ the resulting planar embedded mixed graph.

$(ii)$ We enhance the ILP formulation in [8] to find an upward planar drawing $\Gamma'$ of $G'$ such that the number of bends in the weak quasi-upward planar drawing $\Gamma$ of $G$ is minimized. To this aim, we associate each dummy vertex $d_i$ with a binary variable: if $d_i$ is such that its two incident edges will be both leaving $d_i$ from above or both entering $d_i$ from below in $\Gamma'$, then $d_i$ will correspond to a bend in $\Gamma$ and the associated variable will be set to 1; else, it will be set to 0. Also, since each directed edge of $G$ must have an even number of bends in $\Gamma$ (possibly zero), we add to the ILP model new constraints that guarantee this property. The objective function is defined as the minimum number of dummy vertices that will correspond to bend points in the final drawing.

Concerning $k$, we set its value as the number of bends of a drawing computed by HEURQUASIUPWARDMIXEDREFINED. It is clearly an upper bound to the number of bends of an optimal solution, which guarantees that the number of dummy vertices allocated for each edge $e$ suffices to host all bends along $e$ in an optimal drawing.

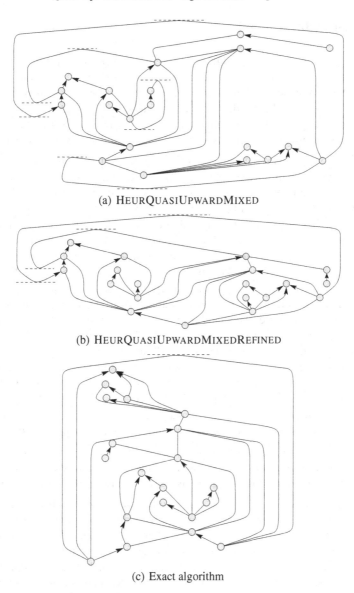

(a) HEURQUASIUPWARDMIXED

(b) HEURQUASIUPWARDMIXEDREFINED

(c) Exact algorithm

**Fig. 4.** Three drawings of the same embedded mixed graph computed with our algorithms (dashed lines are the tangents to the bend points): (a) a drawing with 8 bends; (b) a drawing with 4 bends; (c) a drawing with 1 bend. The graph has 20 vertices and 32 edges, out of which 16 undirected.

We finally remark that our model can be also adapted to compute a strong quasi-upward planar drawing with the minimum number of bends, if such a drawing exists. This is done by constraining the undirected edges to have an even number of bends, as for the directed edges, and suitably redefining the bound $k$.

## 6    Experimental Analysis

We implemented algorithms HEURQUASIUPWARDMIXED and HEURQUASIUPWARD-
MIXEDREFINED in C++, using the GDToolkit library [18]. For solving our ILP model
we used CPLEX with its default setting (http://www.cplex.com/). We ran the algorithms
on a large test suite of 7, 420 random planar embedded mixed graphs. All computations
ran on a common laptop, with an Intel Core i7 - 2.2GHz processor and 6 GB RAM.
Examples of drawings computed with our implemented algorithms are shown in Fig. 4.

**Test suite.** We generated our instances with the same approach used in previous experiments concerned with upward planarity of mixed graphs [8]. Namely, denoting by
$n$ and $m$ the desired number of vertices and edges, and denoting by $p$ the desired percentage of undirected edges, an upward planar embedded digraph with $n$ vertices and
$m$ edges is first generated with the algorithm described in [20]. Then a percentage $p$
of edges is randomly removed from this digraph (with uniform probability distribution)
and then the same number of undirected edges is reinserted, each time selecting at random its two end-vertices; if there already exists in the graph an edge connecting the two
selected vertices, we discard the choice and repeat the selection; this avoids multiple
edges while keeping uniform the random probability distribution, but of course it might
cause a long generation time, especially for high values of $p$ and small values of density
$m/n$. This algorithm gives rise to planar embedded mixed graphs $G = (V, E_d \cup E_u)$
such that $|V| = n$, $|E_d \cup E_u| = m$, $|E_u| = \frac{p \times m}{100}$, and $G_d = (V, E_d)$ is bimodal.

For each pair $\langle n, p \rangle$, where $n \in \{10, 20, 30, \ldots, 100, 200, 300\}$ and $p \in \{20, 50, 80\}$,
we tried to generate 210 different graphs with $n$ vertices, $p\%$ of undirected edges, and
densities ranging from 1.2 to 2.0. All graphs were successfully generated, except those
with 200 or 300 vertices, $p = 80$, and density 1.2; for these instances, the generation
algorithm failed to produce a graph within four hours, thus we stopped it after this time.

**Goals.** We ran the experiments with two main goals in mind: ($i$) Comparing the number
of bends of the drawings computed by the two heuristics with the optimum solutions;
($ii$) Evaluating the running time required by our different algorithmic approaches.

**Results.** Fig. 5 reports the charts that compare the number of bends generated by the
three algorithms. For each pair $\langle n, p \rangle$, the reported values are averaged over all graphs
with $n$ vertices and $p\%$ of undirected edges. Charts (a),(c), and (e) show the ratio between the number of bends generated by the two heuristics and the optimum solutions,
for the different values of $p$, and for those instances on which the optimum is not zero.
It can be seen that HEURQUASIUPWARDMIXEDREFINED outperforms HEURQUASI-
UPWARDMIXED: It generates about 40% of bends less for $p \leq 50$, and about 33%
of bends less for $p = 80$. Also, for $p = 20$ and for $p = 50$ HEURQUASIUPWARD-
MIXEDREFINED exhibits a relatively good approximation factor of the optimum value.
Charts (b), (d), and (f) report the number of solutions with zero bends computed by the
three algorithms, where Opt denotes the ILP based algorithm. HEURQUASIUPWARD-
MIXEDREFINED still behaves much better than HEURQUASIUPWARDMIXED; also,
it achieves the optimum value (i.e., zero bends) on most instances when $p = 20$; for
$p \in \{50, 80\}$, the number of solutions with zero bends is strongly reduced.

Concerning the CPU time, the two heuristics behaved similarly (the two curves overlap) and ran quite fast. They took in the average 1 second on the instances with 300

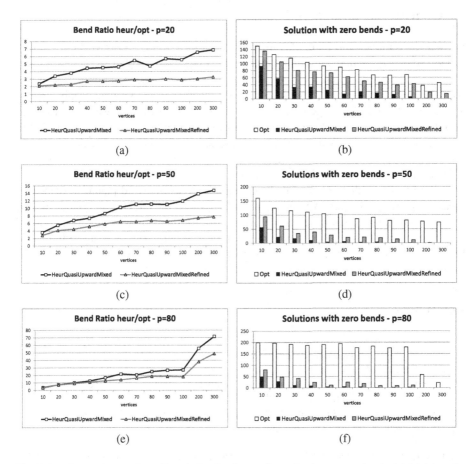

**Fig. 5.** Results on the number of bends for the three values of $p$ (average values). (a),(c),(e): ratio between the number of bends generated by the heuristics and the optimum value, for the instances where the optimum is greater than zero. (b),(d),(f): number of solutions with zero bends computed by the three algorithms, where Opt denotes the ILP based algorithm.

vertices, and their time performance is not affected by the number of undirected edges. Conversely, the ILP based algorithm has a time that increases significantly for growing values of $p$, because for these instances there are many more possible orientations that must be explored by the algorithm in order to find the optimum. In particular, in addition to the charts, we report that the exact algorithm failed to compute the optimum on many instances with 80% of undirected edges, namely 47 instances with 200 vertices and 89 instances with 300 vertices. Finally, in order to better understand how the two heuristics scale with the size of the graph, we ran them on three instances with 3,000 vertices, more than 5,000 edges, and $p = 50$. They took in the average 39 seconds to compute a drawing, and again the number of bends generated by HEURQUASIUP-WARDMIXEDREFINED (370 in the average) was much smaller than that generated by HEURQUASIUPWARDMIXED (558 in the average).

## 7   Conclusions and Open Problems

We addressed the new problem of computing quasi-upward planar drawings of embedded mixed graphs, with the minimum number of bends. We provided fast heuristics and an exact technique based on ILP. The experiments showed that one of our heuristics has good performance in terms of number of bends when applied to mixed graphs having up to 50% of undirected edges. Also, the presented heuristics scale well with the size of the graph (they take less than 40 seconds on graphs with 3,000 vertices and 5,000 edges); the ILP based technique successfully found a solution on all instances with up to 100 vertices, but was unable to find a solution in reasonable time and space for some graphs with 200 and 300 vertices; this further motivates the use of heuristics.

Several research directions related to computing quasi-upward planar drawings of mixed graphs are still open. Among them: ($i$) Devising more effective heuristics, which provide good approximations of the optimum even for mixed graphs that have most of the edges undirected, and investigate the existence of approximation algorithms. ($ii$) Studying the problem in the variable embedding setting. ($iii$) Studying the problem when edge crossings are allowed.

## References

1. Auer, C., Bachmaier, C., Brandenburg, F.J., Gleißner, A.: Classification of planar upward embedding. In: Speckmann, B. (ed.) GD 2011. LNCS, vol. 7034, pp. 415–426. Springer, Heidelberg (2011)
2. Auer, C., Bachmaier, C., Brandenburg, F.J., Gleißner, A., Hanauer, K.: The duals of upward planar graphs on cylinders. In: Golumbic, M.C., Stern, M., Levy, A., Morgenstern, G. (eds.) WG 2012. LNCS, vol. 7551, pp. 103–113. Springer, Heidelberg (2012)
3. Bang-Jensen, J., Gutin, G.: Digraphs: Theory, Algorithms and Applications, 2nd edn. Springer (2009)
4. Bertolazzi, P., Di Battista, G., Didimo, W.: Quasi-upward planarity. Algorithmica 32(3), 474–506 (2002)
5. Bertolazzi, P., Di Battista, G., Liotta, G., Mannino, C.: Upward drawings of triconnected digraphs. Algorithmica 6(12), 476–497 (1994)
6. Bertolazzi, P., Di Battista, G., Mannino, C., Tamassia, R.: Optimal upward planarity testing of single-source digraphs. SIAM J. on Computing 27, 132–169 (1998)
7. Binucci, C., Di Giacomo, E., Didimo, W., Rextin, A.: Switch-regular upward planar embeddings of directed trees. J. of Graph Algorithms and Applications 15(5), 587–629 (2011)
8. Binucci, C., Didimo, W.: Upward planarity testing of embedded mixed graphs. In: Speckmann, B. (ed.) GD 2011. LNCS, vol. 7034, pp. 427–432. Springer, Heidelberg (2011)
9. Binucci, C., Didimo, W., Giordano, F.: Maximum upward planar subgraphs of embedded planar digraphs. Computational Geometry: Theory and Applications 41(3), 230–246 (2008)
10. Binucci, C., Didimo, W., Patrignani, M.: Upward and quasi-upward planarity testing of embedded mixed graphs. Technical report, RT-001-12, DIEI - University of Perugia (2012)
11. Boesch, F., Tindell, R.: Robbins's theorem for mixed multigraphs. American Mathematical Monthly 87(9), 716–719 (1980)
12. Chan, H.: A parameterized algorithm for upward planarity testing. In: Albers, S., Radzik, T. (eds.) ESA 2004. LNCS, vol. 3221, pp. 157–168. Springer, Heidelberg (2004)
13. Chimani, M., Gutwenger, C., Mutzel, P., Wong, H.-M.: Layer-free upward crossing minimization. ACM J. of Experimental Algorithmics 15 (2010)

14. Chimani, M., Gutwenger, C., Mutzel, P., Wong, H.-M.: Upward planarization layout. J. of Graph Algorithms and Applications 15(1), 127–155 (2011)
15. Chimani, M., Zeranski, R.: Upward planarity testing via SAT. In: Didimo, W., Patrignani, M. (eds.) GD 2012. LNCS, vol. 7704, pp. 248–259. Springer, Heidelberg (2013)
16. Chimani, M., Zeranski, R.: Upward planarity testing: A computational study. In: Wismath, S., Wolff, A. (eds.) GD 2013. LNCS, vol. 8242, pp. 13–24. Springer, Heidelberg (2013)
17. Cornelsen, S., Karrenbauer, A.: Accelerated bend minimization. J. of Graph Algorithms and Applications 16(3), 635–650 (2012)
18. Di Battista, G., Didimo, W.: Gdtoolkit. In: Tamassia, R. (ed.) Handbook of Graph Drawing and Visualization. Chapman and Hall/CRC (2013)
19. Di Battista, G., Eades, P., Tamassia, R., Tollis, I.G.: Graph Drawing. Prentice Hall, Upper Saddle River (1999)
20. Didimo, W.: Upward planar drawings and switch-regularity heuristics. J. of Graph Algorithms and Applications 10(2), 259–285 (2006)
21. Didimo, W., Giordano, F., Liotta, G.: Upward spirality and upward planarity testing. SIAM J. on Discrete Mathematics 23(4), 1842–1899 (2009)
22. Didimo, W., Pizzonia, M.: Upward embeddings and orientations of undirected planar graphs. J. of Graph Algorithms and Applications 7(2), 221–241 (2003)
23. Eades, P., Lin, X., Smyth, W.F.: A fast effective heuristic for the feedback arc set problem. Information Processing Letters 47, 319–323 (1993)
24. Eiglsperger, M., Eppinger, F., Kaufmann, M.: An approach for mixed upward planarization. J. of Graph Algorithms and Applications 7(2), 203–220 (2003)
25. Farzad, B., Mahdian, M., Mahmoudian, E., Saberi, A., Sadri, B.: Forced orientation of graphs. Bulletin of Iranian Mathematical Society 32(1), 78–89 (2006)
26. Frati, F., Kaufmann, M., Pach, J., Tóth, C.D., Wood, D.R.: On the upward planarity of mixed plane graphs. In: Wismath, S., Wolff, A. (eds.) GD 2013. LNCS, vol. 8242, pp. 1–12. Springer, Heidelberg (2013)
27. Garey, M.R., Johnson, D.S.: Computers and Intractability: A Guide to the Theory of NP-Completeness. W. H. Freeman and Co. (1979)
28. Garg, A., Tamassia, R.: On the computational complexity of upward and rectilinear planarity testing. SIAM J. on Computing 31(2), 601–625 (2001)
29. Healy, P., Lynch, K.: Fixed-parameter tractable algorithms for testing upward planarity. International Journal of Foundations of Computer Science 17(5), 1095–1114 (2006)
30. Hutton, M.D., Lubiw, A.: Upward planarity testing of single-source acyclic digraphs. SIAM J. on Computing 25(2), 291–311 (1996)
31. Lucchesi, C.L., Younger, D.H.: A minimax theorem for directed graphs. J. London Math. Soc. 17, 369–374 (1978)
32. Mchedlidze, T., Symvonis, A.: Unilateral orientation of mixed graphs. In: van Leeuwen, J., Muscholl, A., Peleg, D., Pokorný, J., Rumpe, B. (eds.) SOFSEM 2010. LNCS, vol. 5901, pp. 588–599. Springer, Heidelberg (2010)
33. Papakostas, A.: Upward planarity testing of outerplanar dags. In: Tamassia, R., Tollis, I.G. (eds.) GD 1994. LNCS, vol. 894, pp. 298–306. Springer, Heidelberg (1995)
34. Stamm, H.: On feedback problems in planar digraphs. In: Möhring, R.H. (ed.) WG 1990. LNCS, vol. 484, pp. 79–89. Springer, Heidelberg (1991)

# On Minimum Average Stretch Spanning Trees in Polygonal 2-Trees*

N.S. Narayanaswamy and G.Ramakrishna

Department of Computer Science and Engineering
Indian Institute of Technology Madras, India
{swamy,grama}@cse.iitm.ac.in

**Abstract.** A spanning tree of an unweighted graph is a *minimum average stretch spanning tree* if it minimizes the ratio of sum of the distances in the tree between the end vertices of the graph edges and the number of graph edges. We consider the problem of computing a minimum average stretch spanning tree in polygonal 2-trees, a super class of 2-connected outerplanar graphs. For a polygonal 2-tree on $n$ vertices, we present an algorithm to compute a minimum average stretch spanning tree in $O(n \log n)$ time. This also finds a minimum fundamental cycle basis in polygonal 2-trees.

## 1 Introduction

Average stretch is a parameter used to measure the quality of a spanning tree in terms of distance preservation, and finding a spanning tree with minimum average stretch is a classical problem in network design. Let $G = (V(G), E(G))$ be an unweighted graph and $T$ be a spanning tree of $G$. For an edge $(u, v) \in E(G)$, $d_T(u, v)$ denotes the distance between $u$ and $v$ in $T$. The average stretch of $T$ is defined as

$$\text{AvgStr}(T) = \frac{1}{|E(G)|} \sum_{(u,v) \in E(G)} d_T(u, v) \qquad (1)$$

A *minimum average stretch spanning tree* of $G$ is a spanning tree that minimizes the average stretch. Given an unweighted graph $G$, the minimum average stretch spanning tree (MAST) problem is to find a minimum average stretch spanning tree of $G$. Due to the unified notation for tree spanners, the MAST problem is equivalent to the problem, MFCB, of finding a minimum fundamental cycle basis in unweighted graphs [16]. Minimum average stretch spanning trees are used to solve symmetric diagonally dominant linear systems [16]. Further, minimum fundamental cycle bases have various applications including determining the isomorphism of graphs, frequency analysis of computer programs, and generation of minimal perfect hash functions (See [4,11] and the references there in]). Due to

---

* Supported by the Indo-Max Planck Centre for Computer Science Programme in the area of *Algebraic and Parameterized Complexity* for the year 2012 - 2013.

S.P. Pal and K. Sadakane (Eds.): WALCOM 2014, LNCS 8344, pp. 310–321, 2014.

these vast applications, finding a minimum average stretch spanning tree is useful in theory and practice. The MAST problem was studied in a graph theoretic game in the context of the $k$-server problem by Alon et al. [1]. The MFCB problem was introduced by Hubika and Syslo in 1975 [12]. The MFCB problem was proved to be NP-complete by Deo et al. [4] and APX-hard by Galbiati et al. [11]. Another closely related problem is the problem of probabilistically embedding a graph into its spanning trees. A graph $G$ is said to be *probabilistically embedded* into its spanning trees with *distortion $t$*, if there is a probability distribution $D$ of spanning trees of $G$, such that for any two vertices the expected stretch of the spanning trees in $D$ is at most $t$. The problem of probabilistically embedding a graph into its spanning trees with low distortion has interesting connections with low average stretch spanning trees.

In the literature, spanning trees with low average stretch has received significant attention in special graph classes such as $k$-outerplanar graphs and series-parallel graphs. In case of planar graphs, Kavitha et al. remarked that the complexity of MFCB is unknown and there is no $O(\log n)$ approximation algorithm [13]. For $k$-outerplanar graphs, the technique of peeling-an-onion decomposition is employed to obtain a spanning tree whose average stretch is at most $c^k$, where $c$ is a constant [7]. In case of series-parallel graphs, a spanning tree with average stretch at most $O(\log n)$ can be obtained in polynomial time (See Section 5 in [8]). The bounds on the size of a minimum fundamental cycle basis is studied in graph classes such as planar, outerplanar and grid graphs [13]. The study of probabilistic embeddings of graphs is discussed in [7,8]. To the best of our knowledge, there is no published work to compute a minimum average stretch spanning tree and minimum fundamental cycle basis in any subclass of planar graphs.

We consider polygonal 2-trees in this work, which are also referred to as polygonal-trees. They have a rich structure that make them very natural models for biochemical compounds, and provide an appealing framework for solving associated enumeration problems.

**Definition 1 ([14]).** *A graph is a polygonal 2-tree if it can be obtained by edge-gluing a set of cycles successively.*

*Edge gluing* on two graphs $G_1$ and $G_2$ results in a graph $G$ such that $V(G) = V(G_1) \cup V(G_2)$, $E(G) = E(G_1) \cup E(G_2)$, $|V(G_1) \cap V(G_2)| = 2$ and $|E(G_1) \cap E(G_2)| = 1$. A graph is a $k$-*gonal tree*, if it can be obtained by edge-gluing a set of cycles of length $k$ successively [14]. For example, a *2-tree* is a 3-gonal tree. The class of polygonal 2-trees is a subclass of planar graphs and it includes 2-connected outerplanar graphs and $k$-gonal trees. 2-trees, in other words 3-gonal trees, are extensively studied in the literature. In particular, previous work on various flavours of counting and enumeration problems on 2-trees is compiled in [10]. Formulas for the number of labeled and unlabeled $k$-gonal trees with $r$ polygons (induced cycles) are computed in [15]. The family of $k$-gonal trees with same number of vertices is claimed as a chromatic equivalence class by Chao and Li, and the claim has been proved by Wakelin and Woodal [14]. The class of polygonal 2-trees is shown to be a chromatic equivalence class by Xu [14]. Further, various subclasses of generalized polygonal 2-trees have been considered, and it has been shown that they also form a chromatic equivalence class [14,17,18]. The enumeration of outerplanar $k$-gonal trees is studied by Harary, Palmer and Read to solve a variant of the cell growth problem [6]. Molecular expansion of the species of outerplanar $k$-gonal trees is shown in [6]. Also outerplanar $k$-gonal trees are of interest in combinatorial chemistry, as the structure of chemical compounds like catacondensed benzenoid hydrocarbons forms an outerplanar $k$-gonal tree.

## 1.1   Our Results

We state our main theorem.

**Theorem 2.** *Given a polygonal 2-tree $G$ on $n$ vertices, a minimum average stretch spanning tree of $G$ can be obtained in $O(n \log n)$ time.*

A quick overview of our approach to solve MAST is presented in **Algorithm 1** below. The detailed implementation is given in Section 4.

---

**Algorithm 1.** An algorithm to find an MAST of a polygonal 2-tree $G$

---

1  $A \leftarrow \emptyset$;
2  **for** *each edge $e \in E(G)$* **do**  $c[e] \leftarrow 0$;
3  **while** $G - A$ *has a cycle* **do**
4      Choose an edge $e$ from $G - A$, such that $e$ belongs to exactly one induced cycle in $G - A$ and $c[e]$ is minimum ;
5      Let $C$ be the induced cycle containing $e$ in $G - A$ ;
6      **for** *each $\hat{e} \in E(C) \setminus \{e\}$* **do**  $c[\hat{e}] \leftarrow c[\hat{e}] + c[e] + 1$;
7      $A \leftarrow A \cup \{e\}$ ;
8  Return $G - A$;

---

Due to the equivalence of MAST and MFCB (shown in Lemma 5), our result implies the following corollary. For a set $\mathcal{B}$ of cycles in $G$, the size of $\mathcal{B}$, denoted by size($\mathcal{B}$), is the number of edges in $\mathcal{B}$ counted according to their multiplicity.

**Corollary 3.** *Given a polygonal 2-tree $G$ on $n$ vertices, a minimum fundamental cycle basis $\mathcal{B}$ of $G$ can be obtained in $O(n \log n + \text{size}(\mathcal{B}))$ time.*

We characterize polygonal 2-trees using a kind of ear decomposition and present the structural properties of polygonal 2-trees that are useful in finding a minimum average stretch spanning tree (In Section 2). We then identify a set of edges in a polygonal 2-tree, called safe edges, whose removal results in a minimum average stretch spanning tree (In Section 3). Finally, we present an algorithm with necessary data-structures to identify the safe set of edges efficiently and compute a minimum average stretch spanning tree in sub-quadratic time (In Section 4).

A graph $G$ can be probabilistically embedded into its spanning trees with distortion $t$ if and only if the multigraph obtained from $G$ by replicating its edges has a spanning tree with average stretch at most $t$ (See [1]). It is easy to observe that, a spanning tree $T$ of $G$ is a minimum average stretch spanning tree for $G$ if and only if $T$ is a minimum average stretch spanning tree for a multigraph of $G$. As a consequence of our result, we have the following corollary.

**Corollary 4.** *For a polygonal 2-tree $G$, let $t$ be the average stretch of a minimum average stretch spanning tree of $G$. Then $G$ can be probabilistically embedded into its spanning trees with distortion $t$.*

## 1.2    Graph Preliminaries

We consider simple, connected, unweighted and undirected graphs. We use standard graph terminology from [20]. Let $G = (V(G), E(G))$ be a graph, where $V(G)$ and $E(G)$ denote the set of vertices and edges, respectively in $G$. We denote $|V(G)|$ by $n$ and $|E(G)|$ by $m$. The union of graphs $G_1$ and $G_2$ is defined as a graph with vertex set $V(G_1) \cup V(G_2)$ and edge set $E(G_1) \cup E(G_2)$ and is denoted by $G_1 \cup G_2$. The intersection of graphs $G_1$ and $G_2$ written as $G_1 \cap G_2$ is a graph with vertex set $V(G_1) \cap V(G_2)$ and edge set $E(G_1) \cap E(G_2)$. The removal of a set $X$ of edges from $G$ is denoted by $G - X$. For a set $X \subset V(G)$, $G[X]$ denotes the induced graph on $X$. An edge $e \in E(G)$ is a *cut-edge* (bridge) if $G - e$ is disconnected. A graph is *2-connected* if it can not be disconnected by removing less than two vertices. A *2-connected component* of $G$ is a maximal 2-connected subgraph of $G$.

Let $T$ be a spanning tree of $G$. An edge $e \in E(G) \setminus E(T)$ is a *non-tree* edge of $T$. For a non-tree edge $(u, v)$ of $T$, a cycle formed by the edge $(u, v)$ and the unique path between $u$ and $v$ in $T$ is referred to as a *fundamental cycle*. For an edge $(u, v) \in E(G)$, *stretch* of $(u, v)$ is the distance between $u$ and $v$ in $T$. The *total stretch* of $T$ is defined as the sum of the stretches of all the edges in $G$. We remark that there are slightly different definitions exit in the literature to refer the average stretch of a spanning tree. We use the definition in Equation 1, presented by Emek and Peleg in [8], to refer the average stretch of a spanning tree. By Proposition 14 in [16], $T$ is a minimum total stretch spanning tree of $G$ if and only if the set of fundamental cycles of $T$ is a minimum fundamental cycle basis of $G$. Then, we can have the following lemma.

**Lemma 5.** *Let $G$ be an unweighted graph and $T$ be a spanning tree of $G$. $T$ is a minimum average stretch spanning tree of $G$ if and only if the set of fundamental cycles of $T$ is a minimum fundamental cycle basis of $G$.*

We use the following notation crucially. A path is a connected graph in which two vertices have degree one and the rest of vertices have degree two. An edge can be considered as a connected graph consisting of single edge.

**Lemma 6.** *Let $G'$ be a 2-connected component in an arbitrary graph $G$ and $T$ be a subgraph of $G$.*

*(a) If $T$ is a spanning tree of $G$, then $T \cap G'$ is a spanning tree of $G'$.*
*(b) If $T$ is a path in $G$, then $T \cap G'$ is a path.*

**Special Graph Classes.** A *partial 2-tree* is a subgraph of a 2-tree. A graph is a *series-parallel* graph, if it can be obtained from an edge, by repeatedly duplicating an edge or replacing an edge by a path. An alternative equivalent definition for series-parallel graphs is given in [9].

## 2   Structural Properties of Polygonal 2-Trees

In this section, we present our key structural result in Lemma 11, which presents crucial structural properties of polygonal 2-trees. This Lemma will be used significantly in proving the correctness of our algorithm. Another major result in this section is Lemma 12, which computes a kind of ear decomposition for polygonal 2-trees. This helps in obtaining an efficient algorithm to solve MAST. The notion of open ear decomposition is well known to characterize 2-connected graphs. An *open ear decomposition* of $G$ is a partition of $E(G)$ into a sequence $(P_0, \ldots, P_k)$ of edge disjoint graphs called as *ears* such that,

1. For each $i \geq 0$, $P_i$ is a path.
2. For each $i \geq 1$, end vertices of $P_i$ are distinct and the internal vertices of $P_i$ are not in $P_0 \cup \ldots \cup P_{i-1}$.

Further, a restricted version of open ear decomposition called *nested ear decomposition* is used to characterize series-parallel graphs [9]. An open ear decomposition $(P_0, \ldots, P_k)$ of $G$ is said to be *nested* if it satisfies the following properties:

1. For each $i \geq 1$, there exists $j < i$, such that the end vertices of path $P_i$ are in $P_j$.
2. Let the end vertices of $P_i$ and $P_{i'}$ are in $P_j$, where $0 \leq j < i, i' \leq k$ and $i \neq i'$. Let $Q_i \subseteq P_j$ be the path between the end vertices of $P_i$ and $Q_{i'} \subseteq P_j$ be the path between the end vertices of $P_{i'}$. Then $E(Q_i) \subseteq E(Q_{i'})$ or $E(Q_{i'}) \subseteq E(Q_i)$ or $E(Q_i) \cap E(Q_{i'}) = \emptyset$.

We define nice ear decomposition to characterize polygonal 2-trees and we show how it helps in efficiently computing the induced cycles. A nested ear decomposition $(P_0, \ldots, P_k)$ is said to be *nice* if it has the following property: $P_0$ is an edge and for each $i \geq 1$, if $x_i$ and $y_i$ are the end vertices of $P_i$, then there is some $j < i$, such that $(x_i, y_i)$ is an edge in $P_j$. Definition 1 naturally gives a nice ear decomposition for polygonal 2-trees. Further, a unique polygonal 2-tree can be constructed easily from a nice ear decomposition. Thus we have the following observation.

**Observation 7.** *A graph $G$ is a polygonal 2-tree if and only if $G$ has a nice ear decomposition.*

In the following lemmas, we present results from the literature that establish polygonal 2-trees as a subclass of 2-connected partial 2-trees, which we formalize in Lemma 11.

**Lemma 8 (Theorem 42 in [2]).** *A graph $G$ is a partial 2-tree if and only if every 2-connected component of $G$ is a series-parallel graph.*

According to Lemma 8, 2-connected series-parallel graphs and 2-connected partial 2-trees are essentially same.

**Lemma 9 (Lemma 1, Lemma 7 and Theorem 1 in [9]).** *A graph $G$ is 2-connected if and only if $G$ has a open ear decomposition in which the first ear is an edge. Further, for a 2-connected series-parallel graph, every open ear decomposition is nested. A graph is series-parallel if and only if it has a nested ear decomposition.*

The above lemma implies that every 2-connected partial 2-tree has a nested ear decomposition *starting* with an edge (first ear is an edge) and vice versa. We strengthen the first part of this result in the following lemma.

**Lemma 10.** *Let $G$ be a 2-connected partial 2-tree. Then there exists a nested ear decomposition $(P_0, \ldots, P_k)$ of $G$, such that $P_0$ is an edge and for each $i \geq 1$, $|E(P_i)| \geq 2$.*

We show in Lemma 12 that a nested ear decomposition as in Lemma 10 is a nice ear decomposition for polygonal 2-trees. From Propositions 1.7.2 and 12.4.2 in [5], partial 2-trees do not contain a $K_4$-subdivision (as a subgraph).

**Lemma 11.** *Let $G$ be a polygonal 2-tree. Then,*
*(a) $G$ is a 2-connected partial 2-tree and $G$ does not contain a $K_4$-subdivision.*
*(b) Any two induced cycles in $G$ share at most one edge and at most two vertices.*
*(c) For $u, v \in V(G)$ such that $(u, v) \notin E(G)$, $G - \{u, v\}$ has at most two components.*

Our algorithm will perform several computations on the induced cycles of a polygonal 2-tree. It is therefore important to obtain the set of induced cycles in a polygonal 2-tree in linear time. We prove this in the following lemma. This is based on a linear-time algorithm for obtaining an open ear decomposition [19].

**Lemma 12.** *Let $G$ be a polygonal 2-tree on $n$ vertices. Let $D$ be a nested ear decomposition of $G$ as in Lemma 10 and $\mathcal{B}$ be the set of induced cycles in $G$. Then $D$ is a nice ear decomposition. Further, $D$ and $\mathcal{B}$ can be computed in linear time and $\text{size}(\mathcal{B})$ is $O(n)$.*

We now present a sufficient condition for a graph to be a polygonal 2-tree.

**Lemma 13.** *If $G$ is a 2-connected partial 2-tree and every two induced cycles in $G$ share at most one edge, then $G$ is a polygonal 2-tree.*

# 3    Structural Properties of MAST in Polygonal 2-Trees

For the rest of the paper, $G$ denotes a polygonal 2-tree. In this section we design an iterative procedure to delete a subset of edges from a polygonal 2-tree, so that the graph on the remaining edges is a minimum average stretch spanning tree. This result is shown in Theorem 18.

*Important Definitions:* We introduce some necessary definitions on polygonal 2-trees. Two induced cycles in $G$ are *adjacent* if they share an edge. An edge in $G$ is *internal* if it is part of at least two induced cycles; otherwise it is *external*. An induced cycle in $G$ is *external* if it has an external edge; otherwise it is *internal*. A fundamental cycle of a spanning tree, created by a non-tree edge is said to be *external* if the associated non-tree edge is external. For a cycle $C$ in $G$, the *enclosure* of $C$ is defined as $G[V(C)]$ and is denoted by $Enc(C)$. A set $A \subseteq E(G)$ consisting of $k$ $(\geq 0)$ edges is said to be an *iterative* set for $G$ if the edges in $A$ can be ordered as $e_1, \ldots, e_k$ such that $e_1$ is external and not a bridge in $G$, and for each $2 \leq i \leq k$, $e_i$ is external and not a bridge in $G - \{e_1, \ldots, e_{i-1}\}$. Let $A$ be an iterative set of edges in $G$. For every edge $(u, v) \in A$, both $u$ and $v$ are not present in the same 2-connected component in $G - A$. We define $\text{bound}(A, G)$ to be the set of external edges in $G - A$ that are not bridges. For an edge $e \in \text{bound}(A, G)$, $G_e$ denotes a 2-connected component in $G - A$ that has $e$.

**Definition 14.** *Let $e \in bound(A, G)$. The support of $e$ is defined as $\{(u, v) \in A \mid$ there is a path $P$ joining $u$ and $v$ in $G - A$ such that $P \cap G_e = e\}$ and is denoted by* Support$(e)$. *The* cost$(e)$ *is defined as* $|\,\text{Support}(e)|$.

In the following lemmas we present a result on the structure of paths connecting the end points of edges in an iterative set $A$. This is useful in setting up the iterative approach for computing a minimum average stretch spanning tree. We apply the necessary properties of polygonal 2-trees (cf. Lemma 11) and sufficient condition for a graph to be a polygonal 2-tree (cf. Lemma 13) in the proofs of the following lemmas.

**Lemma 15.** *Let $A$ be an iterative set of edges for $G$ and $(u, v) \in A$, $P$ be a path joining $u$ and $v$ in $G - A$, $G'$ be a 2-connected component in $G - A$ that has at least two vertices from $P$, and let $P' = P \cap G'$ be a path with end vertices $x$ and $y$. Then the following are true:*
*(a) $(x, y) \in E(G')$.*
*(b) If $P$ is a shortest path, then $P'$ is an edge.*
*(c) Every 2-connected component in $G - A$ is a polygonal 2-tree.*

**Lemma 16.** *Let $A$ be an iterative set of edges for $G$. $(u, v) \in$ Support$(e)$ if and only if there is a shortest path $P$ joining $u$ and $v$ in $G - A$ and $P$ has $e$.*

**Lemma 17.** *Let $T$ be a spanning tree of $G$ and $e$ be an external edge in $G$ such that $e \in E(T)$. For the spanning tree $T$, let $C_{min}$ be the smallest fundamental cycle containing $e$ and let $C_{max}$ be a largest fundamental cycle containing $e$. Let $e'$ and $e''$ be the non-tree edges associated with $C_{min}$ and $C_{max}$, respectively. Then, (a) $e''$ is an external edge (b) $Enc(C_{min}) \subseteq Enc(C_{max})$.*

A set $A$ of edges in $G$ is referred to as a *safe* set for $G$, if $A$ is an iterative set of edges for $G$ and a minimum average stretch spanning tree of $G$ is in $G - A$.

**Theorem 18.** *Let $A$ be a safe set of edges for $G$ such that* $\text{bound}(A, G) \neq \emptyset$. *Let $e$ be an edge in* $\text{bound}(A, G)$ *for which* $\text{cost}(e)$ *is minimum. Then $A \cup \{e\}$ is a safe set for $G$.*

*Proof.* For a safe set $A$, let $T^*$ be a minimum average stretch spanning tree of $G$; that is, $T^* \subset G - A$ as $\text{bound}(A, G) \neq \emptyset$. If $e \notin E(T^*)$, then we are done. Assume that $e \in E(T^*)$. Clearly, $A \cup \{e\}$ is an iterative set for $G$. To show that $A \cup \{e\}$ is a safe set for $G$, we use the technique of cut-and-paste to obtain a spanning tree $T'$ (by deleting the edge $e$ from $T^*$ and adding an appropriately chosen edge $e'$) and show that $\text{AvgStr}(T') \leq \text{AvgStr}(T^*)$.

Let $G_e$ be a 2-connected component in $G - A$ containing $e$ and $G_1, \ldots, G_k$ be the 2-connected components in $G - A$. For clarity, $G_e \in \{G_1, \ldots, G_k\}$. From Lemma 15.(c), $G_e$ is a polygonal 2-tree. For $1 \leq i \leq k$, by Lemma 6.(a), $T_i = T^* \cap G_i$ is a spanning tree of $G_i$. For the spanning tree $T^*$, let $C_{min}$ be the smallest fundamental cycle containing $e$ in $G_e$ and let $C_{max}$ be a largest fundamental cycle containing $e$ in $G_e$. Let $e', e'' \in E(G_e)$ be the non-tree edges associated with $C_{min}$ and $C_{max}$, respectively. From Lemma 17, $e''$ is an external edge in $G_e$ and $Enc(C_{min}) \subseteq Enc(C_{max})$. Let $e' = (x_{min}, y_{min})$, $e'' = (x_{max}, y_{max})$. For a non-tree edge $(u, v)$ in $T^*$, we use $P_{uv}$ to denote the path between $u$ and $v$ in $T^*$ and $C_{uv}$ to denote the fundamental cycle of $T^*$ formed by $(u, v)$. Let $X = \{(u, v) \in E(G) \setminus E(T^*) \mid e \in E(P_{uv}), e' \notin Enc(C_{uv})\}$, $Y = \{(u, v) \in E(G) \setminus E(T^*) \mid e \in E(P_{uv}), e' \in Enc(C_{uv}), (u, v) \neq e'\}$, $Z = \{(u, v) \in E(G) \setminus E(T^*) \mid e \notin E(P_{uv})\}$. The set of non-tree edges in $T^*$ is $X \uplus Y \uplus \{e'\} \uplus Z$. Let $T' = T^* + e' - e$. The set of non-tree edges in $T'$ is $X \uplus Y \uplus Z \uplus \{e\}$. To prove the theorem, we prove the following claims.

**Claim 1:** $X \subseteq A$.
**Claim 2:** $\text{Support}(e) \subseteq X$.
**Claim 3:** $\text{Support}(e'') \subseteq Y$.
**Claim 4:** $X \subseteq \text{Support}(e)$.
**Claim 5:** For every $(u, v) \in Z$, the path between $u$ and $v$ in $T^*$ is in $T'$.

Assuming that the above five claims are true, we complete the proof of the theorem. We know that $\text{cost}(e) \leq \text{cost}(e'')$. As $e$ and $e''$ are in $G_e$, from the definition of Support, we further know that $\text{Support}(e) \cap \text{Support}(e'') = \emptyset$. Therefore, from Claims 2, 3 and 4, it follows that $|X| \leq |Y|$. Since $e', e \in E(C_{min})$, $e \in E(T^*)$ and $e' \notin E(T^*)$, the stretch of $e'$ in $T^*$ is equal to the stretch of $e$ in $T'$. From Claim 5, stretch do not change for the edges in $Z$. For all the edges in $X$, stretch increases by $|C_{min}| - 2$. Further, for all the edges in $Y$, stretch decreases by $|C_{min}| - 2$. Therefore, $\text{AvgStr}(T') \leq \text{AvgStr}(T^*)$. Since $T^*$ is a minimum average stretch spanning tree, $T'$ is also a minimum average stretch spanning tree. Clearly, $T'$ is in $G - (A \cup \{e\})$. Hence $A \cup \{e\}$ is a safe set for $G$.

We now prove the five claims.

**Proof of Claim 1:** On the contrary, assume that $(u, v) \in X$ and $(u, v) \notin A$. To arrive at a contradiction, we show that $e$ is an internal edge. Since $(u, v) \in X$, there is a fundamental cycle $C_{uv}$ of $T^*$ formed by the non-tree edge $(u, v)$ containing $e$. As $(u, v) \notin A$, clearly $(u, v)$ is in $G - A$. Further, $P_{uv}$ is in $G - A$, because $T^* \subset G - A$. So we know that $C_{uv}$ is in $G - A$. If $C_{uv}$ is not in $G_e$, then $G_e \cup C_{uv}$ becomes a 2-connected component in $G - A$, because $G_e$ is in $G - A$, $C_{uv}$ is in $G - A$, and $e$ is both in $G_e$ and $C_{uv}$. But, we know that $G_e$ is a maximal 2-connected subgraph (2-connected component), thereby $C_{uv}$ is in $G_e$. Clearly, $C_{uv}$ and $C_{min}$ are not edge disjoint cycles. If $Enc(C_{min}) \subseteq Enc(C_{uv})$, then either $(u, v) \in Y$ or $(u, v) = e'$, which contradicts the fact that $(u, v) \in X$. Also, $Enc(C_{uv})$ is not contained in $Enc(C_{min})$, because $C_{min}$ is a minimum length induced cycle containing $e$. Therefore, both $C_{min}$ and $C_{uv}$ are not contained in each other. Thus, $e$ is an internal edge in $G - A$. This is a contradiction, as we know that $e$ is external.

**Proof of Claim 2:** Let $(u, v) \in \text{Support}(e)$. In order to prove that $(u, v) \in X$, we show the following: (a) $(u, v) \notin E(T^*)$, (b) $P_{uv}$ has $e$ and (c) $e'$ is not in $Enc(C_{uv})$.

By the definition of $\text{Support}(e)$, $(u, v) \in A$. As $T^* \subset G - A$, it follows that $(u, v) \notin E(T^*)$. By Lemma 16, there is a shortest path $P$ joining $u$ and $v$ in $G - A$ and $P$ has $e$. Let $G'_1, \ldots, G'_r$ be the 2-connected components in $G - A$ containing at least two vertices from $P$. Due to Lemma 15.(b), for each $1 \leq i \leq r$, $P \cap G'_i$ is an edge, say $(x_i, y_i)$. Thus $P \cap G_e$ is $e$. Further, $P$ contains at most one vertex from $e'$, because $e' \in E(G_e)$. The set of edges in $P$ that are cut-edges in $G - A$ are present in $T^*$. Due to Lemma 6.(a), replacing every edge $(x_i, y_i)$ in $P$ by the path between $x_i$ and $y_i$ in $T^*$, $P_{uv}$ is obtained. Since $P \cap G_e$ is $e$ and $e$ is in $T^*$, it implies that $P_{uv}$ has $e$. Thus $P_{uv}$ has $e$, and $e'$ is not in $Enc(C_{uv})$.

**Proof of Claim 3:** Let $(u, v) \in \text{Support}(e'')$. In order to prove that $(u, v) \in Y$, we show the following: (a) $(u, v) \notin E(T^*)$, (b) $P_{uv}$ has $e$ and (c) $e'$ is in $Enc(C_{uv})$.

Because $(u, v) \in A$ and $T^* \subset G - A$, we have $(u, v) \notin E(T^*)$. As $e, e'' \in E(G_e)$, due to Lemma 16, there is a shortest path $P$ joining $u$ and $v$ in $G - A$ and $P$ has $e''$. Let $G'_1, \ldots, G'_r$ be the 2-connected components in $G - A$ such that for each $1 \leq i \leq r$, $P \cap G'_i$ is an edge, say $(x_i, y_i)$, due to Lemma 15.(b). By Lemma 6.(a), we replace every edge $(x_i, y_i)$ in $P$ by the path between $x_i$ and $y_i$ in $T^*$ and obtain the tree path $P_{uv}$. Note that $P \cap G_e$ is $e''$, $e'' = (x_{max}, y_{max})$, and $e''$ in $P$ got replaced with the path between $x_{max}$ and $y_{max}$ in $T^*$. Also, we know that the path between $x_{max}$ and $y_{max}$ in $T^*$ has $e$. Further by Lemma 17.(b), $e'$ is in $Enc(C_{max})$. These observations imply that $P_{uv}$ has $e$ and $Enc(C_{uv})$ contains $e'$.

**Proof of Claim 4:** Let $(u, v) \in X$. By Claim 1, clearly $(u, v) \in A$. Lemma 6.(b) implies that $P_{uv} \cap G_e$ is a path. Let $P' = P_{uv} \cap G_e$ be a path and let $x$ and $y$ be the end vertices of $P'$. If $P'$ is an edge, then the claim holds. On the contrary assume that $P'$ has at least two edges. By Lemma 15.(a), $(x, y) \in E(G)$. Further, $(x, y) \notin E(T^*)$ as it would then form a cycle in the tree. The path $P'$ is strictly contained in the path joining the vertices $x_{min}$ and $y_{min}$ in $T^*$, because $e' \notin Enc(C_{uv})$. Then the fundamental cycle of $T$ formed by $(x, y)$ is of lesser length than the length of $C_{min}$; a contradiction because $C_{min}$ is a minimum length fundamental cycle in $G_e$ containing $e$. Therefore, $P_{uv} \cap G_e$ is $e$. Thus $(u, v) \in \text{Support}(e)$.

**Proof of Claim 5:** Let $(u, v) \in Z$. By the definition of $Z$, clearly $e \notin E(P_{uv})$. It implies that $e' \notin Enc(C_{uv})$ as the path between the end vertices of $e'$ in $T^*$ has $e$. Therefore $P_{uv}$ has at most one end vertex from $e$ and $e'$. Since the symmetric difference of $E(T^*)$ and $E(T')$ is $\{e, e'\}$, the path $P_{uv}$ in $T^*$ remains same in $T'$. Hence the theorem.  □

We now show the termination condition for applying Theorem 18.

**Lemma 19.** *Let $A$ be a safe set of edges for $G$ such that* $\text{bound}(A, G) = \emptyset$. *Then $G - A$ is a minimum average stretch spanning tree of $G$.*

## 4  Computing MAST in Polygonal 2-Trees

In order to obtain a minimum average stretch spanning tree efficiently, we need to efficiently find an edge in $\text{bound}(A, G)$ with minimum cost in every iteration, where $A$ is a safe set for $G$. In this section, we present necessary data-structures, so that a minimum average stretch spanning tree in polygonal 2-trees on $n$ vertices can be computed in $O(n \log n)$ time. This is shown in **Algorithm 2**. For each edge $e \in \text{bound}(A, G)$, we show in Lemma 20, how to compute $cost(e)$ efficiently.

**Notation.** Let $Q$ be a min-heap that supports the following operations: $Q.\text{insert}(x)$ inserts an arbitrary element $x$ into $Q$, $Q.\text{extract-min}()$ extracts the minimum element from $Q$, $Q.\text{decrease-key}(x, k)$ decreases the key value of $x$ to $k$ in $Q$, $Q.\text{delete}(x)$ deletes an arbitrary element $x$ from $Q$. $Q.\text{delete}(x)$ can be implemented by calling $Q.\text{decrease-key}(x, -\infty)$ followed by $Q.\text{extract-min}()$ [3]. For a set $A$ of safe edges for $G$, an induced cycle in $G$ is said to be *processed* if it is not in $G - A$; otherwise it is said to be *unprocessed*. For an edge $e \in E(G)$, we use the sets $Cycles[e]$ and $pCycles[e]$ to store the set of induced cycles and processed induced cycles, respectively containing $e$; $unpCount[e]$ is used to store the number of unprocessed induced cycles containing $e$. For an edge $e \in E(G)$, we use $c[e]$ to store some intermediate values while computing $cost(e)$; whenever $e$ becomes an edge in $\text{bound}(A, G)$, then we make sure that $c[e]$ is $cost(e)$.

**Initialization.** Given a polygonal 2-tree $G$, we first compute the set of induced cycles in $G$. For each induced cycle $C$ in $G$ and for each edge $e \in E(C)$, we insert the cycle $C$ in the set $Cycles[e]$. For each $e \in E(G)$, we perform $unpCount[e] \leftarrow |Cycles[e]|$, $pCycles[e] \leftarrow \emptyset$, $c[e] \leftarrow 0$. We further initialize the set $A$ of safe edges with $\emptyset$. Later, we construct a min-heap $Q$ with the edges $e$ in $\text{bound}(A, G)$ i.e., external edges that are not bridges in $G$, based on $c[e]$.

The **Algorithm 2** maintains the following loop invariants:

---

**Algorithm 2.** An algorithm to find an MAST of a polygonal 2-tree $G$

---

1   Perform the steps described in Initialization ;
2   **while** $Q \neq \emptyset$ **do**
3      $e \leftarrow Q.\texttt{extract-min}()$ ;
4      $A \leftarrow A \cup \{e\}$ ;
5      $C \leftarrow Cycles[e] \setminus pCycles[e]$ ;
6      **for** *each edge* $\hat{e} \in E(C) \setminus \{e\}$ **do**
7          $c[\hat{e}] \leftarrow c[\hat{e}] + c[e] + 1$ ;
8          $pCycles[\hat{e}] \leftarrow pCycles[\hat{e}] \cup C$ ;
9          $unpCount[\hat{e}] \leftarrow unpCount[\hat{e}] - 1$ ;
10          **if** $unpCount[\hat{e}] = 1$ **then** $Q.\texttt{insert}(\hat{e}, c[\hat{e}])$ ;
11          **if** $unpCount[\hat{e}] = 0$ **then** $Q.\texttt{delete}(\hat{e})$ ;

12   Return $G - A$;

---

L1.   The min heap $Q$ only consists of, the set of edges in bound$(A, G)$.
L2.   For an edge $e \in E(G)$, $pCycles[e]$ is the set of processed induced cycles containing $e$ and $unpCount[e]$ is equal to the number of unprocessed induced cycles containing $e$.
L3.   For every edge $e \in$ bound$(A, G)$, $Cycles[e] \setminus pCycles[e]$ is the unique external induced cycle in $G - A$ containing $e$.
L4.   $A$ is a safe set for $G$. (cf. Theorem 18)
L5.   For every edge $e \in$ bound$(A, G)$, $c[e] = \text{cost}(e)$. (cf. Lemma 20)

Let $A_i \subset E(G)$ denote the set of safe edges in $G$ at the end of $i^{th}$ iteration. Let $e$ be an edge extracted from the heap $Q$ in $i^{th}$ iteration and $C$ be the unique induced cycle containing $e$ in $G - A_{i-1}$. That is, $C$ is a cycle in $G - A_{i-1}$ and $C$ is not a cycle in $G - A_i$ as $e$ is added to $A$ in iteration $i$. Then we say that $C$ is processed in iteration $i$ and $e$ is the *destructive* edge for $C$.

**Lemma 20.** *Let* $e \in$ bound$(A_j, G)$. *Let* $C$ *be the unique external induced cycle in* $G - A_j$ *containing* $e$ *and* $C_1, \ldots, C_k$ *be the other induced cycles in* $G$ *containing* $e$. *For* $1 \leq i \leq k$, *let* $e_i$ *be the destructive edge of* $C_i$. *Then* Support$(e) =$ Support$(e_1) \uplus \ldots \uplus$ Support$(e_k) \uplus \{e_1, \ldots, e_k\}$.

The algorithm terminates when $Q$ becomes $\emptyset$, that is, bound$(A, G) = \emptyset$. Then by Lemma 19, $G - A$ is a minimum average stretch spanning tree of $G$.

**Lemma 21.** *For a polygonal 2-tree* $G$ *on* $n$ *vertices,* **Algorithm** *2 takes* $O(n \log n)$ *time.*

*Proof.* The set of induced cycles in $G$ can be obtained in linear time (cf. Lemma 12), thereby line 1 takes linear time. As the size of induced cycles in $G$ is $O(n)$ (Lemma 12), line 5 and lines 7-9 contribute $O(n)$ towards the run time of the algorithm. Also every edge in $G$ gets inserted into the heap $Q$ and gets deleted from $Q$ only once and $|E(G)| \leq 2n - 3$. It takes $O(\log n)$ time for the operations `insert()`,`delete()` and `extract-min()` [3]. Thus Algorithm 2 takes $O(n \log n)$ time.     □

This concludes the presentation of our main result, namely Theorem 2.

# References

1. Alon, N., Karp, R.M., Peleg, D., West, D.: A graph-theoretic game and its application to the k-server problem. SIAM J. of Comput. 24, 78–100 (1995)
2. Bodlaender, H.L.: A partial k-arboretum of graphs with bounded treewidth. Theoretical Computer Science 209(1-2), 1–45 (1998)
3. Cormen, T.H., Leiserson, C.E., Rivest, R.L., Stein, C.: Introduction to Algorithms, 3rd edn. MIT Press (2009)
4. Deo, N., Prabhu, G., Krishnamoorthy, M.S.: Algorithms for generating fundamental cycles in a graph. ACM Transactions on Math. Software 8, 26–42 (1982)
5. Diestel, R.: Graph Theory, 4th edn. Springer (2010)
6. Ducharme, M., Labelle, G., Lamathe, C., Leroux, P.: A classification of outerplanar k-gonal 2-trees. In: 19th Intern. Conf. on FPSAC (2007) (appeared as Poster.)
7. Emek, Y.: k-outerplanar graphs, planar duality, and low stretch spanning trees. Algorithmica 61(1), 141–160 (2011)
8. Emek, Y., Peleg, D.: A tight upper bound on the probabilistic embedding of series-parallel graphs. SIAM J. Discrete Math. 23(4), 1827–1841 (2009)
9. Eppstein, D.: Parallel recognition of series-parallel graphs. Inf. Comput. 98(1), 41–55 (1992)
10. Fowler, T., Gessel, I., Labelle, G., Leroux, P.: The specification of 2-trees. Advances in Applied Mathematics 28(2), 145–168 (2002)
11. Galbiati, G., Rizzi, R., Amaldi, E.: On the approximability of the minimum strictly fundamental cycle basis problem. Discrete Appl. Math. 159(4), 187–200 (2011)
12. Hubicka, E., Syso, M.: Minimal bases of cycles of a graph. In: Recent Advances in Graph Theory, Second Czech Symposium in Graph Theory, pp. 283–293. Academia, Prague (1975)
13. Kavitha, T., Liebchen, C., Mehlhorn, K., Michail, D., Rizzi, R., Ueckerdt, T., Zweig, K.A.: Cycle bases in graphs characterization, algorithms, complexity, and applications. Computer Science Review 3(4), 199–243 (2009)
14. Koh, K.M., Teo, C.P.: Chromaticity of series-parallel graphs. Discrete Mathematics 154(1-3), 289–295 (1996)
15. Labelle, G., Lamathe, C., Leroux, P.: Labelled and unlabelled enumeration of k-gonal 2-trees. J. Comb. Theory, Ser. A 106(2), 193–219 (2004)
16. Liebchen, C., Wünsch, G.: The zoo of tree spanner problems. Discrete Appl. Math. 156, 569–587 (2008)
17. Omoomi, B., Peng, Y.-H.: Chromatic equivalence classes of certain generalized polygon trees, iii. Discrete Mathematics 271(1-3), 223–234 (2003)
18. Peng, Y.-H., Little, C.H.C., Teo, K.L., Wang, H.: Chromatic equivalence classes of certain generalized polygon trees. Discrete Mathematics 172(1-3), 103–114 (1997)
19. Ramachandran, V.: Parallel Open Ear Decomposition with Applications to Graph Biconnectivity and Triconnectivity. Morgan Kaufmann (1992)
20. West, D.B.: Introduction to graph theory, 2nd edn. Prentice-Hall (2001)

# Linear Layouts of Weakly Triangulated Graphs

Asish Mukhopadhyay[1], S.V. Rao[2],
Sidharth Pardeshi[2], and Srinivas Gundlapalli[2]

[1] School of Computer Science, University of Windsor,
Windsor, Ontario N9B 3P4, Canada
[2] Department of Computer Science and Engineering, Indian Institute of Technology,
Guwahati, Assam, India

**Abstract.** A graph $G = (V, E)$ is said to be triangulated if it has no
chordless cycles of length 4 or more. Such a graph is said to be rigid if,
for a valid assignment of edge lengths, it has a unique linear layout and
non-rigid otherwise. Damaschke [7] showed how to compute all linear
layouts of a triangulated graph, for a valid assignment of lengths to the
edges of $G$. In this paper, we extend this result to weakly triangulated
graphs, resolving an open problem. A weakly triangulated graph can be
constructively characterized by a peripheral ordering of its edges. The
main contribution of this paper is to exploit such an edge order to identify
the rigid and non-rigid components of $G$. We first show that a weakly
triangulated graph without articulation points has at most $2^{n_q}$ different
linear layouts, where $n_q$ is the number of quadrilaterals (4-cycles) in
$G$. When $G$ has articulation points, the number of linear layouts is at
most $2^{n_b - 1 + n_q}$, where $n_b$ is the number of nodes in the block tree of $G$
and $n_q$ is the total number of quadrilaterals over all the blocks. Finally,
we propose an algorithm for computing a peripheral edge order of $G$
by exploiting an interesting connection between this problem and the
problem of identifying a two-pair in $\overline{G}$. Using an $\mathcal{O}(n \cdot m)$ time solution
for the latter problem when $G$ has $n$ vertices and $m$ edges, we propose
an $\mathcal{O}(n^2 \cdot m)$ time algorithm for computing its peripheral edge order. For
sparse graphs, the time-complexity can be improved to $\mathcal{O}(m^2)$, using the
concept of handles proposed in [1].

## 1   Introduction

The problem we study in this paper is a restricted version of the point placement
problem for which we seek to determine the locations of a set of distinct points on
a line, uniquely up to translation and reflection, by making the fewest pairwise
distance queries [2]. In the linear layout problem, we are given a set of pairwise
distances (in the form of a graph) and the problem is to determine all possible
placements of the vertices of the graph on a line or their linear layouts. The
linear layout problem, in turn, is a special case of the graph embedding problem.
Given a weighted graph $G = (V, E)$ and a positive integer $k$, an embedding of
$G$ in a $k$-dimensional Euclidean space, $E^k$, is a mapping, $f$, of $V$ into $E^k$ such
that the weight of an edge $e = \{u, v\}$ in $G$ is equal to the Euclidean distance

S.P. Pal and K. Sadakane (Eds.): WALCOM 2014, LNCS 8344, pp. 322–336, 2014.
© Springer International Publishing Switzerland 2014

between $f(u)$ and $f(v)$. For $k = 1$, such an embedding corresponds to a linear layout. Saxe [3] showed that the problem of embedding a weighted (incomplete) graph $G = (V, E)$ in Euclidean $k$-space is strongly $NP$-complete. Indeed, it remains so even when $k = 1$ and the edge weights are restricted to the values $\{1, 2\}$. Barvinok [4] showed that if $G$ is $k$-embeddable for some $k$ then it is $k$-embeddable for $k = \lfloor \sqrt{(8|E| + 1)} - 1)/2 \rfloor$, while Alfakih and Wolkowicz [5] gave an algorithm for constructing such a $k$-embedding. Hastad et al. [6] studied the approximability of the matrix-to-line problem, which is in essence the 1-embeddability problem.

Despite the strongly negative result of Saxe, it is possible to solve this problem in polynomial time for special classes of graphs with well-defined assignment of weights to their edges as we show below.

A graph $G = (V, E)$ is said to be triangulated (also called chordal) if it has no chordless cycle of length 4 or more. Damaschke [7] described an algorithm for generating all linear layouts of a triangulated graph given a valid assignment of lengths, $l$, to the edges of the graph. An assignment of lengths to the edges of $G$ is said to be valid if the distances between the adjacent vertices in a linear placement is consistent with the lengths assigned to the edges of $G$. We indicate this by writing $(G, l)$. It was left as an open problem to extend this algorithm to other classes of graphs with weaker chordality properties. In this paper, we show how to enumerate linear layouts for weakly triangulated (or weakly chordal) graphs.

A graph $G = (V, E)$ is weakly triangulated if neither $G$ nor its complement $\overline{G}$ contains a chordless cycle of 5 or more vertices. A hole in $G$ is an induced cycle on 5 or more vertices and an anti-hole is the complement of a hole. Alternatively, $G$ is weakly triangulated if it does not contain a hole or an anti-hole. Fig. 1(a) shows a weakly triangulated graph; however, the graph in Fig. 1(b) is not a weakly triangulated one as the outer boundary is a chordless 8-cycle. The class of weakly triangulated graphs includes the class of triangulated graphs as well as their complements.

## 2    Preliminaries

Let $G = (V, E)$ be an undirected graph. A component of $G$ is a maximally connected subgraph. A biconnected component is a 2-connected component. When $G$ is connected, it decomposes into a tree of biconnected components called the block tree of the graph. The blocks are joined to each other at shared vertices called cut vertices or *articulation points*. The following constructive characterization, analogous to the perfect elimination ordering of a triangulated graph, is central to our approach.

**Theorem 1.**  [8] *A graph is weakly triangulated iff it can be generated in the following manner:*

1. *Start with an empty graph $G_0$.*
2. *Repeatedly add an edge $e_j$ to the graph $G_{j-1}$ to create the graph $G_j$ such that $e_j$ is not the middle edge of any $P_4$ (a chordless path of 4 vertices) of $G_j$.*

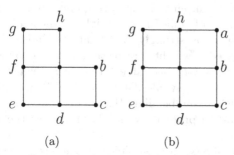

(a)                    (b)

**Fig. 1.** Some example graphs: (a) A weakly triangulated graph and (b) A graph that is not weakly triangulated

An edge of a graph is *peripheral* if it is not the middle edge of any $P_4$. A total order of the $m$ edges $\{e_1, e_2, ..., e_m\}$ of the graph $G$ is a peripheral edge order if for $1 \leq j \leq m$, $e_j$ is peripheral in the graph $G_j = (V, E_j)$, where $E_j = \{e_1, e_2, ..., e_j\}$. Thus the following theorem is equivalent to Theorem 1 [8]:

**Theorem 2.** *A graph is weakly triangulated iff it admits a peripheral edge order.*

The graph of Fig. 1(b), for example, does not admit a peripheral edge order. For a peripheral edge order to exist, the last edge added must be an edge on its outer boundary. Such an edge, when added to the graph, is clearly the middle edge of a $P_4$.

Let $l$ be a valid assignment of lengths to the edges of $G$. This means that there is a placement of the nodes $V$ on a line such that the distance

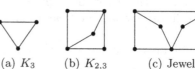

(a) $K_3$     (b) $K_{2,3}$     (c) Jewel     (d) $K_4^-$

**Fig. 2.** Some examples of line rigid graphs

between adjacent nodes is consistent with $l$. We express this by the notation $(G, l)$. By definition $(G, l)$ is said to be line rigid if there is a unique placement up to translation and reflection, while $G$ is said to be line rigid if $(G, l)$ is line rigid for every valid $l$.

Some examples of line rigid graphs are shown in Fig. 2. Except for what is known as the jewel graph (Fig. 2(c)), the rest are also weakly triangulated.

A quadrilateral is not line rigid as an assignment $l$ that makes it a parallelogram has two different layouts (Fig. 3).

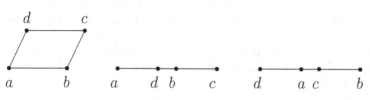

**Fig. 3.** Two different placements of a parallelogram $abcd$

A graph $G$ is *minimally line rigid* if it has no proper induced subgraph that is line rigid. For example, the weakly triangulated graphs $K_{2,3}$ and $K_3$ are minimally line rigid. A subgraph of a graph $G$ is *maximally line rigid* if it has no proper induced supergraph that is line rigid.

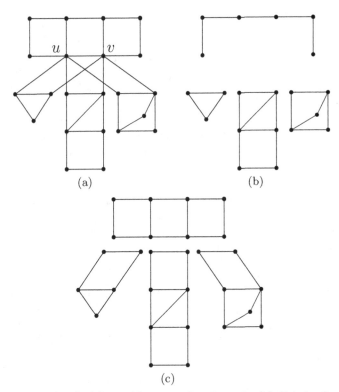

**Fig. 4.** Hinge edge $\{u, v\}$: (a) Weakly triangulated graph, (b) Graph after deleting $u$ and $v$, and (c) Hinge Components

An edge $\{u, v\}$ of $G$ is a *hinge edge* if removal of the vertices $u$ and $v$ and the edges incident on these disconnects $G$ into three or more disjoint components. The edge $\{u, v\}$ in Fig. 4(a) is a hinge edge. Hinge components hanging from the hinge edge $\{u, v\}$ are shown in Fig. 4(c).

## 3  Rigidity Structure

We assume that $G$ is weakly triangulated graph without articulation points or hinge edges and has a valid assignment of edge lengths. Also, we freely use geometric terminology like line segments, triangles and quadrilaterals in lieu of edges, 3-cycles and 4-cycles. Given a peripheral edge order of $G$, we reconstruct $G$ by adding back the edges in reverse peripheral order. During this process, we identity the formation of new rigid and non-rigid components, and the conversion of non-rigid components into rigid ones.

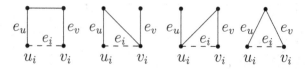

**Fig. 5.** Connectivity of edges $e_u$ and $e_v$

Let $G_i$ be the graph obtained by adding to $G_{i-1}$ the $i$-th peripheral edge $e_i = \{u_i, v_i\}$. The edge $e_i$ does not change the rigidity structure of the graph if it is added to a maximally rigid subgraph of $G_{i-1}$. Otherwise, it changes the rigidity structure of $G_{i-1}$ by way of forming a new rigid or non-rigid component or by changing a non-rigid component to a rigid one.

If only $u_i$ is incident to $G_{i-1}$, then $e_i$ is a dangling edge. Otherwise, $e_i$ joins two vertices of $G_{i-1}$, creating some new cycles. To determine what these are, consider the set of edges $E_u$ and $E_v$ incident respectively on $u_i$ and $v_i$. If $e_u$ is an edge in $E_u$ and $e_v$ an edge in $E_v$ then as $e_i$ is a peripheral edge, either the edges $e_u$ and $e_v$ share a common end point or there is a chord between their end-points as in Fig. 5.

Thus the addition of $e_i$ generates

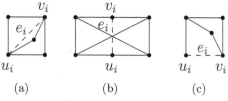

**Fig. 6.** Different length paths between $u_i$ and $v_i$: (a) Triples and pairs of paths of length two combining into a $K_{2,3}$ and a rigid triangulated quadrilateral, (b) Pair of paths of length three makes the configuration rigid before adding $e_i$, and (c) Pair of paths of length three with a common edge forms $K_{2,3}$ with $e_i$

new triangles and quadrilaterals. If there are more than two paths of length two between $u_i$ and $v_i$, excluding the chord $e_i$, each set of three paths group to form a rigid subgraph, $K_{2,3}$ (see Fig. 6(a)). If no two paths are left, $e_i$ gets added to a maximally rigid subgraph and no new rigid or non-rigid components are formed. Otherwise, new rigid or non-rigid components may be formed with the residual set at most two paths of length two between $u_i$ and $v_i$.

Let us examine the collection of quadrilaterals generated, each quadrilateral corresponding to a path of length three between $u_i$ and $v_i$, bounded by $e_i$. In order that pairs of such paths, with no common edge, keep the graph weakly triangulated, there must be chords between two pair of vertices as shown in Fig. 6(b). These give rise to two rigid subgraphs $K_{2,3}$, which makes the configuration rigid before adding $e_i$. Fig. 6(c) shows that if these two paths share an edge, then the addition of $e_i$ creates a $K_{2,3}$. This implies, more rigid or non-rigid components may be formed if there is at most one unpaired path of length 3.

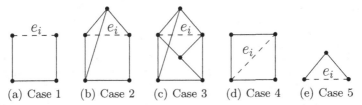

(a) Case 1    (b) Case 2    (c) Case 3    (d) Case 4    (e) Case 5

**Fig. 7.** All combinations of paths of lengths two and three

Combined with the previous step in which we have at most two residual paths of length two, we distinguish among the following 5 cases (see Fig. 7).

Case 1 (zero path of length two and one path of length three): In this case, in conjunction with $e_i$, we generate a quadrilateral.

Case 2 (one path of length two and one path of length three): In this case to maintain the weak triangulation property, there must exist a chord between a pair of vertices as shown in Fig. 7(b) and the addition of $e_i$ to this configuration gives rise to a rigid graph - a $K_{2,3}$ with an additional edge.

Case 3 (two paths of length two and one path of length three): In this case each path of length two together with the path of length three creates a 5-cycle which must be chorded to preserve the weak triangulation property. Since each 5-cycle can be triangulated in four different ways, 16 cases arise (see Fig. 8); of these, two have the configuration shown in Fig. 7(c); the configurations of the remaining 14 cases (marked by ×) are rigid prior to the addition of $e_i$. The former two configurations cannot arise as the complement graph has a chordless 6-cycle. Thus in this case $e_i$ is added to a rigid component.

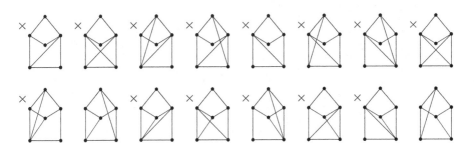

**Fig. 8.** Sixteen different triangulations of two 5-cycles

Case 4 (two paths of length two and zero path of length three): In this case the addition of edge $e_i$ triangulates a quadrilateral, generating a rigid graph.

Case 5 (one path of length two and zero path of length three): In conjunction with $e_i$ we have a triangle, a rigid graph.

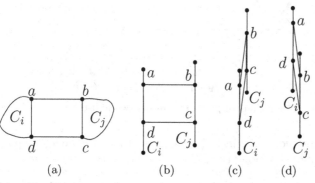

Thus in all cases we generate the rigid components: triangles, triangulated quadrilaterals, $K_{2,3}$'s or a single non-rigid component, viz., a quadrilateral.

Note that a non-rigid quadrilateral remains non-rigid even if all its vertices are parts of maximally rigid components as

**Fig. 9.** Non-rigid quadrilateral remains non-rigid: (a) rigid components $C_i$ and $C_j$ are adjacent to a non-rigid quadrilateral, (b) unique layouts of $C_i$ and $C_j$ are shown as vertical segments, and (c) and (d) are two different layouts of the quadrilateral $abcd$

shown in Fig. 9. This is true even if there are four different rigid components adjacent to a non-rigid quadrilateral, one on each side.

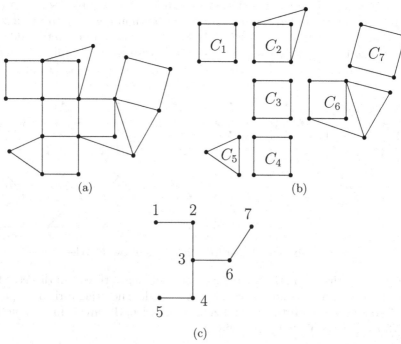

**Fig. 10.** (a) Weakly triangulated graph $G$, (b) Components of $G$, and (c) Rigidity tree of $G$

# 4   Rigidity Tree

By the discussion of the previous section, the only minimally line rigid subgraphs of a weakly triangulated graphs without articulation points and hinge edges are $K_{2,3}$ and $K_3$ and the only non rigid subgraph is a quadrilateral. We use these facts to determine the relationship between maximal rigid components and non-rigid components of a weakly triangulated graphs without articulation points and/or hinge edges. The interesting question is whether these can be found as we construct a weakly triangulated graph in an edge peripheral order.

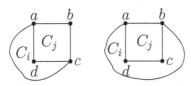

**Fig. 11.** Interaction between a maximally rigid component $C_i$ and a non-rigid component $C_j$

Let $C = \{C_1, C_2, \ldots, C_l\}$ be the set of components of $G$, where each $C_i$, for $1 \leq i \leq l$, is either a quadrilateral or a maximal rigid component of $G$. The rigidity graph $R_G = (V_G, E_G)$ is a graph whose nodes are the components of $G$ and there is an edge connecting two nodes if the corresponding components share an edge. The set of rigid components and rigidity graph of the weakly triangulated graph of Fig. 10(a) are shown in Fig. 10(b) and Fig. 10(c) respectively. We show that $R_G$ is a tree by explaining the interaction between non-rigid components and maximally rigid components.

**Lemma 1.** *No edge is common to more than two rigid components.*

*Proof.* This follows from the assumption that $G$ has no hinge edges.    ∎

**Lemma 2.** *Two quadrilaterals can share at most one edge.*

*Proof.* If two distinct quadrilaterals $C_i$ and $C_j$ share two edges then they form a $K_{2,3}$. This implies that $C_i \cup C_j$ must be a subgraph of a maximally rigid component, which is a contradiction. If three edges are common then the distinct edges of $C_i$ and $C_j$ are parallel. This proves the lemma.    ∎

**Lemma 3.** *No edge is common to two maximally rigid components.*

*Proof.* There cannot be an edge common to two maximally line rigid subgraphs because both have a linear layout on a supporting line of the common edge, fixing the placements of all the vertices. This contradicts the definition of the maximality of the rigid components.    ∎

**Lemma 4.** *A quadrilateral and a maximal rigid component can share at most an edge.*

*Proof.* If more than one edge is common between a maximal rigid component $C_i$ and a quadrilateral $C_j$ (see Fig. 11), then $C_i$ and $C_j$ must be part of the same maximal rigid component. This contradicts the definition of maximality of the rigid component $C_i$.    ∎

**Lemma 5.** *There is no cycle of length two in $R_G$.*

*Proof.* A cycle of length two in $R_G$ between $C_i$ and $C_j$ implies that $C_i$ and $C_j$ have two common edges, which is not possible because of the lemmas 2, 3, and 4.  ∎

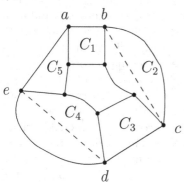

**Fig. 12.** Five components forming cycle

Indeed, we show that there is no cycle of any length in $R_G$.

**Theorem 3.** *The rigidity graph $R_G$ of a weakly triangulated graph without articulation points and hinge edges is a tree.*

*Proof.* Assume there is a 5-cycle in $R_G$. It contains at most two maximally rigid components and at least three non-rigid components, since there is no common edge between any two maximally rigid components. A 5-cycle formed by two maximally rigid components and three quadrilaterals is shown in Fig. 12. This contradicts the definition of a weakly triangulated graph, since there is a cycle $\{a,b\}, \widehat{\{b,c\}}, \{c,d\}, \widehat{\{d,e\}}, \{e,a\}$ with at least five edges, where the notation $\widehat{\{b,c\}}$ denotes a chordless path between $b$ and $c$. This argument extends to all cycles of length $\geq 5$ formed with different types of components.

The nonexistence of 3-cycles and 4-cycles in $R_G$ can be proved by enumerating all possible cycles with different types of components.

All three different types of 3-cycles possible in $R_G$ are shown in Fig. 13. All these 3-cycles violate the definition of a weakly triangulated graph. The case Fig. 13(a) cannot happen because it contains a chordless 6-cycle $\{a,b\}, \{b,c\}, \{c,d\}, \{d,e\}, \{e,f\}, \{f,a\}$ in $G$. The case Fig. 13(b) cannot happen because its complement $\{a,e\}, \{e,c\}, \{c,d\}, \{d,b\}, \{b,f\}, \{f,a\}$ is a 6-cycle in $\overline{G}$. Finally, the case of Fig. 13(c) cannot happen because it contains a chordless cycle $\{a,b\}, \{b,c\}, \{c,d\}, \{d,e\}, \widehat{\{e,a\}}$ in $G$, of length at least five.

All three different types of 4-cycles possible with non-rigid components are shown in Fig. 14. The graph of Fig. 14(a) contains a chordless 8-cycle $\{a,b\}, \{b,c\}, \{c,d\}, \{d,e\}, \{e,f\}, \{f,g\}, \{g,h\}, \{h,a\}$ in $G$, and the graphs of Figs. 14(b) and 14(c) contain a chordless 6-cycle $\{a,b\}, \{b,f\}, \{f,g\}, \{g,h\}, \{h,d\}, \{d,a\}$ The proofs for 4-cycles in $R_G$ formed with non-rigid and maximally rigid components are similar.  ∎

The edges are processed in peripheral edge order to generate the graph and maintain in parallel the rigidity tree of the graph constructed so far. The vertices of the rigidity tree are labeled as either rigid or non-rigid. An *event* occurs when the peripheral edge forms a triangle, quadrilateral, $K_{2,3}$, or splits a quadrilateral. We update the rigidity tree at these events. In all other cases, a dangling edge is formed or joins a pair of vertices of the same rigid component; no update is required for this event.

If the event is a triangle or a quadrilateral formation, insert a new node in the rigidity tree, label the node appropriately, and add the edge to its neighbor (see Fig. 15). For the event $K_{2,3}$ formation or quadrilateral split, change the label of the node to rigid, since the resulting component is rigid (see Fig. 16).

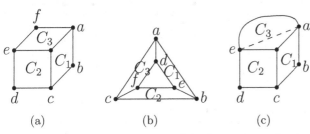

(a)                    (b)                    (c)

**Fig. 13.** Three components cycles: (a) Three non-rigid components cycle with a 6-cycle. (b) Three non-rigid components cycle whose complement is a 6-cycle. (c) Two non-rigid and a maximally rigid component with a minimum cycle length of five.

Whenever a rigid component is formed, merge it with its rigid neighbors, if any. All neighbors of merged nodes are the neighbors of the newly created rigid node. In Figs. 15(b), 16(b), 16(d), and 16(f) the rigid components formed are, in each case, the neighbor of another rigid component.

The number of linear layouts of a weakly triangulated graph without articulation points and hinge edges is $2^{n_q}$, since each non-rigid quadrilateral has at most two distinct layouts, where $n_q$ is number of non-rigid quadrilaterals in the rigidity tree. Moreover, all the linear layouts of the graph can be generated using depth-first or breadth-first traversal of the rigidity tree.

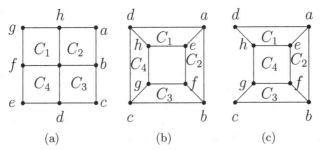

(a)                    (b)                    (c)

**Fig. 14.** Four non-rigid components cycles. (a) Contains an 8-cycle. (b) Contains a 6-cycle. (c) Contains a 6-cycle.

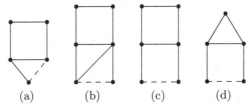

(a)        (b)        (c)        (d)

**Fig. 15.** Two different ways of triangle and non-rigid quadrilateral formations are depicted in (a) and (b), and (c) and (d) respectively

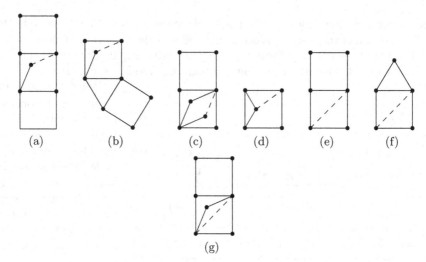

**Fig. 16.** Four different ways of $K_{2,3}$ formation and three different ways of quadrilateral split are depicted in (a) to (d) and (e) to (g) respectively

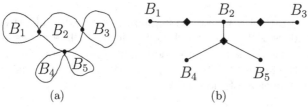

**Fig. 17.** Three different ways of quadrilateral split

When $G$ has articulation points, consider the block tree, $T$, of the graph $G$. If $G$ has $n_b$ blocks then the number of different layouts of the blocks is $2^{n_b-1}$ as each articulation point permits two different placements of a block in a linear layout. Fig. 17(a) shows a graph $G$ with three articulation points and five blocks; Fig. 17(b) shows the corresponding block tree. Fix a placement of the block $B_2$, then relative to it the remaining four blocks each has two different placements and thus 16 different placements.

Divide each block $B_i$ into hinge components if it contains hinge edges as shown in Fig. 4(c). For each hinge components and blocks without hinge edge, construct the rigidity tree from their respective peripheral edge order.

We have already shown that the presence of $q$ quadrilaterals within each block allow for $2^q$ different layouts. Hence we have at most $2^{n_b-1+n_q}$ linear layouts where $n_q$ is the total number of quadrilaterals over all the blocks.

## 5   Computing a Peripheral Edge Order

Vertices $\{x, y\}$ of $G$ is said to be a two-pair, if all chordless paths between $u$ and $v$ are of length two. It was shown in [8] that $\{x, y\}$ is a two-pair in the complement graph $\overline{G}$ if $\overline{xy}$ is a peripheral edge in $G$; furthermore, $(e_1, e_2, e_3 \ldots, e_m)$ is a peripheral edge order of $G$ if and only if $(e_1, e_2, e_3 \ldots, e_m)$ is a two-pair non-edge order (end points of $e_i$ are a two-pairs) of $\overline{G}$. By Theorem 2 stated earlier, we also know that such a peripheral edge order exists for a weakly triangulated graph. The algorithms that we propose will generate the peripheral edge order in reverse; that is, if the output of the algorithm is $(e_m, e_{m-1}, \ldots e_1)$, then the peripheral edge order is $(e_1, e_2, \ldots e_m)$.

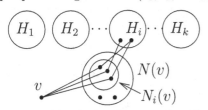

**Fig. 18.** The Arikati-Rangan algorithm in a nutshell

A brute-force approach to obtain the reverse peripheral edge order of $G$ goes thus. We first obtain a two-pair in the complement graph $\overline{G}$ and output the edge joining the corresponding vertices in $G$ as the first peripheral edge. Next, we remove this edge from $G$ and insert it into $\overline{G}$. Repeat these two steps for the updated graphs until $\overline{G}$ is a complete graph, or correspondingly $G$ becomes empty. Then the peripheral edge order is the reverse of the output order of the edges.

To compute a two-pair, we use the $\mathcal{O}(n \cdot m)$ algorithm proposed in [9]. As we shall be referring to some of its details in the subsequent paragraphs, we digress to give a brief description below.

Let $N(v)$ be the neighborhood of a vertex $v$ of $G$. Let $H_1, H_2, \ldots, H_k$ be a decomposition of $H = G - \{v\} - N(v)$ into connected components. The label of a vertex $u \in H$, $label(u)$, is $i$ if it belongs to $H_i$. Let $N_1(v), N_2(v), \ldots, N_k(v)$ be a decomposition of $N(v)$ such that $N_i(v) = \{u | u \in N(v)$ and $u$ is adjacent to some vertex in $H_i\}$ (see Fig. 18).

For $u \in H$, define $NV(u) = \{w | w$ is adjacent to $u$ and in $N(v)\}$. Then $\{u, v\}$ is a two-pair if $NV(u) = N_{label(u)}(v)$.

To continue, since the input graph is $\overline{G}$, the time-complexity for a two-pair computation is in $\mathcal{O}(n^3)$. As $G$ reduces to an empty graph over $m$ iterations, the time complexity of the brute-force algorithm is in $\mathcal{O}(m \cdot n^3)$.

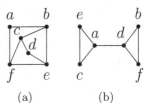

**Fig. 19.** Iteration 1: (a) Graph $G$, (b) its complement

The above brute-force algorithm makes a lot of redundant computations by running the Arikati-Rangan algorithm ab initio in each iteration. To pin down the redundancy precisely, consider a small graph and its complement (Fig. 19).

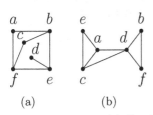

**Fig. 20.** Iteration 2: (a) Graph $G$, (b) its complement

After a first round of the Arikati-Rangan algorithm, if the two-pair returned is $\{c, d\}$, then the updated graphs $G$ and its complement for the next round are shown in Fig. 20. During the first-round (or the $i$-th round, in general) of computations, we need to store the intermediate results of the algorithm so that these can be used in the next round. For example, when we are considering the node $c$ in the complement graph $\overline{G}$, we need to store the components $H_k$ that are produced after removing nodes $c$ and its neighbors (i.e. $a$ and $e$) from the complement graph, so that these can be used in the next round. In this way, all the intermediate results are stored.

It is clear from Fig. 20 that the prinicipal way in which the current round of the two-pair algorithm differs from the previous lies in how we deal with the nodes $c$ and $d$ and the ones adjacent to these. We elaborate on this below.

Consider the $i$-th round of our incremental version of the Arikati-Rangan algorithm. Assume that the edge $\overline{uv}$ was returned by the previous round. The two-pair algorithm considers each node, and computes the neighborhood-deleted components of that node. Depending on the node being processed in the present round, three cases arise:

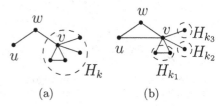

**Fig. 21.** Components $H_k$ of the graph $\overline{G}$: (a) iteration $i$ and (b) iteration $i+1$

**Case 1.** (Node $u$ or $v$ is being processed): In this case, node $u$ ($v$ respectively) has a new neighbor because of the newly-added edge $\overline{uv}$. Also, the components $H_k$ can change (see Fig. 21). For this case, we will just redo the two-pair algorithm for the respective nodes $u$ and $v$.

**Case 2.** (Node $u$ and $v$ are not neighbors of the node $s$ being processed): In this case, the new edge could not have modified the neighbor set of the node being considered. The only change it could have wrought is to join at most two components $H_i$ and $H_j$. If a disjoint set data structure is used, then the union of two sets can be done in $\mathcal{O}(n)$ time. We will also need to modify the set $N_i(v)$ for these two sets $H_i$ and $H_j$, while the neighborhood of all other $H_k$'s remain the same. This takes $\mathcal{O}(n)$ time since the values of $NV(p)$ remain the same for all nodes $p$.

**Case 3.** (Node $u$ or $v$ is a neighbor of the node $s$ being processed:) In this case, the new edge could not have modified the neighbor set of the node being considered. Neither could it modify the components $H_k$'s because when we remove the node and its neighbors, we will get the exact same subgraphs as was considered in the previous iteration. The only change will be in the set $N_i(s)$. Therefore, if the node $u$ is a neighbor of $s$, then the component $H_i$ which contains the node $v$ will now also have $u$ in its set $N_i(v)$. Also, the set $NV(v)$ will now also include the node $u$. Both of these operations can be done in $\mathcal{O}(n)$ time.

If both the nodes $u$ and $v$ are neighbors of $s$, it is easy to see that there will be no change in any of the components or sets. We consolidate all of the above discussion and other details into a formal algorithm described below.

---

**Algorithm 1:** Peripheral-Edge-Ordering

---

**Data**: A graph $G = (V, E)$, and adjacency lists denoted by $Adj(v), v \in V$.
**Result**: A peripheral edge order of the graph, if it exists.
(1) Compute the adjacency matrix of the graph $G = Adj(G)$ and the complement graph $= Adj(\overline{G})$.
(2) **Initialize** the list that will contain the peripheral edge order as $PE = \varnothing$
(3) Run the **two-pair** algorithm on the complement graph $\overline{G}$, let it return the pair $\overline{uv}$.
 Also, while running the algorithm store all intermediate values.
 (in particular, for each node $v$, store the corresponding components $H_k$'s, $NV(u)$, $N_i(u)$ and $label(u)$ for all nodes $u \neq v$).
(4) **Add** the edge $\overline{uv}$ to the list $PE$
(5) **Remove** the edge $\overline{uv}$ from the graph $G$.
(6) **While** $G$ contains some edge **do**
  **(6.1) Update** the adjacency matrix of the complement graph $\overline{G}$ denoted by $Adj(\overline{G})$
  **(6.2) for** $i = 1$ to $n$ **do**
    **(6.2.1) if** node $i$ belongs to Case 1 **then**
       Run the two-pair algorithm for this node and update the various components $H_k$, $NV(u)$, $N_i(u)$ and $label(u)$.
    **(6.2.2) if** node $i$ belongs to Case 2 **then**
       Get the two components in which $u$ and $v$ belong.
       Obtain the set $N_i(v)$ which is the union of these two sets.
       Also make a new set $H_k$ which is a union of these sets.
    **(6.2.3) if** node $i$ belongs to Case 3 **then**
       **Update** the value of $N_i(v)$ and $NV(u)$ for that component in which a new incident edge got added.
    **(6.2.4) for all** $u \in H$ **do**
       **if** $|N_{label(u)}(v)| = |NV(u)|$ **then** declare $(u, v)$ is a two-pair and STOP;
  **(6.3)** Append the edge $\overline{uv}$ to the list $PE$.
  **(6.4)** Remove the edge $\overline{uv}$ from the graph $G$

---

## 5.1   Time Complexity

Steps 2, 4, and 5 take $\mathcal{O}(1)$ time if we are given the adjacency matrix. Step 1 takes $\mathcal{O}(n^2)$ time to compute, and Step 3 takes $\mathcal{O}(n^3)$ time to compute (remember that the complement graph can have as many as $\mathcal{O}(n^2)$ edges).

Each of the *if* conditions in Step 6.2.2 and Step 6.2.3 for case 2 and case 3 take at most $\mathcal{O}(n)$ time. But the *if* condition for case 1 in Step 6.2.1 takes $\mathcal{O}(n^2)$ time (note that we are only executing one iteration of the two-pair algorithm). Also the *for* loop in Step 6.2.4 is executed for each node, i.e. $n$ times. The total time that Step 6.2 takes is $\mathcal{O}(n^2)$ though since case 1 occurs for only two nodes (the ones in which the edge was added), and for the remaining $n - 2$ nodes, it will be $\mathcal{O}(n)$ time. Thus, in total we spend only $\mathcal{O}(n^2)$ time for each iteration. Also, the *while* loop in Step 6 is executed $m$ times, since after each iteration one edge is removed from the graph $G$. Therefore, the time complexity of our algorithm is in $\mathcal{O}(n^2 \cdot m)$.

Hayward et al. [1] proposed an algorithm that employs the interesting concept of handles for recognizing weakly triangulated that runs in $\mathcal{O}(m^2)$ time. We can obtain a peripheral edge order in the same time. When $G$ is sparse, we can use this result to obtain a more efficient algorithm.

# 6   Conclusions

In this paper we have brought together an interesting mix of techniques to solve the problem of generating all linear layouts of weakly triangulated graphs. It would be interesting to explore if these extend to other classes of graphs. Another interesting direction is to consider generating two-dimensional layouts of graphs.

**Acknowledgements.** This work is supported by an NSERC Discovery Grant to the first author. It was done during the first author's sabbatical visit to the Indian Institute of Technology, Guwahati, India. The support provided by the Indian Institute of Technology, Guwahati is gratefully acknowledged.

# References

1. Hayward, R.B., Spinrad, J.P., Sritharan, R.: Improved algorithms for weakly chordal graphs. ACM Trans. Algorithms 3(2) (2007)
2. Alam, M.S., Mukhopadhyay, A.: Improved upper and lower bounds for the point placement problem. Technical report, University of Windsor (2010)
3. Saxe, J.B.: Embeddability of weighted graphs in $k$-space is strongly NP-hard. In: 17th Allerton Conference on Communication, Control and Computing, pp. 480–489 (1979)
4. Barvinok, A.I.: Problems of distance geometry and convex propeties of quadratic maps. Discrete and Computational Geometry 13, 189–202 (1995)
5. Alfakih, A.Y., Wolkowicz, H.: On the embeddability of weighted graphs in euclidean spaces. Technical report, University of Waterloo (1998)
6. Hastad, J., Ivansson, L., Lagergren, J.: Fitting points on the real line and its application to rh mapping. Journal of Algorithms 49, 42–62 (2003)
7. Damaschke, P.: Point placement on the line by distance data. Discrete Applied Mathematics 127(1), 53–62 (2003)
8. Hayward, R.: Generating weakly triangulated graphs. J. Graph Theory 21, 67–70 (1996)
9. Arikati, S.R., Rangan, C.P.: An efficient algorithm for finding a two-pair, and its applications. Discrete Applied Mathematics 31(1), 71–74 (1991)

# Bichromatic Point-Set Embeddings
## of Trees with Fewer Bends
## (Extended Abstract)

Khaled Mahmud Shahriar and Md. Saidur Rahman

Department of Computer Science and Engineering,
Bangladesh University of Engineering and Technology (BUET)
Dhaka-1000, Bangladesh
{khaledshahriar,saidurrahman}@cse.buet.ac.bd

**Abstract.** Let $G$ be a planar graph such that each vertex of $G$ is colored by either red or blue color. Assume that there are $n_r$ red vertices and $n_b$ blue vertices in $G$. Let $S$ be a set of fixed points in the plane such that $|S| = n_r + n_b$ where $n_r$ points in $S$ are colored by red color and $n_b$ points in S are colored by blue color. A bichromatic point-set embedding of $G$ on $S$ is a crossing free drawing of $G$ such that each red vertex of $G$ is mapped to a red point in $S$, each blue vertex of $G$ is mapped to a blue point in $S$, and each edge is drawn as a polygonal curve. In this paper, we study the problem of computing bichromatic point-set embeddings of trees on two restricted point-sets which we call "ordered point-set" and "properly-colored point-set". We show that trees have bichromatic point-set embeddings on these two special types of point-sets with at most one bend per edge and such embeddings can be found in linear time.

**Keywords:** Trees, Bichromatic point-set embedding, Bend, Ordered point-set, Properly-colored point-set.

# 1 Introduction

Let $G = (V, E)$ be a planar graph where $V$ and $E$ are the set of vertices and edges, respectively. Let $V_r$ and $V_b$ be a partition of $V$ such that the vertices in $V_r$ are colored red and the vertices in $V_b$ are colored blue. Let $S$ be a set of points in the plane such that $|S| = |V|$ and $S$ contains $|V_r|$ red points and $|V_b|$ blue points. A *bichromatic point-set embedding* of $G$ on $S$ is a crossing free drawing of $G$ such that each red vertex $v_r \in V_r$ is mapped to a red point $p_r \in S$, each blue vertex $v_b \in V_b$ is mapped to a blue point $p_b \in S$, and each edge of $G$ is drawn as a polygonal curve. The general version of the problem of computing bichromatic point-set embeddings is known as the *k-chromatic point-set embedding* problem where the input graph $G$ and input point-set $S$ are colored using $k$ different colors for $1 \le k \le |V|$. For $k = 1$, Cabello [1] has shown that the problem of determining whether a planar graph $G$ has a monochromatic point-set embedding without bends is *NP*-complete. Hence, researchers have focused

S.P. Pal and K. Sadakane (Eds.): WALCOM 2014, LNCS 8344, pp. 337–348, 2014.

on finding $k$-chromatic point-set embeddings by allowing bends on the edges and consequently, determining the maximum number of bends per edge that is required to compute such drawings. For $k = n$, where $n$ denotes the number of vertices in the given graph, Pach and Wenger [7] have shown that $O(n)$ bends per edge are required for $n$-chromatic point-set embeddings of planar graphs; hence, the number of bends per edge increases linearly with the number of vertices of $G$ for the maximum value of $k$. On the other hand, for $k = 1$, Kaufmann and Wiese [6] have shown that any planar graph has a monochromatic point-set embedding with at most two bends per edge. Surprisingly, for the immediate next value of $k$, i.e., $k = 2$, Di Giacomo et al. [3] have shown that there exists instances of planar graphs that require linear number of bends per edge for bichromatic point-set embeddings. However, there are smaller classes of planar graphs that admit bichromatic point-set embeddings with constant number of bends per edge. Di Giacomo et al. [3] have presented algorithms to compute bichromatic point-set embeddings of paths and cycles with at most one bend per edge and of caterpillars with at most two bends per edge. A more general result by Di Giacomo et al. [2] has shown that every outerplanar graph has a bichromatic point-set embedding with at most 5 bends per edge. Interestingly, it is also possible to find bichromatic point-set embeddings with constant number of bends per edge by working on restricted configurations of point-sets as Di Giacomo et al. [4] have shown that every planar graph has $k$-chromatic point-set embeddings with at most $3k + 7$ bends per edge on ordered point-sets.

Motivated by these results, we have studied the problem of computing bichromatic point-set embeddings of trees, a larger class of planar graphs than caterpillars, with at most one bend per edge on some special types of point-sets. Let the given point set $S$ be such that no two points in $S$ have the same $x$-coordinate. Assume, an ordering $x$-$ord$ of the points in $S$ by increasing $x$-values. $S$ is called an *ordered* point-set when all the points of the same color appear consecutively in $x$-$ord$. $S$ is called a *properly-colored* point-set when no two points of the same color appear consecutively in $x$-$ord$. In this paper, we prove that trees admit bichromatic point-set embeddings on every ordered point-set and every properly-colored point-set with at most one bend per edge and such drawings can be computed in linear time.

The rest of the paper is organized as follows. Section 2 describes some definitions. In Section 3, we prove that trees admit bichromatic point-set embeddings on ordered point-sets with at most one bend per edge. Then we prove the existence of bichromatic point-set embeddings of trees on properly-colored point-sets with at most one bend per edge in Section 4. Finally, Section 5 gives the conclusion and direction for future works.

## 2   Preliminaries

We assume familiarities with basic graph drawing terminology and present only those definitions and known results that are used in the rest of this paper.

Let $G = (V, E)$ be a planar graph. We say $G$ is 2-*colored* if vertices of $G$ are colored either *red* or *blue*. Let $V_r$ and $V_b$ be the set of *red vertices* and *blue vertices* of $G$, respectively. Similarly, we call a set of fixed points $S$ in the Euclidian plane 2-*colored* if points in $S$ are colored either *red* or *blue*. Let $S_r$ and $S_b$ be the set of *red points* and *blue points* of $S$, respectively. We say that $S$ is *compatible* with $G$ if $|V_b| = |S_b|$ and $|V_r| = |S_r|$. We use $c(.)$ to denote the color of a vertex or a point.

In the rest of the paper, we assume that for any given point-set $S$, no two points of $S$ have the same $x$-coordinates (if this is not the case for some point-set $S$, we can rotate the plane to achieve distinct $x$-coordinates for all the points in $S$). This gives an ordering *$x$-ord* of the points in $S$ by increasing $x$-values. We call $S$ an *ordered* point-set when all the points of the same color appear consecutively in *$x$-ord*. It is easy to observe that in an *ordered* point-set red points can be separated from blue points by a line. We call $S$ a *properly-colored* point-set when no two consecutive points in *$x$-ord* have the same color.

We call a 2-colored point-set $\sigma$ an *RB-sequence* when all the points of $S$ are collinear. We call the line $l$ that passes through the points of $\sigma$, the *spine* of $\sigma$. We assume $l$ is always parallel to $x$-axis. For a point $p$ of $\sigma$, we use $next(p)$ to denote the point next to the right of $p$ on the spine $l$ of $\sigma$. A spine defines 2 half planes (called *pages*) sharing line $l$; the top (bottom) half plane is called the *top page* (resp. *bottom page*).

Let $\Gamma$ be a bichromatic point-set embedding of a 2-*colored* tree $G$ on an *RB-sequence* $\sigma$. We say a point $p$ on the spline of $\sigma$ is *accessible from top (bottom) page* in $\Gamma$ if there is no edge $e$ in $\Gamma$ such that $e$ is drawn through the top (resp. bottom) page and $p$ lies between the endpoints of $e$ on $l$. Two points that are accessible from the same page, can be connected with a polygonal chain of at most one bend and without any edge crossing; the proof can be found in [3].

Let $P$ and $Q$ be two 2-colored point-sets. We say $P$ is *chromatic equivalent* to $Q$ if the following two conditions hold: (i) $|P| = |Q|(= n)$, and (ii) $c(p_i) = c(q_i)$ for $0 \leq i \leq n-1$ where $p_0, p_1, \ldots, p_{n-1}$ and $q_0, q_1, \ldots, q_{n-1}$ denote the points in $P$ and $Q$, respectively in the order of increasing $x$-coordinate values.

We have the following lemma that relates bichromatic point-set embeddings on two chromatic equivalent point-sets.

**Lemma 1.** *Let $S$ and $\sigma$ be two chromatic equivalent point-sets. Then, an 1-bend bichromatic point-set embedding of a 2-colored tree $G$ on $S$ can be computed in linear time from an 1-bend bichromatic point-set embedding of $G$ on $\sigma$.*

The proof of this lemma is omitted in this extended abstract.

## 3   Embedding Trees on Ordered Point-Sets

In this section we prove the following theorem.

**Theorem 1.** *Let $G = (V, E)$ be a 2-colored tree. Let $S$ be a 2-colored ordered point-set compatible with $G$. Then $G$ has a bichromatic point-set embedding $\Gamma$ on $S$ with at most one bend per edge. Furthermore, $\Gamma$ can be computed in linear time.*

We give a constructive proof of Theorem 1. Let $\sigma$ be 2-colored RB-sequence chromatic equivalent to $S$. Then $\sigma$ is also an ordered point-set and compatible with $G$. If $G$ has a bichromatic point-set embedding on $\sigma$ with at most one bend per edge, then by Lemma 1, $G$ has an embedding on $S$ with at most one bend per edge. We thus describe, in the rest of this section, an algorithm that finds a bichromatic point-set embedding of $G$ on $\sigma$. We call this algorithm

### Ordered-Sequence-Embedding

Without loss of generality, we assume that the leftmost point in $\sigma$ is red. We assume any red vertex of $G$ as its root and denote it by $v_0$.

We find a bichromatic point-set embedding of $G$ on $\sigma$ by mapping vertices of $G$ on points of $\sigma$ in an incremental way as follows. We first embed the root $v_0$. Then at each subsequent step, we embed a new vertex of $G$ on a point in $\sigma$ in such a way that the resulting drawing maintains some invariants. These invariants ensure that the next vertex can be mapped in the next step without any edge crossings and with at most one bend per edge.

Before proceeding further we need some definitions. Let $\gamma_k$ denote the drawing after some step $k$, $k \geq 0$. For example, Fig. 1(b) shows the drawing $\gamma_4$ of step 4 for the input graph in Fig. 1(a).

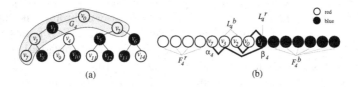

(a)                                        (b)

**Fig. 1.** (a) A 2-colored tree $G$; $G_4$ is shown as the shaded subgraph, and (b) the drawing $\gamma_4$ and the sets $L_4^r$ and $L_4^b$

We denote by $G_k$ the subgraph of $G$ that has been drawn in $\gamma_k$. We call any vertex $v$ in $V(G)\backslash V(G_k)$ an *unmapped* vertex and vertices of $G_k$ *mapped* vertices. A mapped vertex $v$ of $G_k$ is a *live* vertex if it has at least one unmapped neighbor; $v$ is called a *R-live* (*B-live*) vertex if $v$ has at least one unmapped red (resp. blue) neighbor. Thus a vertex can be both a *R-live* and *B-live* vertex if it has unmapped red and blue neighbors.

Let $\sigma_k \subseteq \sigma$ denote the set of points representing the vertices of $G_k$ in $\gamma_k$. We use $p(v)$ to denote the point of $\sigma$ that represents a vertex $v$ of $G$; if $v$ is a R-live (B-live vertex), $p(v)$ is called an *R-live* (*B-live*) point. The set of R-live and B-live points of $\sigma_k$ will be denoted by $L_k^r$ and $L_k^b$, respectively. We use $\alpha_k$ and $\beta_k$ to denote the leftmost and the rightmost points of $\sigma_k$, respectively. We say any point $p$ of $\sigma\backslash\sigma_k$ is a *free* point; thus a free point does not represent any vertex of $G$. The set of free red and blue points of $\sigma\backslash\sigma_k$ will be denoted by $F_k^r$ and $F_k^b$, respectively. For an illustration of the given definitions, see Fig. 1(b).

We call $\gamma_k$, $0\leq k < n$, a *feasible* drawing of $G_k$ if $\gamma_k$ satisfies the following invariants.

**(1)** $G_k$ is connected and $\gamma_k$ represents a bichromatic point-set embedding of $G_k$ on $\sigma_k$ with at most one bend per edge.

**(2)** All points in $F_k^r$ are to the left of $\alpha_k$ and all points in $F_k^b$ are to the right of $\beta_k$.

**(3)** All points in $L_k^r$ are accessible from the bottom page and all points in $L_k^b$ are accessible from the top page.

We now describe Algorithm **Ordered-Sequence-Embedding**.

At step $k = 0$, the root vertex $v_0$ where $c(v_0)$ is red, is mapped to the rightmost free red point $f_r$ of $\sigma$. At any step $k > 0$, we have the following two cases to consider.

**Case 1:** *There is at least one R-live point in* $\sigma_{k-1}$. Let $v$ be the vertex of $G_{k-1}$ which is mapped to the leftmost R-live point $l_r$ of $\sigma_{k-1}$. We take any unmapped red neighbor $u$ of vertex $v$ and map vertex $u$ to the rightmost free red point $f_r$. Then we add the edge $(u, v)$ connecting the points $l_r$ and $f_r$ through the bottom page. As an example, consider Fig. 2(b) which represents the drawing $\gamma_{k-1}$, at some step $k > 0$, for the graph in Fig. 2(a); the shaded subgraph is the graph $G_{k-1}$. Note that, in Fig. 2(b), $p(v_3)$ is the leftmost R-live point and $f_r$ is the rightmost free red point. Vertex $v_7$ is an unmapped red neighbor of vertex $v_3$ in $G_{k-1}$ as can be seen from Fig. 2(a). Hence, we map $v_7$ on $f_r$ and draw the edge $(v_3, v_7)$ as shown in Fig. 2(c).

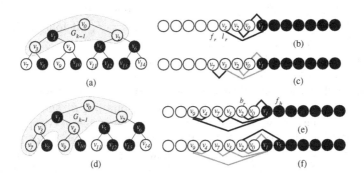

**Fig. 2.** Illustrations of different cases of step $k$ of Algorithm **Ordered-Sequence-Embedding**. (a), (d): A 2-colored tree $G$; $G_{k-1}$ is shown as the shaded subgraph. (b), (e): The drawing $\gamma_{k-1}$. (c), (f): The drawing $\gamma_k$.

**Case 2:** *There is no R-live point in* $\sigma_{k-1}$. In this case, there must be at least one B-live point in $\sigma_{k-1}$; otherwise the drawing process is complete. Let $v$ be the vertex of $G_{k-1}$ which is mapped to the rightmost B-live point $l_b$ of $\sigma_{k-1}$. We take any unmapped blue neighbor $u$ of vertex $v$ and map vertex $u$ to the leftmost free blue point $f_b$. Then we add the edge $(u, v)$ connecting points $l_b$ and $f_b$ through the top page. See Figs. 2(d), 2(e) and 2(f) for an example.

This completes the description of the drawing algorithm. We now prove the following lemma.

**Lemma 2.** $\gamma_k$ *is a feasible drawing of* $G_k$, *for* $0 \le k < n$.

*Proof.* We give an inductive proof.

*Base Case (k =0):* Since the drawing $\gamma_0$ has only one vertex, i.e., the root $v_0$ and no edge, it immediately follows that $\gamma_0$ satisfies Invariants (1)-(3). Hence, $\gamma_0$ is a feasible drawing of $G_0$.

*Induction (k >0):* By induction hypothesis, the drawing $\gamma_{k-1}$ is a feasible drawing of the graph $G_{k-1}$. We now show that the drawing $\gamma_k$ satisfies the given Invariants (1)-(3).

$\gamma_k$ satisfies Invariant (1): First consider Case 1. According to the operation specified, $V(G_k) = V(G_{k-1}) \bigcup \{u\}$. Since vertex $u$ is a neighbor of vertex $v$ in $V(G_{k-1})$ and we draw the edge $(u, v)$ in $\gamma_k$, it follows that $G_k$ is connected. To prove that $\gamma_k$ represents a bichromatic point-set embedding of $G_k$ on $\sigma_k$, we need to show that the edge $(u, v)$ does not create any edge crossing and contains at most one bend. Since $l_r \in L_{k-1}^r$, by Invariant (2), it is accessible from bottom page in $\gamma_{k-1}$. The point $f_r \in F_{k-1}^r$ is to the left of the leftmost point of $\sigma_{k-1}$ (by Invariant (1)); hence, $f_r$ is accessible from both the pages in $\gamma_{k-1}$. Therefore, $f_r$ and $l_r$ can be connected with a polygonal chain through the bottom page that contains at most one bend and does not cross any other edge in $\gamma_{k-1}$.

Now consider Case 2. Using similar arguments as used for Case 1, it can be shown that Invariant (1) holds for Case 2 also.

$\gamma_k$ satisfies Invariant (2): First consider Case 1. By induction hypothesis, points in $F_{k-1}^r$ are to the left of $\alpha_{k-1}$. Since $f_r$ is the rightmost point in $F_{k-1}^r$, it follows that $f_r = \alpha_k$ and $F_k^r = F_{k-1}^r \backslash \{f_r\}$; therefore, points in $F_k^r$ are to the left of $\alpha_k$. On the other hand, $\beta_{k-1} = \beta_k$ and $F_{k-1}^b = F_k^b$. It follows that points in $F_k^b$ are to the right of $\beta_k$.

Now consider Case 2. Using similar arguments as used for Case 1, it can be shown that Invariant (2) holds for Case 2 also.

$\gamma_k$ satisfies Invariant (3): Consider Case 1. We first show that points in $L_k^r$ are accessible from the bottom page. Consider a point $p \in L_k^r$ such that $p$ is not accessible from the bottom page in $\gamma_k$. Let $v_p$ be the vertex of $G$ represented by $p$. Hence, $v_p$ is a R-live vertex in $G_k$. Since $f_r$ is the leftmost point in $\sigma_k$, $f_r$ is accessible from both the pages in $\gamma_k$. It follows that $p \neq f_r$. Since $u$ is mapped on $f_r$, $v_p \neq u$. Then $v_p$ must be a R-live vertex of $G_{k-1}$. It follows that $p$ is a R-live point of $\sigma_{k-1}$. Therefore, by Invariant (2) of induction hypothesis, $p$ is accessible from bottom page in $\gamma_{k-1}$. Consequently, it must be the addition of the edge $(u, v)$ that makes $p$ inaccessible from bottom page in $\gamma_k$. The endpoints of the edge $(u, v)$ are the points $f_r$ and $l_r$; hence, $p$ must lie between $f_r$ and $l_r$. Since $f_r$ is the leftmost point of $\sigma_k$, $f_r$ is to the left of $p$. Now consider the other endpoint $l_r$; either $l_r = p$ or $l_r$ is to the left of $p$ since both $l_r$ and $p$ are in $L_{k-1}^r$ and $l_r$ is the leftmost point of $L_{k-1}^r$. In either case, it follows that $p$ does not lie between $f_r$ and $l_r$, thus the edge $(u, v)$ cannot make $p$ inaccessible from bottom page in $\gamma_k$. Therefore, no point such as $p$ exists. Hence, all points in $L_k^r$ are accessible from the bottom page.

Next, we show that points in $L_k^b$ are accessible from the top page. Consider any point $p \in L_k^b$; either $p = f_r$ or $p \in L_{k_1}^b$. Since $f_r$ is the leftmost point in $\sigma_k$, $f_r$ is accessible from both the pages in $\gamma_k$. If $p \in L_{k_1}^b$, then $p$ is accessible from top page in $\gamma_{k-1}$ by Invariant (2). Since we draw the edge $(u, v)$ through the bottom page, $p$ remains accessible from top page in $\gamma_k$. Thus, all points in $L_k^b$ are accessible from the top page.

Now consider Case 2. Using similar arguments as used for Case 1, it can be shown that Invariant (3) holds for Case 2 also. □

Lemma 2 proves the correctness of Algorithm **Ordered-Sequence-Embedding** to find a bichromatic point-set embedding of $G$ on an ordered RB-sequence $\sigma$ compatible with $G$. Thus, to prove Theorem 1, it remains to show that Algorithm **Ordered-Sequence-Embedding** runs in linear time. We omit the proof in this extended abstract.

# 4    Embedding Trees on Properly-Colored Point-Sets

In this section we prove the following theorem.

**Theorem 2.** *Let $G = (V, E)$ be a 2-colored tree. Let $S$ be a 2-colored point-set such that $S$ is properly-colored and compatible with $G$. Then $G$ has a bichromatic point-set embedding on $S$ with at most one bend per edge. Moreover, such a drawing can be computed in linear time.*

We give a constructive proof of Theorem 2. Let $\Gamma$ be a drawing of $G$ with at most one bend per edge. Let $\sigma$ be the set of points representing the vertices of $G$ in $\Gamma$. If $\sigma$ is chromatic equivalent to $S$, then by Lemma 1, $G$ has an embedding on $S$ with at most one bend per edge. Therefore, we present an algorithm which computes a drawing of $G$ where each edge of $G$ contains at most one bend and the set of points in the drawing is chromatic equivalent to $S$. We call this algorithm **Proper-Sequence-Embedding**. In the rest of this section, we describe Algorithm **Proper-Sequence-Embedding**.

It is implicitly assumed that either there are equal number of red and blue vertices of $G$ or the numbers differ by at most one; otherwise there can be no properly-colored point-set $S$ compatible with $G$. We choose any vertex $v_0$ of $G$ as its root where $v_0$ and the leftmost point of $S$ are of the same color.

The outline of the algorithm is as follows. We start with a point-set $\sigma_0$, which contains a single point $p_0$ where $c(p_0) = c(v_0)$ and map $v_0$ on $p_0$. In subsequent steps, we add new points to the existing point-set and map vertices of $G$ which have not yet been drawn on those points in such a way that the resulting drawing at the end of the each step $k$ $(k > 0)$, satisfies the following two conditions. (i) The set of points $\sigma_k$ is a properly-colored RB-sequence, and (ii) the drawing $\gamma_k$ represents a bichromatic point-set embedding of a connected subgraph $G_k$ of $G$ with at most one bend per edge on a subset of the points in $\sigma_k$; at an intermediate step there may be some points in $\sigma_k$ which do not represent any vertex of $G$ in $\gamma_k$; however our algorithm ensures that no such point exists when the drawing procedure completes. For example, Fig. 3(b) shows the drawing $\gamma_5$ and point-set $\sigma_5$ after some intermediate step $k = 5$ for the input 2-colored tree $G$ in Fig. 3(a); the shaded graph in Fig. 3(a) is the subgraph $G_5$.

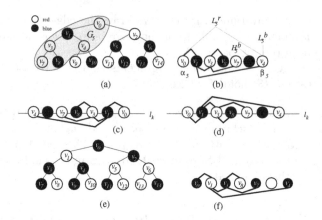

**Fig. 3.** (a) A 2-colored tree $G$. (b) The drawing $\gamma_5$ after step 5 and the sets $L_5^r$, $L_5^b$, $H_5^b$. (c) Drawing obtained after the horizontal flip of the drawing in (b). (d) Drawing obtained after the vertical flip of the drawing in (b). (e) Resulting graph after the inversion of the graph in (a). (f) Resulting drawing after the inversion of the drawing in (b).

For our illustration, we will use the same definitions and notations for un-mapped/mapped vertex, R-live/B-live vertex/point as described in Section 3. Additionally, we use the following definitions and notations. We call any point of $\sigma_k$ that does not represent a vertex of $G$ as a *hole*; a hole can be either red or blue. The set of blue holes and red holes in $\sigma_k$ will be denoted by $H_k^b$ and $H_k^r$, respectively. Hence, $\sigma_k \backslash \{H_k^b \bigcup H_k^r\}$ denotes the set of points in $\sigma_k$ that represent the vertices of $G_k$. We call rotation of $\gamma_k$ by an angle of 180 degree with respect to any line perpendicular to the spine of $\sigma_k$ a *horizontal flip*. As an example, Fig. 3(c) shows the drawing obtained after horizontal flip of the drawing in Fig. 3(b). Likewise, we call rotation of $\gamma_k$ by an angle of 180 degree with respect to the spine of $\sigma_k$ as a *vertical flip*. Fig. 3(d) illustrates vertical flip of the drawing $\gamma_k$ in Fig. 3(b).

We define *inversion* of any 2-colored graph $G$ as changing the color of each of the vertices of $G$ such that each blue vertex of $G$ becomes a red vertex and each red vertex becomes a blue vertex. Similarly inversion of any 2-colored point-set $\sigma$ is defined as changing color of each blue point to red and each red point to blue. One can observe that the point-set obtained after inverting a properly-colored RB-sequence is also a properly-colored RB-sequence. For example, Fig. 3(e) shows the graph obtained after inversion of the graph in Fig. 3(a) and Fig. 3(f) shows the drawing obtained after inversion of the point-set in Fig. 3(b).

We call $\gamma_k$, $0 \leq k < n$, a *feasible* drawing of $G_k$ if $\gamma_k$ satisfies the following invariants.

**(1)** All the R-live points, i.e., points in $L_k^r$ are accessible from the bottom page.

**(2)** All the B-live points, i.e., points in $L_k^b$ are accessible from the top page.

**(3)** All blue holes, i.e., points in $H_k^b$ are accessible from the top page.

**(4)** $\sigma_k$ is a properly-colored RB-sequence that satisfies the following conditions. (i) There is no red hole in $\sigma_k$, i.e., $H_k^r = \phi$; (ii) if the rightmost point of $\sigma_k$ is blue then there is no blue hole in $\sigma_k$, and (iii) there is no B-live point to the left of any blue hole in $\sigma_k$.

**(5)** $G_k$ is connected and $\gamma_k$ represents a bichromatic point-set embedding of $G_k$ on $\sigma_k \backslash H_k^b$ with at most one bend per edge. Moreover, the root $v_0$ of the input graph $G$ is represented by the leftmost point of $\sigma_k$.

For an illustration of the invariants described above, see Fig. 3(b) that shows a *feasible* drawing $\gamma_k$, $k = 5$, of the shaded subgraph $G_k$ in Fig. 3(a).

We now specify the drawing operations performed by Algorithm **Proper-Sequence-Embedding** which ensures that at any step $k$, $\gamma_k$ is a *feasible* drawing of $G_k$.

At step $k = 0$, we take any point $p_0$ on the plane such that $c(p_0) = c(v_0)$ and map the root $v_0$ on $p_0$. The drawing $\gamma_0$ thus obtained has only one vertex $v_0$ and no edge.

At any step $k > 0$, we have the following cases to consider.

**Case 1:** *The rightmost point of $\sigma_{k-1}$ is red, $\sigma_{k-1}$ contains at least one B-live point and does not contain any blue hole.* We add a blue point $p_b$ to the right of $\beta_{k-1}$ on the spine of $\sigma_{k-1}$. Let $v$ be the vertex of $G_{k-1}$ which is mapped to the rightmost B-live point $l_b$ of $\sigma_{k-1}$. We take any unmapped blue neighbor $u$ of vertex $v$ and map $u$ on $p_b$. Then we draw the edge $(v, u)$ connecting the points $l_b$ and $p_b$ through the top page. As an example, consider Fig. 4(b) which represents the drawing $\gamma_{k-1}$ at some step $k$ for the graph in Fig. 4(a), the shaded subgraph is the graph $G_{k-1}$. Note that, in Fig. 4(b), the rightmost point $p(v_7)$ is red and $p(v_1)$ is the rightmost B-live point. $v_4$ is an unmapped blue neighbor of $v_1$ as can be seen from Fig. 4(a). Hence, we add a blue point $p_b$ to the right of $p(v_7)$, map $v_4$ on $p_b$ and draw the edge $(v_1, v_4)$ to obtain the drawing $\gamma_k$ as shown in Fig. 4(c).

**Case 2:** *The rightmost point of $\sigma_{k-1}$ is blue and $\sigma_{k-1}$ contains at least one R-live point.* We add a red point $p_r$ to the right of $\beta_{k-1}$ on the spine of $\sigma_{k-1}$. Let $v$ be the vertex of $G_{k-1}$ which is mapped to the rightmost R-live point $l_r$ of $\sigma_{k-1}$. We take any unmapped red neighbor $u$ of vertex $v$ and map $u$ on $p_r$. Then we draw the edge $(v, u)$ connecting points $l_r$ and $p_r$ through the bottom page. See Figs. 4(d), (e) and (f) for an example.

**Case 3:** *The rightmost point of $\sigma_{k-1}$ is red, $\sigma_{k-1}$ does not contain any B-live point but contains at least one R-live point.* We add a blue point $p_b$ and then a red point $p_r$ to the right of $\beta_{k-1}$ on the spine of $\sigma_{k-1}$. Let $v$ be the vertex of $G_{k-1}$ which is mapped to the rightmost R-live point $l_r$ of $\sigma_{k-1}$. We take any unmapped red neighbor $u$ of vertex $v$ and map $u$ on $p_r$. Then we draw the edge $(v, u)$ connecting points $l_r$ and $p_r$ through the bottom page. See Figs. 4(g), (h) and (i) for an example.

**Case 4:** *The rightmost point of $\sigma_{k-1}$ is red, $\sigma_{k-1}$ contains at least one B-live point and also contains at least one blue hole.* Let $v$ be the vertex mapped to the leftmost B-live point $l_b$ of $\sigma_{k-1}$. Consider any unmapped red neighbor $u$ of vertex $v$. We now apply the drawing procedure separately on the subtree

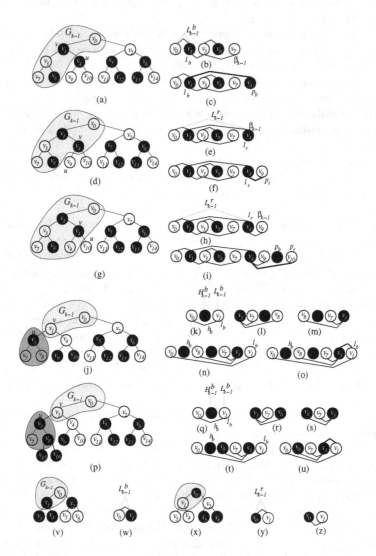

**Fig. 4.** Illustrations for different cases at step $k$ of Algorithm **Proper-Sequence-Embedding**. (a), (d), (g), (j), (p), (v): A 2-colored tree $G$. (b), (e), (h), (k), (q), (w): The drawing $\gamma_{k-1}$. (l), (r) The drawing $\gamma_u$. (m), (s): Resulting drawing after horizontal flip of $\gamma_u$. (n), (t): Resulting drawing after inserting $\gamma_u$ in $\gamma_{k-1}$. (x) Resulting graph after inversion of the graph in (v). (y) Resulting drawing after the inversion of the drawing in (w). (c), (f), (i), (o), (u), (z): The drawing $\gamma_k$.

rooted at vertex $u$ of $G$. In the rest of this paper, we will refer to such recursive application of the drawing procedure as *subtree-embed*. Let $\gamma_u$ be the drawing obtained from such a *subtree-embed* and $G_u$ be the subgraph of $G$ represented by $\gamma_u$. Note that there may be two cases for $G_u$: (i) $G_u$ contains all the vertices of the subtree rooted at $u$ of $G$ if *subtree-embed* terminates as in Case 6, (ii) $G_u$ is a subgraph of the subtree rooted at $u$ of $G$ when *subtree-embed* terminates as in Case 5.

We now merge the drawing $\gamma_u$ with the drawing $\gamma_{k-1}$. First, we flip $\gamma_u$ horizontally. Let $\sigma_u$ denote the point-set in the drawing after the horizontal flip operation. We will prove later that the rightmost point $\beta_u$ of $\sigma_u$ will always represent the vertex $u$. Let $h_b$ be the rightmost point in $H_{k-1}^b$. We insert the drawing $\gamma_u$ between the points $h_b$ and $next(h_b)$ of $\sigma_{k-1}$. Then, if the leftmost point of $\sigma_u$ is blue, we remove the point $h_b$ from the resulting drawing. Finally, we add the edge $(u, v)$ connecting the points $l_b$ and $\beta_u$ through the top page.

For an illustrative example, see Figs. 4(j), (k), (l), (m),(n) and (o). In this example, $G_u$ contains all the vertices of the subtree rooted at $u$. For another example where $G_u$ is a subgraph of the subtree rooted at $u$ of $G$, see Figs. 4(p), (q), (r), (s),(t) and (u).

**Case 5:** *The rightmost point of $\sigma_{k-1}$ is blue and $\sigma_{k-1}$ does not contain any R-live point but contains at least one B-live point.* Here we distinguish two subcases based on whether the drawing algorithm is applied on a subgraph of $G$ or not. Note that, as part of the drawing algorithm on the input graph $G$, when at some intermediate step the resulting drawing matches Case 4, we apply the same algorithm on an unmapped subgraph of $G$ as described in operations for Case 4; we used the term *subtree embed* to denote such recursive step. Thus, the two subcases for Case 5 are as follows.

**Case 5.1:** *In subtree-embed*; if the drawing $\gamma_{k-1}$ is in this state then the drawing process terminates and $\gamma_{k-1}$ is returned as output.

**Case 5.2:** *Not in subtree-embed*; since there are only B-live points in $\sigma_{k-1}$ and the rightmost point of $\sigma_{k-1}$ is blue, it is not possible to map the next unmapped vertex (which is a neighbor of an already mapped vertex) without creating a red hole if we want to maintain that the resulting point-set remains properly-colored. But to maintain Invariant (4), we must ensure that there exists no red hole in the drawing at any intermediate step. Hence, to maintain the desired properties of the drawing, we perform the following operations. First, we invert both $G$ and $\sigma_{k-1}$ and then we flip the drawing $\gamma_{k-1}$ vertically to obtain the drawing $\gamma_k$. For an illustrative example, see Figs. 4(v), (w), (x), (y) and (z).

**Case 6:** *There are no live points in $\sigma_{k-1}$.* Since there is no unmapped vertex to embed, therefore, this case indicates the end of drawing operation.

This completes the description of the Algorithm **Proper-Sequence-Embedding**. To prove Theorem 2, we need to show that Algorithm **Proper-Sequence-Embedding** is correct and runs in linear time. We omit the proof in this extended abstract.

## 5  Conclusion

In this paper, we have shown that trees admit bichromatic point-set embeddings on two special types of point-sets, namely, "ordered" point-sets and "properly-colored" point-sets with at most one bend per edge. It should be mentioned that these results are based on the first author's thesis work [8], and we have noticed that an independent proof of Theorem 1 has appeared in [5] recently.

These results naturally raise some other open problems such as finding other larger classes of outerplanar graphs as well as special configurations of point-sets that admit bichromatic point-set embeddings with at most one bend per edge and exploring 3-chromatic point-set embedding problem with constant number of bends per edge for outerplanar graphs.

**Acknowledgement.** This work is based on an M. Sc. Engineering thesis work [8] done in Bangladesh University of Engineering and Technology (BUET). We thank BUET for its facilities and support.

## References

1. Cabello, S.: Planar embeddability of the vertices of a graph using a fixed point set is np-hard. Journal of Graph Algorithms and Applications 10(2), 353–363 (2006)
2. Di Giacomo, E., Didimo, W., Liotta, G., Meijer, H., Trotta, F., Wismath, S.K.: k-colored point-set embeddability of outerplanar graphs. In: Kaufmann, M., Wagner, D. (eds.) GD 2006. LNCS, vol. 4372, pp. 318–329. Springer, Heidelberg (2007)
3. Di Giacomo, E., Liotta, G., Trotta, F.: On embedding a graph on two sets of points. International Journal of Foundations of Computer Science 17(05), 1071–1094 (2006)
4. Di Giacomo, E., Liotta, G., Trotta, F.: Drawing colored graphs with constrained vertex positions and few bends per edge. In: Hong, S.-H., Nishizeki, T., Quan, W. (eds.) GD 2007. LNCS, vol. 4875, pp. 315–326. Springer, Heidelberg (2008)
5. Frati, F., Glisse, M., Lenhart, W.J., Liotta, G., Mchedlidze, T., Nishat, R.I.: Point-set embeddability of 2-colored trees. In: Didimo, W., Patrignani, M. (eds.) GD 2012. LNCS, vol. 7704, pp. 291–302. Springer, Heidelberg (2013)
6. Kaufmann, M., Wiese, R.: Embedding vertices at points: Few bends suffice for planar graphs. Journal of Graph Algorithms and Applications 6(1), 115–129 (2002)
7. Pach, J., Wenger, R.: Embedding planar graphs at fixed vertex locations. Graphs and Combinatorics 17(4), 717–728 (2001)
8. Shahriar, K.M.: Bichromatic point-set embeddings of trees with fewer bends. M. Sc. Engg. Thesis, Department of CSE, BUET (2008), http://www.buet.ac.bd/library/Web/showBookDetail.asp?reqBookID=66772&reqPageTopBookId=66772

# $\ell_1$-Embeddability of 2-Dimensional $\ell_1$-Rigid Periodic Graphs

## Norie Fu[1,2]

[1] National Institute of Informatics, Japan
[2] JST, ERATO, Kawarabayashi Large Graph Project, Japan
funorie@nii.ac.jp

**Abstract.** The $\ell_1$-*embedding* problem of a graph is the problem to find a map from its vertex set to $\mathbb{R}^d$ such that the length of the shortest path between any two vertices is equal to the $\ell_1$-distance between the mapping of the two vertices in $\mathbb{R}^d$. The $\ell_1$-embedding problem partially contains the shortest path problem since an $\ell_1$-embedding provides the all-pairs shortest paths. While Höfting and Wanke showed that the shortest path problem is NP-hard, Chepoi, Deza, and Grishukhin showed a polynomial-time algorithm for the $\ell_1$-embedding of planar 2-dimensional periodic graphs. In this paper, we study the $\ell_1$-embedding problem on $\ell_1$-*rigid* 2-dimensional periodic graphs, for which there are finite representations of the $\ell_1$-embedding. The periodic graphs form a strictly larger class than planar $\ell_1$-embeddable 2-dimensional periodic graphs. Using the theory of *geodesic fiber*, which was originally proposed by Eon as an invariant of a periodic graph, we show an exponential-time algorithm for the $\ell_1$-embedding of $\ell_1$-rigid 2-dimensional periodic graphs, including the non-planar ones. Through Höfting and Wanke's formulation of the shortest path problem as an integer program, our algorithm also provides an algorithm for solving a special class of *parametric integer programming*.

## 1 Introduction

The $\ell_1$-*embedding* problem of a graph is the problem to find a map from its vertex set to $\mathbb{R}^d$ such that the length of the shortest path between any two vertices is equal to the $\ell_1$-distance between the mapping of the two vertices in $\mathbb{R}^d$. An *n-periodic graph* is an infinite graph which has $\mathbb{Z}^n$ as a subgroup of its automorphism. Although periodic graphs are infinite, every periodic graph can be represented by a finite data structure called a *static graph*, which is formed based on the extraction of a single period. Periodic graphs are used in research to model such things as the structure of crystals [4], very-large-scale integration (VLSI) circuits [11], and systems of uniform recurrence equations [13].

The fundamental problems on periodic graphs have been widely investigated, such as connectivity by Cohen and Megiddo [3] and, planarity by Iwano and Steiglitz [12]. As for the $\ell_1$-embedding problem, motivated by applications in chemistry, Deza, Shtogrin, and Grishukhin [5] computed the $\ell_1$-embedding of the planar 2-periodic graphs in the catalog of tilings made by Chavey [1]. They used

S.P. Pal and K. Sadakane (Eds.): WALCOM 2014, LNCS 8344, pp. 349–360, 2014.

the algorithm for the $\ell_1$-embedding of a (possibly infinite) planar graph proposed by Chepoi, Deza, and Grishukhin [2]. By exploiting the planarity, the algorithm efficiently enumerates all the *convex cuts* on a planar graph, and constructs the $\ell_1$-embedding using them. The theory of the planarity of periodic graphs developed by Iwano and Steiglitz [12] implies that their algorithm runs in a polynomial time on planar 2-periodic graphs.

It is shown by Höfting and Wanke [10] that the shortest path problem is NP-hard even for 2-periodic graphs including non-planar ones. As $\ell_1$-embedding can provide the shortest paths between all pairs of vertices, we can imply from the result that solving $\ell_1$-embedding problem could also be hard. It is not trivial even to show that the problem is computable, since the graph is infinite.

In this paper, we consider the $\ell_1$-embedding problem of an $\ell_1$-rigid 2-periodic graph, which is a 2-periodic graph that admits an essentially unique $\ell_1$-embedding. The problem generalizes the one considered in [2], since the class of $\ell_1$-rigid 2-periodic graph is strictly larger than the class of planar $\ell_1$-embeddable 2-periodic graphs. It is shown that all planar $\ell_1$-embeddable 2-periodic graphs are $\ell_1$-rigid [2], and it is easy to construct a non-planar $\ell_1$-rigid 2-periodic graph. We propose an exponential-time algorithm to solve that problem. The key tools are *geodesic fibers*, which were originally proposed by Eon [7] as topological invariants on periodic graphs. Geodesic fibers are the most fundamental periodic subgraphs of a periodic graph with its vertex set convex. Using the theory of geodesic fibers, we show that convex cuts on $\ell_1$-rigid 2-periodic graphs can be represented as the union of the geodesic fibers. Using this result, we also show an algorithm to enumerate all the convex cuts on an $\ell_1$-rigid 2-periodic graph. This leads to an $O(2^{|\mathcal{E}|}\mathcal{D}^{(|\mathcal{V}|+2)})$-time algorithm for the $\ell_1$-embedding of the periodic graphs, where $|\mathcal{V}|$ (resp. $|\mathcal{E}|$) is the number of the vertices (resp. the edges) and $\mathcal{D}$ is the maximum degree in the static graph.

We note the relationship between the $\ell_1$-embedding of a periodic graph and *parametric integer programming*, which is an important problem which has applications in compiler optimization. Various algorithms for this problem have been proposed [8,14]. However, it seems that an explicit upper bound for the time complexity of the algorithms has not yet been determined. Through Höfting and Wanke's formulation of the shortest path problem as an integer program [10], the $\ell_1$-embedding problem can be interpreted as a special class of parametric integer programming. The computational complexity of the $\ell_1$-embedding problem is left open, but an upper bound for the time complexity is derived.

## 2    Preliminaries

We begin with the definition of *periodic graphs*.

**Definition 1.** *Let $\mathcal{V}$ be a finite set. A locally finite infinite graph $G = (\mathcal{V} \times \mathbb{Z}^n, E)$ with $E \subset (\mathcal{V} \times \mathbb{Z}^n)^2$ is an $n$-periodic graph if, for any edge $((u, \mathbf{y}), (v, \mathbf{z})) \in E$ and any vector $\mathbf{x} \in \mathbb{Z}^n$, $((u, \mathbf{y} + \mathbf{x}), (v, \mathbf{z} + \mathbf{x})) \in E$.*

In this paper, we consider only connected periodic graphs. Every periodic graph has a finite representation called a *static graph*.

**Definition 2.** *For a periodic graph $G = (V \times \mathbb{Z}^n, E)$, the static graph $\mathcal{G}$ of $G$ is the finite graph with the vertex set $V$ constructed in the following manner: For each edge $e \in E$ connecting $(u, \mathbf{y})$ and $(v, \mathbf{z})$ on $G$, add a directed edge from $u$ to $v$ with the label $\mathbf{z} - \mathbf{y}$.*

Conversely, from a given static graph, we can construct the corresponding periodic graph; we say that *a static graph $\mathcal{G}$ generates a periodic graph $G$.* See Fig. 1 (a) and (b) for an example.

Next we briefly review the theory of the $\ell_1$-*embedding* of graphs. For given vertices $v_1$ and $v_2$ of a graph $G$, by $d_G(v_1, v_2)$, we denote the number of the edges in a shortest path between $v_1$ and $v_2$.

**Definition 3.** *A (possibly infinite) graph $G = (V, E)$ is $\ell_1$-embeddable if there exist $d \in \mathbb{N}$ and a map $\phi : V \to \mathbb{R}^d$ such that $d_G(v_1, v_2) = \|\phi(v_1) - \phi(v_2)\|_{\ell_1} = \sum_{k=1}^{d} |\phi_k(v_1) - \phi_k(v_2)|$ with $v_i \in V$, $\phi(v_i) = (\phi_1(v_i), \ldots, \phi_d(v_i))$ $(i = 1, 2)$.*

We call $\phi$ an $\ell_1$-*embedding* of $G$. Note that the set $\mathbb{Z}^d$ is naturally endowed with a $d$-dimensional square lattice, whose path-metric corresponds to the $\ell_1$-distance. The $\ell_1$-embedding of graphs has a deep relationship with cuts in graphs.

**Definition 4.** *The cut semimetric with respect to a vertex set $S$ on a graph, denoted by $\delta(S)$, is the semimetric defined as follows: for two arbitrary vertices $u$ and $v$, $\delta(S)(u, v) = 1$ if $|S \cap \{u, v\}| = 1$ and $\delta(S)(u, v) = 0$ otherwise.*

**Proposition 1 (Proposition 4.2.2, [6]).** *A finite graph $G$ is $\ell_1$-embeddable if and only if there is a set $\mathcal{S}$ of cuts and a set of non-negative reals $\{\lambda_S\}_{(S,\bar{S}) \in \mathcal{S}}$ such that for any two vertices $v_1$ and $v_2$ of $G$,*

$$d_G(v_1, v_2) = \sum_{(S,\bar{S}) \in \mathcal{S}} \lambda_S \delta(S)(v_1, v_2).$$

The decomposition $d_G = \sum_{(S,\bar{S}) \in \mathcal{S}} \lambda_S \delta(S)$ of $d_G$ into a non-negative combination of cut semimetrics is called an $\ell_1$-*decomposition* of $G$.

**Definition 5.** *An $\ell_1$-embeddable (possibly infinite) graph is $\ell_1$-rigid if it admits a unique $\ell_1$-decomposition.*

In the proof of Proposition 1, an $\ell_1$-embedding is constructed from a given $\ell_1$-decomposition. Proposition 1 can be naturally extended to countably infinite graphs, and the same statement holds for them. A subgraph $F$ of a graph $G$ is *geodesically complete* in $G$ if, for any pair of its vertices, $F$ contains all the shortest paths between them in $G$. A vertex set is *convex* if the subgraph induced by it is geodesically complete. A cut $(S, \bar{S})$ of a graph $G$ is *convex* if both $S$ and $\bar{S}$ are convex on $G$. For $\ell_1$-embeddable (countably infinite) graphs, every cut with a non-zero coefficient in the $\ell_1$-decomposition is shown to be convex [6]. Thus, by enumerating all the convex cuts, an $\ell_1$-embedding can be constructed.

A *geodesic fiber*, which was proposed by Eon [7] as an invariant of a periodic graph, is one of the most fundamental geodesically complete subgraphs of a periodic graph. We finish this section by reviewing the theory of geodesic fibers, which we will use in this paper.

**Definition 6 ([7]).** *A pair* $(F, \mathbf{t})$ *of a subgraph* $F$ *of a periodic graph* $G$ *and a vector* $\mathbf{t}$ *is a geodesic fiber if (a) for any edge* $((u, \mathbf{y}), (v, \mathbf{z}))$ *of* $F$, $((u, \mathbf{y} + \mathbf{t}), (v, \mathbf{z} + \mathbf{t}))$ *is also an edge of* $F$, *(b) for any vertex* $(u, \mathbf{y})$ *of* $F$ *and any vector* $\mathbf{s}$ *which is not parallel to* $\mathbf{t}$, $(u, \mathbf{y} + \mathbf{s})$ *is not in* $F$, *(c)* $F$ *is geodesically complete in* $G$, *and (d)* $F$ *is minimal with respect to the conditions of (a), (b) and (c).*

By definition, a geodesic fiber is also a 1-periodic graph. Eon [7] showed that a geodesic fiber of a periodic graph $G$ has a static graph which is a subgraph of a static graph of $G$. See Fig. 1 for an example.

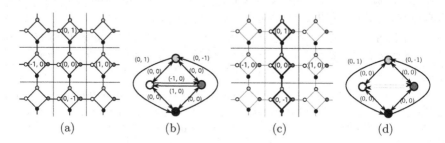

**Fig. 1.** Periodic graph (a) generated by a static graph (b). The bold black lines in (c) indicate a geodesic fiber in the periodic graph. The subgraph (d) of the static graph generates the geodesic fiber indicated by black bold lines in (c).

Eon also proposed an exponential-time algorithm to compute the static graph of a given geodesic fiber. We now give a brief explanation of his algorithm. By the definition of periodic graphs, there is a one-to-one correspondence between directed walks on a static graph and directed walks on the periodic graph generated by it. Given a closed walk $\mathcal{W}$ on a static graph, by repeating $\mathcal{W}$ an infinite number of times, $\mathcal{W}$ lifts to a doubly infinite path on the periodic graph. Such a doubly infinite path is called a *geodesic* if any subpath of it is a shortest path. For a closed walk $\mathcal{W}$ traversing the vertices $(v_1, \mathbf{z}_1), (v_2, \mathbf{z}_2), \ldots, (v_k, \mathbf{z}_k)$ on a static graph, we call the sum $\sum_{i=1}^{k-1}(\mathbf{z}_{i+1} - \mathbf{z}_i)$, denoted by $\mathrm{tran}(\mathcal{W})$, the *transit vector* of $\mathcal{W}$. Clearly, a geodesic fiber $(F, \mathbf{t})$ containing the vertex $(u, \mathbf{y})$ contains all the geodesics lifted from a closed walk $\mathcal{W}$, starting at $u$ and with $\mathrm{tran}(\mathcal{W}) = \mathbf{t}$.

For a vector $\mathbf{t} \in \mathbb{Z}^2$, by $\mathrm{Ext}(\mathbf{t})$ we denote the set of all vectors parallel to $\mathbf{t}$ in $\mathbb{Z}^2$. Obviously, for any $\mathbf{t} \in \mathbb{Z}^2$, there exists a vector $\mathrm{prim}(\mathbf{t})$ such that $\mathrm{Ext}(\mathbf{t}) = \{a \cdot \mathrm{prim}(\mathbf{t}) : a \in \mathbb{Z}\}$. The *reduced length* of a closed walk $\mathcal{W}$ is the ratio $|\mathcal{W}|/|k|$, where $|\mathcal{W}|$ is the number of edges in $\mathcal{W}$ and $k$ is an integer such that $k \cdot \mathrm{prim}(\mathrm{tran}(\mathcal{W})) = \mathrm{tran}(\mathcal{W})$. A closed walk $\mathcal{W}$ starting at $u$ lifts to a geodesic if and only if it is a *cycle*, i.e. it does not pass through the same vertex twice, and it has the shortest reduced length among the closed walks that start at $u$. We note that for a vertex $u$ of a static graph, there does not always exist a closed walk that has the shortest reduced length among the closed walks starting at $u$. The *Kagomé lattice*, which is not $\ell_1$-embeddable [5], is a good example of this.

Following Eon's terminology [7], we say that the geodesic fiber $(F, \mathbf{t})$ runs along the direction $\mathbf{s}$ for a vector $\mathbf{s}$ parallel to $\mathbf{t}$. Basically, for a given vertex

$u$ of a static graph and a vector $\mathbf{t}$, Eon's algorithm computes a static graph of a geodesic fiber running along the direction $\mathbf{t}$ by combining all the cycles that start at $u$, with their transit vectors parallel to $\mathbf{t}$ and with the shortest length. If a closed directed walk $\mathcal{W}$ starting at $u$ consists of more than one cycle with transit vectors not parallel to $\mathrm{tran}(\mathcal{W})$, then a geodesically complete subgraph of $G$ containing the vertices $\{(u, \mathbf{y} + a\mathbf{t}) : a \in \mathbb{Z}\}$ must contain the 2-periodic graph generated by the cycles. Thus, in such a case, there is no geodesic fiber running along the direction $\mathbf{t}$ with its static graph containing $u$.

Finally, we introduce some terminology and results for geodesic fibers.

**Proposition 2 ([7]).** *An $n$-periodic graph admits at least $n$ geodesic fibers in $n$ independent directions.*

Two geodesic fibers $(F_1, \mathbf{t}_1)$ and $(F_2, \mathbf{t}_2)$ are *parallel* if $\mathbf{t}_1$ and $\mathbf{t}_2$ are parallel. If a geodesic lifted from a cycle $\mathcal{C}$ is interrupted by a vertex, then we call each subgraph a *half-geodesic*. If, when following the orientation induced by $\mathcal{C}$, we find that a half-geodesic runs outward from the terminal vertex (resp. towards the terminal vertex), then it is called a *plus* (resp. a *minus*) *half-geodesic*.

**Lemma 1 ([7]).** *Given an infinite geodesically complete subgraph $H$ of a periodic graph $G$, any infinite sequence $(v, \mathbf{z}_1), (v, \mathbf{z}_2), \ldots$ of vertices of $H$ induces at least one half-geodesic with its vertex set contained in $H$.*

**Lemma 2 ([7]).** *Let $\mathcal{C}_1, \mathcal{C}_2$ be cycles on a static graph such that $\mathrm{tran}(\mathcal{C}_1)$ and $\mathrm{tran}(\mathcal{C}_2)$ are parallel. Any geodesically complete subgraph of a periodic graph containing a plus half-geodesic lifted from $\mathcal{C}_1$ and a minus half-geodesic lifted from $\mathcal{C}_2$ also contains the geodesics lifted from $\mathcal{C}_1$ and $\mathcal{C}_2$.*

## 3   Convex Cuts on an $\ell_1$-Embeddable Periodic Graph

In this section, we show a periodic structure of convex cuts of periodic graphs. Throughout the rest of this paper, we sill take $G$ to be an arbitrary connected 2-periodic graph generated by the static graph $\mathcal{G} = (\mathcal{V}, \mathcal{E})$, and we sill take $S$ to be a set of vertices of $G$.

First, we show several properties of convex cuts that hold on all 2-periodic graphs. For a vertex set $U$ of $G$ and a vector $\mathbf{t} \in \mathbb{Z}^2$, denote the vertex set $\{(v, \mathbf{z} + \mathbf{t}) : (v, \mathbf{z}) \in U\}$ by $\mathbf{t}U$. If a set $U$ of vertices in a geodesic fiber $(F, \mathbf{t})$ satisfies the following three conditions, then the subgraph of $F$ induced by $U$ is called a *plus half-geodesic fiber* (resp. a *minus half-geodesic fiber*) of $(F, \mathbf{t})$: (i) $a\mathbf{t}U \subset U$ (resp. $-a\mathbf{t}U$) for all $a \in \mathbb{N}$, (ii) the subgraph of $F$ induced by $U$ does not contain a geodesic, and (iii) for any vertex $v$ of $\mathcal{G}$, $U$ contains at least one vertex corresponding to $v$. For a set $U$ of vertices of $G$ and a subgraph $H$ of $G$, if the vertex set of $H$ is contained in $U$, then we say that $U$ contains $H$.

**Lemma 3.** *Let $(F, \mathbf{t})$ be an arbitrary geodesic fiber on $G$. If $S$ and $\bar{S}$ are both convex, then one of $S$ or $\bar{S}$ contains a plus half-geodesic fiber or a minus half-geodesic fiber of $(F, \mathbf{t})$.*

*Proof.* Let $U$ be a finite set of vertices in $F$ such that for every vertex $u$ of the static graph $\mathcal{F}$ of $(F, \mathbf{t})$, $U$ contains a vertex corresponding to $u$. Since $\mathcal{F}$ is a subgraph of $\mathcal{G}$, for the edges $((u, \mathbf{y}), (v, \mathbf{z}))$ of $F$, $\mathbf{y} - \mathbf{z}$ is bounded. Thus, for a sufficiently large $k \in \mathbb{N}$, the graph obtained by removing the vertex set $\bigcup_{i=0}^{k} itU$ from $(F, \mathbf{t})$ has two connected components. Assume that $at(\bigcup_{i=0}^{k} itU)$ is contained by $S$ for some $a \in \mathbb{Z}$. Let $F_1$ and $F_2$ be the two connected components of the graph obtained by removing $at(\bigcup_{i=0}^{k} itU)$ from $(F, \mathbf{t})$. Since the intersection of two geodesically complete graphs is again geodesically complete, the subgraph of $(F, \mathbf{t})$ induced by the vertices in $\bar{S}$ cannot contain the vertices both from $F_1$ and from $F_2$. Thus, $S$ contains one of $F_1$ and $F_2$, and thus it also contains either a plus half-geodesic or a minus half-geodesic in $(F, \mathbf{t})$. The same argument holds for the case where $at(\bigcup_{i=0}^{k} itU) \subset \bar{S}$.

Assume that $at(\bigcup_{i=0}^{k} itU)$ is not contained by $S$ or $\bar{S}$ for any $a \in \mathbb{Z}$. Let $l$ be the number of cycles that have a transit vector in $\mathrm{Ext}(\mathbf{t})$ in $\mathcal{F}$, and let $L$ be the maximum length of the cycles. There exist vertices $(v, \mathbf{z})$ and $(v, \mathbf{z}')$ of $(F, \mathbf{t})$ such that $d_G((v, \mathbf{z}), (v, \mathbf{z}')) > lL$ and both of them are contained in one of $S$ or $\bar{S}$. Without loss of generality, assume $(v, \mathbf{z}), (v, \mathbf{z}') \in S$. Since the subgraph $F_S$ of $(F, \mathbf{t})$ induced by $S$ is geodesically complete and $d_G((v, \mathbf{z}), (v, \mathbf{z}')) > L$, $F_S$ contains all the shortest paths between $(v, \mathbf{z})$ and $(v, \mathbf{z}')$, including the one lifted from a directed closed walk $C^{(0)}$ in $\mathcal{F}$. For each vertex $u \in C^{(0)}$, there exist two vertices $(u, \mathbf{y}), (u, \mathbf{y}')$ contained in the shortest path lifted from $C^{(0)}$ with $d_G((u, \mathbf{y}), (u, \mathbf{y}')) > (l-1)L$. Again since $F_S$ is geodesically complete and $d_G((u, \mathbf{y}), (u, \mathbf{y}')) > L$, $F_S$ contains all the shortest paths lifted from the cycles $C_1^{(1)}, \ldots, C_{l^{(1)}}^{(1)}$ which intersect $C^{(0)}$ and are contained in $\mathcal{F}$. By recursively enumerating in this way the directed closed walks which lift to the shortest paths contained in $F_S$, we finally obtain a set of directed closed walks. Since $F_S$ is geodesically complete, it also contains all the directed closed walks with zero transit vectors, which provide short cuts. By combining these directed closed walks, we obtain a static graph of a geodesic fiber $(F', \mathbf{s})$ with $\mathbf{s} \in \mathrm{Ext}(\mathbf{s})$. By assumption, $F'$ is properly contained by $F$, contradicting the minimality of $(F, \mathbf{t})$. □

The next lemma follows immediately from Lemma 2.

**Lemma 4.** *Let $(F^{(1)}, \mathbf{t})$ and $(F^{(2)}, \mathbf{s})$ be two parallel geodesic fibers such that for some $a, b \in \mathbb{N}$, $a\mathbf{t} = b\mathbf{s}$, and let $S$ be a convex vertex set of $G$. If the subgraph of $(F^{(1)}, \mathbf{t})$ induced by the intersection of $S$ and the vertex set of $(F^{(1)}, \mathbf{t})$ is a plus half-geodesic fiber (resp. a minus half-geodesic fiber) of $(F^{(1)}, \mathbf{t})$, then the subgraph of $(F^{(2)}, \mathbf{s})$ induced by the intersection of $S$ and the vertex set of $(F^{(2)}, \mathbf{s})$ is also a plus half-geodesic fiber (resp. a minus half-geodesic fiber).*

**Lemma 5.** *If convex sets $S$ and $\bar{S}$ are not empty, then each of $S$ and $\bar{S}$ contains at least one geodesic fiber.*

*Proof.* Suppose to the contrary that $S$ does not contain any geodesic fiber in $G$. By Proposition 2, there is at least one geodesic fiber $(F, \mathbf{t})$ in $G$. Let $\mathfrak{F}$ be the set of all geodesic fibers parallel to $(F, \mathbf{t})$. Without loss of generality, we can assume by Lemma 3 and Lemma 4, that for any $(F', \mathbf{s}) \in \mathfrak{F}$, the intersection of the vertex set of $F'$ and $S$ induces a plus half-geodesic fiber in $(F', \mathbf{s})$. Let $u$ be a vertex of the static graph $\mathcal{G}$ of $G$. For each geodesic fiber $(F', \mathbf{s}) \in \mathfrak{F}$ containing a vertex corresponding to $u$, there exists a vertex $(u, \mathbf{z})$ in $(F', \mathbf{s})$ such that $(u, \mathbf{z} + \mathbf{t})$ is not contained in $S$. By Lemma 1, these vertices induce at least one half-geodesic lifted from a cycle $\mathcal{C}$ and with its vertex set $C_+$ contained in $S$. The vertex set $C$ of the geodesic lifted from $\mathcal{C}$ is also contained in $S$, since otherwise $\bar{S}$ contains the vertex set $C \setminus C_+$ and the vertex set $\mathbf{t}C_+$, but does not contain $C_+$. By Lemma 2, this contradicts the convexity of $\bar{S}$. There does not exist a geodesic fiber containing $C$, since otherwise by Eon's algorithm for geodesic fibers, $S$ must contain a geodesic fiber in order to satisfy convexity. Thus there exists a closed directed walk $\mathcal{C}'$ sharing a common vertex $v$ with $\mathcal{C}$ and consisting of cycles $\mathcal{C}_1, \ldots, \mathcal{C}_k$ with the same reduced length as that of $\mathcal{C}$, such that not all of $\text{tran}(\mathcal{C}_1), \ldots, \text{tran}(\mathcal{C}_k)$ are parallel to $\text{tran}(\mathcal{C})$. By convexity, $S$ contains the 2-periodic graph $H$ generated by $\mathcal{C}'$, and it also contains the translations of $C$ passing through the vertices of $H$ corresponding to $v$. This contradicts the assumption that for each geodesic fiber $(F', \mathbf{s})$ parallel to $(F, \mathbf{t})$ and containing vertices corresponding to $u$, there exists a vertex $(u, \mathbf{z})$ contained in $(F', \mathbf{s})$ such that $(u, \mathbf{z} + \mathbf{t}) \notin S$.     $\square$

Next, we show special properties of $\ell_1$-embeddable periodic graphs.

**Lemma 6.** *If $G$ is $\ell_1$-embeddable and there exists a closed walk $\mathcal{W}$ starting at $u$ with the minimum reduced length among all closed walks starting at $u$ on $\mathcal{G}$, then for each vertex $v$ of $\mathcal{G}$, among the set of all the closed walks starting at $v$ with transit vectors parallel to $\text{tran}(\mathcal{W})$, there exists a closed walk $\mathcal{W}'$ with the minimum reduced length. The reduced length of $\mathcal{W}$ equals that of $\mathcal{W}'$.*

*Proof.* Given two parallel integral vectors $\mathbf{v}$ and $\mathbf{v}'$, by $\text{LCM}(\mathbf{v}, \mathbf{v}')$, we denote the vector $\mathbf{v}'' = k\mathbf{v} = k'\mathbf{v}'$, where $k$ and $k'$ are relatively prime integers. Suppose that for a vertex $v$, there does not exist a closed walk $\mathcal{W}'$ starting at $v$ that has the minimum reduced length among all closed walks starting at $v$, or that the reduced length of $\mathcal{W}'$ is not equal to that of $\mathcal{W}$. We show that there exists a vector $\mathbf{z} \in \mathbb{Z}^d$ such that $d_G((u, a\mathbf{z}), (u, b\mathbf{z})) \neq d_G((v, a\mathbf{z}), (v, b\mathbf{z}))$ for all $a, b \in \mathbb{N}$ with $a \neq b$. First, if such a closed walk $\mathcal{W}'$ exists and $\mathcal{W}$ and $\mathcal{W}'$ have different reduced lengths, then by taking $\text{LCM}(\text{tran}(\mathcal{W}), \text{tran}(\mathcal{W}'))$ as $\mathbf{z}$, the above inequality holds for all $a, b \in \mathbb{N}$ with $a \neq b$. Next, suppose that such a closed walk $\mathcal{W}'$ does not exist and for some $a, b \in \mathbb{N}$, $d_G((u, a \cdot \text{tran}(\mathcal{W}), (u, b \cdot \text{tran}(\mathcal{W})))) = d_G((v, a \cdot \text{tran}(\mathcal{W}), (v, b \cdot \text{tran}(\mathcal{W}))))$. Then there exists a closed walk $\mathcal{W}''$ starting at $v$ with $\text{tran}(\mathcal{W}'')$ parallel to $\text{tran}(\mathcal{W})$ and with the reduced length shorter than the reduced length of $\mathcal{W}$. By taking $\text{LCM}(\text{tran}(\mathcal{W}), \text{tran}(\mathcal{W}''))$ as $\mathbf{z}$, the above inequality holds for all $a, b \in \mathbb{N}$ with $a \neq b$.

Thus, if $\phi : V \times \mathbb{Z}^2 \to \mathbb{Z}^d$ is an $\ell_1$-embedding, then for all $a, b \in \mathbb{Z}$, $\phi((u, a\mathbf{z})) - \phi((v, a\mathbf{z})) \neq \phi((u, b\mathbf{z})) - \phi((v, b\mathbf{z}))$ since, if the equality holds for some $a, b \in \mathbb{Z}$, then $d_G((v, a\mathbf{z}), (v, b\mathbf{z})) = \|\phi((v, a\mathbf{z})) - \phi((v, b\mathbf{z}))\|_{\ell_1} = \|\phi((u, a\mathbf{z})) - \phi((u, b\mathbf{z}))\|_{\ell_1} = d_G((u, a\mathbf{z}), (u, b\mathbf{z}))$. On the other hand, since $d_G((u, a\mathbf{z}), (v, a\mathbf{z})) = d_G((u, b\mathbf{z}),$

$(v, b\mathbf{z}))$ for all $a, b \in \mathbb{Z}$ by the periodicity of $G$, and $d_G((u, a\mathbf{z}), (v, a\mathbf{z}))$ is finite, $\phi((u, a'\mathbf{z})) - \phi((v, a'\mathbf{z})) = \phi((u, b'\mathbf{z})) - \phi((v, b'\mathbf{z}))$ for some $a', b' \in \mathbb{Z}$. This is a contradiction.  □

The property shown to hold on an $\ell_1$-embeddable periodic graph in Lemma 6 is called the *weak coherence*. It is not difficult to construct an exponential-time algorithm to determine if a periodic graph is weakly coherent.

**Lemma 7.** *If $G$ is weakly coherent and a geodesic fiber $(F, \mathbf{t})$ is in $G$, then each vertex in $G$ is contained in some geodesic fiber parallel to $(F, \mathbf{t})$.*

*Proof.* Let $(u, \mathbf{y})$ be a vertex of $G$ contained in $(F, \mathbf{t})$. There is a directed closed walk $C$ beginning at $u$ with $\text{tran}(C) \in \text{Ext}(\mathbf{t})$ and with a length equal to the shortest reduced length among all directed closed walks starting at $u$ on $\mathcal{G}$. By Eon's algorithm for geodesic fibers, if there is no geodesic fiber parallel to $(F, \mathbf{t})$ containing a vertex $(v, \mathbf{z})$ in $G$, then one of the following holds for $\mathcal{G}$: (a) there is no directed closed walk with the minimum reduced length among the closed directed walks starting at $v$ with the transit vector in $\text{Ext}(\mathbf{t})$; or (b) there is such a directed closed walk $C'$ and $C'$ consists of cycles $C'_1, \ldots, C'_k$ in which the individual transit vectors are not all in $\text{Ext}(\mathbf{t})$. By Lemma 6, (a) does not occur. Assume that (b) holds. Since $C'$ has the minimum reduced length, $C'_i$ has the minimum reduced length among the directed closed walks that have transit vectors in $\text{Ext}(\text{tran}(C'_i))$ and that share a vertex with $C'_i$, for $i = 1, \ldots, k$. By Lemma 6, for $i = 1, \ldots, k$, there exists a closed walk $C_i$ starting at $u$ with its transit vector in $\text{Ext}(\text{tran}(C'_i))$ and with the same reduced length as $C'_i$. By combining $C_1, \ldots, C_k$, we can construct a directed closed walk $C''$ with its transit vector in $\text{Ext}(\mathbf{t})$ and with the same reduced length as $C'$. Since, by Lemma 6, $C'$ has the same reduced length as $C$, $C''$ has the same reduced length as $C$. This contradicts the existence of the geodesic fiber $(F, \mathbf{t})$.  □

By Lemmas 3, 4, 5, and 7, the next theorem obviously holds.

**Theorem 1.** *If $G$ is weakly coherent and $S$ and $\bar{S}$ are both convex, then the following two statements hold:*

1. *If $S$ contains a geodesic fiber $(F, \mathbf{t})$, then $S$ and $\bar{S}$ are the union of the vertex sets of the geodesic fibers parallel to $(F, \mathbf{t})$.*
2. *If $S$ does not contain any geodesic fiber parallel to $(F, \mathbf{t})$, then $S$ is the union of the plus half-geodesic fibers (or minus half-geodesic fibers) in $(F', \mathbf{s})$, where $(F', \mathbf{s})$ runs in all geodesic fibers parallel to $(F, \mathbf{t})$.*

# 4    An Algorithm for $\ell_1$-Embedding of an $\ell_1$-Rigid Periodic Graph

Let $\mathcal{S}$ be the collection of all convex cuts in an $\ell_1$-rigid periodic graph $G$, and let $d_G = \sum_{(S, \bar{S}) \in \mathcal{S}} \lambda_S \delta(S)$ be an $\ell_1$-decomposition of $G$. By periodicity, for any vector $\mathbf{t} \in \mathbb{Z}^2$, if $(S, \bar{S}) \in \mathcal{S}$ then $(\mathbf{t}S, \mathbf{t}\bar{S}) \in \mathcal{S}$. Furthermore, by definition, if

$G$ is $\ell_1$-rigid, then $\lambda_S = \lambda_{\mathbf{t}S}$. Thus $\ell_1$-rigidity ensures a finite description of an $\ell_1$-decomposition, and an $\ell_1$-embedding. In this section, we develop an algorithm for the $\ell_1$-embedding of an $\ell_1$-rigid periodic graph, using the property of convex cuts on periodic graphs, that was shown in the previous section.

First, we develop a method to represent a convex cut in a periodic graph with a finite data structure, and we then give an algorithm to enumerate all of the convex cuts. We assume that $G$ is weakly coherent, since otherwise, by Lemma 6, $G$ is not $\ell_1$-embeddable.

Let $q_{\mathcal{G}}$ be the natural projection that maps an edge of $G$ to the corresponding edge of $\mathcal{G}$. For a cut $(S, \bar{S})$, $E(S, \bar{S})$ denotes the edge set $\{(u, v) : u \in S, v \in \bar{S}\}$. By $G_{(S,\bar{S})}$, we denote the periodic graph generated by the static graph $\mathcal{G}_{(S,\bar{S})}$ obtained by removing the set $q_{\mathcal{G}}(E(S, \bar{S}))$ of edges from $\mathcal{G}$. Then $G_{(S,\bar{S})}$ is disconnected, since otherwise removing the set $E(S, \bar{S})$ of edges from $G$ does not yield a cut. By Lemma 7 and Theorem 1, it is not difficult to show that one of the following statements holds: (a) all of the connected components of $G_{(S,\bar{S})}$ are 2-periodic graphs, or (b) all of them are 1-periodic graphs. We call $(S, \bar{S})$ a *2-periodic cut* (resp. a *1-periodic cut*) if (a) holds (resp. (b) holds).

Let $o_{\mathcal{G}}$ be the projection that maps an edge of $\mathcal{G}$ to the set of all corresponding edges of $G$. By Theorem 1, if $(S, \bar{S})$ is a convex cut on $G$ such that $S$ contains a geodesic fiber $(F, \mathbf{t})$, then $q_{\mathcal{G}}(E(S, \bar{S}))$ consists of edges which are not in a static graph of a geodesic fiber parallel to $(F, \mathbf{t})$. The next lemma shows that any 2-periodic convex cut can be uniquely determined by a set of edges on $\mathcal{G}$.

**Lemma 8.** *If $G$ is weakly coherent, then for any 2-periodic convex cut $(S, \bar{S})$ on $G$, $E(S, \bar{S}) = o_{\mathcal{G}}(q_{\mathcal{G}}(E(S, \bar{S})))$.*

*Proof.* By definition, $E(S, \bar{S}) \subset o_{\mathcal{G}}(q_{\mathcal{G}}(E(S, \bar{S})))$. We show $o_{\mathcal{G}}(q_{\mathcal{G}}(E(S, \bar{S}))) \subset E(S, \bar{S})$. The subgraph $G_S$ (resp. $G_{\bar{S}}$) induced by $S$ (resp. $\bar{S}$) is also a connected 2-periodic graph. Thus, any edge $e \in o_{\mathcal{G}}(q_{\mathcal{G}}(E(S, \bar{S})))$ must also be in $E(S, \bar{S})$, since, otherwise, removing $E(S, \bar{S})$ does not make $G$ disconnected. $\square$

To represent a 1-periodic convex cut by a set of edges on $\mathcal{G}$, we need the next.

**Lemma 9.** *If $(S, \bar{S})$ is a convex cut on a weakly coherent periodic graph $G$ and the connected components of $G_{(S,\bar{S})}$ are 1-periodic graphs, then the static graph $\mathcal{G}_{(S,\bar{S})}$ generates a connected 1-periodic graph.*

*Proof.* Suppose that $S$ and $\bar{S}$ contain geodesic fibers running along the direction $\mathbf{t}$. By Theorem 1, there exists a vector $\mathbf{s} \in \mathbb{Z}^2$ such that $\mathbf{s} \notin \mathrm{Ext}(\mathbf{t})$ and the subgraph $G_{S \cap \mathbf{s}\bar{S}}$ of $G$ induced by $S \cap \mathbf{s}\bar{S}$ is a 1-periodic graph which is generated by a subgraph $\mathcal{G}_{S \cap \mathbf{s}\bar{S}}$ of $\mathcal{G}$ with $\mathcal{V}$ as its vertex set. Since $S$ and $\mathbf{s}\bar{S}$ are convex, $G_{S \cap \mathbf{s}\bar{S}}$ is geodesically complete. Thus, $\mathcal{G}_{S \cap \mathbf{s}\bar{S}}$ generates a connected 1-periodic graph. No edge in $q_{\mathcal{G}}(E(S, \bar{S}))$ is contained in $\mathcal{G}_{S \cap \mathbf{s}\bar{S}}$, since, otherwise, $\mathcal{G}_{S \cap \mathbf{s}\bar{S}}$ would generate a connected 2-periodic graph. Thus $\mathcal{G}_{S \cap \mathbf{s}\bar{S}}$ is a subgraph of $\mathcal{G}_{(S,\bar{S})}$, and the lemma follows. $\square$

Lemma 9 means that the vertex set of every connected component of $G_{(S,\bar{S})}$ contains a vertex corresponding to $v$ for any vertex $v$ on $\mathcal{G}$. By Theorem 1, given $G_{(S,\bar{S})}$, $S$ and $\bar{S}$ can be uniquely determined.

Based on the above discussions, by enumerating all subsets of edges on $\mathcal{G}$, we can enumerate all candidates for the convex cuts on $G$ within $O(2^{|\mathcal{E}|})$ time. What remains to be done is to construct an algorithm to determine whether a given cut is convex. The next lemma provides a key for it.

**Lemma 10.** *Let $G$ be an $\ell_1$-rigid periodic graph. Then there is an $\ell_1$-embedding $\phi : \mathcal{V} \times \mathbb{Z}^2 \to \mathbb{Z}^d$ of $G$ such that for all $(u, \mathbf{y}), (v, \mathbf{z}) \in \mathcal{V} \times \mathbb{Z}^2$, $\phi((v, \mathbf{z} + \mathbf{y})) - \phi((v, \mathbf{z})) = \phi((u, \mathbf{y})) - \phi((u, \mathbf{0}))$.*

*Proof.* Let $\mathcal{S}$ be the set of all convex cuts on $G$, and let $\sim$ be an equivalence relation on $\mathcal{S}$ such that $(S, \bar{S}) \sim (S', \bar{S}')$ if there exists a vector $\mathbf{t} \in \mathbb{Z}^2$ satisfying $S = \mathbf{t}S'$ and $\bar{S} = \mathbf{t}\bar{S}'$. In the set $\mathcal{S}/\sim$, there are at most a finite number of equivalence classes $\mathcal{S}_1, \ldots, \mathcal{S}_d$. Let $d_G = \sum_{(S,\bar{S}) \in \mathcal{S}} \lambda_{(S,\bar{S})} \delta(S)$ be the $\ell_1$-decomposition of $G$. Recall that by the definition of $\ell_1$-rigidity, $\lambda_{(S,\bar{S})}$ takes the same value for all convex cuts $(S, \bar{S}) \in \mathcal{S}_i$. Let $\phi : \mathcal{V} \to \mathbb{Z}^d, (v, \mathbf{z}) \mapsto (\phi_1((v, \mathbf{z})), \ldots, \phi_d((v, \mathbf{z})))$ be the $\ell_1$-embedding constructed from the $\ell_1$-decomposition as in the proof of Proposition 1. For a fixed $i \in \{1, \ldots, d\}$, each convex cut in $\mathcal{S}_i$ consists of parallel geodesic fibers that run along the direction $\mathbf{t}$. By construction, it is not difficult to show that, for any $u \in \mathcal{V}$ and $\mathbf{y}, \mathbf{y}' \in \mathbb{Z}^2$, the value $\phi_i((u, \mathbf{y})) - \phi_i((u, \mathbf{y}'))$ is a constant. Thus, for any $(u, \mathbf{y})$ and $(v, \mathbf{z})$, $\phi_i((v, \mathbf{z} + \mathbf{y})) - \phi_i((v, \mathbf{z})) = \phi_i((v, \mathbf{y})) - \phi_i((v, \mathbf{0})) = \phi_i((u, \mathbf{y})) - \phi_i((u, \mathbf{0}))$, and the lemma follows. $\square$

**Corollary 1.** *If $G$ is $\ell_1$-rigid, then for any $u, v \in \mathcal{V}$ and vectors $\mathbf{y}, \mathbf{z} \in \mathbb{Z}^2$, $d_G((u, \mathbf{y}), (u, \mathbf{z})) = d_G((v, \mathbf{y}), (v, \mathbf{z}))$.*

The property shown in Corollary 1 has been defined as *coherence* on general periodic graphs by Fu [9]. An $O(|\mathcal{V}^7||\mathcal{E}|\mathcal{T})$-time algorithm, where $\mathcal{T}$ is the maximum absolute value among the elements of the vectors on the edges, for determining whether a given periodic graph is coherent is also shown by her [9].

**Lemma 11.** *On a coherent periodic graph $G$, there is an algorithm to determine if a given subgraph $H$ of $G$ is geodesically complete on $G$.*

*Proof.* By $G \setminus H$, we denote the graph obtained by removing the vertex set of $H$ from $G$. We show that $H$ is geodesically complete on $G$ if and only if for any two vertices $(u, \mathbf{y})$ and $(v, \mathbf{z})$ in $H$ (resp. in $G \setminus H$) with $d_G((u, \mathbf{y}), (v, \mathbf{z})) \leq |\mathcal{V}| + 2$, any shortest path connecting $(u, \mathbf{y})$ and $(v, \mathbf{z})$ is contained in $H$ (resp. in $G \setminus H$). The enumeration of such shortest paths can be completed in $O(\mathcal{D}^{|\mathcal{V}|+2})$ time, where $\mathcal{D}$ is the maximum degree in $\mathcal{G}$.

Obviously, the latter condition holds when $H$ is geodesically complete. Assume that the latter condition holds but that $H$ is not geodesically complete. Then there exist two vertices $(u, \mathbf{y})$ and $(v, \mathbf{z})$ in $H$, with $d_G((u, \mathbf{y}), (v, \mathbf{z})) > |\mathcal{V}| + 2$, such that a shortest path $P$ that is not contained in $H$ and that connects $(u, \mathbf{y})$ and $(v, \mathbf{z})$ on $G$. Without loss of generality, we can assume that $P$ is the concatenation of an edge $e$ connecting a vertex in $H$ and a vertex $G \setminus H$, a path $P'$ contained in $G \setminus H$ and an edge $e'$ connecting a vertex in $G \setminus H$ and a vertex in $H$. By the definition of periodic graphs, $P'$ corresponds to a walk $\mathcal{P}'$ connecting

$u$ and $v$ on $\mathcal{G}$. The walk $\mathcal{P}'$ decomposes into a (possibly empty) path $\mathcal{Q}$ and cycles $\mathcal{C}_1, \ldots, \mathcal{C}_k$, where each of $\mathcal{Q}, \mathcal{C}_1, \ldots, \mathcal{C}_k$ has a length that is no more than $|\mathcal{V}|$. For each $i = 1, \ldots, k$, let $\mathcal{C}'_i$ be a directed closed walk on $\mathcal{G}$ corresponding to a shortest path from $(v, \mathbf{z})$ to $(u, \mathbf{z} + \text{tran}(\mathcal{C}_i))$. By Corollary 1, $\text{len}(\mathcal{C}'_i) = \text{len}(\mathcal{C}_i)$. Thus, the concatenation of $q(e), \mathcal{Q}, q(e'), \mathcal{C}'_1, \ldots, \mathcal{C}'_k$ also corresponds to a shortest path connecting $(u, \mathbf{y})$ and $(v, \mathbf{z})$. On the other hand, since the concatenation of $q(e), \mathcal{Q}, q(e')$ corresponds to a path on $G$ connecting two vertices of $H$ and with length no more than $|\mathcal{V}| + 2$, by assumption, this cannot be a shortest path. This is a contradiction. $\qquad\square$

Now we can enumerate all the convex cuts on a coherent periodic graph $G$ in $O(2^{|\mathcal{E}|}\mathcal{D}^{|\mathcal{V}|+2})$ time. Finally, we describe an algorithm to compute the $\ell_1$-decomposition $d_G = \sum_{(S,\bar{S}) \in \mathcal{S}} \lambda_S \delta(S)$ of an $\ell_1$-rigid periodic graph $G$. Let $\mathbf{E}$ be the enumerated subsets of $\mathcal{E}$ that represent convex cuts. Then we can write the $\ell_1$-decomposition by

$$d_G = \sum_{\mathcal{E}' \in \mathbf{E}} \sum_{(S,\bar{S}): \text{ the convex cut determined by } \mathcal{E}'} \lambda_{\mathcal{E}'} \delta(S).$$

For each edge $e = ((u, \mathbf{y}), (v, \mathbf{z}))$ in $G$,

$$\sum_{\mathcal{E}' \in \mathbf{E}} \sum_{(S,\bar{S}): \text{ the convex cut determined by } \mathcal{E}' \text{ with } e \in E(S,\bar{S})} \lambda_{\mathcal{E}'} = 1.$$

In this summation, for an edge set $\mathcal{E}' \in \mathbf{E}$, $\lambda_{\mathcal{E}'}$ can appear more than one time because for more than one convex cut $(S, \bar{S})$ determined by $\mathcal{E}'$, $e \in E(S, \bar{S})$ can hold. The number of such convex cuts is equal to the absolute value of the integer $a$ where $a$ is defined using a unit vector $\mathbf{u}^{(i)}$ which is not parallel to the direction of the geodesic fibers contained in $S$ so that the vertex $(u, \mathbf{y}' + a\mathbf{u}^{(i)})$ is on the geodesic fiber running along the direction $\mathbf{t}$ and containing $(u, \mathbf{y})$. For two edges $e$ and $e'$ on $G$, if $q(e) = q(e')$ then these values coincide. For each edge $e \in \mathcal{E}$, we denote this value by $\mu(e)$. Thus, the $\ell_1$-decomposition can be constructed by computing a solution $\{\lambda'_E\}_{\mathcal{E}' \in \mathbf{E}}$ to the linear system

$$\begin{cases} \sum_{\mathcal{E}' \in \mathbf{E}: e \in \mathcal{E}'} \mu(e)\lambda_{\mathcal{E}'} = 1, \forall \text{edge } e \text{ of } \mathcal{G} \\ \lambda'_E \geq 0. \end{cases}$$

By solving a linear program, the solution can be computed in polynomial time with respect to $|\mathbf{E}|$ and $|\mathcal{E}|$. Using the proof of Proposition 1, the $\ell_1$-embedding can also be constructed. Thus, the next theorem holds.

**Theorem 2.** *For a coherent periodic graph $G$ generated by a static graph $\mathcal{G} = (\mathcal{V}, \mathcal{E})$ with maximum degree $\mathcal{D}$, there is an $O(2^{|\mathcal{E}|}\mathcal{D}^{(|\mathcal{V}|+2)})$-time algorithm that can determine if $G$ is $\ell_1$-embeddable, and if it is, it can construct an $\ell_1$-embedding.*

## 5   Concluding Remarks

In this paper, we showed an exponential-time algorithm for the $\ell_1$-embedding of an $\ell_1$-rigid periodic graph. The computational complexity of the $\ell_1$-embedding problem is left open. The characterization shown in this paper for the convex cuts on an $\ell_1$-rigid periodic graph should be of help in investigating this problem. Another interesting open problem is *whether there is an $\ell_1$-embeddable periodic graph which is not $\ell_1$-rigid*. This questions if every $\ell_1$-embeddable periodic graph has a finite representation for its $\ell_1$-embedding, and thus if every the $\ell_1$-embedding of a periodic graph is computable.

# References

1. Chavey, D.: Tilings by regular polygons – II: A catalog of tilings. Computers & Mathematics with Applications 17, 147–165 (1989)
2. Chepoi, V., Deza, M., Grishukhin, V.: Clin d'oeil on $L_1$-embeddable planar graphs. Discrete Applied Mathematics 80(1), 3–19 (1997)
3. Cohen, E., Megiddo, N.: Recognizing properties of periodic graphs. Applied Geometry and Discrete Mathematics 4, 135–146 (1991)
4. Delgado-Friedrichs, O., O'Keeffe, M.: Crystal nets as graphs: Terminology and definitions. Journal of Solid State Chemistry 178, 2480–2485 (2005)
5. Deza, M., Grishukhin, V., Shtogrin, M.: Scale-Isometric Polytopal Graphs in Hypercubes and Cubic Lattices, ch. 9. World Scientific Publishing Company (2004)
6. Deza, M., Laurent, M.: Geometry of Cuts and Metrics. Springer (1997)
7. Eon, J.-G.: Infinite geodesic paths and fibers, new topological invariants in periodic graphs. Acta Crystallographica Section A 63, 53–65 (2007)
8. Feutrier, P.: Parametric integer programming. RAIRO Recherche Opérationnelle 22, 243–268 (1988)
9. Fu, N.: A strongly polynomial time algorithm for the shortest path problem on coherent planar periodic graphs. In: Chao, K.-M., Hsu, T.-S., Lee, D.-T. (eds.) ISAAC 2012. LNCS, vol. 7676, pp. 392–401. Springer, Heidelberg (2012)
10. Höfting, F., Wanke, E.: Minimum cost paths in periodic graphs. SIAM Journal on Computing 24(5), 1051–1067 (1995)
11. Iwano, K., Steiglitz, K.: Optimization of one-bit full adders embedded in regular structures. IEEE Transaction on Acoustics, Speech and Signal Processing 34, 1289–1300 (1986)
12. Iwano, K., Steiglitz, K.: Planarity testing of doubly periodic infinite graphs. Networks 18, 205–222 (1988)
13. Karp, R., Miller, R., Winograd, A.: The organization of computations for uniform recurrence equiations. Journal of the ACM 14, 563–590 (1967)
14. Verdoolaege, S.: barvinok: User guide (2007), http://freshmeat.net/projects/barvinok/

# Author Index